AF477515

The Krzyż Conjecture
Theory and Methods

The Krzyż Conjecture
Theory and Methods

$$f(z) = a_0 + a_1 z + \ldots + a_n z^n + \ldots,$$

$$\max_{f \in B} |a_n| = \max_{f \in B} \frac{f^{(n)}(0)}{n!} = \frac{2}{e}.$$

Equality only for $f(z)$ a rotation of

$$\exp\left(-\frac{1 + z^n}{1 - z^n}\right) = \frac{1}{e} - \frac{2}{e} z^n + \ldots.$$

Ronen Peretz

Ben Gurion University of the Negev, Israel

World Scientific

NEW JERSEY · LONDON · SINGAPORE · BEIJING · SHANGHAI · HONG KONG · TAIPEI · CHENNAI · TOKYO

Published by

World Scientific Publishing Co. Pte. Ltd.

5 Toh Tuck Link, Singapore 596224

USA office: 27 Warren Street, Suite 401-402, Hackensack, NJ 07601

UK office: 57 Shelton Street, Covent Garden, London WC2H 9HE

Library of Congress Cataloging-in-Publication Data
Names: Peretz, Ronen, author.
Title: The Krzyż conjecture : theory and methods / Ronen Peretz, Ben Gurion University of the Negev, Israel.
Description: Hackensack, NJ : World Scientific, [2021] | Includes bibliographical references and index.
Identifiers: LCCN 2020038651 (print) | LCCN 2020038652 (ebook) | ISBN 9789811226373 (hardcover) |
    ISBN 9789811226380 (ebook for institutions) | ISBN 9789811226397 (ebook for individuals)
Subjects: LCSH: Analytic functions.
Classification: LCC QA331 .P379 2021 (print) | LCC QA331 (ebook) | DDC 515/.73--dc23
LC record available at https://lccn.loc.gov/2020038651
LC ebook record available at https://lccn.loc.gov/2020038652

British Library Cataloguing-in-Publication Data
A catalogue record for this book is available from the British Library.

For any available supplementary material, please visit
https://www.worldscientific.com/worldscibooks/10.1142/11993#t=suppl

Desk Editors: Jing Wen Soh/George Vasu

Typeset by Stallion Press
Email: enquiries@stallionpress.com

Dedicated to my beloved ones Noam, Ruth and Amos.

This is Noam

Noam and his mom

About the Author

Ronen Peretz is a faculty member in the department of mathematics at Ben Gurion University of the Negev in Beer Sheva, Israel. His areas of research include Geometric Function Theory, Extremal Problems in Complex Functions Theory, Affine Geometry related to polynomial mappings. In applied mathematics, he is an expert in digital image processing.

Preface

Bounded analytic functions and in particular those that are defined on the open unit disc, U, are among the most important functions in complex function theory of one complex variable. The amount of research on those functions is enormous. It dates back to the early stages of classical function theory and is still vivid and central in modern mathematics. A classical book on this subject is titled "Bounded Analytic Functions" by John B. Garnett. Many general problems on complex analytic functions can be reduced to the bounded analytic functions. Hence, the significance of pursuing further the study of bounded analytic functions. Even if we restrict attention to those bounded analytic functions that are defined on U, we still confront a rich theory with plenty of open problems and conjectures. Those form a great variety: extremal problems, geometric function theoretical problems, boundary behavior, dynamics, topological problems and many others. For example, within the theory of H^p spaces, the three most important spaces are H^1, H^2 (which is a Hilbert space) and the smallest of them all, H^∞, the space of the bounded functions. Classical books on H^p spaces were written by Peter Duren and Paul Koosis. Within H^∞ there are many important sub-spaces. Among them are the so-called inner-functions (defined by Beurling, in his 1949 paper, "On two problems concerning linear transformations in Hilbert space". That paper appeared in the *Acta Mathematica*). In this book we intend to discuss some important problems mostly related to the sub-family of inner functions (within H^∞). The Krzyż conjecture is a famous coefficient problem. It is still unsolved at this time. Let $U = \{z \in \mathbb{C} \,|\, |z| < 1\}$ be the open unit disk, $H(U)$ the algebra of all the holomorphic functions $f : U \to \mathbb{C}$ and $B = \{f \in H(U) \,|\, 0 < |f(z)| < 1, z \in U\}$. For a function $f(z) = a_0 + a_1 z + \cdots + a_n z^n + \cdots \in B$ and a natural number $n \in \mathbb{Z}^+$, the Krzyż conjecture asserts that

$$\max_{f \in B} |a_n| = \max_{f \in B} \frac{f^{(n)}(0)}{n!} = \frac{2}{e}.$$

Moreover, there is equality only for $f(z)$, which is a rotation of the function

$$\exp\left(-\frac{1+z^n}{1-z^n}\right) = \frac{1}{e} - \frac{2}{e}z^n + \cdots.$$

The conjecture is confirmed up to $n \leq 5$. In the typical situation in working coefficient by coefficient, the computations become very complicated. The conjecture was made in 1968 by the Polish mathematician Jan Krzyż in his short article "Coefficient problem for bounded nonvanishing functions". A main reference to the Krzyż conjecture is the beautiful paper "A coefficient problem for bounded nonvanishing functions" by J. A. Hummel, S. Scheinberg and L. Zalcman (1977).

Here is another typical conjecture: We refer to the paper "On the mean boundary behavior and the Taylor coefficients of an infinite Blaschke product" by Patrick Ahern and Hong Oh Kim (1984). Here is the introduction of this nice paper: "If $\{z_n\}$ is a sequence (finite or infinite) of complex numbers of modulus less than 1 such that $\sum(1-|z_n|) < \infty$, then the Blaschke product

$$B(z) = \prod_n \frac{|z_n|}{z_n} \cdot \frac{z_n - z}{1 - \overline{z}_n z},$$

converges uniformly on compact subsets of the unit disc U. We let $\mathcal{B}_\infty$ denote the set of Blaschke products whose zero sequence are infinite ..., **We conjecture that if** $B(z) = \sum_{n=0}^\infty a_n z^n \in \mathcal{B}_\infty$**, then** $\limsup_{n\to\infty} n|a_n| \geq \frac{2}{e}$**."** To introduce another typical problem which is of a different character, we refer to the elegant paper by Kenneth Stephenson (1978), titled "Omitted values of singular inner functions". We quote: "For each inner function f, we define the exceptional set $E(f)$ and the omitted set $O(f)$ as follows:

$$E(f) = \{\alpha \in U \,|\, \Psi_\alpha \circ f \text{ has a non trivial singular factor}\},$$
$$O(f) = \{\alpha \in U \,|\, \Psi_\alpha \circ f \text{ is singular}\}.$$

Here, Ψ_α is the unit disc automorphism

$$\Psi_\alpha(z) = \frac{z - \alpha}{1 - \overline{\alpha}z}.$$

Clearly $O(f) \subseteq E(f)$.

Frostman's construction: A famous theorem of O. Frostman states: *If f is inner, then $E(f)$ is a set of logarithmic capacity zero.* Less well known is his converse (1935): *For each $A \in \mathcal{A}$ (the relatively closed subsets of U of* *logarithmic capacity zero), there is an inner function f such that* $O(f) = E(f) = A$. Frostman's construction appears in the book *The Theory of Cluster Sets* by E. F. Collingwood and A. J. Lohwater. Why are we interested in Stephenson's paper? Because we want to understand the following natural question: If $f(z)$ is an inner function and if $\alpha \in U$, then the zero set of $f(z) - \alpha$, $z \in U$, which is denoted by $Z(f(z) - \alpha)$, is a Blaschke sequence (unless it is an empty set). Here, zeros are counted in a manner such that their multiplicities are taken into account. Does every Blaschke sequence of U have the form $Z(f(z) - \alpha)$? Of course! Just take $f(z) = B(z)$ to be the Blaschke product associated with our Blaschke sequence and $\alpha = 0$. What if we do not allow $f(z)$ to vanish in U? Thus we ask – is it true that every Blaschke sequence of U has the form $Z(S(z) - \alpha)$ for some singular inner function $S(z)$ and some α, $0 < |\alpha| < 1$? We would like to characterize those Blaschke sequences that have the form $Z(S(z) - \alpha)$. The above short presentation gives the taste of the kind of problems we will cover in our book.

Another motivation to write this book is the will to show how a typical research of a mathematician proceeds. Starting from graduate studies and deep into the professional career of the scientist. The frustrations, the hopes that the solution of the "big problem" is around the corner. How that is involved with the personal day-to-day life. It is of interest to youngsters that enter this difficult science to see what might be waiting for them in the future. All in all and in spite of the difficulties, the joy of discovering new theorems is worth it all.

Ronen Peretz
December 2020

Contents

Contents

Chapter 1

Introduction

This book is about one of the beautiful topics in mathematics. It describes an ongoing research on bounded analytic functions which are defined on the unit disc $\{z \in \mathbb{C} \,|\, |z| < 1\}$, which will be denoted by U or by $\mathbb{D}$. This is a very active topic that belongs to the theory of complex analysis in a single complex variable. Complex analysis is one of the classical chapters in mathematics. It contains the analytic theory of functions, the geometric function theory among other theoretical areas, as well as applications. There are a lot of applications. Some originate in other fields of mathematics: geometry, topology, arithmetic and number theory in general, algebra, and the list goes on. Other applications originate in other scientific and engineering disciplines: physics, dynamical systems, electrical engineering, and the list goes on here as well. As is frequently the case in mathematical research, the fuel that advances the research includes open problems to be solved. These are more than just a set of difficult problems, Riddles. They were raised based on the insight of excellent scientists, experts in their area of research, that were able to point to their significance. It might be that solving those open problems clarifies a still vague part of the theory and deepens our understanding of the topic. But on some occasions the wisdom of asking those particular questions leads the mathematicians, that try to solve them, to discover new theories that were hidden before. Mathematical research developed traditional working tools to communicate those open problems within the community of the researchers. Mathematicians used to ask their colleagues challenging questions that were sent in the past, in the mail. Those were simply detailed handwritten letters. Mathematicians in the various fields of research very carefully maintained lists of problems that included also the source of the problems, who asked, why did he ask, dates of certain breakthroughs, usually partial solutions, etc.

The establishment of scientific conferences is also a standard way to communicate new problems as well as advancements that were made. Scientific conferences are just one way of arranging meetings among mathematicians. Other common ways include various types of visits of a scientist in a host research institute of his colleagues. These visits are the sabbaticals, the invited lectures, the invited courses or mini-courses, etc.

It turns out that such meetings of mathematicians from different geographical locations have the magical ability to advance the scientific research. These scientific trips are not just a fulfillment of tourism. Mathematicians need to talk about their research with one another as a tool to make scientific progress. They sometimes are motivated by a very high competitive personality and a strong will to "win the race" (the first who solves is the winner). Thus, these scientific gatherings are also social events but sometimes they turn into anti-social events because of disputes.

Coming back to the topics covered in this book, they were born within the theory of one of the most important function spaces of analytic functions, the so-called $\mathbb{H}^p(U)$, $p > 0$ functions. The letter $\mathbb{H}$ is to honor the British mathematician Godfrey Harold **Hardy** from Cambridge. Most notable are the following three spaces: $\mathbb{H}^1(U)$, $\mathbb{H}^2(U)$, which is a Hilbert space, and the smallest of them all $\mathbb{H}^\infty(U)$. The theory of Hardy spaces, $\mathbb{H}^p(U)$, was initiated by Hardy and his friend at Cambridge, John Edensor Littlewood but it was further developed by lots of mathematicians over the years. It is still an active field of research and naturally there are still important unsolved problems in this area. Here is one such an open problem:

The Krzyż conjecture is a famous coefficient problem. It is still unsolved. Let $U = \{z \in \mathbb{C} \,|\, |z| < 1\}$ be the open unit disk and $H(U)$ the algebra of all the holomorphic functions $f : U \to \mathbb{C}$ and $B = \{f \in H(U) \,|\, 0 < |f(z)| < 1,\ z \in U\}$. For a function $f(z) = a_0 + a_1 z + \cdots + a_n z^n + \cdots$ and a natural number $n \in \mathbb{Z}^+$ the Krzyż conjecture asserts that

$$\max_{f \in B} |a_n| = \max_{f \in B} \left| \frac{f^{(n)}(0)}{n!} \right| = \frac{2}{e}.$$

Moreover, there is equality only for $f(z)$ which is a rotation of the function

$$\exp\left(-\frac{1 + z^n}{1 - z^n} \right) = \frac{1}{e} - \frac{2}{e} z^n + \cdots .$$

The conjecture is confirmed up to $n \leq 5$. As the typical situation in working coefficient by a coefficient, the computations become very complicated. The conjecture was made in 1968 by the Polish mathematician Jan Krzyż in his short article [61]. A main reference is still the beautiful paper [49].

As one can immediately observe, the condition $f(z) \neq 0$ for all $|z| < 1$ makes the problem nonlinear. One might call it log-linear but the standard functional analytic methods that work so well on linear functionals do not work here. The problem has already attracted the attention of many mathematicians who tried to solve it. Some developed ingenious ideas and methods in order to tackle the problem. In some sense, those ideas and methods are the most important issues and not the knowledge of the actual answer to the Krzyż problem. Is it really important that the exact coefficient bound is $\frac{2}{e}$ and not $\frac{2}{e} + 10^{-255}$? Or, maybe that the bound is independent of n for some mysterious reason?

This book includes much more than just a review on the Krzyż conjecture. A large portion of the book reviews many more topics that are related or at least remind one of the conjecture, but really form a core part of the theory of the $\mathbb{H}^\infty(U)$ space. The fact that we have a factorization theorem of $\mathbb{H}^p$ functions into the product of a Blaschke product, a singular inner function and an outer function, led us to investigate each one of these three types of factors. For example, one of Beurling's famous theorems on the shift operator related this very famous problem (on invariant spaces of the shift operator) to inner functions. In fact, it was Beurling who coined the term inner function. These are the functions whose $\mathbb{H}^\infty$ canonical factorization contains no outer part. In other words, a function $g(z)$ is an inner function if $|g(z)| \leq 1$ for all $|z| < 1$, and which has a unimodular radial limit a.e. on the unit circle $\mathbb{T}$.

The book will include topics on inner functions within the context of problems that are different from the Krzyż conjecture.

The way the author conducts his research is the routine of reading more and more papers and books on the topics. Sometimes, the choice of concentrating on a certain paper is not at all clear. Why is that relevant to the solution of the Krzyż conjecture? This is the part of the mathematician, which is based on the so-called mathematical intuition. Sometimes, this leads nowhere — the intuition was faulty. At the end of the day, the mathematician might get lucky and discover an interesting new fact. That is the reward. The author has made an effort to convey in this book his research habits. Mathematics has the reputation of a very formal and ordered science. Dry — some will say. Not lively. Well, it is so. But at the same time it is not. This other part is usually hidden very carefully by mathematicians. They write down in their papers only the final results, and the proofs are as short as possible. Almost cryptic. The way to achieve those results, though, is vivid and messy and filled (usually) with trials and errors. Mathematicians are human beings and so their psychology (sometimes it is a very interesting psychology) plays a vital role in the way they conduct the research. The reader of this book will quickly notice that some of the problems tackled appear in many chapters scattered across the book. The author intentionally did not just write down the final and most general result on a problem. This was not the purpose of the book. This is not how a mathematician discovers his theorems. The chapters of the book are ordered chronologically. This means that the new results that are proved in the book are recorded from author's handwritten research notes exactly in the calendar order. The book forms a research diary that indicates how the results that are presented were discovered, at least in what timely order. A mathematician works on a problem till either he solves it (a rare event) or till he is stuck and hits a dead end (that happens to the author frequently). He then moves to another problem on his list. Usually, the next problem is very different from the one he just left. This will open a new chapter in this book. Eventually, after enough time had elapsed, a new idea comes up and our mathematician goes back to the problem he abandoned and makes a small progress. A new chapter will describe the progress just made on the old problem. And so on goes our book-diary. What triggered the new idea on the old problem? Probably, something from the current chapter that dealt with another problem. It is not always clear what exactly was that trigger. It involves the psychology of the mathematician (the author). Sometimes, the author was puzzled. "How did I miss that before?!". Also, the research habits are of interest. Because of (old) age, the author writes his research notes with a pen on a paper. He fills long notebooks with his mathematics. Only later when he decides that a result is worth publishing, he types that part electronically. Why to write twice? Psychology.

The author dedicates this book to three persons. These are his grand children. At this time, these are three babies, two of which cannot yet speak (an adult language). It did not bother the author at all, who conducted long mathematical monologs while he was a babysitter. In particular, Noam heard a huge amount of mathematics while playing with grandpa. Did they understand? Of course not. They cannot even speak (at least the two young ones). This, I think, is better than talking to yourself, or making strange movements with your hands while walking in the street without any visible reason.

The book contains a lot of mathematics. Most of it is a review of results of many mathematicians whose papers were studied by the author. In those reviews, there are a lot of quotations but often the author expanded computations or arguments for him to understand better. But there are new results due to the author that appear in intermediate chapters. These appear (to the author's best knowledge) for the first time in print. Thus, an expert on these topics should benefit from reading these parts.

To sum up, an important objective in writing this book is to show that mathematicians are human beings that try to solve problems. Most of the time, they are stuck or they make mistakes. From time to time, they actually discover something new that is worth a scientific publication. There is nothing mysterious or divine about this particular brand of scientists. It is very important for our students to remember that!

It is appropriate to mention here the late Professor Wolfgang Heinrich Johannes Fuchs from Cornell. He passed away on February 24, 1997. He was a referee of one of my Ph.D. papers, which was badly written. He decided to expose himself to the author (the only time I encountered such a noble gesture) to invite me to Cornell to talk about my results and revise the paper. Professor Fuchs and his wife Dorothy hosted me for a week or so in their house, one of the most important events in my career. One evening we were walking around in Ithaca (a beautiful small city) and I dared asking Professor Fuchs: "What is the most important thing you accomplished in your career?". I expected to hear on one of the many important theorems he proved. He did not hesitate even for a second! His answer was "My students!". I never forgot that remarkable, simple and very human answer. That is for me the model of a true mathematician.

Chapter 2

A Partial Summary of What We Know: The Direct Approach

It was proved in [49] that the extremal function for a_N, $N \in \mathbb{Z}^+$ is of the following form:

$$F(z) = \exp\left(-\sum_{j=1}^{A} \lambda_j \frac{1 + k_j z}{1 - k_j z}\right),$$

where $1 \le A \le N$, $\lambda_j > 0$, $|k_j| = 1$, $j = 1, \ldots, A$. We have

$$\frac{1 + kz}{1 - kz} = (1 + kz)(1 + (kz) + (kz)^2 + \cdots) = 1 + 2((kz) + (kz)^2 + \cdots),$$

here $|k| = 1$ and $|z| < 1$. Hence,

$$\sum_{j=1}^{A} \lambda_j \frac{1 + k_j z}{1 - k_j z} = \sum_{j=1}^{A} \lambda_j \left\{1 + 2\sum_{m=1}^{\infty} (k_j z)^m\right\}$$

$$= \sum_{j=1}^{A} \lambda_j + 2\sum_{m=1}^{\infty} z^m \left(\sum_{j=1}^{A} \lambda_j k_j^m\right).$$

We recall the recursion that one obtains from exponentiation of power series. See [21], especially Chapter 4 (pp. 118–140) (for Grunsky inequalities, and the Goluzin–Lebedev inequalities), Chapter 5 (pp. 142–187) (exponentiation of the Grunsky inequalities), and Chapter 7 (pp. 214–231) (integral means). Namely, if $\sum_{n=0}^{\infty} \beta_n z^n = \exp(\sum_{k=1}^{\infty} \alpha_k z^k)$, then

$$\beta_0 = 1 \quad \text{and} \quad \beta_n = \frac{1}{n}\sum_{k=1}^{n} k\alpha_k \beta_{n-k}, \quad n = 1, 2, 3, \ldots.$$

We will use the notation $t = \sum_{j=1}^{A} \lambda_j > 0$. Then

$$e^t F(z) = \sum_{n=0}^{\infty} \gamma_n z^n = e^t \exp\left(-\sum_{j=1}^{A} \lambda_j \frac{1+k_j z}{1-k_j z}\right) = \exp\left(\sum_{k=1}^{\infty} \alpha_k z^k\right),$$

where

$$\left(t - \sum_{j=1}^{A} \lambda_j\right) - 2\sum_{m=1}^{\infty} z^m \left(\sum_{j=1}^{A} \lambda_j k_j^m\right) = \sum_{m=1}^{\infty} \alpha_m z^m,$$

hence, $\alpha_m = -2\sum_{j=1}^{A} \lambda_j k_j^m$. Using the recursion, we obtain

$$\gamma_n = -\frac{2}{n} \sum_{k=1}^{n} k \left(\sum_{j=1}^{A} \lambda_j k_j^k\right) \gamma_{n-k}.$$

If $\beta_n = e^{-t}\gamma_n$ are the coefficients of $F(z)$, then

$$\beta_n = -\frac{2}{n} \sum_{k=1}^{n} k \left(\sum_{j=1}^{A} \lambda_j k_j^k\right) \beta_{n-k},$$

where $\beta_0 = e^{-t}$, $n = 1, 2, \ldots$, $1 \le A \le N$, $\lambda_j > 0$, $|k_j| = 1$. Let us fix $t > 0$ and assume that $t = \sum_{j=1}^{A} \lambda_j$. We use the recursion above and write down the first few coefficients of $F(z)$ explicitly:

$$\beta_0 = \exp\left(-\sum_{j=1}^{A} \lambda_j\right), \quad \beta_1 = -2\left(\sum_{j=1}^{A} \lambda_j k_j\right) \exp\left(-\sum_{j=1}^{A} \lambda_j\right),$$

$$\beta_2 = -\left\{-2\left(\sum_{j=1}^{A} \lambda_j k_j\right)^2 + 2\left(\sum_{j=1}^{A} \lambda_j k_j^2\right)\right\} \exp\left(-\sum_{j=1}^{A} \lambda_j\right),$$

$$\beta_3 = -\frac{2}{3}\left\{2\left(\sum_{j=1}^{A} \lambda_j k_j\right)^3 - 6\left(\sum_{j=1}^{A} \lambda_j k_j\right)\left(\sum_{j=1}^{A} \lambda_j k_j^2\right)\right.$$

$$\left. + 3\left(\sum_{j=1}^{A} \lambda_j k_j^3\right)\right\} \exp\left(-\sum_{j=1}^{A} \lambda_j\right), \ldots.$$

Having these explicit expressions, let us use them in order to try and compute the largest values of their absolute values. We will see that already for $N = 2$ the computation becomes tedious.

$N = 1$: Then $A = 1$ and we need to estimate the absolute value of $\beta_1 = -2te^{-t}$.

$N = 2$: Let us take in the explicit formula for β_2, $A = 2$ but keep in mind that $\lambda_1, \lambda_2 \geq 0$ (and not $\lambda_1, \lambda_2 > 0$). We have $|k_1| = |k_2| = 1$, $\lambda_1 + \lambda_2 = t \geq 0$. Then $\beta_2 = 2\{(\lambda_1 k_1 + \lambda_2 k_2)^2 - (\lambda_1 k_1^2 + \lambda_2 k_2^2)\}e^{-t}$. We may assume that $k_1 = 1$, $k_2 = e^{i\theta}$. We denote

$$I(\theta) = (\lambda_1 + \lambda_2 e^{i\theta})^2 - (\lambda_1 + \lambda_2 e^{2i\theta})$$
$$= (\lambda_1^2 - \lambda_1) + 2\lambda_1\lambda_2 e^{i\theta} + (\lambda_2^2 - \lambda_2)e^{2i\theta}.$$
$$\Phi(\theta) = |I(\theta)|^2 = (\lambda_1^2 - \lambda_1)^2 + 4\lambda_1^2\lambda_2^2 + (\lambda_2^2 - \lambda_2)^2$$
$$+ 4\lambda_1\lambda_2\{(\lambda_1^2 - \lambda_1) + (\lambda_2^2 - \lambda_2)\}\cos\theta$$
$$+ 2(\lambda_1^2 - \lambda_1)(\lambda_2^2 - \lambda_2)\cos 2\theta.$$

In order to find the values of the extrema θ's, we differentiate $\Phi(\theta)$ and equate to zero.

$$-\Phi'(\theta) = 4\lambda_1\lambda_2\{(\lambda_1^2 - \lambda_1) + (\lambda_2^2 - \lambda_2)\}\sin\theta$$
$$+ 4(\lambda_1^2 - \lambda_1)(\lambda_2^2 - \lambda_2)\sin 2\theta = 0.$$

One possibility is $\sin\theta = 0$. Another possibility, when $\sin\theta \neq 0$, is

$$\lambda_1\lambda_2\{((\lambda_1^2 - \lambda_1) + (\lambda_2^2 - \lambda_2)) + 2(\lambda_1 - 1)(\lambda_2 - 1)\cos\theta\} = 0.$$

The possibilities are $(\lambda_1 = 0)$ or $(\lambda_2 = 0)$ or $\lambda_1\lambda_2 \neq 0$ and $((\lambda_1^2 - \lambda_1) + (\lambda_2^2 - \lambda_2)) + 2(\lambda_1 - 1)(\lambda_2 - 1)\cos\theta = 0$. The third possibility splits to $(\lambda_1 = 1$ or $\lambda_2 = 1)$ or $\lambda_1\lambda_2(\lambda_1 - 1)(\lambda_2 - 1) \neq 0$ and

$$\cos\theta = -\frac{(\lambda_1^2 - \lambda_1) + (\lambda_2^2 - \lambda_2)}{2(\lambda_1 - 1)(\lambda_2 - 1)} = -\frac{1}{2}\left(\frac{\lambda_1}{\lambda_2 - 1} + \frac{\lambda_2}{\lambda_1 - 1}\right).$$

If $\sin\theta = 0$, then we obtain $|I(\theta)| = |(\lambda_1 \pm \lambda_2)^2 - t^2|$. The case where

$$\cos\theta = -\frac{(\lambda_1^2 - \lambda_1) + (\lambda_2^2 - \lambda_2)}{2(\lambda_1 - 1)(\lambda_2 - 1)},$$

is more complicated, and leads to

$$|I(\theta)| = |\lambda_1 - \lambda_2| \frac{|1 - t|^{3/2}}{|\lambda_1 - 1)(\lambda_2 - 1)|^{1/2}}.$$

That should be consistent with

$$|\cos \theta| = \left| \frac{(\lambda_1^2 - \lambda_1) + (\lambda_2^2 - \lambda_2)}{2(\lambda_1 - 1)(\lambda_2 - 1)} \right| \leq 1.$$

This demonstrates how complicated a straightforward computation of coefficient by coefficient can be. Thus, it is probable that such a "brute force" attack will not lead anywhere.

Chapter 3

Indirect Approach: Representations of $R(z)$

We begin with a note on the special rational function that determines the structure of the extremals $F(z)$ for the Krzyż conjecture.

Proposition 3.0.1. *If $\lambda_1, \lambda_2, \ldots, \lambda_n > 0$, $k_1, k_2, \ldots, k_n \in \mathbb{C}$ satisfy both $|k_j| = 1$, $j = 1, \ldots, n$ and $i \neq j \Rightarrow k_i \neq k_j$, then the rational function*

$$R(z) = \lambda_1 \frac{1 + k_1 z}{1 - k_1 z} + \cdots + \lambda_n \frac{1 + k_n z}{1 - k_n z},$$

has n poles at the points $\overline{k}_1, \overline{k}_2, \ldots, \overline{k}_n$ and exactly n zeros on $|z| = 1$ and only on that circle.

Proof. If $F(z) = \exp(-R(z))$, then because in U we have

$$\Re\left(\frac{1 + k_j z}{1 - k_j z}\right) > 0$$

and because $\lambda_j > 0$, it follows that $\Re(-R(z)) \leq 0$ in U and so $|F(z)| \leq 1$ in U. If $R(z_0) = 0$ for some $z_0 \in U$, then $F(z_0) = 1$ and the maximum principle implies that $F(z) \equiv 1$, i.e., $R(z) \equiv 0$ which contradicts $R(0) = \sum_{j=1}^{n} \lambda_j > 0$. This proves that $R(z) \neq 0$ for all $z \in U$. Similarly, if we define $G(z) = \exp(R(z))$, then because of

$$\Re\left(\frac{1 + k_j z}{1 - k_j z}\right) < 0 \quad \text{for all } |z| > 1,$$

and because $\lambda_j > 0$, it follows that $\Re(R(z)) \leq 0$ and hence $|G(z) \leq 1$ for all $|z| > 1$. If $R(z_0) = 0$ for some $|z_0| > 1$, then $G(z_0) = 1$. We note that $G(\infty) = \exp(-\sum_{j=1}^{n} \lambda_j) < 1$. So, the maximum principle implies that $G(z) \equiv 1$, i.e., $R(z) \equiv 0$ in $|z| > 1$. This, however, contradicts $G(\infty) < 1$. $\qquad\square$

Proposition 3.0.1 tells us that given a fixed sequence $k_1, \ldots, k_n$, such that $|k_j| = 1$, $j = 1, \ldots, n$ and such that for $i \neq j$ we have $k_i \neq k_j$ then there are n functions $l_1, \ldots, l_n$ defined on $(\mathbb{Z}^+)^n$, such that $|l_j(\lambda_1, \ldots, \lambda_n)| = 1$, $j = 1, \ldots, n$ and such that

$$\lambda_1 \frac{1 + k_1 z}{1 - k_1 z} + \lambda_2 \frac{1 + k_2 z}{1 - k_2 z} + \cdots + \lambda_n \frac{1 + k_n z}{1 - k_n z}$$

$$= \left(\sum_{j=1}^{n} \lambda_j \right) \left(\frac{(1 + l_1 z)(1 + l_2 z) \cdots (1 + l_n z)}{(1 - k_1 z)(1 - k_2 z) \cdots (1 - k_n z)} \right). \qquad (3.0.1)$$

For example, if $n = 1$, then $l_1(\lambda_1) \equiv k_1$ a constant.

Problem: To give a characterization of all the n-tuples $(l_1, \ldots, l_n)$ that correspond to $(k_1, \ldots, k_n; l_1, \ldots, l_n)$ so that the identity in equation (3.0.1) will be satisfied. Is it true that $(k_1, \ldots, k_n; l_1, \ldots, l_n)$ determines a unique set $\{l_1, \ldots, l_n\}$ of numbers on ∂U that satisfy equation (3.0.1)? If not, then up to what is it unique?

Example 3.0.2. Let us consider the case $n = 2$. We have $k_1 \neq k_2$, $|k_1| = |k_2| = 1$ and $\lambda_1, \lambda_2 > 0$. Then our defining equation for l_1, l_2, (3.0.1) is

$$\lambda_1 \frac{1 + k_1 z}{1 - k_1 z} + \lambda_2 \frac{1 + k_2 z}{1 - k_2 z} \equiv (\lambda_1 + \lambda_2) \frac{(1 + l_1 z)(1 + l_2 z)}{(1 - k_1 z)(1 - k_2 z)}.$$

This could be written in terms of convex combination:

$$\left(\frac{\lambda_1}{\lambda_1 + \lambda_2} \right) \frac{1 + k_1 z}{1 - k_1 z} + \left(\frac{\lambda_2}{\lambda_1 + \lambda_2} \right) \frac{1 + k_2 z}{1 - k_2 z} \equiv \frac{(1 + l_1 z)(1 + l_2 z)}{(1 - k_1 z)(1 - k_2 z)}.$$

We have the real part conditions:

$$\Re \frac{(1 + l_1 z)(1 + l_2 z)}{(1 - k_1 z)(1 - k_2 z)} \text{ is } \begin{cases} > 0 & \text{for } |z| < 1 \\ < 0 & \text{for } |z| > 1 \end{cases}.$$

We know that the necessary and sufficient condition for that is the existence of an holomorphic function $w(z)$, $z \in U$ that satisfies the following:

$$\frac{(1 + l_1 z)(1 + l_2 z)}{(1 - k_1 z)(1 - k_2 z)} = \frac{1 + w(z)}{1 - w(z)},$$

where $|w(z)| \leq |z|$ for $z \in U$ and where $|w(z)| > 1$ for $|z| > 1$. The equation $\frac{1+w}{1-w} = L$ implies that $w = \frac{L-1}{L+1}$. In our case, this gives the following identity:

$$w(z) = \frac{(l_1 + l_2 + k_1 + k_2) + (l_1 l_2 - k_1 k_2)z}{2 + (l_1 + l_2 - k_1 - k_2)z + (l_1 l_2 + k_1 k_2)z^2} \cdot z.$$

This implies the following restrictions:

$$\left| \frac{(l_1 + l_2 + k_1 + k_2) + (l_1 l_2 - k_1 k_2)z}{2 + (l_1 + l_2 - k_1 - k_2)z + (l_1 l_2 + k_1 k_2)z^2} \cdot z \right| \text{ is } \begin{cases} \leq 1 & \text{for } |z| < 1 \\ > 1 & \text{for } |z| > 1 \end{cases}.$$

That pair of conditions on the absolute value is equivalent to the pair of conditions on $\Re \frac{(1+l_1 z)(1+l_2 z)}{(1-k_1 z)(1-k_2 z)}$. We could also argue as follows: The conformal mapping $\frac{(1+l_1 z)(1+l_2 z)}{(1-k_1 z)(1-k_2 z)}$ is covering $\{w \in \mathbb{C} \mid \Re w > 0\}$ exactly twice (needs to be proved). Hence, the mapping $U \to U$ given by

$$\left\{ \frac{(1+l_1 z)(1+l_2 z)}{(1-k_1 z)(1-k_2 z)} - 1 \right\} \Big/ \left\{ \frac{(1+l_1 z)(1+l_2 z)}{(1-k_1 z)(1-k_2 z)} + 1 \right\}, \quad z \in U$$

covers U exactly twice. Recall that $\frac{w-1}{w+1}$, $w \in \{\xi \in \mathbb{C} \mid \Re \xi > 0\}$ is a conformal mapping of the right half plan $\{\xi \in \mathbb{C} \mid \Re \xi > 0\}$ onto U. But the only mappings $U \to U$ that cover U exactly twice are Blaschke products of degree 2, i.e.,

$$e^{i\alpha} \left(\frac{z - \alpha_1}{1 - \overline{\alpha}_1 z} \right) \left(\frac{z - \alpha_2}{1 - \overline{\alpha}_2} \right), \quad |\alpha_1|, |\alpha_2| < 1, \quad 0 \leq \alpha < 2\pi.$$

So,

$$\left\{ \frac{(1+l_1 z)(1+l_2 z)}{(1-k_1 z)(1-k_2 z)} - 1 \right\} \Big/ \left\{ \frac{(1+l_1 z)(1+l_2 z)}{(1-k_1 z)(1-k_2 z)} + 1 \right\}$$
$$= e^{i\alpha} \left(\frac{z - \alpha_1}{1 - \overline{\alpha}_1 z} \right) \left(\frac{z - \alpha_2}{1 - \overline{\alpha}_2} \right), \tag{3.0.2}$$

for appropriate α, α_1, α_2. This could also be written as follows:

$$\frac{(1+l_1 z)(1+l_2 z)}{(1-k_1 z)(1-k_2 z)} = \left\{ 1 + e^{i\alpha} \left(\frac{z - \alpha_1}{1 - \overline{\alpha}_1 z} \right) \left(\frac{z - \alpha_2}{1 - \overline{\alpha}_2} \right) \right\} \Big/$$
$$\left\{ 1 - e^{i\alpha} \left(\frac{z - \alpha_1}{1 - \overline{\alpha}_1 z} \right) \left(\frac{z - \alpha_2}{1 - \overline{\alpha}_2} \right) \right\}.$$

Note that if we substitute $z = 0$ into equation (3.0.2), we obtain: $0 = e^{i\alpha}\alpha_1\alpha_2$. Thus, we assume $\alpha_1 = 0$ and get

$$\left\{\frac{(1+l_1z)(1+l_2z)}{(1-k_1z)(1-k_2z)} - 1\right\} \Big/ \left\{\frac{(1+l_1z)(1+l_2z)}{(1-k_1z)(1-k_2z)} + 1\right\}$$

$$= e^{i\alpha}z\left(\frac{z-\alpha_2}{1-\overline{\alpha_2}}\right). \tag{3.0.3}$$

From equation (3.0.3), we get

$$\frac{(l_1+l_2+k_1+k_2)+(l_1l_2-k_1k_2)z}{2+(l_1+l_2-k_1-k_2)z+(l_1l_2+k_1k_2)z^2} = e^{i\alpha}\left(\frac{z-\alpha_2}{1-\overline{\alpha_2}}\right).$$

Hence, the denominator on the left-hand side must be linear, i.e., $l_1l_2 + k_1k_2 = 0$. The zero of the numerator is

$$\alpha_2 = \frac{l_1+l_2+k_1+k_2}{-2k_1k_2},$$

and the pole is

$$(\overline{\alpha_2})^{-1} = \frac{2}{k_1+k_2-l_1-l_2}.$$

So, we get the following equations:

$$\begin{cases} l_1l_2 + k_1k_2 = 0 \\ l_1 + l_2 + k_1 + k_2 = -2k_1k_2\alpha_2, \\ -l_1 - l_2 + k_1 + k_2 = 2\overline{\alpha_2} \end{cases}$$

where $|k_1| = |k_2| = 1$ are given and we need $|l_1| = |l_2| = 1$ and $|\alpha_2| < 1$.

Remark 3.0.3. We can state a rather surprising principle of N-to-1 which links that geometric property in simply connected domains in $\mathbb{C}$, to the very simple arithmetic property of multiplying N automorphisms. The key is two-folded: (1) The Riemann mapping theorem and (2) The role of finite Blaschke products. We start with the following well-known result.

Theorem (The multiplicity of finite Blaschke Products). *A mapping $F : U \to U$ is an holomorphic N to 1 if and only if it is a Blaschke product of degree N. This means that*

$$F(z) = e^{i\alpha}\prod_{j=1}^{N}\left(\frac{z-\alpha_j}{1-\overline{\alpha}_jz}\right), \quad z \in U,$$

for some $0 \le \alpha < 2\pi$ and some α_j in U, $|\alpha_j| < 1$, $j = 1,\ldots,N$.

We easily conclude the following principle.

Theorem (The principle of N to 1 in simply connected domains).
(1) *Let D be a simply connected domain in $\mathbb{C}$ which is different from $\mathbb{C}$. A mapping $F : D \to D$ is an holomorphic N to 1 if and only if it is conjugate to a Blaschke product of degree N by a Riemann mapping $G : U \to D$. This means that*

$$F(z) = G\left(e^{i\alpha} \prod_{j=1}^{N} \left(\frac{G^{-1}(z) - \alpha_j}{1 - \overline{\alpha}_j G^{-1}(z)} \right) \right), \quad z \in D,$$

for some $0 \le \alpha < 2\pi$ and some α_j in U, $|\alpha_j| < 1$, $j = 1, \ldots, N$.
  (2) *A mapping $F : \mathbb{C} \to \mathbb{C}$ is an holomorphic N to 1 if and only if F is a polynomial $F \in \mathbb{C}[z]$ of degree $N = \deg F$.*

Remark 3.0.4. We recall (see [49]) that the extremal function for $|a_N|$ has the following form:

$$F(z) = \exp\left(-\sum_{j=1}^{A} \lambda_j \frac{1 + k_j z}{1 - k_j z} \right),$$

where $1 \le A \le N$, $\lambda_j > 0$ and $|k_j| = 1$ for $j = 1, \ldots, A$ and the k_j's are different from one another. We have

$$\frac{1 + k_j z}{1 - k_j z} = \frac{1 - |z|^2}{|1 - k_j z|^2} + 2i \frac{\Im\{k_j z\}}{|1 - k_j z|^2}.$$

If $k_j = e^{i\theta_j}$ and $z = re^{i\theta}$ for some $0 \le \theta_j, \theta < 2\pi$ and $0 \le r < 1$, then

$$\frac{1 + k_j z}{1 - k_j z} = \frac{1 - r^2}{|1 - re^{i(\theta + \theta_j)}|^2} + 2i \frac{r \sin(\theta + \theta_j)}{|1 - re^{i(\theta + \theta_j)}|^2}.$$

For $\theta = -\theta_j$ this gives $\frac{1+r}{1-r} \to +\infty$ when $r \to 1^-$. We conclude that $F(z)$ tends to exactly A zeros at the points $\overline{k}_j$, $j = 1, \ldots, A$ when $z \to \overline{k}_j$ from within U, radially. On the other hand, $|F(z)| \to 1$ at exactly A points $-\overline{l}_j$, $|l_j| = 1$, $j = 1, \ldots, A$. So, we conclude that $F(z)$ has exactly A zero points and these are $\overline{k}_j$, $j = 1, \ldots, A$. It has at least A one points and these are $-\overline{l}_j$, $j = 1, \ldots, A$. Thus, $F(z)$ has A maximal domains that tile up U. Each maximal domain contains on its boundary 2 adjacent numbers $\overline{k}_j$ and $-\overline{l}_j$. Also, we note that $R(z)$ maps $|z| = 1$ on the imaginary axis and hence, $F(\partial U) = \partial U$ except for the A zero points $\overline{k}_j$, $j = 1, \ldots, A$. It is natural to inquire about the automorphic functions of $F(z)$, i.e., those functions (typically multi-valued) $\phi(z)$ that satisfy the automorphic functional equation $F(z) = F(\phi(z))$. Clearly, for that to be satisfied, we need $R(\phi(z)) = R(z) + 2\pi i k$ for some $k \in \mathbb{Z}$. Thus,

$$\sum_{j=1}^{A} \lambda_j \frac{1 + k_j \phi(z)}{1 - k_j \phi(z)} = \sum_{j=1}^{A} \lambda_j \frac{1 + k_j z}{1 - k_j z} + 2\pi i k.$$

we already know that $R(z)$ is exactly A to 1 and so we expect at least A different solutions $\phi(z)$ for a given k. Indeed, since we expect geometrically at least A maximal domains of F that tile up U, we know in advance that we will be able to find at least A^2 solutions $\phi(z)$, that map the maximal domain Ω_i onto the maximal domain Ω_j where $i, j = 1, \ldots, A$ ($i = j$ is, of course allowed, the identity $\phi(z) = z$ does the A cases $i = j$. So, the different $\phi(z)$'s are at most $A^2 - A + 1$ in number). However, the total number of maximal domains is infinite.

Example 3.0.5. Let us consider the simplest case, $A = 1$. We assume $F(z) = \exp\left(-\frac{1+z}{1-z}\right)$. Then the automorphic function $\phi_k(z)$ is defined by

$$\frac{1 + \phi_k}{1 - \phi_k} = \frac{1 + z}{1 - z} + 2\pi i k.$$

$$\phi_k(z) = \frac{(1 - \pi i k)z + \pi i k}{-\pi i k z + (1 + \pi i k)}.$$

We note that

$$\begin{vmatrix} 1 - \pi i k & \pi i k \\ -\pi i k & 1 + \pi i k \end{vmatrix} = 1,$$

so, $\phi_k(U) = U$ conformally. Also, $\phi_k(1) = 1 \ \forall k \in \mathbb{Z}$ so all the maximal domains include on their boundary $\partial \Omega_k$ the 0 point $z = 1$ of $F(z)$. Also, $\phi_k \circ \phi_l = \phi_{k+l}$ so $\mathrm{Aut}(F)$ is isomorphic (as an abstract group) to $\mathbb{Z}$ and a generator is given by

$$\phi_1(z) = \frac{(1 - \pi i)z + \pi i}{-\pi i z + (1 + \pi i)},$$

or by

$$\phi_{-1}(z) = \frac{(1 + \pi i)z - \pi i}{\pi i z + (1 - \pi i)}.$$

The tiling of U is determined by arcs that start at the point $z = 1$.

Chapter 4

Indirect Approach: Properties of the Extremal Functions

Let us denote the extremal function for n ($n \in \mathbb{Z}^+$) by: $f(z) = a_0 + a_1 z + \cdots + a_n z^n + \cdots$.

Proposition 4.0.6. $2|a_0| \leq |a_n|$ and $|a_0| \leq \sqrt{2} - 1$.

Proof. Let $\epsilon > 0$. Let $F_\epsilon(z) = \exp\left(-\epsilon \frac{1+kz^n}{1-kz^n}\right)$, where $|k| = 1$. Then:

$$F_\epsilon(z) = 1 - \epsilon \frac{1 + kz^n}{1 - kz^n} + O(\epsilon^2) = (1 - \epsilon) - 2\epsilon(kz^n + k^2 z^{2n} + \cdots) + O(\epsilon^2).$$

Hence,

$$(f \cdot F_\epsilon)(z) = (1-\epsilon)(a_0 + a_1 z + \cdots + a_{n-1} z^{n-1}) + ((1-\epsilon)a_n - 2\epsilon k a_0)z^n + \cdots + O(\epsilon^2).$$

But $f \cdot F_\epsilon \in B$ and f is extremal for n, and hence,

$$|(1 - \epsilon)a_n - 2k\epsilon a_0 + O(\epsilon^2)| \leq |a_n|,$$

for any $\epsilon > 0$ and any k, $|k| = 1$. Hence, we conclude that $(1 - \epsilon)|a_n| + 2\epsilon|a_0| + O(\epsilon^2) \leq |a_n|$. So, $2|a_0| + O(\epsilon) \leq |a_n|$ and taking the limit when $\epsilon \to 0^+$ we obtain $2|a_0| \leq |a_n|$. To prove the second inequality, we will make a use in the well-known inequality: $|a_n| \leq 1 - |a_0|^2$. Combining this with the first inequality (that was just proved), we get $2|a_0| \leq 1 - |a_0|^2$. Hence $|a_0| \leq \sqrt{2} - 1$. $\square$

Similarly, we can obtain more inequalities: Let $\epsilon > 0$. This time, we define the variational multiplier by $F_\epsilon(z) = \exp\left(-\epsilon \frac{1+kz}{1-kz}\right)$ where $|k| = 1$. We have the following expansion:

$$F_\epsilon(z) = (1 - \epsilon) - 2\epsilon(kz + k^2 z^2 + \cdots) + O(\epsilon^2).$$

Multiplying, we get

$$(f \cdot F_\epsilon)(z) = (1 - \epsilon)a_0 + ((1 - \epsilon)a_1 - 2\epsilon k a_0)z$$
$$+ \cdots + ((1 - \epsilon)a_n - 2\epsilon(k a_{n-1} + k^2 a_{n-2}$$
$$+ \cdots + k^n a_0))z^n + \cdots + O(\epsilon^2).$$

Hence, since $f \cdot F_\epsilon \in B$ and since f is extremal for n,

$$|(1 - \epsilon)a_n - 2\epsilon(k a_{n-1} + k^2 a_{n-2} + \cdots + k^n a_0) + O(\epsilon^2)| \leq |a_n|,$$

for each $\epsilon > 0$ and each k, $|k| = 1$. Let us denote $S = k a_{n-1} + k^2 a_{n-2} + \cdots + k^n a_0$. Then we have $|(1 - \epsilon)a_n - 2\epsilon S + O(\epsilon^2)|^2 \leq |a_n|^2$. So,

$$(1 - \epsilon)^2|a_n|^2 - 4\epsilon(1 - \epsilon)\Re\{S\bar{a}_n\} + O(\epsilon^2) \leq |a_n|^2,$$

or $-4\epsilon(1 - \epsilon)\Re\{S\bar{a}_n\} + O(\epsilon^2) \leq (2 - \epsilon)\epsilon|a_n|^2$. If we divide by ϵ and after that take the limit when $\epsilon \to 0^+$ we get

$$-4\Re\{(k a_{n-1} + k^2 a_{n-2} + \cdots + k^n a_0)\bar{a}_n\} \leq 2|a_n|^2.$$

This can be written as

$$\Re\{|a_n|^2 + 2(k a_{n-1} + k^2 a_{n-2} + \cdots + k^n a_0)\bar{a}_n\} \geq 0.$$

Here, $k = e^{i\theta}$, $0 \leq \theta < 2\pi$ is arbitrary. Thus, the left-hand side is a non-negative trigonometric polynomial. This inequality can be re-written as follows:

$$\Re\left\{1 + 2\left(\left(\frac{a_{n-1}}{a_n}\right)k + \left(\frac{a_{n-2}}{a_n}\right)k^2 + \cdots + \left(\frac{a_0}{a_n}\right)k^n\right)\right\} \geq 0. \qquad (4.0.1)$$

We note that the last inequality includes as a special case the first inequality of Proposition 4.0.6. Namely, let us denote the n'th root of the unity by $e_n = \exp(2\pi i/n)$. Then

$$\sum_{k=1}^{n}\left(e_n^k\right)^j = \begin{cases} 0, & j = 1, 2, \ldots, n - 1 \\ n, & j = n \end{cases}. \qquad (4.0.2)$$

Using equation (4.0.1), we get

$$\Re\left\{1 + 2\left(\left(\frac{a_{n-1}}{a_n}\right)(e_n^l k) + \left(\frac{a_{n-2}}{a_n}\right)(e_n^l k)^2 + \cdots + \left(\frac{a_0}{a_n}\right)(e_n^l k)^n\right)\right\} \geq 0,$$

for $l = 1, 2, \ldots, n$. We add these n inequalities:

$$\Re\left\{n + 2\left(\left(\frac{a_{n-1}}{a_n}\right)k\left(\sum_{l=1}^{n}e_n^l\right) + \left(\frac{a_{n-2}}{a_n}\right)k^2\left(\sum_{l=1}^{n}(e_n^l)^2\right)\right.\right.$$
$$\left.\left. + \cdots + \left(\frac{a_0}{a_n}\right)k^n\left(\sum_{l=1}^{n}(e_n^l)^n\right)\right)\right\} \geq 0.$$

Using equation (4.0.2),

$$\Re\left\{n + 2\left(\frac{a_0}{a_n}\right)k^n n\right\} \geq 0,$$

which gives $1 \geq 2\left|\frac{a_0}{a_n}\right|$. This is the first inequality of Proposition 4.0.6. Another conclusion follows from the theorem of Caratheodory on the coefficients of functions in $H(U)$ that have non-negative real part, namely

Theorem (Caratheodory). *Let* $g(z) = 1 + b_1 z + b_2 z^2 + \cdots \in H(U)$ *satisfy* $\Re\{g(z)\} \geq 0 \ \forall\, |z| < 1$. *Then* $|b_k| \leq 2$, $k = 1, 2, 3, \ldots$.

In our case, we can take

$$g(z) = 1 + 2\left(\left(\frac{a_{n-1}}{a_n}\right)z + \left(\frac{a_{n-2}}{a_n}\right)z^2 + \cdots + \left(\frac{a_0}{a_n}\right)z^n\right),$$

and conclude that $2\left|\frac{a_{n-j}}{a_n}\right| \leq 2$ for $j = 1, 2, \ldots, n$. Hence, $|a_{n-j}| \leq |a_n|$ for $j = 1, 2, \ldots, n$. Next, we state and prove a generalization of the first inequality in Proposition 4.0.6.

Proposition 4.0.7. *Let* $f(z) = a_0 + a_1 z + \cdots + a_n z^n + \cdots \in B$ *be extremal for* $n \in \mathbb{Z}^+$. *Then for any* $l = 1, \ldots, n$, *we have the inequality:*

$$\Re\left\{1 + 2\sum_{j=1}^{[n/l]}\left(\frac{a_{n-l\cdot j}}{a_n}\right)k^j\right\} \geq 0,$$

for any $k = e^{i\theta} \in \mathbb{T}$, $0 \leq \theta < 2\pi$.

Proof. For a given $\epsilon > 0$, we define

$$F_\epsilon(z) = \exp\left(-\epsilon\frac{1 + kz^l}{1 - kz^l}\right) = 1 - \epsilon\left(\frac{1 + kz^l}{1 - kz^l}\right) + O(\epsilon^2)$$

$$= (1 - \epsilon) - 2\epsilon\sum_{j=1}^{\infty}(kz^l)^j + O(\epsilon^2).$$

On multiplication of $f(z)$ by $F_\epsilon(z)$, we obtain

$$(f \cdot F_\epsilon)(z) = \cdots + \left\{ (1 - \epsilon)a_n - 2\epsilon \sum_{j=1}^{[n/l]} k^j a_{n-l\cdot j} \right\} z^n + \cdots .$$

We note that if $l \cdot j + x = n$ then $x = n - l \cdot j$ so necessarily $n - l \cdot j \geq 0$, i.e., $n \geq l \cdot j$ so $j \leq \frac{n}{l}$. But j is an integer and hence $j \leq \left[\frac{n}{l}\right]$. Let us denote $S = \sum_{j=1}^{[n/l]} a_{n-l\cdot j} k^j$. Then we have

$$|(1 - \epsilon)a_n - 2\epsilon S + O(\epsilon^2)| \leq |a_n|.$$

This follows because $f \cdot F_\epsilon \in B$ and f is extremal for n. Squaring the last inequality, we obtain

$$|a_n|^2(1 - \epsilon)^2 - 4\epsilon(1 - \epsilon)\Re\{\bar{a}_n \cdot S\} + O(\epsilon^2) \leq |a_n|^2.$$

This could be written as follows: (letting $\epsilon \to 0^+$)

$$\Re\left\{ |a_n|^2 + 2\bar{a}_n \sum_{j=1}^{[n/l]} a_{n-l\cdot j} k^j \right\} \geq 0.$$

So,

$$\Re\left\{ 1 + 2 \sum_{j=1}^{[n/l]} \left(\frac{a_{n-l\cdot j}}{a_n} \right) \cdot k^j \right\} \geq 0.$$

$\square$

Remark 4.0.8. Thus, Proposition 4.0.7 constructs n non-negative trigonometric polynomials of the degrees $\left[\frac{n}{l}\right]$, $l = 1, 2, 3, \ldots, n$. Any convex combination of these polynomials will also be non-negative trigonometric polynomials normalized (too) to have the free term 1. What are these degrees? Here are the list of the first 14 values of n.

$n = 1$: 1
$n = 2$: 2, 1
$n = 3$: 3, 1, 1
$n = 4$: 4, 2, 1, 1
$n = 5$: 5, 2, 1, 1, 1
$n = 6$: 6, 3, 2, 1, 1, 1
$n = 7$: 7, 3, 2, 1, 1, 1, 1

$n = 8$: 8, 4, 2, 2, 1, 1, 1, 1
$n = 9$: 9, 4, 3, 2, 1, 1, 1, 1, 1
$n = 10$: 10, 5, 3, 2, 2, 1, 1, 1, 1, 1
$n = 11$: 11, 5, 3, 2, 2, 1, 1, 1, 1, 1, 1
$n = 12$: 12, 6, 4, 3, 2, 2, 1, 1, 1, 1, 1, 1
$n = 13$: 13, 6, 4, 3, 2, 2, 1, 1, 1, 1, 1, 1
$n = 14$: 14, 7, 4, 3, 2, 2, 2, 1, 1, 1, 1, 1, 1, 1.

It is not difficult to see that the degrees equal 1 for $n/2$ if n is even and $(n+1)/2$ values of l if n is odd. Hence, we obtain

Proposition 4.0.9. $2|a_j| \leq |a_n|$ *for* $j = 0, 1, \ldots, (n/2) - 1$ *if n is even, but for* $j = 1, 2, \ldots, (n+1)/2 - 1$ *if n is odd.*

Proposition 4.0.10. *Let* $f(z) = a_0 + a_1 z + \cdots + a_n z^n + \cdots \in B$ *be extremal for* $n \in \mathbb{Z}^+$. *Then for* $l = 1, 2, \ldots, n$, *we have* $H_l(z) \in \mathbb{C}[z]$, $\deg H_l(z) = [\frac{n}{l}]$, $H_l(z) \neq 0 \; \forall |z| < 1$, $H_l(0) > 0$ *and*

$$\Re\left\{ 1 + 2 \sum_{j=1}^{[n/l]} \left(\frac{a_{n-l \cdot j}}{a_n} \right) k^j \right\} = |H_l(k)|^2 \quad \forall |k| = 1.$$

Moreover, if $a_0, a_1, \ldots, a_n \in \mathbb{R}$ *then* $H_l(x) \in \mathbb{R}[x]$.

Proof. This follows by the Fejér–Riesz representation theorem of non-negative trigonometric polynomials. See [73, pp. 22–28] and in particular Theorem 2.3.3 (p. 25). Also see [85, pp. 72–78] and in particular number 43 (p. 77). $\square$

It will be convenient to write the following real non-negative polynomials

$$\Re\left\{ 1 + 2 \sum_{j=1}^{[n/l]} \left(\frac{a_{n-l \cdot j}}{a_n} \right) k^j \right\},$$

in the form of a linear combination over the real field of the functions: $\{1, \cos\theta, \sin\theta, \cos 2\theta, \sin 2\theta, \ldots\}$. In our case, $k^j = e^{ij\theta} = \cos j\theta + i \sin j\theta$. We note that for any $\alpha, \beta \in \mathbb{R}$, we have $\Re\{(\alpha + i\beta)k^j\} = \alpha \cos j\theta - \beta \sin j\theta$. In our case, $\alpha + i\beta = a_{n-l \cdot j}/a_n$ and hence, $\alpha = \Re\{a_{n-l \cdot j}/a_n\}$, $\beta = \Im\{a_{n-l \cdot j}/a_n\}$. So, by the above,

$$\Re\left\{ \left(\frac{a_{n-l \cdot j}}{a_n} \right) k^j \right\} = \left\{ \Re\left(\frac{a_{n-l \cdot j}}{a_n} \right) \right\} \cos j\theta - \left\{ \Im\left(\frac{a_{n-l \cdot j}}{a_n} \right) \right\} \sin j\theta.$$

We arrive at the following formula:

$$\Re\left\{1 + 2\sum_{j=1}^{[n/l]}\left(\frac{a_{n-l\cdot j}}{a_n}\right)k^j\right\}$$

$$= 1 + 2\sum_{j=1}^{[n/l]}\left(\left\{\Re\left(\frac{a_{n-l\cdot j}}{a_n}\right)\right\}\cos j\theta - \left\{\Im\left(\frac{a_{n-l\cdot j}}{a_n}\right)\right\}\sin j\theta\right).$$

This expresses our real non-negative trigonometric polynomial in the standard form:

$$\lambda_0 + \lambda_1\cos\theta + \mu_1\sin\theta + \cdots + \lambda_{[n/l]}\cos\left[\frac{n}{l}\right]\theta + \mu_{[n/l]}\sin\left[\frac{n}{l}\right]\theta,$$

where

$$\lambda_j = 2\Re\left(\frac{a_{n-l\cdot j}}{a_n}\right), \quad \mu_j = -2\Im\left(\frac{a_{n-l\cdot j}}{a_n}\right), \quad j = 0, 1, \ldots, \left[\frac{n}{l}\right],$$

here, $\lambda_0 = 1$ and $\mu_0 = 0$. In particular, for $j > 0$, we have

$$\lambda_j^2 + \mu_j^2 = 4\left|\frac{a_{n-l\cdot j}}{a_n}\right|^2.$$

Proposition 4.0.11. *Let $f(z) = a_0 + a_1 z + \cdots + a_n z^n + \cdots \in B$ be extremal for $n \in \mathbb{Z}^+$. Then:*

(1) *For $l = 1, \ldots, n$, we have the following inequality:*

$$\Re\left\{1 + 2\sum_{j=1}^{[n/l]}\left(\frac{a_{n-l\cdot j}}{a_n}\right)k^j\right\} \le \left[\frac{n}{l}\right] + 1,$$

    for any $|k| = 1$.

(2) *We have the following inequality:*

$$\left|\frac{a_{n-[n/l]\cdot l}}{a_n}\right| \le \frac{1}{2}.$$

(3) *We have the following inequality:*

$$\left|\frac{a_{n-l}}{a_n}\right| \le \cos\left(\frac{\pi}{[n/l]+2}\right).$$

(4) *We have the following inequality:*

$$\max_{|k|=1} \Re \left\{ 1 + 2 \sum_{j=1}^{[n/l]} \left(\frac{a_{n-l\cdot j}}{a_n} \right) k^j \right\} \geq 2 \left| \frac{a_{n-[n/l]\cdot l}}{a_n} \right|.$$

Proof. (1) Follows by number 50 in [85, p. 78]. (2) Follows by number 51 in [85, p. 79]. (3) Follows by number 52 in [85, p. 79]. (4) Follows by number 60 in [85, p. 80], taking into an account that for a real non-negative trigonometric polynomial (as is our case) $g(\theta)$, we have $0 \leq |m| = \min_\theta g(\theta) \leq M = \max_\theta g(\theta)$. $\square$

Remark 4.0.12. We would like to expand formulas for integral valued norms of a real non-negative trigonometric polynomial. To start with, for such a non-negative trigonometric polynomial $T(\theta)$, we have for any $1 \leq p < \infty$:

$$\|T\|_p = \left(\frac{1}{2\pi} \int_0^{2\pi} |T(\theta)|^p d\theta \right)^{1/p} = \left(\frac{1}{2\pi} \int_0^{2\pi} T(\theta)^p d\theta \right)^{1/p}.$$

We can write $T(\theta) = \lambda_0 + \lambda_1 \cos\theta + \mu_1 \sin\theta + \cdots + \lambda_n \cos n\theta + \mu_n \sin n\theta$ using $\cos m\theta = \frac{1}{2}(e^{im\theta} + e^{-im\theta})$, $\sin m\theta = \frac{1}{2i}(e^{im\theta} - e^{-im\theta})$. We get

$$T(\theta) = \frac{1}{2} \sum_{m=0}^{n} ((\lambda_m - i\mu_m)e^{im\theta} + (\lambda_m + i\mu_m)e^{-im\theta})$$

$$= b_{-n}e^{-in\theta} + b_{-(n-1)}e^{-(n-1)\theta} + \cdots + b_0 + b_1 e^{i\theta} + \cdots + b_n e^{in\theta},$$

where $b_m = \bar{b}_{-m} = \frac{1}{2}(\lambda_m - i\mu_m)$, $m = -n, \ldots, n$, $m \neq 0$ and $b_0 = \lambda_0$. We get: $\|T\|_1 = \lambda_0 = b_0$ and

$$\|T\|_2 = \left(\sum_{m=-n}^{n} |b_m|^2 \right)^{1/2} = \left\{ \lambda_0^2 + \frac{1}{2}((\lambda_1^2 + \mu_1^2) + \cdots + (\lambda_n^2 + \mu_n^2)) \right\}^{1/2}.$$

In general, for $p \in \mathbb{Z}^+$, we use the multinomial expansion and the orthogonality of $\{\ldots, e^{-2i\theta}, e^{-i\theta}, 1, e^{i\theta}, e^{2i\theta}, \ldots\}$ and obtain

$$\|T\|_p = \left(\frac{1}{2\pi} \int_0^{2\pi} T(\theta)^p d\theta \right)^{1/p}$$

$$= \left(\sum_{\substack{p_{-n} + \cdots + p_n = p \\ -np_{-n} + \cdots + np_n = 0}} \binom{p}{p_{-n} \cdots p_0 \cdots p_n} b_{-n}^{p_{-n}} \cdots b_0^{p_0} \cdots b_n^{p_n} \right)^{1/p}.$$

In our case, we computed above

$$\lambda_j = 2\Re\left(\frac{a_{n-l\cdot j}}{a_n}\right), \quad \mu_j = -2\Im\left(\frac{a_{n-l\cdot j}}{a_n}\right),$$

$$j = 1,\ldots,\left[\frac{n}{l}\right], \ \lambda_0 = b_0 = \|T\|_1, \ \mu_0 = 0.$$

So

$$b_j = \frac{1}{2}(\lambda_j - i\mu_j) = \left(\frac{a_{n-l\cdot j}}{a_n}\right), \quad b_{-j} = \bar{b}_j = \overline{\left(\frac{a_{n-l\cdot j}}{a_n}\right)}.$$

From this, we get the expansion we wanted for

$$\left\|\Re\left\{1 + 2\sum_{j=1}^{[n/l]}\left(\frac{a_{n-l\cdot j}}{a_n}\right)k^j\right\}\right\|_p.$$

We quote a result of Zygmund. In [107, p. 244], we find the following:
Let

$$\left|\frac{1}{2}a_0 + \sum_{k=1}^{n}(a_k\cos k\theta + b_k\sin k\theta)\right| \leq 1.$$

Then

$$\frac{1}{2}|a_0| + \sum_{k=1}^{n}(|a_k| + |b_k|) \leq A\sqrt{n},$$

where A is a numerical constant.

Conclusion:

$$1 + 2\sum_{j=1}^{[n/l]}\left(\left|\Re\left(\frac{a_{n-l\cdot j}}{a_n}\right)\right| + \left|\Im\left(\frac{a_{n-l\cdot j}}{a_n}\right)\right|\right)$$

$$\leq A\cdot\sqrt{\left[\frac{n}{l}\right]}\cdot\max_{\theta}\Re\left\{1 + 2\sum_{j=1}^{[n/l]}\left(\frac{a_{n-l\cdot j}}{a_n}\right)k^j\right\},$$

for any $l = 1,2,\ldots,n$. Next, we recall the following result:

Theorem (Fejes, L. [29]). *For any trigonometric polynomial of degree n, and for $3m > n$,*

$$\frac{1}{2\pi}\int_0^{2\pi}\left|\sum_{\nu=0}^{n}(a_\nu\cos\nu\theta + b_\nu\sin\nu\theta)\right|d\theta \geq \frac{2}{\pi}\sqrt{a_m^2 + b_m^2}.$$

The constant $\frac{2}{\pi}$ is best possible and the inequality is, in general, not true for $3m \leq n$.

We note that if (as in our case) the trigonometric polynomial is non-negative, then the above Fejes result becomes

$$a_0 \geq \frac{2}{\pi} \sqrt{a_m^2 + b_m^2} \quad \text{for } 3m > n.$$

In our particular case, the trigonometric polynomial is

$$\Re \left\{ 1 + 2 \sum_{j=1}^{[n/l]} \left(\frac{a_{n-l \cdot j}}{a_n} \right) k^j \right\},$$

$$\lambda_j = 2\Re \left(\frac{a_{n-l \cdot j}}{a_n} \right), \quad \mu_j = -2\Im \left(\frac{a_{n-l \cdot j}}{a_n} \right),$$

$$j = 1, 2, \ldots, \left[\frac{n}{l} \right] \quad \text{and } \lambda_0 = 1,$$

so that, we have $\sqrt{\lambda_j^2 + \mu_j^2} = 2 \left| \frac{a_{n-l \cdot j}}{a_n} \right|$. We get

$$1 \geq \frac{2}{\pi} \cdot 2 \cdot \left| \frac{a_{n-l \cdot j}}{a_n} \right| \quad \text{for } 3j > \left[\frac{n}{l} \right].$$

Proposition 4.0.13. *For the extremal function $f(z) = a_0 + a_1 z + \cdots + a_n z^n + \cdots \in B$ for $n \in \mathbb{Z}^+$ in the Krzyż conjecture we, have*

$$|a_{n-l \cdot j}| \leq \frac{\pi}{4} |a_n| \quad \text{for } 3j > \left[\frac{n}{l} \right].$$

In particular, we have

$$|a_k| \leq \frac{\pi}{4} |a_n| \quad \text{for } k < \frac{2}{3} n.$$

This result extends the indices domain in Proposition 4.0.9. We would like to extend the estimate (3) of Proposition 4.0.11 from $k = 1$ to any k in the admissible range. Let τ_n^* be be the set of all the non-negative trigonometric polynomials of degree at most n with a prescribed constant term,

$$T(\theta) = 1 + a_1 \cos \theta + b_1 \sin \theta + \cdots + a_n \cos n\theta + b_n \sin n\theta \geq 0.$$

The authors of [99] and [27] found an estimate for $\sqrt{a_k^2 + b_k^2}$, where k is an arbitrary integer between 1 and n.

Theorem (Szegö, G., [99], Egerváry, E. und Szász, O., [27]). *If $T \in \tau_n^*$, then*

$$\sqrt{a_k^2 + b_k^2} \leq \cos\left(\frac{\pi}{[n/k]+2}\right) \quad (1 \leq k \leq n).$$

The equality is attained only for the non-negative polynomials:

$$T^*(\theta) = r(\theta)\left\{1 + \frac{2}{p+2}\sum_{\nu=1}^{p}\left((p-\nu+1)\cos\nu\alpha \right.\right.$$
$$\left.\left. +\frac{\sin(\nu+1)\alpha}{\sin\alpha}\right)\cos\nu m(\theta - \psi)\right\},$$

where $r(\theta)$ is an arbitrary non-negative trigonometric polynomial from τ_q^*, $\alpha = \frac{\pi}{p+2}$, $p = \left[\frac{n}{k}\right]$, $n = pk + q$ $(0 \leq q < k)$ and ψ is an arbitrary constant. The bounds above are sharp.

An intermediate consequence is the following generalization of Proposition 4.0.11(3).

Proposition 4.0.14. *Let $f(z) = a_0 + a_1 z + \cdots + a_n z^n + \cdots \in B$ be extremal for $n \in \mathbb{Z}^+$. Then for $l = 1, \ldots, n$, $1 \leq k \leq [n/l]$, we have*

$$\left|\frac{a_{n-l\cdot k}}{a_n}\right| \leq \cos\left(\frac{\pi}{[[n/l]/k]+2}\right).$$

Another estimate is the following:

Proposition 4.0.15. *Let $f(z) = a_0 + a_1 z + \cdots + a_n z^n + \cdots \in B$ be extremal for $n \in \mathbb{Z}^+$. Then*

$$\left|\frac{d}{d\theta}\Re\left\{1 + 2\sum_{j=1}^{[n/l]}\left(\frac{a_{n-l\cdot j}}{a_n}\right)e^{i\theta j}\right\}\right| \leq \sqrt{\frac{[n/l]+1}{2}\binom{[n/l]+2}{3}},$$

for $l = 1, \ldots, n$.

Proof. The following theorem was proved in [27]: If $T \in \tau_n^*$, then

$$|T'(\theta)| \leq \sqrt{\frac{n+1}{2}\binom{n+2}{3}}.$$

Our proposition follows with

$$T(\theta) = \Re\left\{1 + 2\sum_{j=1}^{[n/l]}\left(\frac{a_{n-l\cdot j}}{a_n}\right)e^{i\theta j}\right\},$$

and $\left[\frac{n}{l}\right]$ replacing n. $\qquad\square$

Remark 4.0.16. We note that

$$\frac{d}{d\theta}\Re\{(\alpha + i\beta)e^{i\theta j}\} = -j\Im\{(\alpha + i\beta)e^{i\theta j}\}.$$

So, the inequality in Proposition 4.0.15 could be written as follows:

$$\left|\Im\left\{\sum_{j=1}^{[n/l]}j\left(\frac{a_{n-l\cdot j}}{a_n}\right)e^{i\theta j}\right\}\right| \leq \frac{1}{2}\cdot\sqrt{\frac{[n/l]+1}{2}\left(\frac{[n/l]+2}{3}\right)}.$$

Chapter 5

Corona

For reference, we will use [59], in particular, Chapter XI: Wolff's proof of the Corona theorem (pp. 252–262).

Theorem (Carleson, 1962 [12]). *Let $f_1, \ldots, f_n \in H_\infty$. Suppose $\|f_k\|_\infty \leq 1$ for $k = 1, \ldots, n$ and that for some $\delta > 0$ we have $\sup_k |f_k(z)| > \delta$ for all z, $|z| < 1$. There is a number $M(\delta, n)$ depending only on δ and n such that $g_1 f_1 + g_2 f_2 + \cdots + g_n f_n \equiv 1$ on $\{|z| < 1\}$ with some functions $g_k \in H_\infty$ satisfying $\|g_k\|_\infty \leq M(\delta, n)$.*

Remark 5.0.17. Sometimes, the assumption "$\sup_k |f_k(z)| > \delta$ for all z, $|z| < 1$" is replaced by "$|f_1(z)| + \cdots + |f_n(z)| > \epsilon > 0$ for all z, $|z| < 1$". This is clear with the choice $\delta = \frac{1}{n}\epsilon$. For with that choice of δ if the assumption $\sup_k |f_k(z)| > \delta$ is violated at z_0, $|z_0| < 1$, then it means that $|f_k(z_0)| < \delta$ for $k = 1, \ldots, n$ and hence, $|f_1(z_0)| + \cdots + |f_n(z_0)| < n\delta = \epsilon$, which violates the second assumption.

How is the Corona Theorem relevant to the Krzyż conjecture? Let

$$F(z) = \exp\left(-\sum_{j=1}^{A} \lambda_j \frac{1 + k_j z}{1 - k_j z}\right)$$

be an extremal function for $n \in \mathbb{Z}^+$. Thus, $\lambda_j > 0$, $|k_j| = 1$ for $j = 1, \ldots, A$ and where if $1 \leq j_1 \neq j_2 \leq A$, then $k_{j_1} \neq k_{j_2}$, and where $A \leq n$. Then the pair $\frac{1}{2}(F + f)$ and f have no common zero in U for any $f \in H_\infty$, such that $\|f\|_\infty \leq 1$. Also $\|\frac{1}{2}(F + f)\|_\infty \leq 1$. Hence,

$$\left|\frac{1}{2}(F(z) + f(z))\right| + |f(z)| > 0 \quad \text{for } z, |z| < 1.$$

However, this is short of the assumption needed in the Corona Theorem (with $n = 2$, $f_1 = \frac{1}{2}(F + f)$ and $f_2 = f$). It may happen that

$$\inf_{|z|<1} \left(\left| \frac{1}{2}(F(z) + f(z)) \right| + |f(z)| \right) = 0.$$

We note that the function $F(z)$ tends radially to 0 exactly at the A points $\{\overline{k}_1, \ldots, \overline{k}_A\}$ on $|z| = 1$. If also $f(\overline{k}_{j_0}) = 0$ for one of these points, then $\frac{1}{2}(F(\overline{k}_{j_0}) + f(\overline{k}_{j_0})) = 0$. Let us call a pair $(f_1(z), f_2(z)) \in H^2_\infty$, a Corona pair, if $\|f_1\|_\infty \leq 1$ and $\|f_2\|_\infty \leq 1$ and there is an $\epsilon > 0$ such that $|f_1(z)|+|f_2(z)| > \epsilon$ for all $|z| < 1$. Equivalently, $\sup_{k=1,2} |f_k(z)| > \frac{1}{2}\epsilon = \delta > 0$. We can easily construct Corona pairs from extremal functions to the Krzyż conjecture for $n \in \mathbb{Z}^+$. Let us take

$$F_1(z) = \exp\left(-\sum_{j=1}^{A} \lambda_j \frac{1 + k_j z}{1 - k_j z} \right), \quad F_2(z) = \exp\left(-\sum_{j=1}^{B} \omega_j \frac{1 + \xi_j z}{1 - \xi_j z} \right),$$

$(A, B \leq n)$ where $\{\overline{k}_1, \ldots, \overline{k}_A\} \cap \{\overline{\xi}_1, \ldots, \overline{\xi}_B\} = \emptyset$.

Example 5.0.18. Here are few examples of Corona pairs:

(1) (F_1, F_2).
(2) $(F_1(z), F_1(kz))$ where $\{\overline{k}_1, \ldots, \overline{k}_A\} \cap \{\overline{kk}_1, \ldots, \overline{kk}_A\} = \emptyset$.
(3) $\left(\frac{1}{2}(F_1 + f), f \right)$, where $f \in H_\infty$, $\|f\|_\infty \leq 1$ and $f(z)$ is analytic in each of $\{\overline{k}_1, \ldots, \overline{k}_A\}$ and $f(\overline{k}_j) \neq 0$ for $j = 1, \ldots, A$.

Chapter 6

Indirect Approach: More Properties of the Extremal Functions

We continue our discussion from Chapter 4. We first point at a useful inequality.

Proposition 6.0.19. *If* $|z|$, $|w| \leq 1$ *and* $0 < \alpha < 1$, *then*

$$\left| \frac{\alpha z + (1 - \alpha)w + zw}{(1 - \alpha)z + \alpha w + 1} \right| \leq 1.$$

This becomes an equality if and only if $|z| = |w| = 1$.

Proof.   Let us denote

$$A = \alpha \left(\frac{1 - z}{1 + z} \right), \quad B = (1 - \alpha) \left(\frac{1 - w}{1 + w} \right).$$

By the assumption $|z|$, $|w| \leq 1$, $0 < \alpha < 1$, it follows that $\Re\{A\} \geq 0$, $\Re\{B\} \geq 0$ and hence $\Re\{A + B\} \geq 0$. We note that for $z = w = 0$, we have $A + B = 1$. Hence, there exists a ξ such that

$$|\xi| \leq 1 \quad \text{and} \quad A + B = \frac{1 - \xi}{1 + \xi}.$$

Solving for ξ yields

$$\xi = \frac{\alpha z + (1 - \alpha)w + zw}{(1 - \alpha)z + \alpha w + 1}.$$

So, $|\xi| \leq 1$ proves our inequality. This becomes an equality if and only if $\Re\{A + B\} = 0$. This is equivalent to $\Re\{A\} = \Re\{B\} = 0$, which is equivalent to $1 - |z|^2 = 1 - |w|^2 = 0$. $\qquad \square$

Next, we prove a result that roughly says that if we knew how to replace the real part $\Re\{\cdot\}$, in the inequality of Proposition 4.0.7 (and for just one case, namely $l = 1$) by the absolute value, $|\cdot|$, we had a proof of the Krzyż conjecture. Thus, it is a type of a reduction theorem. It is going to be the first of a list of reductions we will find for the Krzyż conjecture.

Theorem 6.0.20. *Let* $f(z) = a_0 + \cdots + a_n z^n + \cdots \in B$ *be an extremal function for* $n \in \mathbb{Z}^+$. *If it were true that for all* k, $|k| = 1$ *we had*

$$2|k a_{n-1} + k^2 a_{n-2} + \cdots + k^n a_0| \leq |a_n|,$$

then $|a_0| \leq 1/e$ *and if we had equality,* $|a_0| = 1/e$, *then*

$$a_1 = a_2 = \cdots = a_{n-1} = 0 \quad \text{and} \quad |a_n| = \frac{2}{e}.$$

In particular, the Krzyż conjecture would follow.

Remark 6.0.21. We note that we actually proved the following:

$$-2\Re\left\{\bar{a}_n\left(k a_{n-1} + k^2 a_{n-2} + \cdots + k^n a_0\right)\right\} \leq |a_n|^2,$$

for all k, $|k| = 1$. See equation (4.0.1) or Proposition 4.0.7 in the case $l = 1$. Thus, the assumption in Theorem 6.0.20 is that this inequality is valid even when $\Re\{\cdot\}$ is replaced by $|\cdot|$. So, had we been able to prove the following:

$$2\sqrt{(\Re\{\bar{a}_n(k a_{n-1} + \cdots + k^n a_0)\})^2 + (\Im\{\bar{a}_n(k a_{n-1} + \cdots + k^n a_0)\})^2} \leq |a_n|^2,$$

the Krzyż conjecture would have followed.

A proof of Theorem 6.0.20. By [49], there are non-negative numbers $\lambda_j \geq 0$ and points on the unit circle, k_j, $|k_j| = 1$, $j = 1, \ldots, n$, so that

$$f(z) = \exp\left(-\sum_{j=1}^{n} \lambda_j \frac{1 + k_j z}{1 - k_j z}\right),$$

where we assumed that $a_0 = \exp(-\sum_{j=1}^{n} \lambda_j) > 0$ is positive. By the exponentiation principle, we obtain the standard recursion, which for the value $n \in \mathbb{Z}^+$ gives

$$n a_n = -2 \sum_{k=1}^{n} k \left(\sum_{j=1}^{n} \lambda_j k_j^k\right) a_{n-k} = -2 \sum_{j=1}^{n} \lambda_j k_j \left(\sum_{k=1}^{n} k k_j^{k-1} a_{n-k}\right). \tag{6.0.1}$$

By Bernstein–Szegö Inequality (Theorem 5.1.3 in [7, p. 232]) or even just by its corollary (Corollary 5.1.6 in [7, p. 233]), we have

$$|a_{n-1} + 2k_j a_{n-2} + 3k_j^2 a_{n-3} + \cdots + n k_j^{n-1} a_0| \leq n\|a_{n-1}k + a_{n-2}k^2 + \cdots + a_0 k^n\|_\infty,$$

where the norm $\|\cdot\|_\infty$ is taken on the unit circle $|k| = 1$. Moreover, there is an equality, if and only if $a_{n-1} = \cdots = a_1 = 0$. So, by equation (6.0.1), we have

$$n|a_n| \leq 2 \sum_{j=1}^{n} \lambda_j |a_{n-1} + 2k_j a_{n-2} + \cdots + nk_j^{n-1} a_0|$$

$$\leq 2 \sum_{j=1}^{n} \lambda_j n \|a_{n-1}k + \cdots + a_0 k^n\|_\infty.$$

By the assumption $2\|a_{n-1}k + \cdots + a_0 k^n\|_\infty \leq |a_n|$ and hence $n|a_n| \leq n|a_n|(\sum_{j=1}^{n} \lambda_j)$. Thus, we proved that $1 \leq \sum_{j=1}^{n} \lambda_j$ and so

$$a_0 = \exp\left(-\sum_{j=1}^{n} \lambda_j\right) \leq \frac{1}{e}$$

as was to be asserted. Now, we treat the case of equality which by Bernstein–Szegö is equivalent to $a_1 = a_2 = \cdots = a_{n-1} = 0$. Proving now the Krzyż conjecture is a well-known routine. We will recall that for the sake of convenience: In the case of equality, we have

$$h(z) = \frac{1}{f(z)} = \frac{1}{a_0 + a_n z^n + \cdots} = b_0 + b_n z^n + \cdots,$$

where $b_0 = 1/a_0$ and $b_n = -a_n/a_0^2$. Now $h(z) \in H(U)$ and clearly $|h(z)| > 1$ for all $z \in U$. We define the analytic branch:

$$g(z) = \log h(z) = d_0 + d_n z^n + \cdots, \quad \text{where} \quad d_0 > 0.$$

If $g(z) = u(z) + iv(z)$, then $u(z) = \Re\{g(z)\} = \log|h(z)| > 0$, and so if we define

$$F(z) = \frac{1}{z^{n-1}}\left(\frac{g(z) - d_0}{g(z) + d_0}\right),$$

then $F(z) \in H(U)$ and satisfies $F(0) = 0$ and $|F(z)| < 1$ for

$$|F(z)|^2 = \frac{1}{|z^2|^{n-1}} \frac{(u - d_0)^2 + v^2}{(u + d_0)^2 + v^2}.$$

By the Schwarz Lemma: $|F'(0)| \leq 1$. But $|F'(0)| = |d_n/(2d_0)|$ and we deduce that $|d_n| \leq 2|d_0|$. We recall that $d_0 = \log b_0$, $d_n = b_n/b_0$ and hence, $|b_n| \leq 2|b_0| \log|b_0|$. This means that

$$\left|\frac{a_n}{a_0^2}\right| \leq 2\frac{1}{|a_0|} \log \frac{1}{|a_0|}.$$

Hence, $|a_n| \leq -2|a_0| \log |a_0|$. The maximum of $-t \log t$ in $0 \leq t \leq 1$ is attained for only $t = 1/e$ and its value is also $1/e$. Hence, $|a_n| \leq 2/e$. However, the extremal function for n,

$$\exp \left(-\frac{1 - z^n}{1 + z^n} \right),$$

satisfies the equality $|a_n| = 2/e$ and the Krzyż conjecture follows. $\qquad\square$

Here is another similar variant of the same type of a theorem.

Theorem 6.0.22. *Let $f(z) = a_0 + \cdots + a_n z^n + \cdots \in B$ be an extremal function for $n \in \mathbb{Z}^+$. Then*

(i) *If $n = 2p+1$ is an odd natural number and if $a_1 = a_3 = \cdots = a_{2p-1} = 0$, then $|a_0| \leq 1/e$ and equality occurs if and only if $a_n = 2/e$.*

(ii) *If $n = 2p$ is an even natural number and if $a_2 = a_4 = \cdots = a_{2p-2} = 0$, then $|a_0| \leq 1/e$ and equality occurs if and only if $a_n = 2/e$.*

Proof. (i) We recall (see Remark 6.0.21) that we proved the following inequality:

$$-2\Re\{\bar{a}_n(k a_{n-1} + k^2 a_{n-2} + \cdots + k^n a_0)\} \leq |a_n|^2$$

for all k, $|k| = 1$. By our assumptions on the coefficients of $f(z)$ in (i), we get

$$|2\Re\{\bar{a}_n(k a_{n-1} + k^2 a_{n-2} + \cdots + k^n a_0)\}| \leq |a_n|^2$$

for all k, $|k| = 1$. We can assume that

$$f(z) = \exp \left(-\sum_{j=1}^{N} \lambda_j \frac{1 + k_j z}{1 - k_j z} \right),$$

where $1 \leq N \leq n$, $\lambda_j > 0$, $|k_j| = 1$. Differentiation of the last identity gives

$$\sum_{m=1}^{\infty} m a_m z^{m-1} = -\sum_{j=1}^{N} \lambda_j \frac{2k_j}{(1 - k_j z)^2} \sum_{m=0}^{\infty} a_m z^m.$$

By equating coefficients, this proves the standard recursion:

$$(m+1)a_{m+1} = -2\sum_{j=1}^{N} \lambda_j \left\{ \sum_{l=0}^{m} (l+1)k_j^{l+1} a_{m-l} \right\}, \quad m = 0, 1, 2, \ldots, a_0$$

$$= \exp \left(-\sum_{j=1}^{N} \lambda_j \right). \tag{6.0.2}$$

We denote $p(z) = -2\overline{a_n}\sum_{m=1}^{N} a_{n-m}z^m$. Using equation (6.0.2) with $m = n - 1$ and multiplying it by $\overline{a_n}$, we get

$$n|a_n|^2 = \sum_{j=1}^{N} \lambda_j k_j p'(k_j) \leq \sum_{j=1}^{N} \lambda_j |p'(k_j)| \leq$$

(by the Bernstein–Szegö inequality)

$$\leq \sum_{j=1}^{N} \lambda_j \cdot n \cdot \max_{|k|=1} |\Re p(k)| \leq \sum_{j=1}^{N} \lambda_j n|a_n|^2.$$

Thus, $1 \leq \sum_{j=1}^{N} \lambda_j$ or (equivalently) $|a_0| \leq 1/e$. To have equality, we must have equality in the Bernstein–Szegö inequality, so $p(z) = -2\overline{a_n}a_0 z^n$ and hence, $a_1 = a_2 = \cdots = a_{n-1} = 0$. We must have equality in the inequality above, as mentioned in Remark 6.0.21. So, $|a_n| = 2|a_0| = 2/e$.

(ii) has a similar proof to that of (i). $\qquad\square$

Let us continue extracting properties of the extremal function for $n \in \mathbb{Z}^+$. Here is a simple example:

Proposition 6.0.23. *Let $f(z) = a_0 + \cdots + a_n z^n + \cdots \in B$ be extremal for $n \in \mathbb{Z}^+$. Let $k > 1$ be any integer. Then*

$$\left| \sum_{\substack{i_1,\ldots,i_{k-1} \geq 0 \\ i_1 + \cdots + i_{k-1} \leq n}} a_{i_1}\ldots a_{i_{k-1}} a_{n-i_1-i_2-\cdots-i_{k-1}} \right| \leq |a_n|.$$

Proof. We just have to note that $f(z)^k \in B$ and that the coefficient of z^n in the power series expansion of $f(z)^k$ is on the left-hand side of our inequality (the expression under the absolute value sign). $\qquad\square$

In fact, we can extend the claim in Proposition 6.0.23 to any positive power of $f(z)$, namely Proposition 6.0.6.

Proposition 6.0.24. *Let $f(z) = a_0 + \cdots + a_n z^n + \cdots \in B$ be extremal for $n \in \mathbb{Z}^+$. Let $p > 0$ be any positive number. Then*

$$\lim_{r \to 1^-} \left| \left(\frac{1}{2\pi}\right) \int_0^{2\pi} \exp(p \log f(re^{i\theta})) e^{-in\theta} d\theta \right| \leq |a_n|.$$

Proof. This follows by $f(z)^p \in B$ and the fact that the integral times $1/(2\pi)$ under the absolute value sign tend to the coefficient of z^n in $f(z)^p$. $\qquad\square$

More interesting is the vanishing of the two neighboring coefficients of a_n in $f(z)$.

Proposition 6.0.25. *Let $f(z) = a_0 + \cdots + a_n z^n + \cdots \in B$ be extremal for $n \in \mathbb{Z}^+$, where $n \geq 2$. Then $a_{n-1} = a_{n+1} = 0$.*

Proof. Let $\epsilon \in U$. Then

$$\frac{z - \epsilon}{1 - \bar{\epsilon}z} = (z - \epsilon)\left(1 + \sum_{j=1}^{\infty}(\bar{\epsilon}z)^j\right) = z - \epsilon + \bar{\epsilon}z^2 + O(|\epsilon|^2) \text{ for } |\epsilon| \to 0.$$

For any $k \in \mathbb{Z}^+$, we have

$$(z - \epsilon + \bar{\epsilon}z^2)^k = -k\epsilon z^{k-1} + z^k + k\bar{\epsilon}z^{k+1} + O(|\epsilon|^2).$$

We expand $f((z - \epsilon)/(1 - \bar{\epsilon}z))$ to its power series in z and compute the coefficient of z^n up to $O(|\epsilon|^2)$. Since $f((z - \epsilon)/(1 - \bar{\epsilon}z)) \in B$, this coefficient has an absolute value which is not larger than $|a_n|$. The contribution to z^n clearly comes from the following three elements:

$$a_{n-1}(-(n-1)\epsilon z^{n-2} + z^{n-1} + (n-1)\bar{\epsilon}z^n) + a_n(-n\epsilon z^{n-1} + z^n + n\bar{\epsilon}z^{n+1})$$

$$+ a_{n+1}(-(n+1)\epsilon z^n + z^{n+1} + (n+1)\bar{\epsilon}z^{n+2})$$

$$= \cdots + ((n-1)\bar{\epsilon}a_{n-1} + a_n - (n+1)\epsilon a_{n+1})z^n + \cdots .$$

We conclude that when $\epsilon \to 0$ ($\epsilon \in \mathbb{C}$), we approach asymptotically the following inequality:

$$|(n-1)\bar{\epsilon}a_{n-1} + a_n - (n+1)\epsilon a_{n+1}| \leq |a_n|.$$

We distinguish among the following three cases.

Case 1: $|(n-1)a_{n-1}| < |(n+1)a_{n+1}|$.

Case 2: $|(n-1)a_{n-1}| = |(n+1)a_{n+1}|$.

Case 3: $|(n-1)a_{n-1}| > |(n+1)a_{n+1}|$.

These are all the possibilities.

Case 1: We pick $\arg \epsilon$ so that $|a_n - (n+1)\epsilon a_{n+1}| = |a_n| + (n+1)|\epsilon||a_{n+1}|$. Then by geometry or the triangle inequality, we get

$$|(n-1)\bar{\epsilon}a_{n-1} + a_n - (n+1)\epsilon a_{n+1}| \geq |a_n| + (n+1)|\epsilon||a_{n+1}| - (n-1)|\epsilon||a_{n-1}| > |a_n|,$$

which is a contradiction. The strict second inequality is because we are in Case 1.

Case 3: This is similarly impossible. This leaves us with **Case 2:**

In this case, the choice of $\arg \epsilon$ as in Case 1 gives us $\arg \epsilon = \arg \bar{\epsilon}$, so that we have cancellation $(n-1)\bar{\epsilon}a_{n-1} - (n+1)\epsilon a_{n+1} = 0$. For any $w \in \mathbb{C}^{\times}$, we have $\arg \overline{w} = -\arg w$. In our case $w = \epsilon$, we get $\arg \epsilon = 0 \,(\mathrm{mod}\, 2\pi)$. Geometrically, this means that $\epsilon > 0$ is a real and positive number (an horizontal vector pointing to the right). Suppose that instead of $0 \,(\mathrm{mod}\, 2\pi)$ we take $\arg \epsilon > 0$ and small. Then $\arg \bar{\epsilon}$ is the negative of that. But then, the total vector $\overline{v} = (n-1)\bar{\epsilon}a_{n-1} + a_n - (n+1)\epsilon a_{n+1}$ is longer than a_n, unless $a_{n-1} = a_{n+1} = 0$. The proof is now completed. $\square$

Remark 6.0.26. In the discussion of Case 2, we assumed that $a_n > 0$, as indeed we may (for simplicity).

Proposition 6.0.27. *Let* $f(z) = a_0 + \cdots + a_n z^n + \cdots \in B$ *be extremal for* $n \in \mathbb{Z}^+$. *Then we have*

$$\begin{cases} |a_j| \leq |a_n|, & j = 1, \ldots, n-1 \\ 2|a_0| \leq |a_n| \end{cases}.$$

Proof. We consider the polynomial: $p(z) = |a_n|^2 + 2\bar{a}_n(a_{n-1}z + \cdots + a_0 z^n)$. Then $\pm \Re\left\{p(z)\right\}$ are harmonic functions in $\overline{U}$. By equation (4.0.1) or Proposition 4.0.7 in the case $l = 1$, we have $0 \leq \Re\{p(z)\}$ for $|z| = 1$. By a theorem of C. Caratheodory for the coefficients in the power series expansion of functions in $H(U)$ with a non-negative real part, we get

$$\left|\frac{2\bar{a}_n a_j}{|a_n|^2}\right| \leq 2, \quad j = 1, \ldots, n-1.$$

Hence, $|a_j| \leq |a_n|$ for $j = 1, \ldots, n-1$. The second inequality $2|a_0| \leq |a_n|$ was proved in Proposition 4.0.1. We will now give another proof of that same inequality using properties of harmonic functions. We have $p(z) = 2\bar{a}_n a_0(z - \alpha 1)\ldots(z - \alpha_n)$. We saw above (the maximum principle) that $\Re\{p(z)\} > 0$ (notice! strict inequality holds) in U and hence, $|\alpha_j| \geq 1$ for $j = 1, \ldots, n$. By Vieté, we deduce our result:

$$\left|\frac{|a_n|^2}{2\bar{a}_n a_0}\right| = |\alpha_1 \ldots \alpha_n| \geq 1. \qquad \square$$

Next, we will prove a list of identities and inequalities that relate the $2n$ parameters in the representation of the extremal function for $n \in \mathbb{Z}^+$, as given in [49].

Proposition 6.0.28. *Let* $f(z) = a_0 + \cdots + a_n z^n + \cdots \in B$ *be extremal for* $n \in \mathbb{Z}^+$. *Assuming* $a_0 > 0$, *there are* $\lambda_j \geq 0$ *and* k_j, $|k_j| = 1$, $j = 1, \ldots, n$

such that (see [49])

$$f(z) = \exp\left(-\sum_{j=1}^{n} \lambda_j \frac{1 + k_j z}{1 - k_j z}\right).$$

Then we have the following:

(1) $\Re\{\bar{a}_n(a_n + 2(k_j a_{n-1} + k_j^2 a_{n-2} + \cdots + k_j^n a_0))\} = 0, \quad j = 1, 2, \ldots, n.$

(2) $\Im\{\bar{a}_n(k_j a_{n-1} + 2k_j^2 a_{n-2} + \cdots + nk_j^n a_0)\} = 0, \quad j = 1, 2, \ldots, n.$

(3) $\Re\{\bar{a}_n(k_j a_{n-1} + 2^2 k_j^2 a_{n-2} + \cdots + n^2 k_j^n a_0)\} \geq 0, \quad j = 1, 2, \ldots, n.$
    If there is an equality for some j, then

$$\Im\{\bar{a}_n(k_j a_{n-1} + 2^3 k_j^2 a_{n-2} + \cdots + n^3 k_j^n a_0)\} = 0, \quad j = 1, 2, \ldots, n \text{ and}$$

$$\Re\{\bar{a}_n(k_j a_{n-1} + 2^4 k_j^2 a_{n-2} + \cdots + n^4 k_j^n a_0)\} \geq 0, \quad j = 1, 2, \ldots, n.$$

(4) *Let us denote*

$$d_{lj} = 4\Re\{\bar{a}_n[(k_j + k_l)a_{n-1} + \cdots + (k_j^n + 2k_j^{n-1}k_l + \cdots + 2k_j k_l^{n-1} + k_l^n)a_0]\}$$

$$+ 8\Re\{(k_j a_{n-1} + \cdots + nk_j^n a_0)\overline{(k_l a_{n-1} + \cdots + k_l^n a_0)}\}, \quad l, j = 1, 2, \ldots, n.$$

Then $\det(d_{lj})_{l,j=1}^{n} \geq 0.$
If the inequality is strict (not equality), then

$$2\Re\{\bar{a}_n(k_j a_{n-1} + \cdots + nk_j^n a_0)\} < |a_n|^2 - 4|k_j a_{n-1} + \cdots + k_j^n a_0|^2,$$

$$j = 1, 2, \ldots, n.$$

Proof.   We have the following identity:

$$a_n = \frac{1}{2\pi}\int_0^{2\pi} \exp\left(-\sum_{m=1}^{n} \lambda_m \frac{1 + k_m e^{i\theta}}{1 - k_m e^{i\theta}}\right) e^{-in\theta}\, d\theta.$$

We consider $|a_n|^2$ as a function of the $2n$ variables $(\lambda_1, \ldots, \lambda_n; k_1, \ldots, k_n)$. This function attains its maximal value for certain values of the $2n$ variables. A necessary condition for the function $|a_n|^2$ to attain its maximal value is that

$$\frac{\partial |a_n|^2}{\partial \lambda_j} = 0, \quad j = 1, 2, \ldots, n.$$

We compute the first-order partial derivatives of $|a_n|^2$ with respect to the λ_j's:

$$\frac{\partial |a_n|^2}{\partial \lambda_j} = -\left(\frac{1}{2\pi}\right)^2 \Re\left\{\int_0^{2\pi}\left(\frac{1 + k_j e^{i\theta}}{1 - k_j e^{i\theta}}\right)\exp\left(-\sum_{m=1}^{n}\lambda_m \frac{1 + k_m e^{i\theta}}{1 - k_m e^{i\theta}}\right)e^{-in\theta}\,d\theta\right.$$

$$\left. \times \overline{\int_0^{2\pi}\exp\left(-\sum_{m=1}^{n}\lambda_m \frac{1 + k_m e^{i\theta}}{1 - k_m e^{i\theta}}\right)e^{-in\theta}\,d\theta}\right\}.$$

We have

$$\frac{1+k_j z}{1-k_j z} = 1 + 2(k_j z + k_j^2 z^2 + k_j^3 z^3 + \cdots) \text{ and } \exp\left(-\sum_{m=1}^{n} \lambda_m \frac{1+k_m e^{i\theta}}{1-k_m e^{i\theta}}\right)$$

$$= a_0 + a_1 z + \cdots.$$

and hence, we get

$$\frac{1}{2\pi}\int_0^{2\pi} \left(\frac{1+k_j e^{i\theta}}{1-k_j e^{i\theta}}\right) \exp\left(-\sum_{m=1}^{n} \lambda_m \frac{1+k_m e^{i\theta}}{1-k_m e^{i\theta}}\right) e^{-in\theta}\, d\theta$$

$$= a_n + 2(k_j a_{n-1} + k_j^2 a_{n-2} + \cdots + k_j^n a_0).$$

Thus, the condition $\partial|a_n|^2/\partial\lambda_j = 0$ is

$$\Re\{\bar{a}_n(a_n + 2(k_j a_{n-1} + k_j^2 a_{n-2} + \cdots + k_j^n a_0))\} = 0, \quad j = 1, 2, \ldots, n.$$

On the other hand, we proved (equation (4.0.1) or Proposition 4.0.7 with $l = 1$) that

$$\Re\{\bar{a}_n(a_n + 2(k a_{n-1} + k^2 a_{n-2} + \cdots + k^n a_0))\} \geq 0 \text{ for } |k| = 1.$$

Hence, the zeros $k_1, \ldots, k_n$ of this trigonometric polynomial have even (positive) order. Hence, the derivatives of this polynomial at $k_1, \ldots, k_n$ are zero. This proves the following identities:

$$\Im\{\bar{a}_n(k_j a_{n-1} + 2k_j^2 a_{n-2} + \cdots + nk_j^n a_0)\} = 0, \quad j = 1, 2, \ldots, n.$$

The next derivative at those points, $k_1, k_2, \ldots, k_n$ is non-negative. If this derivative vanishes at one of these points, then the third derivative at that same point is 0 and the fourth must be non-negative. Hence, we proved that

$$\Re\{\bar{a}_n(k_j a_{n-1} + 2^2 k_j^2 a_{n-2} + \cdots + n^2 k_j^n a_0)\} \geq 0, \quad j = 1, 2, \ldots, n,$$

and if there is an equality for one of the j's, then

$$\Im\{\bar{a}_n(k_j a_{n-1} + 2^3 k_j^2 a_{n-2} + \cdots + n^3 k_j^n a_0)\} = 0,$$

and

$$\Re\{\bar{a}_n(k_j a_{n-1} + 2^4 k_j^2 a_{n-2} + \cdots + n^4 k_j^n a_0)\} \geq 0.$$

So far, we considered only the condition $\partial|a_n|^2/\partial\lambda_j = 0$, $j = 1, 2, \ldots, n$. However, since we are at a local maximum, we must have the following: let us denote

$$\Delta = \det\left(\frac{\partial^2 |a_n|^2}{\partial\lambda_l \partial\lambda_j}\right)_{l,j=1}^{n},$$

then $\Delta \geq 0$ and if $\Delta > 0$ then $\partial^2 |a_n|^2 / \partial \lambda_j^2 < 0$, $j = 1, 2, \ldots, n$. The computation of the second-order derivatives gives

$$\frac{\partial^2 |a_n|^2}{\partial \lambda_l \partial \lambda_j} = d_{lj} = 4\Re\{\bar{a}_n[(k_j + k_l)a_{n-1}$$

$$+ \cdots + (k_j^n + 2k_j^{n-1}k_l + \cdots + 2k_j k_l^{n-1} + k_l^n)a_0]\}$$

$$+ 8\Re\{(k_j a_{n-1} + \cdots + nk_j^n a_0)\overline{(k_l a_{n-1} + \cdots + k_l^n a_0)}\}.$$

For the diagonal elements $l = j$, we get

$$\frac{\partial^2 |a_n|^2}{\partial \lambda_j^2} = 4\Re\{\bar{a}_n(k_j a_{n-1} + \cdots + nk_j^n a_0)\}$$

$$- 2|a_n|^2 + 8|k_j a_{n-1} + \cdots + k_j^n a_0|^2.$$

The proof of Proposition 6.0.28 is now completed. $\square$

Proposition 6.0.29. *Let $f(z) = a_0 + \cdots + a_n z^n + \cdots \in B$ be extremal for $n \in \mathbb{Z}^+$. Let us assume that (see [49])*

$$f(z) = \exp\left(-\sum_{j=1}^{n} \lambda_j \frac{1 + k_j z}{1 - k_j z}\right),$$

where $\lambda_j > 0$ and $|k_j| = 1$ for $j = 1, 2, \ldots, n$, and where if $1 \leq j \neq l \leq n$ then $k_j \neq k_l$ (so we insist that the Radon measure has exactly n atoms, the maximal number). Then for any k, $|k| = 1$, we have

$$\frac{|a_n|^2 |(k - k_1)(k - k_2)\ldots(k - k_n)|^2}{2 + |\sum_{i_1=1}^{n} k_{i_1}|^2 + \cdots + |\sum_{1 \leq i_1 < \cdots < i_{n-1} \leq n} k_{i_1} k_{i_2} \ldots k_{i_{n-1}}|^2}$$

$$= \Re\{\bar{a}_n(a_n + 2(ka_{n-1} + k^2 a_{n-2} + \cdots + k^n a_0))\}.$$

Proof. We refer to E.3, C on p. 85 of [7], or to p. 5 of the classical book [100]. We find there the following theorem:

Theorem ([7, 100]). *If $g(\theta)$ is a trigonometric polynomial with real coefficients of degree n, such that $g(\theta) \geq 0$ for all θ, then there exists an algebraic polynomial of degree n, $q(z) \in \mathbb{C}[z]$, $\deg q = n$ so that $g(\theta) = |q(\theta)|^2$. If $g(\theta) \not\equiv 0$, then we can choose such a $q(z)$, so that $q(z) \neq 0$ for all $z \in U$ and $q(0) > 0$ in which case $q(z)$ is uniquely determined. Also, if $g(\theta)$ is an even function, then we can choose $q(z) \in \mathbb{R}[z]$.*

We take $g(\theta) = \Re\{\bar{a}_n(a_n + 2(e^{i\theta}a_{n-1} + \cdots + e^{in\theta}a_0))\}$. Then $g(\theta) \geq 0$ for all θ (by equation (4.0.1) or Proposition 4.0.7 in the case $l = 1$). So, there exists a polynomial $q(z) \in \mathbb{C}[z]$ with $\deg q = n$ that has no zero in U and $q(0) > 0$ such that

$$|q(k)|^2 = \Re\{\bar{a}_n(a_n + 2(ka_{n-1} + \cdots + k^n a_0))\},$$

for all k, $|k| = 1$. By the assumptions of our proposition (Proposition 6.0.29) and by Proposition 6.0.28(1), we conclude that $q(k_j) = 0$, $j = 1, 2, \ldots, n$. So, these are all the zeros of $q(z)$ and hence, there exists a constant $c \in \mathbb{C}^\times$ so that: $q(z) = c(z - k_1)(z - k_2) \ldots (z - k_n)$. Hence,

$$q(z) = c[z^n - (k_1 + \cdots + k_n)z^{n-1}$$
$$+ (k_1 k_2 + \cdots + k_{n-1} k_n)z^{n-2} - \cdots + (-1)^n k_1 k_2 \ldots k_n].$$

We compute the constant c by integration of $|q|^2$:

$$\frac{1}{2\pi} \int_0^{2\pi} |q(e^{i\theta})|^2 d\theta = |c|^2(1 + |k_1 + \cdots + k_n|^2 + \cdots + |k_1 k_2 \ldots k_n|^2).$$

On the other hand,

$$\frac{1}{2\pi} \int_0^{2\pi} |q(e^{i\theta})|^2 d\theta$$
$$= \frac{1}{2\pi} \int_0^{2\pi} \Re\{\bar{a}_n(a_n + 2(e^{i\theta} a_{n-1} + \cdots + e^{in\theta} a_0))\} d\theta = |a_n|^2.$$

Hence,

$$|c|^2 = \frac{|a_n|^2}{2 + |\sum_{i_1=1}^n k_{i_1}|^2 + \cdots + |\sum_{1 \leq i_1 < \cdots < i_{n-1} \leq n} k_{i_1} k_{i_2} \ldots k_{i_{n-1}}|^2}.$$

The proof of Proposition 6.0.29 is now completed. $\qquad\square$

Definition 6.0.30. Let $z_1, \ldots, z_n \in \mathbb{C}$ be n complex numbers. Let j be an integer such that $1 \leq j \leq n$. The j'th symmetric function at $(z_1, \ldots, z_n)$ is denoted and defined by $S_j(z_1, \ldots, z_n) = \sum_{1 \leq i_1 < \cdots < i_j \leq n} z_{i_1} z_{i_2} \ldots z_{i_j}$.

In our case, we will mostly evaluate the symmetric functions at $(k_1, \ldots, k_n)$ where for any k_i, $|k_i| = 1$, $i = 1, \ldots, n$. These will be n of the parameters of the extremal function for $n \in \mathbb{Z}^+$. Thus, we will be able to think of $(k_1, \ldots, k_n)$ as fixed. In this case, we will shorten our notation for the symmetric functions as follows: $S_j = S_j(k_1, \ldots, k_n)$, $1 \leq j \leq n$.

Remark 6.0.31. Using Definition 6.0.30, we may express $|c|^2$ in the proof of Proposition 6.0.29 as follows:

$$|c|^2 = \frac{|a_n|^2}{1 + |S_1|^2 + |S_2|^2 + \cdots + |S_n|^2} = \frac{|a_n|^2}{2 + |S_1|^2 + \cdots + |S_{n-1}|^2}.$$

Also, we note that

$$|S_n| = |k_1 \ldots k_n| = 1, \quad \overline{S}_n = \frac{1}{S_n}, \quad \overline{S}_j = \overline{S}_n \cdot S_{n-j},$$

$$|S_j|^2 = S_j \cdot \overline{S}_j = S_j \cdot S_{n-j} \cdot \overline{S}_n.$$

Hence,

$$|c|^2 = \frac{|a_n|^2}{2 + (S_1 \cdot S_{n-1} + S_2 \cdot S_{n-2} + \cdots + S_{n-1} \cdot S_1)\overline{S}_n}$$

$$= \frac{S_n \cdot |a_n|^2}{2S_n + S_1 \cdot S_{n-1} + S_2 \cdot S_{n-2} + \cdots + S_{n-1} \cdot S_1}.$$

Proposition 6.0.32. *Let $f(z) = a_0 + \cdots + a_n z^n + \cdots \in B$ be extremal for $n \in \mathbb{Z}^+$. Let us assume that (see [49]):*

$$f(z) = \exp\left(-\sum_{j=1}^{n} \lambda_j \frac{1 + k_j z}{1 - k_j z}\right),$$

where $\lambda_j > 0$ and $|k_j| = 1$ for $j = 1, \ldots, n$ and where $1 \le j \ne l \le n$ implies that $k_j \ne k_l$, then we have the following $(n+1)$ identities:

$$\begin{cases} |c|^2(1 + |S_1|^2 + |S_2|^2 + \cdots + |S_n|^2) & = & |a_n|^2 \\ |c|^2(\overline{S}_1 + S_1\overline{S}_2 + S_2\overline{S}_3 + \cdots + S_{n-1}\overline{S}_n) & = & -\overline{a}_n a_{n-1} \\ |c|^2(\overline{S}_2 + S_1\overline{S}_3 + S_2\overline{S}_4 + \cdots + S_{n-2}\overline{S}_n) & = & \overline{a}_n a_{n-2} \\ \vdots & \vdots & \vdots \\ |c|^2(\overline{S}_{n-2} + S_1\overline{S}_{n-1} + S_2\overline{S}_n) & = & (-1)^{n-2}\overline{a}_n a_2 \\ |c|^2(\overline{S}_{n-1} + S_1\overline{S}_n) & = & (-1)^{n-1}\overline{a}_n a_1 \\ |c|^2\overline{S}_n & = & (-1)^n\overline{a}_n a_0 \end{cases}$$

We also have the following n identities:

$$\begin{cases} -a_1/a_0 & = & 2S_1 \\ a_2/a_0 & = & 2S_2 + S_1^2 \\ -a_3/a_0 & = & 2S_3 + 2S_1 S_2 \\ a_4/a_0 & = & 2S_4 + 2S_1 S_3 + S_2^2 \\ \vdots & \vdots & \vdots \\ (-1)^{n-1}a_{n-1}/a_0 & = & 2S_{n-1} + 2S_1 S_{n-2} + 2S_2 S_{n-3} + \cdots \\ (-1)^n a_n/a_0 & = & 2S_n + 2S_1 S_{n-1} + 2S_2 S_{n-2} + \cdots \end{cases}$$

Finally, we also have the following n equations:

$$\sum_{j=1}^{n} k_j^m = m \sum_{j=1}^{n} \lambda_j k_j^m, \quad m = 1, 2, \ldots, n.$$

Proof. We note the following two facts:

(1) If for all $|k| = 1$ we have $\Re\{\alpha_0 + \alpha_1 k + \cdots + \alpha_n k^n\} = \Re\{\beta_0 + \beta_1 k + \cdots + \beta_n k^n\}$, where $\alpha_0, \ldots, \alpha_n, \beta_0, \ldots, \beta_n \in \mathbb{C}$, then for all $|k| = 1$, we have $\Re\{\alpha_j k^j\} = \Re\{\beta_j k^j\}$, $j = 1, 2, \ldots, n$.

(2) If $j > 0$ is an integer and if for all $|k| = 1$ we have $\Re\{\alpha k^j\} = \Re\{\beta k^j\}$, then $\alpha = \beta$.

We use these two facts and compare coefficients in the identity: $|q(k)|^2 = \Re\{\bar{a}_n(a_n + 2(a_{n-1}k + \cdots + a_0 k^n))\}$. Looking at the calculations in the proof of Proposition 6.0.29, we get

$$|q(k)|^2 = |c|^2 (k^n - (k_1 + \cdots + k_n)k^{n-1} + \cdots + (-1)^n k_1 k_2 \ldots k_n)$$

$$\times (\bar{k}^n - (\bar{k}_1 + \cdots + \bar{k}_n)\bar{k}^{n-1} + \cdots + (-1)^n \bar{k}_1 \bar{k}_2 \ldots \bar{k}_n)$$

$$= |c|^2 \left((1 + |S_1|^2 + \cdots + |S_n|^2) - 2\Re\{(S_1 + \bar{S}_1 S_2 + \cdots + \bar{S}_{n-1} S_n)\bar{k}\} \right.$$

$$\left. + \cdots + (-1)^n 2\Re\{S_n \bar{k}^n\} \right).$$

We compare the coefficients in the last expression to those in the trigonometric polynomial $\Re\{\bar{a}_n(a_n + 2(a_{n-1}k + \cdots + a_0 k^n))\}$ and obtain the first set of $(n+1)$ identities. To obtain the second set of n identities, we divide each one of the first identities $1, 2, \ldots, n$ of the first set, by the last, $(n+1)$st, identity noting that $\bar{a}_n a_0 \neq 0$. The result is as follows:

$$-a_1/a_0 = (\bar{S}_{n-1} + S_1 \bar{S}_n)/\bar{S}_n$$

$$a2/a0 = (\bar{S}_{n-2} + S_1 \bar{S}_{n-1} + S_2 \bar{S}_{n-2})/\bar{S}_n$$

$$\vdots \quad \cdots$$

$$(-1)^n a_n/a_0 = (1 + S_1 \bar{S}_1 + S_2 \bar{S}_2 + \cdots + S_n \bar{S}_n)/\bar{S}_n.$$

But $S_j/S_n = \bar{S}_{n-j}$, $j = 1, 2, \ldots, n-1$ and so the right-hand sides of the above n identities become

$$-a_1/a_0 = 2S_1$$

$$a_2/a_0 = 2S_2 + S_1^2$$

$$\vdots \quad \vdots$$

$$(-1)^n a_n/a_0 = 2S_n + 2S_1 S_{n-1} + 2S_2 S_{n-2} + \cdots$$

and we proved the validity of the second set of n identities. Finally, we show how to derive the equations:

$$\sum_{j=1}^{n} k_j^m = m \sum_{j=1}^{n} \lambda_j k_j^m, \quad m = 1, 2, \ldots, n.$$

By exponentiation, we obtained the recursion:

$$a_m = -\frac{2}{m} \sum_{k=1}^{m} k \left(\sum_{j=1}^{n} \lambda_j k_j^k \right) a_{m-k}, \quad m = 1, 2, \ldots$$

For $m = 1$: $a_1 = -2(\sum_{j=1}^{n} \lambda_j k_j) a_0 = -2(\sum_{j=1}^{n} k_j) a_0$, where the right-hand side comes from the first equation in the set of n identities we proved, i.e., $-a_1/a_0 = 2S_1$. Hence $\sum_{j=1}^{n} k_j = \sum_{j=1}^{n} \lambda_j k_j$.

For $m = 2$: $a_2 = (2(\sum_{j=1}^{n} \lambda_j k_j)^2 - 2(\sum_{j=1}^{n} \lambda_j k_j^2)) a_0 = (2(\sum k_{j_1} k_{j_2}) + (\sum k_j)^2) a_0$. We now use the case $m = 1$:

$$2 \left(\sum_{j=1}^{n} \lambda_j k_j^2 \right) = 2 \left(\sum_{j=1}^{n} \lambda_j k_j \right)^2 - 2 \left(\sum k_{j_1} k_{j_2} \right) - \left(\sum k_j \right)^2$$

$$= 2 \left(\sum k_j \right)^2 - 2 \left(\sum k_{j_1} k_{j_2} \right) - \left(\sum k_j \right)^2$$

$$= \left(\sum k_j \right)^2 - 2 \left(\sum k_{j_1} k_{j_2} \right) = \sum_{j=1}^{n} k_j^2.$$

For $m = 3$,

$$a_3 = -\frac{2}{3} \left(2 \left(\sum \lambda_j k_j \right)^3 - 6 \left(\sum \lambda_j k_j \right) \left(\sum \lambda_j k_j^2 \right) + 3 \left(\sum \lambda_j k_j^3 \right) \right) a_0$$

$$= - \left(2 \left(\sum k_{j_1} k_{j_2} k_{j_3} \right) + 2 \left(\sum k_j \right) \left(\sum k_{j_1} k_{j_2} \right) \right) a_0.$$

we now use the cases $m = 1, 2$:

$$2 \left(\sum \lambda_j k_j^3 \right) = 2 \left(\sum k_{j_1} k_{j_2} k_{j_3} \right) + 2 \left(\sum k_j \right) \left(\sum k_{j_1} k_{j_2} \right)$$

$$+ 4 \left(\sum \lambda_j k_j \right) \left(\sum \lambda_j k_j^2 \right) - \frac{4}{3} \left(\sum \lambda_j k_j \right)^3$$

$$= 2 \left(\sum k_{j_1} k_{j_2} k_{j_3} \right) + 2 \left(\sum k_j \right) \left(\sum k_{j_1} k_{j_2} \right)$$

$$+ 2 \left(\sum k_j \right) \left(\sum k_j^2 \right) - \frac{4}{3} \left(\sum k_j \right)^3 = \frac{2}{3} \left(\sum k_j^3 \right),$$

and hence $3(\sum_{j=1}^{n} \lambda_j k_j^3) = \sum_{j=1}^{n} k_j^3$.

In general, we use induction on m. Assuming we proved the equations for $1, 2, \ldots, m-1$, where $-1 \leq m-1 < n$, then

$$a_m = -\frac{2}{m} \sum_{k=1}^{m-1} k \left(\sum_{j=1}^{n} \lambda_j k_j^k \right) a_{m-k} - 2 \left(\sum_{j=1}^{n} \lambda_j k_j^n \right) a_0$$

(using the inductive hypothesis)

$$= -\frac{2}{m} \sum_{k=1}^{m-1} \left(\sum_{j=1}^{n} k_j^k \right) a_{m-k} - 2 \left(\sum_{j=1}^{n} \lambda_j k_j^n \right) a_0$$

$$= -\frac{2}{m} \sum_{k=1}^{m-1} \left(\sum_{j=1}^{n} k_j^k \right) (-1)^{m-k}(2S_{m-k} + 2S_1 S_{m-k-1} + \cdots) - 2 \left(\sum_{j=1}^{n} \lambda_j k_j^m \right) a_0.$$

The last equality follows by the second set of n identities we already proved above. It follows by the same set of identities, that $a_m = (-1)^m(2S_m + 2S_1 S_{m-1} + \cdots)a_0$. From the last two equations, we obtain

$$m \left(\sum_{j=1}^{n} \lambda_j k_j^m \right) = (-1)^m \left(\frac{m}{2} \right) (2S_m + 2S_1 S_{m-1} + \cdots)$$

$$- \sum_{k=1}^{m-1} (-1)^{m-k} \left(\sum_{j=1}^{n} k_j^k \right) (2S_{m-k} + 2S_1 S_{m-k-1} + \cdots),$$

and to conclude the inductive proof,

$$\left(\sum_{j=1}^{m} k_j^m \right) + \sum_{k=1}^{m-1} (-1)^{m-k} \left(\sum_{j=1}^{n} k_j^k \right) (2S_{m-k} + 2S_1 S_{m-k-1} + \cdots)$$

$$+ (-1)^m \left(\frac{m}{2} \right) (2S_m + 2S_1 S_{m-1} + \cdots) = 0. \qquad \square$$

By now, we have enough information to prove the following reduction theorem for the Krzyż conjecture.

Theorem 6.0.33. *Let* $f(z) = a_0 + \cdots + a_n z^n + \cdots \in B$ *be extremal for* $n \in \mathbb{Z}^+$. *Let us assume that* ([49]):

$$f(z) = \exp \left(-\sum_{j=1}^{n} \lambda_j \frac{1 + k_j z}{1 - k_j z} \right),$$

where $\lambda_j > 0$ and $|k_j| = 1$ for $j = 1, \ldots, n$ and where if $1 \le j \neq l \le n$, then $k_j \neq k_l$. Then $2|a_0| \le |a_n|$ and if $2|a_0| = |a_n|$, then $|a_n| = 2/e$ and the Krzyż conjecture follows.

Proof. We consider the first set of $n+1$ identities in Proposition 6.0.32. By dividing the first identity: $|c|^2(1+|S_1|^2+|S_2|^2+\cdots+|S_n|^2) = |a_n|^2$ by the last $|c|^2 = (-1)^n \overline{a}_n a_0$, we obtain $a_n = (-1)^n S_n(2+|S_1|^2+|S_2|^2+\cdots+|S_{n-1}|^2)a_0$, which implies that $|a_n| \ge 2|a_0|$, an inequality we already proved twice, before. Equality, though, is the interesting new part. Namely, there is an equality $|a_n| = 2|a_0|$ if and only if $S_1 = S_2 = \cdots = S_{n-1} = 0$. This is equivalent to $a_1 = a_2 = \cdots = a_{n-1} = 0$, using (for example) the second set of n identities in Proposition 6.0.32. But this implies as in the proof of Theorem 6.0.20 that $a_n = 2/e$ and $|a_0| = 1/e$ of course. $\qquad\square$

Remark 6.0.34. We clarify how Theorem 6.0.20 was used in the final part of the proof of Theorem 6.0.33. We are in the situation in which $a_1 = \cdots = a_{n-1} = 0$. The assumption in Theorem 6.0.20 is that $2|ka_{n-1}+k^2 a_{n-2}+\cdots+k^n a_0| \le |a_n|$, for all k, $|k| = 1$. Thus, in our case, this assumption reduces to $2|k^n a_0| \le |a_n|$, i.e., $2|a_0| \le |a_n|$, which of course is true.

Proposition 6.0.35. *Let $f(z) = a_0 + \cdots + a_n z^n + \cdots \in B$ be extremal for $n \in \mathbb{Z}^+$. Let us assume that ([49]):*

$$f(z) = \exp\left(-\sum_{j=1}^{n} \lambda_j \frac{1 + k_j z}{1 - k_j z}\right),$$

where $\lambda_j > 0$ and $|k_j| = 1$ for $j = 1, \ldots, n$ and where if $1 \le j \neq l \le n$, then $k_j \neq k_l$. Then we have the following identity:

$$\overline{S}_1 + S_1\overline{S}_2 + S_2\overline{S}_3 + \cdots + S_{n-1}\overline{S}_n = 2S_{n-1} + 2S_1 S_{n-2} + 2S_2 S_{n-3} + \cdots = 0.$$

Proof. By Proposition 6.0.32, we have $|c|^2(\overline{S}_1 + S_1\overline{S}_2 + \cdots + S_{n-1}\overline{S}_n) = -\overline{a}_n a_{n-1}$ and $(-1)^{n-1}a_{n-1}/a_0 = 2S_{n-1} + 2S_1 S_{n-2} + 2S_2 S_{n-3} + \cdots$. By Proposition 6.0.25, $a_{n-1} = 0$. The result follows. $\qquad\square$

Proposition 6.0.36. *Let $f(z) = a_0 + \cdots + a_n z^n + \cdots \in B$ be extremal for $n \in \mathbb{Z}^+$. Let us assume that ([49]):*

$$f(z) = \exp\left(-\sum_{j=1}^{n} \lambda_j \frac{1 + k_j z}{1 - k_j z}\right),$$

where $\lambda_j > 0$ and $|k_j| = 1$ for $j = 1, \ldots, n$ and where if $1 \leq j \neq l \leq n$, then $k_j \neq k_l$. Then we have the following identities:

$$\lambda_j \times (k_1 k_2 \ldots k_n) \prod_{1 \leq i < l \leq n} (k_l - k_i)$$

$$= \begin{vmatrix} k_1 & k_2 & \ldots & k_{j-1} & (k_1 + k_2 + \cdots + k_n) & k_{j+1} & \ldots & k_n \\ k_1^2 & k_2^2 & \ldots & k_{j-1}^2 & (1/2)(k_1^2 + k_2^2 + \cdots + k_n^2) & k_{j+1}^2 & \ldots & k_n^2 \\ \vdots & \vdots & \vdots & \vdots & \vdots & \vdots & \vdots & \vdots \\ k_1^n & k_2^n & \ldots & k_{j-1}^n & (1/n)(k_1^n + k_2^n + \cdots + k_n^n) & k_{j+1}^n & \ldots & k_n^n \end{vmatrix}, \quad j = 1, 2, \ldots, n.$$

In particular, all the quotients of the $n \times n$ determinants by the finite products $(k_1 k_2 \ldots k_n) \prod_{1 \leq i < l \leq n} (k_l - k_i)$ are positive numbers.

Proof. By Proposition 6.0.32, we have

$$\sum_{j=1}^{n} \lambda_j k_j^m = \frac{1}{m} \left(\sum_{j=1}^{n} k_j^m \right), \quad m = 1, 2, \ldots, n.$$

This is a linear system of equations. There are n equations in the n unknowns $\lambda_1, \lambda_2, \ldots, \lambda_n$. The determinant of the coefficients is a multiple of the Vandermonde on $k_1, k_2, \ldots, k_n$, namely

$$\begin{vmatrix} k_1 & k_2 & \ldots & k_n \\ k_1^2 & k_2^2 & \ldots & k_n^2 \\ \vdots & \vdots & \vdots & \vdots \\ k_1^n & k_2^n & \ldots & k_n^n \end{vmatrix} = (k_1 k_2 \ldots k_n) \prod_{1 \leq i < l \leq n} (k_l - k_i).$$

This determinant is different from zero because our assumptions imply that the unit numbers $k_1, k_2, \ldots, k_n$ are different from one another. So, there is a unique solution of the linear system. It can be represented by Cramer's rule. The proof is now completed. $\square$

Now, we can prove one more reduction theorem for the Krzyż conjecture, namely it is sufficient to prove that $\lambda_1 = \cdots = \lambda_n$.

Theorem 6.0.37. *Let $f(z) = a_0 + \cdots + a_n z^n + \cdots \in B$ be extremal for $n \in \mathbb{Z}^+$. Let us assume that ([49])*

$$f(z) = \exp \left(\sum_{j=1}^{n} \lambda_j \frac{1 + k_j z}{1 - k_j z} \right),$$

where $\lambda_j > 0$ and $|k_j| = 1$ for $j = 1, \ldots, n$ and where if $1 \le j \ne l \le n$, then $k_j \ne k_l$. If $\lambda_1 = \cdots = \lambda_n = \lambda > 0$, then $\lambda = 1/n$, there is a real number α such that $e^{i\alpha} k_j = \exp(2\pi(j-1)i/n)$, $j = 1, \ldots, n$ and

$$f(z) = \exp\left(-\frac{1 + e^{i\theta} z^n}{1 - e^{i\theta} z^n}\right),$$

for some real θ, and $|a_n| = 2/e$.

Proof. By the assumption $\lambda_1 = \cdots = \lambda_n = \lambda > 0$, and by the third set of identities in Proposition 6.0.32, we get the following homogeneous system: $\sum_{j=1}^{n}(1 - \lambda m)k_j^m = 0$ for $m = 1, \ldots, n$. If for each integer m, $1 \le m \le n$, we have $1 - \lambda \cdot m \ne 0$, then our system is the following set of n homogeneous equations: $\sum_{j=1}^{n} k_j^m = 0$, $m = 1, \ldots, n$, which says that the first n moments of $(k_1, \ldots, k_n)$ vanish. We view these equations as n linear homogeneous equations in the n unknowns $k_1, \ldots, k_n$. Then the determinant of the coefficient matrix is the Vandermonde $\prod_{1 \le i < j \le n}(k_i - k_j)$, which, by our assumption $j \ne l \Rightarrow k_j \ne k_l$, is not 0. So, the unique solution is the trivial solution $k_1 = \cdots = k_n = 0$, which is a clear contradiction. We conclude that there is an integer l, $1 \le l \le n$ such that $\lambda = 1/l$ in which case the l'th equation in our system is the trivial equation $0 = 0$. For each $m \ne l$, we get again the moment equals zero equation:

$$0 = \sum_{j=1}^{n}(1 - \lambda m)k_j^m = \left(1 - \frac{m}{l}\right)\sum_{j=1}^{n} k_j^m.$$

Thus, the m'th moment $M_m := \sum_{j=1}^{n} k_j^m = 0$. The only non-zero moment is the l'th: $M_l = \sum_{j=1}^{n} k_j^l \ne 0$. We recall our exponentiation recursion on the coefficients of the extremal $f(z) = a_0 + \cdots + a_n z^n + \cdots$, namely

$$a_m = -\frac{2}{m}\sum_{k=1}^{m} k\left(\sum_{j=1}^{n} \lambda_j k_j^k\right) a_{m-k} = -\frac{2}{m}\sum_{k=1}^{m}\left(\sum_{j=1}^{n} k_j^k\right) a_{m-k},$$

(by $k\sum_{j=1}^{n} \lambda_j k_j^k = \sum_{j=1}^{n} k_j^k$.) Since the only non-zero moment is the l'th, this becomes

$$a_m = -\frac{2}{m}\left(\sum_{j=1}^{n} k_j^l\right) a_{m-l} = -\frac{2}{m} \cdot M_l \cdot a_{m-l}.$$

We conclude that if $l \nmid n$ (l is not a divisor of n), then $a_n = 0$. This is because the series of coefficients is

$$a_0, a_1 = 0, \ldots, a_l = \left(-\frac{2}{l}\right) M_l a_0, a_{l+1} = 0, \ldots, a_{2l}$$

$$= \left(-\frac{2}{l}\right)\left(-\frac{2}{2l}\right) M_l^2 a_0, \ldots, a_n = 0.$$

That contradicts the extremality of $f(z)$. Hence, $l \mid n$ and $\lambda = 1/l$ must be one of the reciprocals of divisors of n. So, $\lambda \in \{1/n, \ldots, 1\}$. Also, we note that for $\lambda = 1/l$, we have the following value of the free term of $f(z)$:

$$a_0 = f(0) = \exp\left(-\sum_{j=1}^{n} \lambda_j\right) = \exp\left(-\frac{n}{l}\right),$$

where $n/l \in \mathbb{Z}^+$ (because $l \mid n$). Thus, $a_0 \in \{e^{-1}, \ldots, e^{-n}\}$ (the exponents are all the negative divisors of n). Our aim is to prove that $l = n$. If indeed $\lambda = 1/n$, then by the recursion, we get

$$a_1 = a_2 = \cdots = a_{n-1} = 0 \text{ and } a_n = -\frac{2}{n}\left(\sum_{j=1}^{n} k_j^n\right) e^{-1}.$$

We know by Theorem 6.0.20 that in this case $|a_n| = 2 \cdot e^{-1}$ and the Krzyż conjecture follows. Let us now take for l a value which is another divisor of n, i.e., different from n itself, and see why this leads to a contradiction. For example, $l = 1$. Then $a_0 = \exp(-n/l) = e^{-n}$. We have the following system:

$$\begin{cases} k_1^2 + \cdots + k_n^2 = 0 \\ \qquad \vdots \\ k_1^n + \cdots + k_n^n = 0 \end{cases},$$

and our recursion leads to

$$a_j = (-1)^j \frac{2^j}{j!}(k_1 + \cdots + k_n)^j e^{-n}.$$

We use the following trick: We define the following monic polynomial of degree n,

$$p_n(z) = (z - k_1)(z - k_2)\ldots(z - k_n) = z^n + \alpha_{n-1} z^{n-1} + \cdots + \alpha_1 z + \alpha_0.$$

In fact, the n lowest coefficients of $p_n(z)$ are the n signed symmetric functions of $(k_1, \ldots, k_n)$. Thus,

$$\alpha_{n-1} = -S_1 = -(k_1 + \cdots + k_n),$$

$$\alpha_{n-2} = \sum_{1 \leq j_1 < j_2 \leq n} k_{j_1} k_{j_2}, \ldots, \alpha_0 = (-1)^n k_1 k_2 \ldots k_n.$$

Then we have $p_n(k_j) = k_j^n + \alpha_{n-1} k_j^{n-1} + \cdots + \alpha_1 k_j + \alpha_0 = 0$ for $j = 1, \ldots, n$. Let us add these n identities:

$$\sum_{j=1}^n p_n(k_j) = \left(\sum_{j=1}^n k_j^n \right) + \alpha_{n-1} \left(\sum_{j=1}^n k_j^{n-1} \right) + \cdots$$

$$+ \alpha_1 \left(\sum_{j=1}^n k_j \right) + n\alpha_0 = 0.$$

But in our case $\lambda = 1$, so the last $n - 1$ moments vanish, $\sum_{j=1}^n k_j^k = 0$ for $k = 2, 3, \ldots, n$ and the identity we had becomes much simpler:

$$\alpha_1 \left(\sum_{j=1}^n k_j \right) + n\alpha_0 = 0. \text{ Thus } n = -\frac{\alpha_1}{\alpha_0} \left(\sum_{j=1}^n k_j \right).$$

On the other hand, the α_j's are the corresponding signed symmetric functions of $(k_1, \ldots, k_n)$, that is,

$$-\frac{\alpha_1}{\alpha_0} = -\frac{-(-1)^{n-1}(\sum k_{j_1} \ldots k_{j_{n-1}})}{(-1)^n (k_1 \ldots k_n)} = \overline{\left(\sum_{j=1}^n k_j \right)}.$$

We calculate that $n = |\sum_{j=1}^n k_j|^2$ or equivalently $|\sum_{j=1}^n k_j| = \sqrt{n}$. Coming back to our recursion, we get

$$a_n = (-1)^n \frac{2^n}{n!} (k_1 + \cdots + k_n)^n e^{-n},$$

and hence,

$$|a_n| = \frac{2^n n^{n/2}}{n!} e^{-n}.$$

We can observe that the sequence of absolute values of the coefficients of the extremal is monotonic decreasing. For,

$$\left\{ \frac{2^{k+1}(k+1)^{(k+1)/2} e^{-(k+1)}}{(k+1)!} \right\} \Big/ \left\{ \frac{2^k k^{k/2} e^{-k}}{k!} \right\} = \left(\frac{2}{e} \right) \frac{(k+1)^{(k+1)/2}}{k^{k/2}(k+1)}$$

$$= \left(\frac{2}{e} \right) \left(1 + \frac{1}{k} \right)^{k/2} \frac{1}{\sqrt{k+1}} < \frac{2}{\sqrt{e(k+1)}} < 1.$$

Thus, the sequence $\{|a_k|\} = \{2^k k^{k/2} e^{-k}/k!\}$ is a strictly decreasing sequence. Its largest element is the first, $k = 1$, and its value is $2/e$. We conclude that if $n > 1$ and $\lambda = 1$, we get $|a_n| < 2/e$. But the bounded non-vanishing function

$$\exp\left(-1 \cdot \frac{1 + z^n}{1 - z^n}\right),$$

attains the value $2/e$ which gives us the desired contradiction. Similarly, if l is any divisor of n, which is different from n, we will get the contradiction $|a_n| < 2/e$. To conclude our proof of Theorem 6.0.37, we know by now that $\lambda = 1/n$ and we have the following system:

$$\begin{cases} k_1 + \cdots + k_n = 0 \\ k_1^2 + \cdots + k_n^2 = 0 \\ \qquad\vdots \\ k_1^{n-1} + \cdots + k_n^{n-1} = 0 \end{cases}.$$

We repeat our trick with the monic polynomial:

$$p_n(z) = (z - k_1)\ldots(z - k_n) = z^n + \alpha_{n-1}z^{n-1} + \cdots + \alpha_0.$$

We sum up $\sum_{j=1}^{n} p_n(k_j) = 0$ and obtain $k_1^n + \cdots + k_n^n = ne^{i\beta}$ for some real β. We define $\alpha = -\beta/n$ and get that $e^{i\alpha}k_j$ are (for $j = 1, \ldots, n$) the n roots of unity of order n (up to order). Thus,

$$\sum_{j=1}^{n} \lambda_j \frac{1 + k_j z}{1 - k_j z} = \frac{1}{n} \sum_{j=1}^{n} \frac{\overline{k}_j + z}{\overline{k}_j - z} = \frac{1 + e^{i\theta} z^n}{1 - e^{i\theta} z^n},$$

for some fixed real θ. So,

$$f(z) = \exp\left(-\frac{1 + e^{i\theta} z^n}{1 - e^{i\theta} z^n}\right),$$

and

$$a_n = -\frac{2}{n}\left(\sum_{j=1}^{n} k_j^n\right) e^{-1} = -\frac{2}{e} e^{i\beta}.$$

So, finally, indeed $|a_n| = 2/e$. $\qquad\square$

Remark 6.0.38. Let

$$f(z) = \exp\left(-\sum_{j=1}^{n} \lambda_j \frac{1 + k_j z}{1 - k_j z}\right) \in B,$$

be extremal for $n \in \mathbb{Z}^+$, where $\lambda_j > 0$, $|k_j| = 1$ for $j = 1, \ldots, n$. Then

$$\max\{2|a_0|, |a_1|, \ldots, |a_{n-1}|\} \leq |a_n| \leq 2\sum_{j=0}^{n-1} |a_j|.$$

A proof of the remark. The left-hand inequality is true even without our extra assumptions on $f(z) = a_0 + \cdots + a_n z^n + \cdots$, $|z| < 1$ (by Proposition 6.0.27). On the other hand, we have

$$
na_n = -2 \sum_{k=1}^{n} k \left(\sum_{j=1}^{n} \lambda_j k_j^k \right) a_{n-k} = -2 \sum_{k=1}^{n} \left(\sum_{j=1}^{n} k_j^k \right) a_{n-k}.
$$

The last step follows by the third set of identities in Proposition 6.0.32. Hence,

$$
n|a_n| \leq 2 \sum_{k=1}^{n} \left| \sum_{j=1}^{n} k_j^k \right| |a_{n-k}| \leq 2n \sum_{k=1}^{n} |a_{n-k}|. \qquad \square
$$

Chapter 7

Indirect Approach: Estimates of Growth and Distortion

Proposition 7.0.39. *Let* $f(z) \in B = \{g \in H(U) \,|\, 0 < |g(z)| \leq 1,\ \text{for}\ |z| < 1\}$. *Then*

(1)

$$|f'(z)| \leq \frac{2}{e \cdot (1 - |z|^2)}, \quad z \in U.$$

(2)

$$\left| \frac{1}{2}(1 - |z|^2)f''(z) - \bar{z}f'(z) \right| \leq \frac{2}{e \cdot (1 - |z|^2)}, \quad z \in U.$$

Alternatively,

$$\left| zf''(z) - \frac{2|z|^2}{1 - |z|^2}f'(z) \right| \leq \frac{4|z|}{e \cdot (1 - |z|^2)^2}, \quad z \in U.$$

(3)

$$|f(Re^{i\theta}) - f(\pm re^{i\theta})| \leq \frac{1}{e}\log\left(\frac{1 + R}{1 - R}\right)\left(\frac{1 - r}{1 + r}\right),$$

$$0 < r < R < 1,\ 0 \leq \theta < 2\pi.$$

(4)

$$(1 - |z|^2)|f'(z)| \leq -2|f(z)|\log|f(z)|, \quad z \in U.$$

53

Proof. All these facts are well known and elementary. We will just write down proofs along the lines of the theory of S (as in [49]). For any $z, \xi \in U$, we have

$$f\left(\frac{\xi + z}{1 + \overline{z}\xi}\right) \in B, \quad \text{as a function of } \xi.$$

$$f\left(\frac{\xi + z}{1 + \overline{z}\xi}\right) = f(z) + (1 - |z|^2)f'(z)\xi + \frac{1}{2}((1 - |z|^2)^2 f''(z)$$

$$- 2\overline{z}(1 - |z|^2)f'(z)\xi^2 + \cdots,$$

and hence $(1 - |z|^2)|f'(z)| \leq 2/e$ which is (1), and also

$$\frac{1}{2}|(1 - |z|^2)^2 f''(z) - 2\overline{z}(1 - |z|^2)f'(z)| \leq \frac{2}{e} \quad \text{which is (2).}$$

By (1),

$$|f(z) - f(0)| \leq \int_0^{|z|} |f'(re^{i\theta})| dr \leq \frac{2}{e} \int_0^{|z|} \frac{dr}{1 - r^2} = \frac{1}{e} \log\left(\frac{1 + |z|}{1 - |z|}\right),$$

where $\theta = \arg z$. (3) follows. Finally,

$$f\left(\frac{\xi + z}{1 + \overline{z}\xi}\right)^{-1} = \frac{1}{f(z)} - \frac{(1 - |z|^2)f'(z)}{(f(z))^2}\xi + \cdots$$

and by a well-known estimate,

$$\frac{(1 - |z|^2)|f'(z)|}{|f(z)|^2} \leq 2\frac{1}{|f(z)|} \log \frac{1}{|f(z)|},$$

which is (4). $\qquad\qquad\qquad\qquad\qquad\qquad\qquad\qquad\qquad\qquad\qquad\qquad\Box$

Remark 7.0.40. We used as in the parallel theory for S, the coefficients estimates for $f(z) = a_0 + a_1 z + a_2 z^2 + \cdots \in B$, $|a_1|, |a_2| \leq 2/e$.

Remark 7.0.41. It is always a bit surprising to see the wonders of cancellations in singular integrals. For $f(z) = a_0 + \cdots + a_n z^n + \cdots \in B$, an extremal for $n \in \mathbb{Z}^+$, we know by [49], that we may assume that $f(z)$ has the following representation:

$$f(z) = \exp\left(-\sum_{j=1}^{n} \lambda_j \frac{1 + k_j z}{1 - k_j z}\right), \quad z \in U,$$

where $\lambda_j \geq 0$ and $|k_j| = 1$ for $j = 1, \ldots, n$. Hence, using the integral representations for the coefficients of $f(z)$ and of $f'(z)$, we obtain

$$a_n = \lim_{r \to 1^-} \left(\frac{1}{2\pi}\right) \int_{|z|=r} \frac{1}{z^n} \exp\left(-\sum_{j=1}^{n} \lambda_j \frac{1+k_j z}{1-k_j z}\right) dz,$$

$$na_n = \lim_{r \to 1^-} \left(\frac{1}{2\pi}\right) \int_{|z|=r} \frac{1}{z^n} \exp\left(-\sum_{j=1}^{n} \lambda_j \frac{1+k_j z}{1-k_j z}\right) \times \sum_{j=1}^{n} \frac{2k_j \lambda_j}{(1-k_j z)^2} dz.$$

So, multiplying the bounded integrand of the first integral by the rational function,

$$\sum_{j=1}^{n} \frac{2k_j \lambda_j}{(1-k_j z)^2},$$

that has double poles at $\overline{k}_j$ whenever $\lambda_j > 0$, did not elevate the size of the integral by too much. It only multiplied it by n.

Chapter 8

When Will Many of the Coefficients $a_1, \ldots, a_{n-1}$ in the Extremal Vanish?

The Krzyż conjecture asserts for the problem $\max_B |a_n|$ that the solutions are functions of z^n, $f(z) = g(z^n)$. This means that $a_1 = \cdots = a_{n-1} = 0$. Moreover, we saw in Theorems 6.0.20, 6.0.33 and 6.0.37 how the vanishing of these $n-1$ coefficients is sufficient to solve the Krzyż conjecture.

Theorem 8.0.42. *Let us fix a positive number $t > 0$ and n numbers of modulus 1, $k_1, \ldots, k_n$, $|k_j| = 1$, $j = 1, \ldots, n$. We assume that the $k_1, \ldots, k_n$ are different from one another. Then there are n non-negative numbers $\lambda_1, \ldots, \lambda_n \geq 0$ such that $\sum_{j=1}^n \lambda_j = t$, and there are n signs, $(-1)^{a_1}, \ldots, (-1)^{a_n}$ such that*

$$\exp\left(-\sum_{j=1}^n \lambda_j \frac{1 + (-1)^{a_j} k_j z}{1 - (-1)^{a_j} k_j z} \right) = a_0 + a_1 z + \cdots + a_n z^n + \cdots,$$

where if $n = 2k+1$, then $a_1 = a_3 = \cdots = a_{2k-1} = 0$, and if $n = 2k$ then $a_1 = a_3 = \cdots = a_{2k-3} = 0$.

Proof. We will make use of the Borsuk–Ulam Theorem. See [70, Chapter 9, p. 356], [41, pp. 176, 38, 32], and [20, p. 367]. Here is a standard statement of this theorem.

Theorem (Borsuk–Ulam). *For every $n \geq 0$, the following statements are equivalent and true.*

(1) *For every continuous mapping $f : S^n \to \mathbb{R}^n$, there exists a point $x \in S^n$ with $f(x) = f(-x)$. Here, $S^n = \{y \in \mathbb{R}^{n+1} \,|\, \|y\|_2 = 1\}$ is the n-dimensional unit sphere centered at the origin.*

57

(2) *For every continuous mapping $f : S^n \to \mathbb{R}^n$, which is antipodal (satisfies $f(-x) = -f(x)$), there exists a point $x \in S^n$ satisfying $f(x) = 0$.*

(3) *There is no continuous antipodal mapping $S^n \to S^{n-1}$.*

(4) *There is no continuous mapping $f : B^n \to S^{n-1}$ that is antipodal on the boundary satisfying $f(-x) = -f(x)$ for all $x \in S^{n-1} = \partial B^n$ (thus, $B^n = \{y \in \mathbb{R}^n \mid \|y\|_2 < 1\}$, the n-dimensional unit ball centered at the origin).*

We will change (within our proof) for convenience the set of notations in the statement of our Theorem 8.0.42. We define the following set of mappings in B.

$$E = \left\{ \exp\left(-\sum_{j=1}^n \lambda_j^2 \frac{1 + \operatorname{sgn}(\lambda_j)k_j z}{1 - \operatorname{sgn}(\lambda_j)k_j z} \right) \,\bigg|\, \sum_{j=1}^n \lambda_j^2 = t \right\}.$$

As before, we assume the generic power series expansion for each mapping in E:

$$\exp\left(-\sum_{j=1}^n \lambda_j^2 \frac{1 + \operatorname{sgn}(\lambda_j)k_j z}{1 - \operatorname{sgn}(\lambda_j)k_j z} \right) = a_0 + a_1 z + \cdots + a_n z^n + \cdots \in E \subset B.$$

Here, $a_0 = \exp(-\sum_{j=1}^n \lambda_j^2) = e^{-t}$. Let us first assume that $n = 2k + 1$. We define the following mapping:

$$T : S^{n-1}(t) \to \mathbb{R}^{n-1} = \mathbb{R}^{2k},$$

$$T((\lambda_1, \ldots, \lambda_n)) = (\Re(a_1), \Im(a_1), \ldots, \Re(a_{2k-1}), \Im(a_{2k-1})).$$

Only odd indices are used on the right-hand side, $1, 3, \ldots, 2k - 1$. Then by the following formula

$$a_l = \lim_{r \to 1^-} \left(\frac{1}{2\pi} \right) \int_{|z|=r} \frac{1}{z^{l+1}} \exp\left(-\sum_{j=1}^n \lambda_j^2 \frac{1 + \operatorname{sgn}(\lambda_j)k_j z}{1 - \operatorname{sgn}(\lambda_j)k_j z} \right) dz,$$

it follows that T is a continuous mapping. Since T depends only on the odd indexed coefficients, the signs $\operatorname{sgn}(\lambda_j)$ for odd j are sensitive to ± 1, and so T is antipodal. By the Borsuk–Ulam Theorem, there exists a point $(\lambda_1, \ldots, \lambda_n) \in S^{n-1}(t)$ such that $T((\lambda_1, \ldots, \lambda_n)) = (0, 0, \ldots, 0)$. This means that $\Re(a_1) = \Im(a_1) = \Re(a_3) = \Im(a_3) = \cdots = \Re(a_{2k-1}) = \Im(a_{2k-1}) = 0$. Hence, $a_1 = a_3 = \cdots = a_{2k-1} = 0$.

Let us now turn to the case where $n = 2k$. We define the following mapping:

$$T : S^{n-1}(t) \to \mathbb{R}^{n-1} = \mathbb{R}^{2k-1},$$

$$T((\lambda_1, \ldots, \lambda_n)) = (\Re(a_1), \Im(a_1), \ldots, \Re(a_{2k-3}), \Im(a_{2k-3}), \Re(a_{2k-1})),$$

and by the Borsuk–Ulam Theorem, we obtain a point $(\lambda_1, \ldots, \lambda_n) \in S^{n-1}(t)$ such that the image function satisfies $a_1 = a_3 = \cdots = a_{2k-3} = \Re(a_{2k-1}) = 0$. $\square$

Remark 8.0.43. It is clear that not for every choice of n points on the unit circle, k_j, $|k_j| = 1$, $j = 1, 2, \ldots, n$, we can find positive numbers $\lambda_1, \ldots, \lambda_n > 0$ for which all the $n - 1$ coefficients vanish, $a_1 = a_2 = \cdots = a_{n-1} = 0$. Even a dimension argument hints on that impossibility because the aggregate of points $(\lambda_1^2, \ldots, \lambda_n^2)$ with $\sum_{j=1}^{n} \lambda_j^2 = t$ is $(n-1)$-dimensional (real dimensions), but the desired conclusion $a_1 = a_2 = \cdots = a_{n-1} = 0$ is $2 \cdot (n-1)$-dimensional. Thus, the points k_j, $|k_j| = 1$, $j = 1, \ldots, n$ must lie in very special types of locations on $\mathbb{T} = \{z \in \mathbb{C} \,|\, |z| = 1\}$. So, our next task is to find those very special configurations of $k_1, \ldots, k_n$ on $\mathbb{T}$. It turns out that this task is not too difficult. The characterization of the geometric configurations is not completely simple though.

Theorem 8.0.44. *Let us fix a positive number $t > 0$. We consider functions of the form given in* [49]:

$$f(z) = \exp\left(-\sum_{j=1}^{n} \lambda_j \frac{1 + k_j z}{1 - k_j z} \right),$$

where $\lambda_j \geq 0$ and $|k_j| = 1$ for $j = 1, \ldots, n$. Assuming that $f(z) = a_0 + \cdots + a_n z^n + \cdots$ there is a unique non-negative vector $(\lambda_1, \ldots, \lambda_n)$ such that $\sum_{j=1}^{n} \lambda_j = t$ and $a_1 = a_2 = \cdots = a_{n-1} = 0$ if and only if the n unit numbers $k_1, k_2, \ldots, k_n$ satisfy the following n inequalities:

$$\prod_{\substack{l=1 \\ l \neq j}}^{n} \left(1 - \frac{k_j}{k_l} \right) > 0, \quad j = 1, 2, \ldots, n.$$

Proof. The conditions to be fulfilled $a_0 = e^{-t}$, $a_1 = a_2 = \cdots = a_{n-1} = 0$ are equivalent to the following system:

$$
\begin{cases}
\lambda_1 + \lambda_2 + \cdots + \lambda_n & = t \\
\lambda_1 k_1 + \lambda_2 k_2 + \cdots + \lambda_n k_n & = 0 \\
\lambda_1 k_1^2 + \lambda_2 k_2^2 + \cdots + \lambda_n k_n^2 & = 0 \\
\quad \vdots & \quad \vdots \; \vdots \\
\lambda_1 k_1^{n-1} + \lambda_2 k_2^{n-1} + \cdots + \lambda_n k_n^{n-1} & = 0
\end{cases}.
$$

We view this as a system of n linear equations in the n unknowns $\lambda_1, \lambda_2, \ldots, \lambda_n$, but the solutions have to be real and non-negative. The determinant of the coefficients is the Vandermonde on $(k_1, k_2, \ldots, k_n)$:

$$
\begin{vmatrix}
1 & 1 & \cdots & 1 \\
k_1 & k_2 & \cdots & k_n \\
\vdots & \vdots & \vdots & \vdots \\
k_1^{n-1} & k_2^{n-1} & \cdots & k_n^{n-1}
\end{vmatrix} = \prod_{j>l} (k_j - k_l),
$$

which is non-zero when all the $k_1, k_2, \ldots, k_n$ are different from one another. Assuming a single solution, we can write it down explicitly using Cramer's formula. For example,

$$
\lambda_1 = \begin{vmatrix}
t & 1 & \cdots & 1 \\
0 & k_2 & \cdots & k_n \\
\vdots & \vdots & \vdots & \vdots \\
0 & k_2^{n-1} & \cdots & k_n^{n-1}
\end{vmatrix} \Bigg/ \prod_{j>l} (k_j - k_l) = \frac{t k_2 \ldots k_n \prod_{j>l>1} (k_j - k_l)}{\prod_{j>l} (k_j - k_l)}
$$

$$
= \frac{t k_2 k_3 \ldots k_n}{(k_2 - k_1)(k_3 - k_1) \ldots (k_n - k_1)}.
$$

Similarly,

$$
\lambda_2 = \frac{-t k_1 k_3 k_4 \ldots k_n}{(k_2 - k_1)(k_3 - k_2)(k_4 - k_2) \ldots (k_n - k_2)} \qquad \text{etc.}
$$

$$
\lambda_n = \frac{(-1)^{n-1} t k_1 k_2 \ldots k_{n-1}}{(k_n - k_1)(k_n - k_2) \ldots (k_n - k_{n-1})}.
$$

So, the unique solution is given by

$$
\lambda_j = t \Bigg/ \prod_{\substack{l=1 \\ l \neq j}}^{n} \left(1 - \frac{k_j}{k_l} \right), \quad j = 1, 2, \ldots, n.
$$

We need, though, a non-negative solution and this is equivalent to

$$\prod_{\substack{l=1 \\ l \neq j}}^{n} \left(1 - \frac{k_j}{k_l}\right) > 0, \quad j = 1, 2, \ldots, n.$$

$\square$

An obvious question to ask is the following: What are the geometric configurations of $\{k_1, k_2, \ldots, k_n\}$ that correspond to the n conditions given in Theorem 8.0.44? That is, what are the possible arrangements of $\{k_1, k_2, \ldots, k_n\}$ on the unit circle $\mathbb{T} = \{z \in \mathbb{C} \,|\, |z| = 1\}$ that correspond to the following conditions:

$$\prod_{\substack{l=1 \\ l \neq j}}^{n} \left(1 - \frac{k_j}{k_l}\right) > 0, \quad j = 1, 2, \ldots, n.$$

The answer is not surprising. It is what we expect according to the Krzyż conjecture, namely the points $\{k_1, k_2, \ldots, k_n\}$ are evenly spread over $\mathbb{T}$. Thus, we can rephrase Theorem 8.0.44 in geometric terms as follows:

Theorem 8.0.45. *Let us fix a positive number $t > 0$. We consider functions of the form given in* [49]:

$$f(z) = \exp\left(-\sum_{j=1}^{n} \lambda_j \frac{1 + k_j z}{1 - k_j z}\right),$$

where $\lambda_j \geq 0$ and $|k_j| = 1$ for $j = 1, \ldots, n$. Assuming that $f(z) = a_0 + \cdots + a_n z^n + \cdots$ there is a unique non-negative vector $(\lambda_1, \ldots, \lambda_n)$ such that $\sum_{j=1}^{n} \lambda_j = t$ and $a_1 = a_2 = \cdots = a_{n-1} = 0$ if and only if there is a real number α, so that after arranging the indices properly we will have: $k_j = e^{i\alpha} \cdot e^{2\pi ij/n}$, $j = 1, 2, \ldots, n$. Moreover, the unique solution is given by

$$\lambda_j = \frac{t}{n}, \quad j = 1, \ldots, n,$$

and the function determined is

$$f(z) = \exp\left(-t \frac{1 + e^{in\alpha} z^n}{1 - e^{in\alpha} z^n}\right).$$

Proof. The problem is linear algebraic in which we have a system of n linear equations in the n unknowns $\alpha_j = \arg k_j$, $j = 1, \ldots, n$. Each equation is equivalent to one of the conditions in Theorem 8.0.44,

$$\prod_{\substack{l=1 \\ l \neq j}}^{n} \left(1 - \frac{k_j}{k_l}\right) > 0, \quad j = 1, 2, \ldots, n.$$

The rank of the coefficients matrix is $n - 1$, which generates a 1-parameter family of solutions, hence, the origin of α in the statement. The system is not homogeneous and the free terms are determined mod 2π. However, once we remember that $0 \leq \alpha_j = \arg k_j < 2\pi$, $j = 1, \ldots, n$, we manage to control the mod 2π ambiguity. To get a feeling of the situation, let us analyze in full details the first two cases. Before doing that, let us first generate the linear equations. We consider one of the n conditions

$$\prod_{\substack{l=1 \\ l \neq j}}^{n} \left(1 - \frac{k_j}{k_l}\right) > 0.$$

We have

$$\arg \prod_{\substack{l=1 \\ l \neq j}}^{n} \left(1 - \frac{k_j}{k_l}\right) \equiv 0 \ (\mathrm{mod} \ 2\pi).$$

$$\arg \prod_{\substack{l=1 \\ l \neq j}}^{n} \left(1 - \frac{k_j}{k_l}\right) = \sum_{\substack{l=1 \\ l \neq j}}^{n} \arg(k_l - k_j) - \arg k_l.$$

We denote $\alpha_j = \arg k_j$ for the choice $0 \leq \alpha_j < 2\pi$. Looking at the geometric picture, we find

$$\arg(k_l - k_j) = \pi + \alpha_j + \frac{\pi - (\alpha_j - \alpha_l)}{2} = \frac{1}{2}(3\pi + \alpha_j + \alpha_l).$$

$$\arg \prod_{\substack{l=1 \\ l \neq j}}^{n} \left(1 - \frac{k_j}{k_l}\right) = \sum_{\substack{l=1 \\ l \neq j}}^{n} \frac{1}{2}(3\pi + \alpha_j + \alpha_l) - \alpha_l$$

$$= \frac{1}{2} \sum_{\substack{l=1 \\ l \neq j}}^{n} (3\pi + \alpha_j - \alpha_l).$$

So, the j'th linear equation is

$$\frac{3}{2}\pi(n - 1) + \frac{1}{2} \sum_{\substack{l=1 \\ l \neq j}}^{n} (\alpha_j - \alpha_l) \equiv 0 \ (\mathrm{mod} \ 2\pi).$$

We note that since

$$\sum_{j=1}^{n} \sum_{\substack{l=1 \\ l \neq j}}^{n} (\alpha_j - \alpha_l) = 0,$$

the equations without the free terms add to zero and hence the rank of the coefficients matrix is less than n. In fact, we can show that the rank is $n-1$.

The case $n = 2$: $(3/2)\pi + (1/2)(\alpha_2 - \alpha_1) = 2\pi k$ hence $(1/2)(\alpha_2 - \alpha_1) = (2k - 3/2)\pi$. We remember that $0 \leq \alpha_1, \alpha_2 < 2\pi$ and so necessarily $k = 1$, $\alpha_2 - \alpha_1 = \pi$, which is the claim.

The case $n = 3$:

$$
\begin{cases}
\dfrac{3}{2} \cdot 2\pi + \dfrac{1}{2}((\alpha_1 - \alpha_2) + (\alpha_1 - \alpha_3)) = 2\pi k_1 \\[2mm]
\dfrac{3}{2} \cdot 2\pi + \dfrac{1}{2}((\alpha_2 - \alpha_1) + (\alpha_2 - \alpha_3)) = 2\pi k_2 \\[2mm]
\dfrac{3}{2} \cdot 2\pi + \dfrac{1}{2}((\alpha_3 - \alpha_1) + (\alpha_3 - \alpha_2)) = 2\pi k_3
\end{cases}
\cdot
$$

So,

$$
\begin{cases}
2\alpha_1 - \alpha_2 - \alpha_3 = 2 \cdot (2k_1 - 3)\pi \\
-\alpha_1 + 2\alpha_2 - \alpha_3 = 2 \cdot (2k_2 - 3)\pi \\
-\alpha_1 - \alpha_2 + 2\alpha_3 = 2 \cdot (2k_3 - 3)\pi
\end{cases}
\cdot
$$

The rank indeed is $3 - 1 = 2$. Remembering that $0 \leq \alpha_1, \alpha_2, \alpha_3 < 2\pi$ and using α_3 as a free parameter, we get

$$
\begin{cases}
2\alpha_1 - \alpha_2 = \pm 2\pi + \alpha_3 \\
-\alpha_1 + 2\alpha_2 = \pm 2\pi + \alpha_3
\end{cases}
\cdot
$$

So,

$$
\begin{cases}
3\alpha_1 = 2 \cdot (\pm 2\pi) + 3\alpha_3 \\
3\alpha_2 = 2 \cdot (\pm 2\pi) + 3\alpha_3
\end{cases}
\quad (k = 1).
$$

This gives the unknowns $\alpha_1, \alpha_2, \alpha_3$ the possible values $0, (2/3)\pi, (4/3)\pi$ up to a translation $\bmod 2\pi$. This concludes the claim's proof for this case, and it is clear that this works for any value of n. Finally, we can compute the unique solution $(\lambda_1, \ldots, \lambda_n)$. All we need is to substitute the values for $k_1, k_2, \ldots, k_n$ into Cramer's formula given at the end of the proof of Theorem 8.0.44:

$$
\begin{cases}
\lambda_1 = \dfrac{t k_2 k_3 \ldots k_n}{(k_2 - k_1)(k_3 - k_1) \ldots (k_n - k_1)} = t/n \\[4mm]
\lambda_2 = \dfrac{-t k_1 k_3 k_4 \ldots k_n}{(k_2 - k_1)(k_3 - k_2)(k_4 - k_2) \ldots (k_n - k_2)} = t/n \\[4mm]
\vdots \quad \vdots \quad \vdots \qquad\qquad\qquad\qquad\qquad\qquad \vdots \quad \vdots \\[2mm]
\lambda_n = \cdots \qquad\qquad\qquad\qquad\qquad\qquad = t/n
\end{cases}
\quad (k_j = e^{i\alpha} e^{2\pi i j/n}).
$$

We concluded the proof of Theorem 8.0.45.   □

Let us make a summary of the possible reductions we proved so far for the Krzyż conjecture. We think of an extremal for $n \in \mathbb{Z}^+$ given in two forms:

$$f(z) = \exp\left(-\sum_{j=1}^{n} \lambda_j \frac{1+k_j z}{1-k_j z}\right) = a_0 + \cdots + a_n z^n + \cdots \in B,$$

where $\lambda_j > 0$ and $|k_j| = 1$, $j = 1, \ldots, n$. Then each one of the following implies the validity of the Krzyż conjecture:

(1) $2|ka_{n-1} + k^2 a_{n-2} + \cdots + k^n a_0| \leq |a_n|$ for all k, $|k| = 1$.
     (Theorem 6.0.20)
(2) $2|a_0| = |a_n|$.
     (Theorem 6.0.33)
(3) $\lambda_1 = \lambda_2 = \cdots = \lambda_n$.
     (Theorem 6.0.37)
(4) $a_1 = a_2 = \cdots = a_{n-1} = 0$.
     (The proof of Theorem 6.0.20)
(5)

$$\prod_{\substack{l=1 \\ l \neq j}} \left(1 - \frac{k_j}{k_l}\right) > 0 \quad \text{for } j = 1, 2, \ldots, n.$$

     (Theorem 8.0.44)
(6) There is a real number $\alpha \in \mathbb{R}$ such that $k_j = e^{i\alpha} \cdot e^{2\pi i j/n}$, $j = 1, 2, \ldots, n$.
     (Theorem 8.0.45)

What do we know in relation to the above reductions?

(1') $\Re\{|a_n|^2 + 2(a_{n-1}k + a_{n-2}k^2 + \cdots + a_0 k^n)\bar{a}_n\} \geq 0$ for all k, $|k| = 1$ and more inequalities of this type (Equation (4.0.1) and Proposition 4.0.7).
(2')

$$\begin{cases} 2|a_0| \leq |a_n| \\ |a_j| \ \leq |a_n| \end{cases}$$

     (Proposition 6.0.28)

$$2|a_j| \leq |a_n|,$$

for $j = 0, 1, \ldots, (n/2) - 1$ if n is even, and for $j = 0, 1, \ldots, ((n+1)/2) - 1$ if n is odd (Proposition 4.0.7).

(4$'$) We can find $\lambda_1, \ldots, \lambda_n \geq 0$ such that if $n = 2k+1$ then $a_1 = a_3 = \cdots = a_{2k-1} = 0$, and if $n = 2k$ then $a_1 = a_3 = \cdots = a_{2k-3} = \Re\{a_{2k-1}\} = 0$ (Theorem 8.0.42).

Remark 8.0.46. Regarding reduction (1) and the currently known situation (1$'$), we have (that is well known) that

$$|a_n| \leq 2 \cdot \left(\sum_{j=1}^{n} \lambda_j \right) \|a_{n-1}k^{n-1} + \cdots + a_0\|_\infty,$$

for the extremal of n. Also, $(2/e) \leq |a_n|$. Hence, for all k, $|k| = 1$, we have

$$\frac{1}{e} \leq \left(\sum_{j=1}^{n} \lambda_j \right) \|a_{n-1}k^{n-1} + \cdots + a_0\|_\infty.$$

(4$''$) $a_{n-1} = a_{n+1} = 0$.
(Proposition 6.0.25)

Chapter 9

More on the System $\sum_{j=1}^{n} k_j^m = m \sum_{j=1}^{n} \lambda_j k_j^m$, $m = 1, \ldots, n$

Assuming there is an extremal for $n \in \mathbb{Z}^+$ which has a form as in [49]:

$$f(z) = \exp\left(-\sum_{j=1}^{n} \lambda_j \frac{1 + k_j z}{1 - k_j z}\right)$$

where $\lambda_j > 0$ and $|k_j| = 1$ for $j = 1, \ldots, n$ we saw (the third system of identities in Proposition 6.0.32) that

$$\sum_{j=1}^{n} k_j^m = m \sum_{j=1}^{n} \lambda_j k_j^m, \quad m = 1, \ldots, n.$$

This gives us n equations for the n complex unknowns k_j, $j = 1, \ldots, n$ and the n positive unknowns λ_j, $j = 1, \ldots, n$. We get n more equations by conjugation and keeping in mind that: $\overline{\lambda_j} = \lambda_j$ and $\overline{k_j} = 1/k_j$ (for $\lambda_j \in \mathbb{R}$, $|k_j| = 1$):

$$\sum_{j=1}^{n} \frac{1}{k_j^m} = m \sum_{j=1}^{n} \lambda_j \frac{1}{k_j^m}, \quad m = 1, \ldots, n.$$

These $2n$ equations make it possible to solve the Krzyż coefficients problem for $m = 2, 3$.

The case $n = 2$:
We will use only 2 (out of the 4) equations we have:

$$\begin{cases} k_1 + k_2 = \lambda_1 k_1 + \lambda_2 k_2 \\ 1/k_1 + 1/k_2 = \lambda_1/k_1 + \lambda_2/k_2 \end{cases}.$$

Multiplying the second equation by $k_1 k_2$ we get $k_1 + k_2 = \lambda_1 k_2 + \lambda_2 k_1$. Subtracting from the first equation gives $0 = (\lambda_1 - \lambda_2)(k_1 - k_2)$, which implies (since $k_1 \neq k_2$) that $\lambda_1 = \lambda_2$ which proves (Theorem 6.0.37) that $|a_2| = 2/e$.

The case $n = 3$:

We take 2 (out of the 6) equations:

$$\begin{cases} k_1 + k_2 + k_3 = \lambda_1 k_1 + \lambda_2 k_2 + \lambda_3 k_3 \\ 1/k_1 + 1/k_2 + 1/k_3 = \lambda_1/k_1 + \lambda_2/k_2 + \lambda_3/k_3 \end{cases}.$$

We divide the first equation by k_1 and multiply the second equation by k_1 and subtract. This gives the first equation in the following system. Similarly, we get the second and the third equations:

$$\left(\frac{k_2}{k_1} - \frac{k_1}{k_2} \right) + \left(\frac{k_3}{k_1} - \frac{k_1}{k_3} \right) = \lambda_2 \left(\frac{k_2}{k_1} - \frac{k_1}{k_2} \right) + \lambda_3 \left(\frac{k_3}{k_1} - \frac{k_1}{k_3} \right)$$

$$\left(\frac{k_1}{k_2} - \frac{k_2}{k_1} \right) + \left(\frac{k_3}{k_2} - \frac{k_2}{k_3} \right) = \lambda_1 \left(\frac{k_1}{k_2} - \frac{k_2}{k_1} \right) + \lambda_3 \left(\frac{k_3}{k_2} - \frac{k_2}{k_3} \right).$$

$$\left(\frac{k_1}{k_3} - \frac{k_3}{k_1} \right) + \left(\frac{k_2}{k_3} - \frac{k_3}{k_2} \right) = \lambda_1 \left(\frac{k_1}{k_3} - \frac{k_3}{k_1} \right) + \lambda_2 \left(\frac{k_2}{k_3} - \frac{k_3}{k_2} \right)$$

We add these 3 equations and obtain

$$0 = (\lambda_2 - \lambda_1) \left(\frac{k_2}{k_1} - \frac{k_1}{k_2} \right) + (\lambda_3 - \lambda_1) \left(\frac{k_3}{k_1} - \frac{k1}{k_3} \right) + (\lambda_3 - \lambda_2) \left(\frac{k_3}{k_2} - \frac{k_2}{k_3} \right).$$

Similarly,

$$0 = (\lambda_2 - \lambda_1) \left(\frac{k_2^2}{k_1^2} - \frac{k_1^2}{k_2^2} \right) + (\lambda_3 - \lambda_1) \left(\frac{k_3^2}{k_1^2} - \frac{k1^2}{k_3^2} \right) + (\lambda_3 - \lambda_2) \left(\frac{k_3^2}{k_2^2} - \frac{k_2^2}{k_3^2} \right),$$

and

$$0 = (\lambda_2 - \lambda_1) \left(\frac{k_2^3}{k_1^3} - \frac{k_1^3}{k_2^3} \right) + (\lambda_3 - \lambda_1) \left(\frac{k_3^3}{k_1^3} - \frac{k1^3}{k_3^3} \right) + (\lambda_3 - \lambda_2) \left(\frac{k_3^3}{k_2^3} - \frac{k_2^3}{k_3^3} \right).$$

This is a linear homogeneous system in the unknowns $(\lambda_2 - \lambda_1)$, $(\lambda_3 - \lambda_1)$, $(\lambda_3 - \lambda_2)$. We now compute the determinant of the coefficients matrix:

$$\Delta = \begin{vmatrix} k_2/k_1 - k_1/k_2 & k_3/k_1 - k_1/k_3 & k_3/k_2 - k_2/k_3 \\ k_2^2/k_1^2 - k_1^2/k_2^2 & k_3^2/k_1^2 - k_1^2/k_3^2 & k_3^2/k_2^2 - k_2^2/k_3^2 \\ k_2^3/k_1^3 - k_1^3/k_2^3 & k_3^3/k_1^3 - k_1^3/k_3^3 & k_3^3/k_2^3 - k_2^3/k_3^3 \end{vmatrix}$$

$$= \left(\frac{k_2}{k_1} - \frac{k_1}{k_2} \right) \left(\frac{k_3}{k_1} - \frac{k_1}{k_3} \right) \left(\frac{k_3}{k_2} - \frac{k_2}{k_3} \right) \left(\frac{k_3}{k_1} - \frac{k_2}{k_1} \right) \left(\frac{k_3}{k_2} - \frac{k_2}{k_1} \right)$$

$$\times \left(\frac{k_3}{k_2} - \frac{k_3}{k_1} \right) \left(1 - \frac{k_1}{k_3} \right) \left(1 - \frac{k_1^2}{k_2 k_3} \right) \left(1 - \frac{k_1 k_2}{k_3^2} \right).$$

We recall that if $l \neq j$ then $k_l \neq k_j$. Hence,

(1) If $\Delta \neq 0$, then the system has a unique solution, the trivial one. Hence, $\lambda_1 = \lambda_2 = \lambda_3 = 0$ and $|a_3| = 2/e$.

(2) If $\Delta = 0$, we distinguish different cases:

(a) There are $l \neq j$ such that $k_l = -k_j$. For example, we take $(l, j) = (1, 2)$. Then

$$\begin{cases} k_3 = 7(\lambda_2 - \lambda_1) + \lambda_3 k_3 \\ 1/k_3 = (\lambda_2 - \lambda_1) + \lambda_3/k_3 \end{cases},$$

where we assumed without losing the generality that $k_1 = -1$, $k_2 = 1$. Hence, $k_3^2 \neq 1$. We subtract the second equation from the first equation:

$$k_3 - \frac{1}{k_3} = \lambda_3 \left(k_3 - \frac{1}{k_3} \right),$$

and by $k_3^2 \neq 1$, we deduce that $\lambda_3 = 1$ and $\lambda_2 - \lambda_1 = 0$. We remember that $k_3^3 = 3(\lambda_2 - \lambda_1) + 3\lambda_3 k_3^3 = 3k_3^3$ and so $k_3 = 0$, which is a contradiction.

(b) For $1 \leq s, t, u \leq 3$ distinct from one another, we have $k_s^2 = k_t k_u$. Without losing the generality, we assume that $k_1 = 1$, $k_2 = \alpha$, $k_3 = \alpha^2$. We have

$$\begin{cases} 1 + \alpha + \alpha^2 = \lambda_1 + \lambda_2 \alpha + \lambda_3 \alpha^2 \\ 1 + 1/\alpha + 1/\alpha^2 = \lambda_1 + \lambda_2/\alpha + \lambda_3/\alpha^2 \end{cases}.$$

We multiply the second equation by α^2 and subtract the resulting equation from the first equation: $0 = (1 - \alpha^2)(\lambda_1 - \lambda_3)$. It follows by $k_3 \neq k_1$ that $1 - \alpha^2 \neq 0$. Hence, $\lambda_1 = \lambda_3$. Let us denote $\lambda = \lambda_1 = \lambda_3$ and $\delta = \lambda_2$ and the system becomes

$$\begin{cases} 1 + \alpha + \alpha^2 = \lambda(1 + \alpha^2) + \delta\alpha \\ 1 + \alpha^2 + \alpha^4 = 2\lambda(1 + \alpha^4) + \delta\alpha^2 \\ 1 + \alpha^3 + \alpha^6 = 3\lambda(1 + \alpha^6) + \delta\alpha^3 \end{cases}.$$

The three equations with $1/\alpha$ instead of α are exactly the same as those above. We recall that $|\alpha| = 1$, so $\alpha = e^{i\theta}$ $(0 < \theta < 2\pi)$. We obtain

$$\begin{cases} 2\cos\theta + 1 = 2\lambda\cos\theta + \delta \\ 2\cos 2\theta + 1 = 4\lambda\cos 2\theta + 2\delta \\ 2\cos 3\theta + 1 = 6\lambda\cos 3\theta + 3\delta \end{cases}.$$

Denoting $X = \cos\theta$, the last system can be written as

$$\begin{cases} 2X + 1 = 2\lambda X + \delta \\ 2(2X^2 - 1) + 1 = 4\lambda(2X^2 - 1) + 2\delta \\ 2X(4X^2 - 3) + 1 = 6\lambda X(4X^2 - 3) + 3\delta \end{cases}.$$

We multiply the first equation by 2 and subtract from the second equation: $(X + 1/2)(X - 3/2) = 2\lambda(X - 1)(X + 1/2)$. If $X + 1/2 = 0$, we get from the original system $\lambda = \delta = 1/3$, so $|a_3| = 2/e$. If $X + 1/2 \neq 0$, then $X - 3/2 = 2\lambda(X - 1)$. We multiply the first equation by 3 and subtract the result from the third equation: $8X^3 - 12X - 2 = 24\lambda(X - 1)(X + 1)$. From the last two equations, we can eliminate λ and get $2X^3 - 3X^2 - 3X + 1 = 0$. This equation has 3 real roots: $-1 < X_0 < -1/2 < 1/4 < X_1 < 1/2 < 1 < X_2$. So X_2 is not possible $(X = \cos\theta)$. The two other roots X_0, X_1 are possible and generate positive (λ_0, δ_0) and (λ_1, δ_1). We can verify numerically that for these two solutions, we get $|a_3| < 2/e$. However, we have enough information to verify that for the extremal function $\lambda \neq \delta$ is an impossibility: assuming $e^t f(z) = a_0 + a_1 z + a_2 z^2 + \cdots$. we get by the exponential recursion for the coefficients

$$\begin{cases} a_0 = 1 \\ a_1 = -(\lambda + \delta\alpha + \lambda\alpha^2) \\ a_2 = 2\{(\lambda + \delta\alpha + \lambda\alpha^2) - (\lambda + \delta\alpha^2 + \lambda\alpha^4)\} \\ a_3 = (-2/3)\{2(\lambda + \delta\alpha + \lambda\alpha^2)^3 - 6(\lambda + \delta\alpha + \lambda\alpha^2)(\lambda + \delta\alpha^2 + \lambda\alpha^4) \\ \qquad + 3(\lambda + \delta\alpha^3 + \lambda\alpha^6)\} \end{cases}.$$

These give us

$$\begin{cases} a_0 = 1 \\ a_1 = -2(1 + \alpha + \alpha^2) \\ a_2 = 4\alpha(1 + \alpha + \alpha^2) \\ a_3 = (2/3)(1 - 11\alpha^3 + \alpha^6) \end{cases}.$$

Hence, the polynomial we obtain from the variation in equation (4.0.1) becomes

$$p(k) = \Re\{\bar{a}_3(2(k^3 + k^2 a_1 + k a_2) + a_3)\} = \Re\left\{\frac{2}{3}\bar{\alpha}^3(2\cos 3\theta - 11)\right.$$

$$\left. \times \left(\frac{2}{3}\alpha^3(2\cos 3\theta - 11) + 8\alpha^2(2\cos\theta + 1)k - 4\alpha(2\cos\theta + 1)k^2 + 2k^3\right)\right\}.$$

We recall that necessarily $p(\alpha) = 0$ (by Proposition 6.0.28(1)), hence,

$$\frac{2}{3}(2\cos 3\theta - 11)\left(\frac{2}{3}(2\cos 3\theta - 11) + 8(2\cos\theta + 1) - 4(2\cos\theta + 1) + 2\right) = 0.$$

So, $(2\cos 3\theta - 11)(2\cos 3\theta + 12\cos\theta - 2) = 0$. But $2\cos 3\theta - 11 < 0$ and so, $2\cos 3\theta + 12\cos\theta - 2 = 0$, i.e., $4X^3 + 3X - 1 = 0$ ($X = \cos\theta$). However, we recall that we have proved before that $2X^3 - 3X^2 - 3X + 1 = 0$ and when we add the last two equations, we get $3X^2(2X - 1) = 0$. This means that $X_0 = 0$ and $X_1 = X_2 = 1/2$, which is a contradiction. So, the case $X + 1/2 \neq 0$ is impossible. We proved that we always have $\lambda_1 = \lambda_2 = \lambda_3$ and so $|a_3| = 2/e$.

We see that proving the Krzyż conjecture for small values of n becomes more and more complicated if it is done as in this chapter.

Chapter 10

On $\arg a_n$

Proposition 10.0.47. *Let $f(z) = a_0 + \cdots + a_n z^n + \cdots \in B$ be extremal for $n \in \mathbb{Z}^+$. We assume that $a_0 > 0$. Then*

(1) *The following two polynomials have no zeros in $|z| < 1$: $a_n + a_{n-1}z + \cdots + a_0 z^n$, and $a_n + 2(a_{n-1}z + \cdots + a_0 z^n)$.*

(2) *Consider the form of the extremal $f(z)$ given in [49]:*

$$\exp\left(-\sum_{j=1}^{n}\lambda_j \frac{1 + k_j z}{1 - k_j z}\right),$$

where $\lambda_j \geq 0$, $|k_j| = 1$, $j = 1, \ldots, n$. Then $X(k_j) = a_{n-1}k_j + a_{n-2}k_j^2 + \cdots + a_0 k_j^n$ are co-linear points and for $1 \leq l \neq j \leq n$, we have

$$\arg(X(k_j) - X(k_l)) = \arg a_n + \frac{\pi}{2} \quad \text{or} \quad \arg(X(k_l) - X(k_j)) = \arg a_n + \frac{\pi}{2}.$$

In other words, the vectors $X(k_j) - X(k_l)$ and a_n are orthogonal.

Proof.

(1) Let us denote $p(z) = |a_n|^2 + 2\bar{a}_n(a_{n-1}z + \cdots + a_0 z^n)$. By equation (4.0.1) or Proposition 4.0.7 in the case $l = 1$, we have $0 \leq \Re\{p(z)\}$ for $|z| < 1$. Hence, there is $q(z) \in H(U)$, $q(0) = 0$, $|q(z)| < 1$ for $|z| < 1$, so that

$$p(z) = |a_n|^2 \cdot \left(\frac{1 + q(z)}{1 - q(z)}\right).$$

Hence, solving for $q(z)$:

$$q(z) = \frac{p(z) - |a_n|^2}{p(z) + |a_n|^2} = \frac{a_{n-1}z + \cdots + a_0 z^n}{a_n + a_{n-1}z + \cdots + a_0 z^n}.$$

73

We just proved that for all $|z| < 1$, we have

$$\left| \frac{a_{n-1}z + \cdots + a_0 z^n}{a_n + a_{n-1}z + \cdots + a_0 z^n} \right| < 1.$$

Hence, by the Rouche Theorem, the following polynomials have the same number of zeros in $|z| < 1$:

$$a_n, \quad a_n + a_{n-1}z + \cdots + a_0 z^n, \quad a_n + 2(a_{n-1}z + \cdots + a_0 z^n).$$

(2) By Proposition 6.0.28(1), we have $\Re\{k_j\} = 0$, $j = 1, \ldots, n$. We note that

$$\Re\{p(z)\} = |a_n|^2 \cdot \left(\frac{1 - |q(z)|^2}{|1 - q(z)|^2} \right),$$

and hence, $|q(k_j)| = 1$, $j = 1, \ldots, n$ (and $q(k_j) \neq 1$). Thus,

$$|a_n + a_{n-1}k_j + \cdots + a_0 k_j^n| = |a_{n-1}k_j + \cdots + a_0 k_j^n|.$$

Using the notation in the statement of our theorem, this is equivalent to $|a_n + X(k_j)| = |X(k_j)|$. Let us sketch a_n and $X(k_j)$ as planar vectors so that the vertical segment to a_n that starts at the origin is in fact the vertical segment that ends in the mid point of a_n. Then necessarily the vector that starts at the origin and ends at the starting point of a_n is $X(k_j)$ (because $|X(k_j)| = |a_n + X(k_j)|$). This concludes the proof of part (2).

$\square$

Chapter 11

One More Variation

Proposition 11.0.48. *Let $f(z) = a_0 + \cdots + a_n z^n + \cdots \in B$ be extremal for $n \in \mathbb{Z}^+$. We assume that $a_0 > 0$. Let $-\log f(z) = \alpha_0 + \alpha_1 z + \cdots$ where the branch we take satisfies $\alpha_0 > 0$ $(a_0 = e^{-\alpha_0})$. Then for any angle β, $(-\pi/2) \leq \beta \leq (\pi/2)$, we have*

$$\Re\{\bar{a}_n e^{i\beta}(2\alpha_0\alpha_1 a_{n-1} + (2\alpha_0\alpha_2 + \alpha_1^2)a_{n-2}$$

$$+ \cdots + (2\alpha_0\alpha_n + 2\alpha_1\alpha_{n-1} + \cdots)a_0)\} \leq 0.$$

Equivalently,

$$\bar{a}_n(2\alpha_0\alpha_1 a_{n-1} + (2\alpha_0\alpha_2 + \alpha_1^2)a_{n-2}$$

$$+ \cdots + (2\alpha_0\alpha_n + 2\alpha_1\alpha_{n-1} + \cdots)a_0) \leq 0.$$

Proof. We introduce an appropriate variation that will enable us derive our inequalities. If $p(z) \in H(U)$ satisfies the two conditions: $0 \leq \Re\{p(z)\}$ and $p(0) = \lambda$, then there is a function $q(z) \in H(U)$ so that $|q(z)| \leq |z|$ and so that

$$p(z) = \lambda \cdot \left(\frac{1 + q(z)}{1 - q(z)}\right), \quad z \in U.$$

Solving for $q(z)$:

$$q(z) = \frac{p(z) - \lambda}{p(z) + \lambda}, \quad z \in U.$$

if $\alpha \in \mathbb{R}$ and if r, $0 \leq r \leq 1$, we denote $\delta = re^{i\alpha}$. We define a new function:

$$p_\delta(z) = \lambda \cdot \left(\frac{1 + \delta \cdot q(z)}{1 - \delta \cdot q(z)} \right), \quad z \in U.$$

Then $p_\delta(z) \in H(U)$, $0 \leq \Re\{p_\delta(z)\}$, and $p_\delta(0) = \lambda$. In this sense, $p_\delta(z)$ is an admissible function. We now estimate the difference $p(z) - p_\delta(z)$:

$$p - p_\delta = \lambda \left(\frac{1+q}{1-q} - \frac{1+\delta q}{1-\delta q} \right) = 2\lambda(1-\delta) \frac{q}{(1-q)(1-\delta q)}$$

$$= 2\lambda(1-\delta) \frac{(p-\lambda)/(p+\lambda)}{(1 - (p-\lambda)/(p+\lambda))(1 - \delta(p-\lambda)/(p+\lambda))}$$

$$= (1-\delta)\left\{ \frac{p^2 - \lambda^2}{(1-\delta)p + (1+\delta)\lambda} \right\}.$$

We come to our function $f \in B$ that satisfies $0 < |f(z)| < 1$, $z \in U$ and $f(0) = a_0 > 0$. There is a function $p(z) \in H(U)$ such that $0 \leq \Re\{p(z)\}$, and $p(0) = \lambda$, and also $f(z) = e^{-p(z)}$, $a_0 = e^\lambda$. For $\delta = re^{i\alpha}$, $0 \leq r \leq 1$, the following function is an admissible function: $f_\delta(z) = e^{-p_\delta(z)}$. Our computation implies the following:

$$f_\delta(z) = e^{-p_\delta} = e^{p - p_\delta} e^{-p} = \exp\left\{ (1-\delta) \frac{p^2 - \lambda^2}{(1-\delta)p + (1+\delta)\lambda} \right\} \cdot f.$$

We have

$$\exp\left\{ (1-\delta) \frac{p^2 - \lambda^2}{(1-\delta)p + (1+\delta)\lambda} \right\}$$

$$= 1 + (1-\delta) \frac{p^2 - \lambda^2}{(1-\delta)p + (1+\delta)\lambda} + O((1-\delta)^2)$$

$$= 1 + (1-\delta) \frac{p^2 - \lambda^2}{(1+\delta)\lambda} + O((1-\delta)^2).$$

Thus,

$$f_\delta(z) = \left\{ 1 + \left(\frac{1-\delta}{1+\delta} \right) \cdot \frac{1}{\lambda} \cdot (p^2 - \lambda^2) + O((1-\delta)^2) \right\} \cdot f(z).$$

We recall that $f(z) = \sum_{j=0}^\infty a_j z^j$, $p(z) = \sum_{j=0}^\infty \alpha_j z^j$ where $\lambda = \alpha_0$, $a_0 = e^{-\lambda}$. Using that, we get

$$f_\delta(z) = \left\{ 1 + \frac{1}{\alpha_0} \left(\frac{1-\delta}{1+\delta} \right) (2\alpha_0\alpha_1 z + (2\alpha_0\alpha_2 + \alpha_1^2)z^2 + \cdots) \right\}$$

$$\times \sum_{j=0}^\infty a_j z^j + O((1-\delta)^2).$$

From here, we compute the coefficient $A_n(\delta)$ of z^n:

$$A_n(\delta) = a_n + \frac{1}{\alpha_0}\left(\frac{1-\delta}{1+\delta}\right)(2\alpha_0\alpha_1 a_{n-1} + (2\alpha_0\alpha_2 + \alpha_1^2)a_{n-2}$$

$$+ \cdots + (2\alpha_0\alpha_n + 2\alpha_1\alpha_{n-1} + \cdots)a_0) + O((1-\delta)^2),$$

where $\delta = re^{i\alpha}$, $0 \leq r \leq 1$. By extremality, we have $|A_n(\delta)| \leq |a_n|$, $\forall\,\delta$. The above computation gives

$$\Re\left\{\left(\frac{1-\delta}{1+\delta}\right)\frac{\overline{a}_n}{\alpha_0}(2\alpha_0\alpha_1 a_{n-1} + \cdots + (2\alpha_0\alpha_n + 2\alpha_1\alpha_{n-1} + \cdots)a_0)\right\}$$

$$+ O((1-\delta)^2) \leq 0,$$

for all $\delta = re^{i\alpha}$, $0 \leq r \leq 1$. This implies that for every angle β, $(-\pi/2) < \beta < (\pi/2)$, we can find a δ such that $1 - \delta = \epsilon e^{i\beta}$, $\epsilon \to 0^+$ and hence,

$$\Re\left\{\left(\frac{\epsilon e^{i\beta}}{2 - \epsilon e^{i\beta}}\right)\frac{\overline{a}_n}{\alpha_0}(2\alpha_0\alpha_1 a_{n-1} + \cdots + (2\alpha_0\alpha_n + 2\alpha_1\alpha_{n-1} + \cdots)a_0)\right\}$$

$$+ O(\epsilon^2) \leq 0.$$

Now, we divide by $\epsilon > 0$ and take the limit as $\epsilon \to 0^+$. $\qquad\square$

Chapter 12

l^p Versions of the Krzyż Conjecture

We now point at a family of problems that resemble the problem that lies at the heart of the Krzyż conjecture. The nonlinear feature of being non-vanishing will remain as is. The difference will be in the norms. We will be dealing with l^p spaces.

Definition 12.0.49. We will denote $l^p(U) = \{f(z) = \sum_{j=0}^{\infty} a_j z^j \in H(U) \mid \sum_{j=0}^{\infty} |a_j|^p < \infty\}$. The fixed parameter, p, will usually be in $1 \leq p$. We also may take $p = \infty$, by defining $l^\infty(U) = \{f(z) = \sum_{j=0}^{\infty} a_j z^j \in H(U) \mid \sup_j |a_j| < \infty\}$. The unit (open) balls are as usual: $B(l^p(U)) = \{f \in l^p(U) \mid \|f\|_{l^p} < 1\}$. If $f(z) = \sum_{j=0}^{\infty} a_j z^j \in l^p(U)$, then we used the standard notation:

$$\|f\|_{l^p} = \begin{cases} \left(\displaystyle\sum_{j=0}^{\infty} |a_j|^p\right)^{1/p} & \text{if } 1 \leq p < \infty \\ \displaystyle\sup_j |a_j| & \text{if } p = \infty \end{cases} .$$

If we fix p, $1 \leq p \leq \infty$, and $n \in \mathbb{Z}^+ \cup \{0\}$, then the corresponding Krzyż problem is

$$A_n(p) = \sup_{\begin{cases} f(z) \in B(l^p(U)) \\ f(z) \neq 0 \; \forall z \in U \end{cases}} |a_n|.$$

Remark 12.0.50. (1) We clearly have $A_n(p) \leq 1$ for all $n \in \mathbb{Z}^+ \cup \{0\}$ and for all $1 \leq p \leq \infty$. This is true even if the condition $f(z) \neq 0 \; \forall z \in U$ is dropped. It follows by $\forall f \in B(l^p(U)) \; \forall n \in \mathbb{Z}^+ \cup \{0\}$, $|a_n| \leq \|f\|_{l^p} \leq 1$.

(2) If there exists an extremal function $f(z) = \sum_{j=0}^{\infty} a_j z^j$ for $A_n(p)$, $(p < \infty)$, then $\sum_{j=0}^{\infty} |a_n|^p = 1$. For if $\alpha^p = \sum_{j=0}^{\infty} |a_n|^p < 1$, then clearly $|\alpha| > 0$ and the dilated function $(1/\alpha)f(z)$, is admissible for $A_n(p)$, and its $(n+1)$'st coefficient $(1/\alpha)a_n$ is larger than a_n in absolute value $((1/|\alpha|) > 1)$, which contradicts the extremality of $f(z)$.

(3) The problem $A_n(2)$ is more general than the Krzyż conjecture that takes into an account only the functions $f(z) \in H(U)$, for which $0 < |f(z)| \le 1 \ \forall z \in U$. For: $\sum_{j=0}^{\infty} |a_j|^2 = (1/(2\pi)) \int_0^{2\pi} |f(e^{i\theta})|^2 d\theta \le 1$. Thus, the Krzyż set of admissible functions is a subset of $\{g \in B(l^2(U)) \,|\, g(z) \ne 0 \ \forall z \in U\}$. Hence, if $|a_n|$ is a solution for the Krzyż problem for $n \in \mathbb{Z}^+$, then $|a_n| \le A_n(2)$. This might be significant if we could show that $\sup_n A_n(2) < 1$. For it provides a uniform upper bound, which is strictly less than 1, for all the coefficients in Krzyż admissible functions, $0 < |f(z)| \le 1$.

(4) Solving $A_n(p)$ is not hard for certain values of p. Here is an example: $A_n(1) = 1/2$, with the extremal function $f(z) = (1/2)(1 + z^n)$.

Proof. Let us suppose that $a_n = 1 - \epsilon$. If for all z, $|z| = 1$, the following inequality were true:

$$\left| \frac{\sum_{j=0,j\ne n}^{\infty} a_j z^j}{a_n z^n} \right| < 1,$$

then by the Rouche Theorem, the functions $a_n z^n$ and $f(z)$ had the same number of zeros in U. But $n \ge 1$ while $f(z) \ne 0 \ \forall z \in U$. We conclude that there is a z_0, $|z_0| = 1$ such that

$$\left| \frac{\sum_{j=0,j\ne n}^{\infty} a_j z_0^j}{a_n z_0^n} \right| \ge 1.$$

But by the triangle inequality,

$$\left| \frac{\sum_{j=0,j\ne n}^{\infty} a_j z_0^j}{a_n z_0^n} \right| \le \frac{\epsilon}{1-\epsilon},$$

and we conclude that $\epsilon/(1 - \epsilon) \ge 1$, i.e., $\epsilon \ge (1/2)$ and hence $a_n = 1 - \epsilon \le (1/2)$. The function $(1/2)(1 + z^n)$ is admissible for $A_n(1)$ and hence shows that $A_n(1) = (1/2)$. $\qquad\square$

Let us formally store the fourth remark in the following:

Proposition 12.0.51.

(1) $A_n(1) = 1/2$ *and an extremal function is* $f(z) = (1/2)(1 + z^n)$.
(2) $A_n(\infty) = 1$ *and an extremal function is* $f(z) = 1 + z^n$.

Problem 12.0.52. Let us denote the set of admissible functions for $A_n(p)$ by $Kr(p)$. Thus, $Kr(p) = \{f \in H(U) \,|\, \|f\|_{l^p}^p = \sum_{j=0}^{\infty} |a_j|^p \leq 1, \, f(z) \neq 0 \,\forall\, z \in U\}$. Let us denote by C_n the functional on $H(U)$ that assigns to every function in $H(U)$, its $(n+1)$'st coefficient in its power series expansion:

$$C_n : H(U) \to \mathbb{C}, \quad C_n(g) = \frac{g^{(n)}(0)}{n!}.$$

We know that (Proposition 12.0.51) $\max_{g \in Kr(1)} C_n(g) = 1/2$ and $\max_{g \in Kr(\infty)} C_n(g) = 1$. Could we develop an interpolation theory of operators (not on linear spaces) that will give us an estimate for $\max_{g \in Kr(2)} C_n(g)$? As mentioned before, an affirmative answer might lead to a uniform bound strictly less than 1 in the Krzyż conjecture. We mention that at the end of the paper [49] the authors brought up that idea of nonlinear interpolation, but they have chosen the setting of $\mathbb{H}_p(U)$ spaces and not $l^p(U)$ spaces. They suggested a conjecture for the solution of the problem, and the extremal functions there are quite complicated: A kind of a hybrid between the conjectured extremal in the Krzyż conjecture and a polynomial in a certain power that comes in as a factor. The author cannot understand the origin of this impressive hybrid. On the other hand, in the $l^p(U)$ setting, the extremals might be simpler. At least by Proposition 12.0.51 at the end values, we are dealing with polynomials with only two terms.

Problem 12.0.53. We do not even have a conjecture for $A_n(p)$ and not for corresponding extremals. Looking at the extremal functions $(1/2)(1+z^n)$ and $1 + z^n$ for $A_n(1)$ and for $A_n(\infty)$, respectively, we might have thought that we are lucky and try $(1/2)^{1/p}(1 + z^n)$ as an extremal for $A_n(p)$ and $A_n(p) = (1/2)^{1/p}$. This cannot be true for we have (owing to the Krzyż conjecture): $A_n(2) \geq (2/e) > (1/\sqrt{2})$ for all $n \in \mathbb{Z}^+$. We remark that $A_n(2) > (1/2)^{1/2}$ is easy to verify without using facts from the theory developed to solve the Krzyż conjecture. For instance,

$$A_n(2) \geq \sqrt{\frac{2}{3}} > \left(\frac{1}{2}\right)^{1/2}, \quad n > 1,$$

because $\sqrt{1/6}(1 + 2z^n + z^{2n}) \in Kr(2)$. It is worth observing that $\sqrt{2/3} > 2/e$ is to be expected because of the relations between the Krzyż conjecture and

$A_n(2)$. So, at this point, we might conjecture that

$$A_0(2) = 1 \quad \text{and} \quad A_n(2) = \left(\frac{2}{3}\right)^{1/2}, \quad \text{if } n = 2, 3, 4, \ldots.$$

If true, then we have values for $A_n(p)$ for $p = 1, 2, \infty$ ($p = 2$ is merely a speculation) and we might aim at a simple formula that will model that. For example, is the following true? For $1 \le p \le \infty$ and for $n \in \mathbb{Z}^+$,

$$A_n(p) = \left(\frac{p}{p+1}\right)^{1/p}.$$

Unfortunately, this nice conjecture cannot be true. The reason is that the problem $A_n(2)$ is identical with the extension of the Krzyż conjecture to $\mathbb{H}_2(U)$ (Parseval). The conjecture at the end of [49] gives for $p = 2$ a different value than the above. For example, the value in [49] is not an algebraic number unlike $\sqrt{2/3}$.

Remark 12.0.54. (1) If $f, g \in Kr(2)$ then $\sqrt{f \cdot g} \in Kr(2)$.

Proof. $f, g \in Kr(2) \Rightarrow f(z), g(z) \ne 0 \ \forall z \in U$ and

$$(1/(2\pi)) \int_0^{2\pi} |f(e^{i\theta})|^2 d\theta, \quad (1/(2\pi)) \int_0^{2\pi} |g(e^{i\theta})|^2 d\theta \le 1.$$

Hence, $f(z) \cdot g(z) \ne 0 \ \forall z \in U$ and by the Cauchy–Schwarz inequality:

$$\frac{1}{2\pi} \int_0^{2\pi} \left| \sqrt{f(e^{i\theta})g(e^{i\theta})} \right|^2 d\theta = \frac{1}{2\pi} \int_0^{2\pi} \left| f(e^{i\theta})g(e^{i\theta}) \right| d\theta$$

$$\le \sqrt{\frac{1}{2\pi} \int_0^{2\pi} |f(e^{i\theta})|^2 d\theta} \cdot \sqrt{\frac{1}{2\pi} \int_0^{2\pi} |g(e^{i\theta})|^2 d\theta} \le 1.$$

$\square$

(2) If $f \in Kr(2)$, $g \in B$, then $f \cdot g \in Kr(2)$.

Proof. $f \in Kr(2)$, $g \in B \Rightarrow f(z), g(z) \ne 0 \ \forall z \in U$ and $(1/(2\pi)) \int_0^{2\pi} |f(e^{i\theta})|^2 d\theta \le 1$, $|g(z)| \le 1 \ \forall z \in U$. Hence, $f(z) \cdot g(z) \ne 0 \ \forall z \in U$ and

$$\frac{1}{2\pi} \int_0^{2\pi} |f(e^{i\theta})g(e^{i\theta})|^2 d\theta \le \frac{1}{2\pi} \int_0^{2\pi} |f(e^{i\theta})|^2 d\theta \le 1.$$

$\square$

Or, simply use $B \subseteq Kr(2)$ and (1) above.

In particular, the multiplicative variations that worked for the Krzyż problem apply also here. For example, if $f(z) = a_0 + \cdots + a_n z^n + \cdots$ is extremal for $A_n(2)$, then by Remark 12.0.54(2) above (the current), for any $\epsilon > 0$ and any k, $|k| = 1$, we have

$$f(z) \cdot \exp\left(-\epsilon \frac{1+kz}{1-kz}\right) \in Kr(2).$$

As with the Krzyż problem, we compare the coefficient of this last function with a_n, as $\epsilon \to 0^+$ and get equation analog to equation (4.0.1):

$$\Re\{|a_n|^2 + 2\overline{a}_n(a_{n-1}k + a_{n-2}k^2 + \cdots + a_0 k^n)\} \geq 0 \quad \forall\, |k| = 1,$$

and similarly for other results that we derived in the setting of Krzyż problem from the theory of non-negative trigonometric polynomials. For example, also for $A_n(2)$, we have $2|a_0| \leq |a_n|$. Indeed, in our (faulty) conjecture $A_n(p) = (p/(p+1))^{1/p}$, we get for the case $p = 2$, $\sqrt{2/3}$ and we saw at least one extremal function, namely

$$\frac{1}{\sqrt{6}}(1+z^n)^2 = \frac{1}{\sqrt{6}} + \frac{2}{\sqrt{6}}z^n + \frac{1}{\sqrt{6}}z^{2n}.$$

Thus, here, $a_0 = 1/\sqrt{6}$ and $a_n = 2/\sqrt{6} = 2a_0$ as expected. The principle that we have equality $|a_n| = 2|a_0|$ for the extremal seems to be in common for the Krzyż conjecture theory and for $A_n(2)$. Of course, our conjecture for $A_n(2)$ is faulty, but even if we did not know that, we have no explanation for this doubling principle, $|a_n| = 2|a_0|$. If this principle holds true for $A_n(p)$, it seems to appear for different reasons in the Krzyż conjecture and in $A_n(p)$ (even in $A_n(2)$).

Another principle that might be true for the families of the Krzyż problem type is the principle of a uniform value. This means (if true) that the value of $A_n(2)$ is independent of n ($2/e$ for the Krzyż conjecture and something like $\sqrt{2/3}$ for $A_n(2)$). One more principle to inquire about is that for different values of n, we obtain different extremal functions. However, all these functions originate from a single function, $\phi(z)$ (the extremal for $n = 1$) by subordination, $\phi(z^n)$ for the n'th coefficient. In particular, if this subordination principle is true, it explains the principle of the vanishing of many coefficients in the extremal function. For example, $a_1 = a_2 = \cdots = a_{n-1} = 0$. For instance, we have for the Krzyż conjecture and for $A_n(2)$, respectively,

$$\phi(z) = \exp\left(-\frac{1+z}{1-z}\right) \quad \text{and} \quad \phi(z) = \frac{1}{\sqrt{6}}(1+z)^2.$$

So far for letting our imagination go loose....

Chapter 13

The Structure of Functions in $Kr(2)$

We recall that $Kr(2) = \{f(z) = \sum_{j=0}^{\infty} a_j z^j \in H(U) \,|\, \sum_{j=0}^{\infty} |a_j|^2 \leq 1$ and $f(z) \neq 0 \,\forall\, |z| < 1\}$. We can re-write this definition in terms of Hilbert space

$$\mathbb{H}_2(U) = \left\{ f(z) \in H(U) \,\Big|\, \|f\|_2^2 = \frac{1}{2\pi} \int_0^{2\pi} |f(e^{i\theta})|^2 d\theta < \infty \right\},$$

namely

$$Kr(2) = \left\{ f \in \mathbb{H}_2(U) \,\big|\, 0 < |f(z)| \,\forall\, |z| < 1, \ \|f\|_2 \leq 1 \right\}.$$

In other words, $Kr(2)$ is the set of all the non-vanishing functions in the closed unit ball of the Hilbert space $\mathbb{H}_2(U)$, i.e., $\overline{B}_{\mathbb{H}_2(U)}$. We will refer to [59]; Chapter IV (pp. 65–86) describes the canonical structure of the functions in $\mathbb{H}_p$. This structure is the well-known factorization into a product of inner and outer functions.

Theorem ([59], p. 76). *Let $F(z) \not\equiv 0$ belong to $\mathbb{H}_p$, $p > 0$. Let $B(z)$ be the Blaschke product consisting of the zeros of $F(z)$. Then there is a singular measure σ on $[-\pi, \pi]$ with*

$$F(z) = B(z) \exp\left(-\frac{1}{2\pi} \int_{-\pi}^{\pi} \frac{e^{it} + z}{e^{it} - z} d\sigma(t) \right) \cdot e^{ic}$$

$$\cdot \exp\left(\frac{1}{2\pi} \int_{-\pi}^{\pi} \frac{e^{it} + z}{e^{it} - z} \log |F(e^{it})| dt \right),$$

for $|z| < 1$, where c is a real constant.

Proposition 13.0.55. $Kr(2)$ *is composed of functions:*

$$F(z) = \exp\left(-\frac{1}{2\pi}\int_{-\pi}^{\pi}\frac{e^{it}+z}{e^{it}-z}d\sigma(t)\right)\cdot e^{ic}$$

$$\cdot\exp\left(\frac{1}{2\pi}\int_{-\pi}^{\pi}\frac{e^{it}+z}{e^{it}-z}\log|F(e^{it})|dt\right),$$

such that

$$\frac{1}{2\pi}\int_{0}^{2\pi}|F(e^{it})|^2dt \leq 1.$$

Remark 13.0.56. We note that this is the same as in [49] representation of non-vanishing functions in the closed unit ball of $\mathbb{H}_\infty(U)$, $\overline{B}_{\mathbb{H}_\infty(U)}$. In the representation, the outer factor

$$\exp\left(\frac{1}{2\pi}\int_{-\pi}^{\pi}\frac{e^{it}+z}{e^{it}-z}\log|F(e^{it})|dt\right),$$

is missing for the Krzyż conjecture.

Chapter 14

Is the Conjecture We Made for $A_n(p)$ a Good One?

No! It might have worked for $p = 2$, but certainly not for all $1 < p$. The way we arrived to conjecture that $A_n(2)$ is $\sqrt{2/3}$ was by normalizing the quadratic $(1+z)^2 = 1+2z+z^2$, which does not vanish in U, to get a quadratic that belongs to $Kr(2)$. So, we divided by the square root of the sum of the squares of the coefficients: $\sqrt{1^2 + 2^2 + 1^2} = \sqrt{6}$. The largest coefficient we get is $2/\sqrt{6} = \sqrt{2/3}$. The nice thing about the simple expression $(p/(p+1))^{1/p}$ is that it also gives the correct values for $p = 1$ $(1/2)$ and for $p = \infty$ (1). On the other hand, this simple expression does not respect the same normalization procedure for other values of p. We might want to repeat the procedure: Take our quadratic $(1 + z)^2 = 1 + 2z + z^2$ and normalize it into $Kr(p)$ by dividing by the p'th root of the sum of the p-powers of the coefficients: $(1^p + 2^p + 1^p)^{1/p} = (2 + 2^p)^{1/p}$. This leads to the largest value:

$$\frac{2}{(2 + 2^p)^{1/p}} = \left(\frac{2^p}{2 + 2^p}\right)^{1/p}.$$

Also, this simple expression gives the correct values at $p = 1$ and at $p = \infty$. Clearly, we want to compare the two expressions:

$$\left(\frac{p}{p+1}\right)^{1/p} \quad \text{and} \quad \left(\frac{2^p}{2^p + 2}\right)^{1/p}.$$

Of course, we take the larger. We can verify that

$$\left(\frac{p}{p+1}\right)^{1/p} \leq \left(\frac{2^p}{2^p + 2}\right)^{1/p} \quad \text{for } p \geq 2.$$

Also,

$$\left(\frac{p}{p+1}\right)^{1/p} \geq \left(\frac{2^p}{2^p+2}\right)^{1/p} \quad \text{for } 1 \leq p \leq 1 - \left(\frac{\log\log 2}{\log 2}\right).$$

In fact, the second inequality is true for $1 \leq p \leq 2$. We adjust accordingly our conjecture: For $1 \leq p \leq \infty$ and for $n \in \mathbb{Z}^+$,

$$A_n(p) = \left(\frac{2^p}{2^p+2}\right)^{1/p} = \left(\frac{2^{p-1}}{2^{p-1}+1}\right)^{1/p}.$$

The problems $A_n(p)$ for most values of p (say for $p \neq 1, \infty$) do not seem *a priori* to be easier than the Krzyż conjecture. One thing that is not clear to work is the property of being closed to multiplication by functions of B, i.e., $f \in Kr(p)$, $g \in B \Rightarrow f \cdot g \in Kr(p)$. We saw that it is valid for $p = 2$ but we were able to prove that by switching from the $l^2(U)$ characterization of $Kr(2)$ to the $\mathbb{H}_2(U)$ characterization. In the latter, the property (closedness for B-multiplication) is obvious because the characterization is via an L^2 integral on $\mathbb{T}$ and we have $|g| \leq 1$ and the triangle inequality. The last tool (integration) is not available in the discrete $l^2(U)$ description. For values $p \neq 2$, we do not have the equivalence between $l^p(U)$ (discrete) and some L^q (integration with the Lebesgue measure on $\mathbb{T}$). These arguments lead us towards a different family of coefficient problems, which seem more tractable than the discrete l^p problems. The disadvantage is that for the $A_n(p)$ problems we have a conjecture $A_n(p) = (2^p/(2^p+2))^{1/p}$ that is supported by a very simple and discrete construction:

$$\frac{(1+z)^2}{(2^p+2)^{1/p}}.$$

We are very interested in the special case $A_n(2)$ (is it $\sqrt{2/3}$? We know that the answer is negative!), and fortunately this particular case belongs to both families $A_n(p)$ and the new family which we will describe in the following chapter.

Chapter 15

$\mathbb{H}_p$ Versions of the Krzyż Conjecture

Again, the nonlinear feature of being non-vanishing will remain as is. The difference will be in the norms. In fact, here, we will arrive at a more familiar zone for a complex analyst.

Definition 15.0.57. For a fixed $1 \le p \le \infty$ and $n \in \mathbb{Z}^+$ we consider the problem:

$$B_n(p) := \sup\left\{ |a_n| \,\Big|\, f(z) = \sum_{j=0}^{\infty} a_j z^j \in \mathbb{H}_p(U),\ \|f\|_{\mathbb{H}_p} \le 1,\ f(z) \ne 0 \,\forall\, z \in U \right\}.$$

Here, as usual, we have

$$\|f\|_{\mathbb{H}_p} = \lim_{r \to 1^-} \left(\frac{1}{2\pi} \int_0^{2\pi} |f(re^{i\theta})|^p \right)^{1/p},$$

and where,

$$\mathbb{H}_p(U) = \left\{ f \in H(U) \,\big|\, \|f\|_{\mathbb{H}_p} < \infty \right\}.$$

The unit $\mathbb{H}_p(U)$ ball is

$$B(\mathbb{H}_p(U)) = \left\{ f \in \mathbb{H}_p(U) \,\big|\, \|f\|_{\mathbb{H}_p} < 1 \right\}.$$

the closed unit $\mathbb{H}_p$ ball is

$$\overline{B}(\mathbb{H}_p(U)) = \left\{ f \in \mathbb{H}_p(U) \,\big|\, \|f\|_{\mathbb{H}_p} \le 1 \right\}.$$

So, $B_n(p)$ is the n'th coefficient problem on the nonlinear family of non-vanishing $\mathbb{H}_p(U)$ functions in the closed unit $\mathbb{H}_p(U)$ ball. For $p = \infty$, this is exactly the original Krzyż problem. So, it nicely embeds into the family $B_n(p)$ as $B_n(\infty)$. Also, as mentioned in Chapter 14, we have $A_n(2) = B_n(2)$, which means that we can try, for example,

A conjecture on $B_n(2)$: For $n \in \mathbb{Z}^+$, $B_n(2) = \sqrt{2/3}$ and an extremal function is $f_n(z) = (1/\sqrt{6})(1 + z^n)^2$.

Of course, one can change the data and the extremal according to what is suggested at the end of [49].

The Krzyż conjecture is $B_n(\infty) = 2/e$ and the extremal functions are the rotations of

$$f_n(z) = \exp\left(-\frac{1 + z^n}{1 - z^n}\right).$$

It will be convenient to give a special notation to the space of functions underlying $B_n(p)$:

$$B_p = \left\{f \in \mathbb{H}_p(U) \,\middle|\, \|f\|_{\mathbb{H}_p} \leq 1, \ f(z) \neq 0 \,\forall\, |z| < 1\right\}.$$

Hence, we can write the formula:

$$B_n(p) = \sup\left\{|a_n| \,\middle|\, f(z) = \sum_{j=0}^{\infty} a_j z^j \in B_p\right\}$$

or

$$B_n(p) = \sup\left\{\left|\frac{f^{(n)}(0)}{n!}\right| \,\middle|\, f \in B_p\right\}.$$

Our first observation is that multiplication by a function in $B = B_\infty$ of a function in B_p gives a function in B_p.

Proposition 15.0.58. *If $f \in B_p$, $g \in B = B_\infty$ then $f \cdot g \in B_p$. Alternatively, $B \cdot B_p = B_p$.*

Proof. $f \in B_p$, $g \in B \Rightarrow f(z), g(z) \neq 0$, $z \in U$ and $\|f\|_{\mathbb{H}_p}, \|g\|_{\mathbb{H}_\infty} \leq 1$. $\Rightarrow f(z) \cdot g(z) \neq 0 \,\forall\, |z| < 1$ and

$$\|fg\|_{\mathbb{H}_p} = \left(\frac{1}{2\pi}\int_0^{2\pi} |f(e^{i\theta})g(e^{i\theta})|^p d\theta\right)^{1/p} \leq \left(\frac{1}{2\pi}\int_0^{2\pi} |f(e^{i\theta})|^p d\theta\right)^{1/p} \leq 1,$$

where in the one before the last inequality, we used $|g(e^{i\theta})| \leq 1$ a.e., θ. So, $B_p \cdot B \subseteq B_p$. Taking $g = 1 \in B$, we get equality. $\qquad\square$

We extend Proposition 13.0.55.

Proposition 15.0.59. *B_p consists of all the functions $F(z)$ in $\mathbb{H}_p(U)$ (here, we can even have $0 < p < \infty$) such that there is a singular measure σ on $[0, 2\pi]$ with*

$$F(z) = \exp\left(\frac{1}{2\pi}\int_0^{2\pi} \frac{e^{i\theta}+z}{e^{i\theta}-z}d\sigma(\theta)\right)\cdot e^{ic}\cdot\exp\left(\frac{1}{2\pi}\int_0^{2\pi} \frac{e^{i\theta}+z}{e^{i\theta}-z}\log\left|F(e^{i\theta})\right|d\theta\right),$$

for $|z| < 1$, where c is a real constant, and such that

$$\frac{1}{2\pi}\int_0^{2\pi} |F(e^{i\theta})|^p d\theta \leq 1.$$

Proof. By the theorem on factorization in [59, p. 76]. $\square$

We automatically gain all the results that follow by variations of multiplying by

$$\exp\left(-\epsilon\frac{1+kz}{1-kz}\right)$$

for $\epsilon > 0$ and $|k| = 1$. For instance, $2|a_0| \leq |a_n|$.

Chapter 16

The $A_n(2) = B_n(2)$ Problem

We recall that

$$Kr(2) = \left\{ f \in H(U) \mid f(z) = \sum_{j=0}^{\infty} a_j z^j \neq 0 \,\forall\, |z| < 1, \ \sum_{j=0}^{\infty} |a_j|^2 \leq 1 \right\},$$

and $A_n(2) = \sup\{|a_n| \mid \sum_{j=0}^{\infty} a_j z^j \in Kr(2)\}$. We also denote

$$B_n(2) = \sup\left\{ |a_n| \mid f(z) = \sum_{j=0}^{\infty} a_j z^j \in \mathbb{H}_2(U), \ f(z) \neq 0 \,\forall\, z \in U, \ \|f\|_{\mathbb{H}_2} \leq 1 \right\}.$$

Clearly, $A_n(2) = B_n(2) \leq 1$. We conjectured at first that $A_n(2) = \sqrt{2/3}$, where an extremal function is given by the polynomial $(1/\sqrt{6})(1+z^n)^2$. This conjecture should be replaced by the one in [49], $A_n(2) = \sqrt{2/e}$, which is larger than $\sqrt{2/3}$, but this is of no significance to our discussion here. In this chapter, we will demonstrate how one might go about resolving $A_n(2)$ or equivalently $B_n(2)$. We recall that $A_n(1) = 1/2$ with the extremal function $(1/2)(1 + z^n)$, while $A_n(\infty) = 1$ with the extremal function $1 + z^n$. We wondered if there might exist a kind of interpolation of operators theory in the nonlinear setting of the spaces $Kr(p)$, $1 \leq p \leq \infty$. The advantage of the problem $A_n(2)$ on the other $A_n(p)$ is that we can either work within $Kr(2)$ with $\| \sum_{j=0}^{\infty} a_j z^j \|_{l^2(U)} = (\sum_{j=0}^{\infty} |a_j|^2)^{1/2}$ or within $\mathbb{H}_2(U)$ with (in fact the same norm) $\|f\|_{\mathbb{H}_2} = \lim_{r\to 1^-} ((1/(2\pi)) \int_0^{2\pi} (|f(re^{i\theta})|^2 d\theta)^{1/2}$.

Remark 16.0.60. For a general $1 < p \leq \infty$, the problems $A_n(p)$ and $B_n(p)$ differ a lot from one another. The norm in $A_n(p)$ is very simply related to the coefficients of $f(z) = \sum_{k=0}^{\infty} a_k z^k$. Namely, $\|f\|_{l^p} = (\sum_{k=0}^{\infty} |a_k|^p)^{1/p}$. On the other hand, the norm in $B_n(p)$ is not (usually) easily expressible in terms of these coefficients. However, there is an exception for a particular sequence of p. This originates in the simple observation that $\|f\|_{\mathbb{H}_p} = \|f^{p/2}\|_{\mathbb{H}_2}^{2/p} = \|f^{p/2}\|_{l^2}^{2/p}$. Thus, for even natural numbers $p = 2m$, $m \in \mathbb{Z}^+$ we get again a tame relation between $\|f\|_{\mathbb{H}_{2N}}$ and the coefficients. In fact, it is best to take powers of 2 for p. I learnt this clever observation when I was a student reading a beautiful paper of Rogosinski on subordination. So, $\|f\|_{\mathbb{H}_{2N}} = \|f^N\|_{l^2}^{1/N}$. It is reasonable to expect that when $N \to \infty$ we arrive at the classical Krzyż problem.

Remark 16.0.61. If $f(z) \neq 0 \, \forall \, |z| < 1$, then there is a $g(z) = u(z) + iv(z) \in H(U)$ such that $f(z) = \exp(g(z))$. Assuming $\|f\|_{\mathbb{H}_2} = 1$ is in fact the following equation:

$$1 = \lim_{r \to 1^-} \frac{1}{2\pi} \int_0^{2\pi} |f(re^{i\theta})|^2 d\theta = \lim_{r \to 1^-} \frac{1}{2\pi} \int_0^{2\pi} \exp(2u(re^{i\theta})) d\theta.$$

We can assume (multiplying by an appropriate $e^{i\alpha}$) that $a_n \in \mathbb{R}$ and $a_n \geq 0$, in which case

$$a_n = \lim_{r \to 1^-} \frac{1}{2\pi} \int_0^{2\pi} f(re^{i\theta}) e^{-in\theta} d\theta$$

$$= \lim_{r \to 1^-} \frac{1}{2\pi} \int_0^{2\pi} \exp(u(re^{i\theta})) \cos(v(re^{i\theta}) - n\theta) d\theta.$$

So, a reformulation of $A_n(2)$ is

$$A_n(2) = \sup \lim_{r \to 1^-} \frac{1}{2\pi} \int_0^{2\pi} \exp(u(re^{i\theta})) \cos(v(re^{i\theta}) - n\theta) d\theta,$$

where the supremum is taken over all harmonic conjugates u, v in $|z| < 1$ for which

$$\lim_{r \to 1^-} \frac{1}{2\pi} \int_0^{2\pi} \exp(2u(re^{i\theta})) d\theta = 1.$$

Since the harmonic conjugate $v(z)$ of $u(z)$ in U can be constructed from $u(z)$ (up to an additive constant), the above gives a reformulation of $A_n(2)$ as a variational problem with a side condition, on the space of all the harmonic functions $u(z)$ in (the simply connected domain) U. Thus, maybe, variational techniques can be invoked. For example, we might want to expand the appropriate Euler–Lagrange equations for $A_n(2)$.

Owing to an experience the author had while he was an undergraduate student, due to a faulty theorem of Morris Marden he used (and thought that he proved the Krzyż conjecture), here is a "lemma" that could have been nice to have.

Lemma 16.0.62. Really a speculation! *There exists an absolute constant c, $0 < c < 2$ such that if $p(z) = a_0 + a_1 z + \cdots + a_n z^n$ is a complex polynomial, such that $\sum_{j=0}^{n} |a_j|^2 \leq 1$ and which does not vanish in $|z| < 1$, then there exists a sequence $\epsilon_0, \ldots, \epsilon_n$ such that:*

(1)

$$\frac{\max_k |\epsilon_k|}{\min_k |\epsilon_k|} \leq c.$$

(2) *The polynomial $q(z) = \epsilon_0 a_0^2 + \epsilon_1 a_1^2 z + \cdots + \epsilon_n a_n^2 z^n$ does not vanish in $|z| < 1$.*

Proof. We have none! We do not know if the claim is true! $\square$

So, why is it nice to have? Here is a reason.

Proposition 16.0.63. *Let $f(z) = a_0 + a_1 z + \cdots \in H(U)$ and satisfy:*

(1) $f(z) \neq 0 \, \forall \, |z| < 1$.
(2) $\sum_{j=0}^{\infty} |a_j|^2 \leq 1$.

Then if $k \geq 1$, we have the estimate

$$|a_k| \leq \sqrt{\frac{c}{2}},$$

where $c < 2$ is the absolute constant of Lemma 16.0.62.

A proof conditioned in the validity of Lemma 16.0.62.
Without losing the generality, let us assume that $f(z)$ is a polynomial of degree n: $f(z) = a_0 + a_1 z + \cdots + a_n z^n = p(z)$. We achieve that by an argument that combines a truncation (far enough) of the holomorphic $f(z)$ followed by a dilation using a factor close to 1. By Lemma 16.0.62, there exists a polynomial: $q(z) = \epsilon_0 a_0^2 + \epsilon_1 a_1^2 z + \cdots + \epsilon_n a_n^2 z^n$ which does not vanish in U and the maximal quotient of the multiplier sequence is bounded from above by $c < 2$, i.e.,

$$\frac{\max_k |\epsilon_k|}{\min_k |\epsilon_k|} \leq c.$$

By the assumption (2) in our proposition, we have $\sum_{k=0}^{n} |\epsilon_k a_k^2| \leq \max_k |\epsilon_k|$. Hence, by our solution of $A_k(1)$, we conclude that for $k \geq 1$, we have $|\epsilon_k a_k^2| \leq (1/2) \max_j |\epsilon_j|$. Hence,

$$|a_k| \leq \sqrt{\frac{1}{2} \frac{\max_j |\epsilon_j|}{\epsilon_k}} \leq \sqrt{\frac{1}{2} \frac{\max_j |\epsilon_j|}{\min_j |\epsilon_j|}} \leq \sqrt{\frac{c}{2}}.$$

$\square$

This is an elegant proof of a uniform bound less than 1 in the Krzyż conjecture, but it is conditioned in the validity of Lemma 16.0.62.

Remark 16.0.64. To conclude the personal experience of the author due to Marden's theorem: Marden's result was corrected and the correct version appeared in [84], 1992. However, the correct version was not enough to get a result such as Lemma 16.0.62 and make a progress on the Krzyż conjecture (getting a uniform upper bound which is less than 1). At any rate, the author was a very excited undergraduate student for couple of days when he thought he had a significant result on the Krzyż conjecture. The way he discovered that something was wrong was that using the faulty theorem, he achieved the bound $\sqrt{1/2}$ for the Krzyż conjecture. That, of course, was too good to be true (compared to $2/e$).

The case $n = 2$.

We will give two solutions, and we insist on bringing the longer one too, because we can extend it later to $n > 2$. If $p(z) = a_0 + a_1 z + a_2 z^2$ does not vanish in $|z| < 1$ and $|a_0|^2 + |a_1|^2 + |a_2|^2 = 1$, then $\max\{|a_1|, |a_2|\} = \sqrt{2/3}$.

Proof. By Rouche's Theorem, there is a $|z_0| = 1$ such that

$$\frac{|a_0 + a_1 z_0 + a_2 z_0^2 - a_j z_0^j|}{|a_j z_0^j|} \geq 1, \quad j \in \{1, 2\}.$$

Let us assume, for example, that $\max\{|a_1|, |a_2|\} = |a_1|$. We take $j = 1$ and deduce that

$$\frac{|a_0 + a_1 z_0 + a_2 z_0^2 - a_1 z_0|}{|a_1 z_0|} \geq 1,$$

hence, $|a_0 + a_2 z_0^2| \geq |a_1|$ and by the triangle inequality $|a_0| + |a_2| \geq |a_1|$. We denote $x = |a_0|$, $y = |a_1|$, $z = |a_2|$, then we have to solve for $\max y$ in

$$\begin{cases} x, y, z & \geq 0 \\ x^2 + y^2 + z^2 = 1 \\ z \leq y & \leq x + z \end{cases}.$$

Thus, $y = (1 - x^2 - z^2)^{1/2}$ and so $z \le y$ implies that $2z^2 \le 1 - x^2$. Similarly $y \le x + z$ implies that $(1/2) \le x^2 + xz + z^2$. We can sketch the domain $D : x, z \ge 0, x^2 + 2z^2 \le 1, 2(x^2 + xz + z^2) \ge 1$ and we have to solve $\max_{(x,z)\in\overline{D}}\{1 - x^2 - z^2\}$ in $\overline{D}$. It is attained on the boundary. So,

(i) $x^2 + 2z^2 = 1 \Rightarrow y^2 = 1 - x^2 - z^2 = z^2 \le (1/2) \Rightarrow y \le (1/\sqrt{2})$.

(ii) $2(x^2 + xz + z^2) = 1$. We use Lagrange multipliers: We form $F = 1 - x^2 - z^2 + \lambda(2(x^2 + xz + z^2) - 1)$ and use the necessary conditions:

$$\frac{\partial F}{\partial x} = \frac{\partial F}{\partial y} = 0,$$

to get

$$\begin{cases} (2\lambda - 1)x + \lambda z = 0 \\ \lambda x + (2\lambda - 1)z = 0 \end{cases},$$

and since $(x, z) \ne (0,0)$ we deduce that $2\lambda - 1 = \pm\lambda \Rightarrow x = z = (1/\sqrt{6})$ and $y = \sqrt{2/3}$.

$\square$

Here is a second (shorter) solution: Once more, by Rouche's Theorem,

$$\frac{(1 - |a_1|^2)^{1/2}\sqrt{2}}{|a_1|} = \frac{(|a_0|^2 + |a_2|^2)^{1/2}\sqrt{2}}{|a_1|} \ge \frac{|a_0 + a_2 z_0^2|}{|a_1|} \ge 1.$$

The one before the last inequality follows by Cauchy–Schwarz inequality, and the last inequality follows by Rouche's Theorem. So, $2(1 - |a_1|^2) \ge |a_1|^2$ and finally, $|a_1| \le \sqrt{2/3}$.

The case $n = 3$.

If $p(z) = a_0 + a_1 z + a_2 z^2 + a_3 z^3$ does not vanish in U and $|a_0|^2 + |a_1|^2 + |a_2|^2 + |a_3|^2 = 1$, then

$$\max\{|a_1|, |a_2|, |a_3|\} = \frac{\sqrt{3}}{2} = \sqrt{\frac{3}{4}}.$$

Proof. By Rouche's Theorem, there is a $|z_0| = 1$ such that

$$\frac{|p(z_0) - a_j z_0^j|}{|a_j z_0^j|} \ge 1, \quad j \in \{1, 2, 3\}.$$

Let us assume, for example, that $\max\{|a_1|, |a_2|, |a_3|\} = |a_2|$ and we get $|a_0| + |a_1| + |a_3| \geq |a_2|$ (taking $j = 2$). We will denote $x = |a_0|$, $y = |a_1|$, $z = |a_2|$, $w = |a_3|$, then we need to solve for $\max z$ in

$$\begin{cases} x, y, z, w & \geq 0 \\ x^2 + y^2 + z^2 + w^2 = 1 \\ y & \leq z \\ w & \leq z \\ z & \leq x + y + w \end{cases} .$$

Thus, $z^2 = 1 - x^2 - y^2 - w^2$. By $y^2 \leq z^2 \Rightarrow x^2 + 2y^2 + w^2 \leq 1$. By $w^2 \leq z^2 \Rightarrow x^2 + y^2 + 2w^2 \leq 1$. Finally, by $z^2 \leq (x+y+w)^2 \Rightarrow 1 - x^2 - y^2 - w^2 \leq (x^2 + y^2 + w^2 + 2xy + 2xw + 2yw) \Rightarrow 1 \leq 2(x^2 + y^2 + w^2 + xy + xw + yw)$. We got a domain D:

$$\begin{cases} x^2 + 2y^2 + w^2 \leq 1 \\ x^2 + y^2 + 2w^2 \leq 1 \\ 1 & \leq 2(x^2 + y^2 + w^2 + xy + xw + yw) \\ 0 & \leq x, y, w \end{cases} .$$

We need to solve in $\overline{D}$, $\max_{(x,y,w)\in\overline{D}}\{1 - x^2 - y^2 - w^2\}$. The maximum is attained on ∂D.

(i) $x^2 + 2y^2 + w^2 = 1 \Rightarrow z^2 = -y^2 + 2y^2 = y^2 \leq (1/\sqrt{2})^2$.
(ii) $x^2 + y^2 + 2w^2 = 1 \Rightarrow z^2 = w^2 \leq (1/\sqrt{2})^2$.
(iii) $2(x^2 + y^2 + w^2 + xy + xw + yw) = 1$. We use Lagrange multipliers and form

$$F = (1 - x^2 - y^2 - w^2) + \lambda(2(x^2 + y^2 + w^2 + xy + xw + yw) - 1).$$

The necessary conditions are $F_x = F_y = F_z = 0$. We get

$$\begin{cases} (2\lambda - 1)x + \lambda y + \lambda w = 0 \\ \lambda x + (2\lambda - 1)y + \lambda w = 0 \\ \lambda x + \lambda y + (2\lambda - 1)w = 0 \end{cases} .$$

The homogeneous system has non-trivial solutions and hence,

$$\begin{vmatrix} 2\lambda - 1 & \lambda & \lambda \\ \lambda & 2\lambda - 1 & \lambda \\ \lambda & \lambda & 2\lambda - 1 \end{vmatrix} = 0.$$

So, $(\lambda - 1)^2(4\lambda - 1) = 0$. $\lambda = 1$ is impossible because $x + y + w = 0$ for $x, y, w \geq 0$ gives $x = y = w = 0$, $z = 1$ (so the trivial solution — because of positivity). $\lambda = (1/4) \Rightarrow x = y = w = (1/\sqrt{12})$, $z^2 = 1 - x^2 - y^2 - w^2 = (3/4)$, so $z = \sqrt{3/4}$. We conclude that our polynomial $p(z) = a_0 + a_1 z + a_2 z^2 + a_3 z^3$ satisfies $p(z) \neq 0 \, \forall \, |z| < 1$ and $|a_0| = |a_1| = |a_3| = (1/3)|a_2| \neq 0$.

Claim: No such $p(z)$ exists.

A proof of the claim.
We need to show that for any $|\epsilon_0| = |\epsilon_1| = |\epsilon_2| = 1$, the polynomial $z^3 + 3\epsilon_2 z^2 + \epsilon_1 z + \epsilon_0$ must have a zero in U. Suppose that for a certain $|\epsilon_0| = |\epsilon_1| = |\epsilon_2| = 1$, we have

$$z^3 + 3\epsilon_2 z^2 + \epsilon_1 z + \epsilon_0 = (z - \alpha)(z - \beta)(z - \gamma),$$

where $|\alpha|, |\beta|, |\gamma| \geq 1$. Then

$$z^3 + 3\epsilon_2 z^2 + \epsilon_1 z + \epsilon_0 = z^3 - (\alpha + \beta + \gamma)z^2 + (\alpha\beta + \alpha\gamma + \beta\gamma)z - \alpha\beta\gamma.$$

So, $\epsilon_2 = -\alpha - \beta - \gamma$, where $|\epsilon_2| = 1$, $|\alpha|, |\beta|, |\gamma| \geq 1$ imply that $|\alpha| = |\beta| = |\gamma| = 1$ (by $-\alpha\beta\gamma = \epsilon_0$) and $\arg \alpha = \arg \beta = \arg \gamma = \pi + \arg \epsilon_2$. Thus, we get $\alpha = \beta = \gamma = -\epsilon_2$. But this implies that $\alpha\beta + \alpha\gamma + \beta\gamma = 3\epsilon_2^2$. So $|\alpha\beta + \alpha\gamma + \beta\gamma| = 3$ which contradicts $\alpha\beta + \alpha\gamma + \beta\gamma = \epsilon_1$. $\qquad\square$

Thus, the bound $\sqrt{3/4}$ is not admissible! This is consistent with the conjecture in [49] which speculates that $A_n(2) = B_n(2) = \sqrt{2/e}$. The bound $\sqrt{3/4}$ is larger. It shows that our method outlined above needs an upgrade.

Remark 16.0.65. The choice of the maximal coefficient in $p(z)$ we made seems to be random. For example, in the case $n = 2$, we decided arbitrarily that $\max\{|a_1|, |a_2|\} = |a_1|$. The point is that other choices are impossible. For if $\max\{|a_1|, |a_2|\} = |a_2|$, then we will eventually have to see why the polynomial $1 + z + 2z^2 = 2(z^2 + (1/2)z + (1/2)) = 2(z - \alpha)(z - \beta)$ must clearly have a zero in U. For if $|\alpha|, |\beta| \geq 1$, then $|\alpha\beta| \geq 1$, but $\alpha\beta = (1/2)$.

Remark 16.0.66. In constructing the domain D over which we maximize, there is really only one boundary component which is important. This component is important because it encodes the requirement that $p(z) \neq 0 \, \forall \, |z| < 1$, which is the nonlinear condition in the Krzyż type problems. We used Rouche's Theorem to get this component, namely there exists a z_0, $|z_0| = 1$ such that: $|(p(z_0) - a_j z_0^j)/(a_j z_0^j)| \geq 1$. In the case $n = 2$, this gave us $|a_0 + a_2 z_0^2| \geq |a_1|$. In the case $n = 3$, this gave us

$|a_0 + a_1 z_0 + a_3 z_0^3| \geq |a_2|$. We then used the triangle inequality to simplify matters by getting rid of z_0. For $n = 2$, $|a_0| + |a_2| \geq |a_1|$, i.e., $x + z \geq y$. For $n = 3$, $|a_0| + |a_1| + |a_3| \geq |a_2|$, i.e., $x + y + w \geq z$. Both gave us a quadratic component, using $y^2 = 1 - x^2 - z^2$ for $n = 2$ and using $z^2 = 1 - x^2 - y^2 - w^2$ for $n = 3$.

However, we might have used the Cauchy–Schwarz inequality instead of the triangle inequality. For $n = 2$, $|a_1|^2 \leq |a_0 + a_2 z_0^2|^2 \leq (|a_0|^2 + |a_2|^2)(1+1) = 2(|a_0|^2 + |a_2|^2)$, i.e., $y^2 \leq 2(x^2 + z^2) \Rightarrow 1 - x^2 - z^2 \leq 2(x^2 + z^2) \Rightarrow 1 \leq 3(x^2 + z^2) \Rightarrow (1/3) \leq x^2 + z^2$ and on the boundary $x^2 + z^2 = (1/3)$, so $y^2 = 1 - x^2 - z^2 = (2/3)$ and we get immediately $y = \sqrt{2/3}$. There is no need for Lagrange multipliers.

Similarly, for $n = 3$, $|a_2|^2 \leq |a_0 + a_1 z_0 + a_3 z_0^3|^2 \leq (|a_0|^2 + |a_1|^2 + |a_3|^3) \cdot 3$, i.e., $z^2 \leq 3(x^2 + y^2 + w^2) \Rightarrow 1 - x^2 - y^2 - w^2 \leq 3(x^2 + y^2 + w^2) \Rightarrow 1 \leq 4(x^2 + y^2 + w^2)$ and on the boundary $x^2 + y^2 + w^2 = (1/4)$. So $z^2 = 1 - x^2 - y^2 - w^2 = 1 - (1/4) = (3/4)$ and we get immediately $z = \sqrt{3/4}$, there is no need for Lagrange multipliers.

Remark 16.0.67. We see how easy our method (which only gives rough bounds) works. We demonstrated it for $n = 2, 3$. For $n = 3$, we got no solution eventually. The reason is that the polynomial $p(z)$ we get must have a zero in U. It proves though that $A_3(2) < \sqrt{3/4}$. Why did that happen? Maybe the use we made in Rouche's Theorem is not strong enough to ensure the existence of a non-vanishing polynomial solution in U. Or, maybe our approximation to the true domain D, using the triangle inequality or the Cauchy–Schwarz inequality is too coarse. So, let us try to figure out in the case $n = 3$, what is exactly the domain D. To start with, the dimension is one higher: For the variables are $(x, y, z, w) = (a_0, a_1, a_2, a_3)$, λ (the multiplier), θ the argument of $z_0 = e^{i\theta}$ of Rouche's Theorem. We have $|x|^2 + |y|^2 + |z|^2 + |w|^2 = 1$ and we want to maximize $|z|$, $\max |z| = \max \sqrt{1 - |x|^2 - |y|^2 - |w|^2}$, but the domain D is more complicated because the variables are complex and there is an extra real variable θ. The component that expresses Rouche's Theorem is given as before by the inequality $|x + yz_0 + wz_0^3|^2 \geq |z|^2$. If we expand this, we get

$$2\left(|x|^2 + |y|^2 + |w|^2 + \Re(\bar{x}ye^{i\theta} + \bar{x}we^{3i\theta} + \bar{y}we^{2i\theta})\right) \geq 1.$$

If we use polar representation: $x = ae^{i\alpha}$, $y = be^{i\beta}$, $z = ce^{i\gamma}$, $w = d \cdot e^{i\delta}$ the inequality becomes

$$2(a^2 + b^2 + d^2 + ab\cos(-\alpha+\beta+\theta) + ad\cos(-\alpha+\delta+3\theta) + bd\cos(-\beta+\delta+2\theta)) \geq 1.$$

So, for the Lagrange multiplier on that component of the boundary, we form

$$F = 1 - a^2 - b^2 - d^2 + \lambda \cdot (2(a^2 + b^2 + d^2$$
$$+ ab\cos(-\alpha + \beta + \theta) + ad\cos(-\alpha + \delta + 3\theta)$$
$$+ bd\cos(-\beta + \delta + 2\theta)) - 1).$$

The variables are $a, b, d, \lambda, \alpha, \beta, \delta, \theta$. That gives us some idea on the complications we had put aside using the triangle inequality or the Cauchy–Schwarz inequality.

We end this chapter by outlining the first (coarse) approach in the general case. We have enough experience from our treatments of the cases $n = 2, 3$. Suppose that $p(z) = a_0 + a_1 z + \cdots + a_n z^n$ does not vanish in $|z| < 1$, and that $|a_0|^2 + \cdots + |a_n|^2 = 1$. We would like to compute $\max\{|a_1|, |a_2|, \ldots, |a_n|\}$. Let us assume, for example, that $\max\{|a_1|, |a_2|, \ldots, |a_n|\} = |a_n|$. It is probably more likely that the maximum will be around the central index, but we just demonstrate. We will use the following notation: $x_1 = |a_0|$, $x_2 = |a_1|, \ldots, x_n = |a_{n-1}|, y = |a_n|$. Then we need to compute $\max y$ in the domain which is defined by

$$\begin{cases} x_1, \ldots, x_n, y & \geq & 0 \\ x_1^2 + \cdots + x_n^2 + y^2 & = & 1 \\ x_j^2 \leq y^2, & 1 \leq j \leq n & \text{recall that max} \\ & & \{|a_1|, |a_2|, \ldots, |a_n|\} = |a_n| \\ y & \leq & x_1 + \cdots + x_n \\ \text{the triangle inequality or} \\ y^2 & \leq & n \cdot (x_1^2 + \cdots + x_n^2) \\ \text{Cauchy–Schwarz inequality} \end{cases}$$

We note that (as we already experienced) if the Cauchy–Schwarz inequality is used (instead of the triangle inequality), then $1 - x_1^2 - \cdots - x_n^2 \leq n \cdot (x_1^2 + \cdots + x_n^2)$ and hence $x_1^2 + \cdots + x_n^2 = 1/(n+1)$. So,

$$y = (1 - x_1^2 - \cdots - x_n^2)^{1/2} = \left(1 - \frac{1}{n+1}\right)^{1/2} = \sqrt{\frac{n}{n+1}}.$$

That was easily computed, and no Lagrange multipliers were needed. On the other hand, if we had used the triangle inequality, the computation would have been longer, and using Lagrange multipliers we would end up with the same estimate for $\max y$. We leave the elementary details to the reader. The result is that $\lambda = 1/(n+1)$ (the case $\lambda = 1$ does not give the maximum). We get

$$x_1 = x_2 = \cdots = x_n = \frac{1}{\sqrt{n(n+1)}}, \quad y = \sqrt{\frac{n}{n+1}}.$$

A poor result indeed for $y \to 1$ as $n \to \infty$. The only case of interest is $n = 2$ when we get $\sqrt{2/3}$. In this case, we actually have an extremal $(1/\sqrt{6})(1 + z^n)^2$. Finally, we point out that in order to get the above bound in the shortest path, we can use **the following better approach:**

Suppose that $p(z) = a_0 + a_1 z + \cdots + a_n z^n$ does not vanish in $|z| < 1$ and that $|a_0|^2 + \cdots + |a_n|^2 = 1$. Then by Rouche's Theorem there is a $|z_0| = 1$ such that

$$\frac{|p(z_0) - a_k z_0^k|}{|a_k z_0^k|} \geq 1,$$

where $k \geq 1$ is a fixed index. So,

$$1 \leq \frac{(1 - |a_k|^2)^{1/2} n^{1/2}}{|a_k|},$$

and

$$|a_k| \leq \sqrt{\frac{n}{n+1}}.$$

Short and elegant. For equality to hold it is necessary that the two vectors $(a_0, \ldots, a_{k-1}, a_{k+1}, \ldots, a_n)$, $(1, z_0, \ldots, z_0^{k-1}, z_0^{k+1}, \ldots, z_0^n)$ are proportional. On the other hand, we know that when equality occurs, we have

$$|a_0| = |a_1| = \cdots = |a_{k-1}| = |a_{k+1}| = \cdots = |a_n| \quad \text{and} \quad |a_k| = n|a_0|.$$

Note that only when $n = 2$, we get an instance of the doubling principle, $|a_k| = 2|a_0|$. So, the question of equality is reduced to: Is there an $|\epsilon| = 1$ such that $z^n + \cdots + z^{k+1} + n\epsilon z^k + z^{k-1} + \cdots + z + 1$ does not vanish in $|z| < 1$. In fact, we should examine that in general when each coefficient is given an arbitrary argument. We always have

$$|e^{in\theta} + \cdots + e^{i(k+1)\theta} + e^{i(k-1)\theta} + \cdots + e^{i\theta} + 1| < n, \qquad (16.0.1)$$

unless $\theta = 0$ (we consider only $0 \le \theta < 2\pi$). If it were possible for $r < 1$ close to 1 to have

$$|r^n e^{in\theta} + \cdots + r^{k+1} e^{i(k+1)\theta} + r^{k-1} e^{i(k-1)\theta} + \cdots + r e^{i\theta} + 1| < nr^k,$$

then by Rouche's Theorem, the polynomial $z^n + \cdots + z^{k+1} + \epsilon n z^k + z^{k-1} + \cdots + z + 1$ would have a zero in $|z| < r$. We have to consider only $\theta = 0$ (by equation (16.0.1)). We need $f(r) = r^n + \cdots + r^{k+1} - nr^k + r^{k-1} + \cdots + r + 1 > 0$ for $r < 1$ close to 1. Since $f(1) = 0$ we need $f'(r) = nr^{n-1} + \cdots + (k+1)r^k - nkr^{k-1} + (k-1)r^{k-2} + \cdots + 1$ to be < 0 at $r = 1$. So $n + (n-1) + \cdots + (k+1) - nk + (k-1) + \cdots + 1 < 0$, and hence $(1/2)n(n+1) < nk + k = (n+1)k$ thus $(1/2)n < k$. If we choose $f(z) = z^n + \cdots + z^{k+1} + nz^k + z^{k-1} + \cdots + z + 1$ then $f(1) = 2n > 0$, $f(0) = 1 > 0$. So, for $f(-1)$ to be ≥ 0, we must have k even.

Chapter 17

The $B_n(p)$ Problem — Some Background

We will refer to [6, 8, 49]. In [8], the authors denote by B_p the set of all non-vanishing $\mathbb{H}_p$ functions $f(z) = a_0 + a_1 z + a_2 z^2 + \cdots$ with $\|f\|_{\mathbb{H}_p} \leq 1$. The authors of [49] conjectured that

$$\sup_{B_p} |a_n| = \left(\frac{2}{e}\right)^{1/q}, \quad \forall n \geq 1, \tag{17.0.1}$$

where $1 < p < \infty$ and $(1/p) + (1/q) = 1$. If true, the bound is attained by

$$H_n(z) = \left(\frac{(1 + z^n)^2}{2}\right)^{1/p} \left(\exp\left(\frac{z^n - 1}{z^n + 1}\right)\right)^{1/q},$$

and its rotations $e^{i\nu} H_n(e^{i\mu} z)$, where $\nu, \mu \in \mathbb{R}$. Till 1988, the only evidence supporting (17.0.1) was given in [6] where the conjecture was verified for $n = 1$ and for arbitrary $n \geq 2$ provided that $a_m = 0$ for all $1 \leq m < ((n+1)/2)$. In [8], the conjecture was proved for $n = 2$ and $n = 3$ for a certain natural subclass of non-vanishing $\mathbb{H}_p$ functions. So, the conjecture in [49] is that $A_n(2) = B_n(2) = \sqrt{2/e}$ and it is attained on rotations of

$$\left(\frac{(1 + z^n)^2}{2}\right)^{1/2} \left(\exp\left(\frac{z^n - 1}{z^n + 1}\right)\right)^{1/2}.$$

This is in contrast to our speculation: $A_n(2) = B_n(2) = \sqrt{2/3}$ with an extremal $(1/\sqrt{6})(1 + z^n)^2$. However, it might make sense to ask for the solution in the subfamily of $B_2 = Kr(2)$, which is composed of polynomials only, up to (including) a certain fixed degree.

If f is a non-vanishing $\mathbb{H}_p$ function, then as is well known,

$$f(z) = e^{i\lambda} \Omega(z) I(z),$$

105

where $\Omega(z)$ and $I(z)$ are the outer and the inner factors of f, respectively, and where $I(z)$ does not have a Blaschke product part (just the part which is the integral of a singular measure), and where $\lambda \in \mathbb{R}$. For $n \geq 2$, $B_p(n)$ denotes in [8] those functions $f \in B_p$ of the form given above with $\Omega'(0) = \cdots = \Omega^{(n-1)}(0) = 0$. One can check that $H_n \in B_p(n)$. Here (as before), H_n is the conjectured extremal for $B_n(p)$ in [49], for each $n \geq 2$. Also the classes $B_p(n)$ are nested in B_p. The main result in [8] is as follows.

Theorem 1 ([8]). *Let $1 < p < \infty$. (1) If $f \in B_p(2)$, then $|a_2| \leq (2/e)^{1/q}$. (2) If $f \in B_p(3)$, then $|a_3| \leq (2/e)^{1/q}$.*

Equality holds only for $H_2(z)$ and $H_3(z)$, respectively, and their rotations.

Chapter 18

One More Elementary Variation
for the Krzyż Conjecture

This chapter is really just one remark that we want to record before diving into a new set of ideas that make an extensive use of approximation by finite Blaschke products. That new method will occupy the following four chapters.

Let

$$f(z) = \exp\left(-\sum_{j=1}^{n} \lambda_j \frac{1 + k_j z}{1 - k_j z}\right)$$

$$= \sum_{m=0}^{\infty} a_m z^m, \quad \lambda_j > 0, \ |k_j| = 1, \ 1 \le j \le n,$$

be a function in B that maximizes $|a_n|$. For each $|k| = 1$, let us define

$$g(z) = \frac{1}{2}\log 2 - \frac{1}{2}\log(1 - kz) = \frac{1}{2}\log 2 - \frac{1}{2}\sum_{m=1}^{\infty} \frac{k^m}{m} z^m, \quad |z| < 1.$$

Then

$$\Re\{g(z)\} = \frac{1}{2}\log 2 - \frac{1}{2}\log|1 - kz| > \frac{1}{2}\log 2 - \frac{1}{2}\log 2 = 0, \quad |z| < 1.$$

So, for $\epsilon > 0$, $F(z) = f(z)\exp(-\epsilon g(z))$ is in B, and so it is a variation of $f(z)$. We can see that

$$g(z) = \frac{1}{2}\log 2 - \frac{1}{2}\log(1 - kz) = \frac{1}{2}\log\left(\frac{2}{1 - kz}\right) = \log\left(\frac{2}{1 - kz}\right)^{1/2}.$$

107

Hence,

$$F(z) = f(z) \left(\frac{2}{1-kz} \right)^{-\epsilon/2} = \left(\sum_{m=0}^{\infty} a_m z^m \right)$$

$$\times \left(1 - \epsilon \left(\frac{1}{2} \log 2 + \frac{1}{2} \sum_{m=1}^{\infty} \frac{k^m}{m} z^m \right) \right) + O(\epsilon^2).$$

We denote $F(z) = \sum_{m=0}^{\infty} b_m z^m$. By extremality of $f(z)$, $|b_n|^2 \leq |a_n|^2$ and in that inequality letting $\epsilon \to 0^+$, we get

$$\Re \left\{ \overline{a_n} \left((\log 2) \cdot a_n + a_{n-1} k + a_{n-2} \frac{k^2}{2} + \cdots + a_0 \frac{k^n}{n} \right) \right\} \geq 0,$$

for any $|k| = 1$. Substituting $k = k_j$ and adding for $j = 1, \ldots, n$, we get

$$-n|a_n|^2 \log 2 \leq \Re \left\{ \overline{a_n} \left(a_{n-1} \sum_{j=1}^{n} k_j + a_{n-2} \sum_{j=1}^{n} \frac{k_j^2}{2} + \cdots + a_0 \sum_{j=1}^{n} \frac{k_j^n}{n} \right) \right\}.$$

$$(18.0.1)$$

On the other hand, for each $1 \leq j \leq n$, we have

$$\Re\{\overline{a_n}(a_n + 2a_{n-1}k_j + \cdots + 2a_0 k_j^n)\} = 0, \quad \text{so}$$

$$\Re\{\overline{a_n}(a_n \lambda_j + 2a_{n-1}\lambda_j k_j + \cdots + 2a_0 \lambda_j k_j^n)\} = 0,$$

and adding these, we obtain

$$\Re \left\{ \overline{a_n} \left(a_n \sum_{j=1}^{n} \lambda_j + 2a_{n-1} \sum_{j=1}^{n} \lambda_j k_j + \cdots + 2a_0 \sum_{j=1}^{n} \lambda_j k_j^n \right) \right\} = 0.$$

Using our identities (Proposition 6.0.32) $\sum_{j=1}^{n} \lambda_j k_j^m = (1/m) \sum_{j=1}^{n} k_j^m$, we obtain

$$-\frac{1}{2}|a_n|^2 \sum_{j=1}^{n} \lambda_j = \Re \left\{ \overline{a_n} \left(a_{n-1} \sum_{j=1}^{n} k_j + a_{n-2} \sum_{j=1}^{n} \frac{k_j^2}{2} + \cdots + a_0 \sum_{j=1}^{n} \frac{k_j^n}{n} \right) \right\}.$$

$$(18.0.2)$$

Comparing equations (18.0.1) and (18.0.2) we deduce that $-n|a_n|^2 \log 2 \leq -(1/2)|a_n|^2 \sum_{j=1}^{n} \lambda_j$. So $-2n \log 2 \leq -\sum_{j=1}^{n} \lambda_j$. But $a_0 = \exp(-\sum_{j=1}^{n} \lambda_j)$ and hence $4^{-n} \leq a_0$. Not an exciting lower bound!

Chapter 19

The Blaschke Products Approach to the Krzyż Conjecture

In this chapter, we will indicate a method of approximating the largest function (in coefficientwise sense) in $B = \{f \in H(U) \mid 0 < |f(z)| < 1\}$ by finite Blaschke products, uniformly on compact subsets of U. Then, using those approximations, we get information on

$$A_n = \max_{f(z) = \sum_{j=0}^{\infty} a_j z^j \in B} |a_n|,$$

i.e., on the Krzyż conjecture. The idea seems at first to be bizarre. For any function, $f \in B$ has no zeros in U. It is an inner function without the Blaschke product factor, and so out of the standard $\mathbb{H}_\infty$ factorization as a product $e^{i\lambda} \cdot Bl \cdot I \cdot \Omega$ where Bl is the Blaschke product that is formed with the zeros of the $\mathbb{H}_\infty$ function, $\lambda \in \mathbb{R}$, I the non-vanishing inner factor and Ω the outer function, only $e^{i\lambda} \cdot I$ remains. Yet, we intend to construct a sequence $B_m(z)$ of Blaschke products such that $\lim_{m \to \infty} B_m(z) = e^{i\lambda} \cdot I(z) = f(z)$, (here, we assume $\|f\|_\infty = 1$) uniformly on compact subsets of U. Note that this implies, by the Hurwitz Theorem, that the zero sets $Z(B_m(z))$ of those Blaschke products run away to ∂U, i.e., $\forall 0 < r < 1$, $\exists N(r) > 0$ such that $\forall m > N(r)$, $Z(B_m(z)) \cap \{|z| \leq r\} = \emptyset$. We start with a very preliminary such an approximation theorem. Much stronger and accurate theorems are known. Later, we will bring and use them.

Theorem 19.0.68. *Let $f \in B$, $\|f\|_\infty = 1$. There exists a sequence of finite Blaschke products $B_m(z)$ such that $\lim B_m = e^{i\lambda} \cdot f$, for some $\lambda \in \mathbb{R}$, uniformly on compact subsets of $|z| < 1$. Moreover, we can arrange it so that the cardinals $|Z(B_m)|$ of the zero sets tend to infinity, and $\forall m$, $B_m(0) \neq 0$.*

Proof. We sketch the construction of the sequence $B_m(z)$ of those Blaschke products. We start with the following elementary computation: given $|\alpha| < 1$,

$$\frac{z - \alpha}{1 - \overline{\alpha}z} = -\alpha + (1 - |\alpha|^2)z + (1 - |\alpha|^2)\overline{\alpha}z^2 + \cdots,$$

and more generally:

$$\left(\frac{z - \alpha}{1 - \overline{\alpha}z}\right)^m = (-\alpha)^m + m(-\alpha)^{m-1}(1 - |\alpha|^2)z + \cdots.$$

Making α dynamic by choosing $-\alpha = 1 - (1/m)$, we get

$$\lim_{m \to \infty} \left(\frac{z - \alpha}{1 - \overline{\alpha}z}\right)^m = \frac{1}{e} + \frac{2}{e}z + \cdots.$$

These numbers (the coefficients) look familiar to anyone who studies the Krzyż conjecture. In fact, we have uniformly on compact subsets of U:

$$\lim_{m \to \infty} \left(\frac{z + \left(1 - \frac{1}{m}\right)}{1 + \left(1 - \frac{1}{m}\right)z}\right)^m = \lim_{m \to \infty} \left(1 - \frac{1}{m}\left(\frac{1 - z}{1 + z - \frac{z}{m}}\right)\right)^m$$

$$= \exp\left(-\frac{1 - z}{1 + z}\right).$$

Finally, f is an inner function that is non-vanishing in U, and $\|f\|_\infty = 1$. So, there is a singular measure $\sigma \geq 0$ on $[-\pi, \pi]$ such that

$$f(z) = \exp\left(-\frac{1}{2\pi} \int_{-\pi}^{\pi} \frac{e^{it} + z}{e^{it} - z} d\sigma(t)\right) \cdot e^{ic},$$

where $c \in \mathbb{R}$ and where

$$\left\|\exp\left(-\frac{1}{2\pi} \int_{-\pi}^{\pi} \frac{e^{it} + z}{e^{it} - z} d\sigma(t)\right)\right\|_\infty = 1.$$

See the theorem of [59, p. 76]. Now, we approximate this exponential by finite Blaschke products. This is because we can use in each step only finitely many factors of the form:

$$\left(\frac{z - \alpha_k}{1 - \overline{\alpha_k}z}\right)^{m_k}.$$

$\square$

Remark 19.0.69. By the Hurwitz Theorem (on the zero sets of compactly convergent sequences of holomorphic functions), we deduce that if

$$B_m(z) = \prod_{j=1}^{n_m} \left(\frac{z - \alpha_{mj}}{1 - \overline{\alpha_{mj}} z} \right),$$

then for every $0 < \epsilon < 1$, there is an $N(\epsilon)$ such that $1 - \epsilon < |\alpha_{mj}| < 1$, $1 \leq j \leq n_m$ for $m > N(\epsilon)$.

To solve in affirmative the Krzyż conjecture, we can start by using the following equivalent form of the conjecture.

Theorem 19.0.70. *The Krzyż conjecture is true for $n \in \mathbb{Z}^+$ if and only if the following holds true: Fix an $\epsilon > 0$ and an $n \in \mathbb{Z}^+$. There is a function $C(n, \epsilon)$ depending on (n, ϵ) such that for every*

$$B_m(z) = \prod_{j=1}^{m} \left(\frac{z - \alpha_j}{1 - \overline{\alpha_j} z} \right),$$

that satisfies: $1 - \epsilon \leq |\alpha_j| < 1$, $1 \leq j \leq m$, $B_m(z) = \sum_{k=0}^{\infty} a_k z^k$, we have $|a_k| \leq C(n, \epsilon) < 1$. Furthermore, $\lim_{\epsilon \to 0+} C(n, \epsilon) = 2/e$.

Remark 19.0.71. We neglected $C(0, \epsilon)$ for it is clear that $C(0, \epsilon) = 1$ for any $0 < \epsilon < 1$.

This gives an alternative proof to the known proofs of the Krzyż conjecture for $n = 1$:

$$B_m(z) = \prod_{j=1}^{m} \left(\frac{z - \alpha_j}{1 - \overline{\alpha_j} z} \right) = \prod_{j=1}^{m} ((-\alpha_j) + (1 - |\alpha_j|^2)z + \cdots) = \sum_{k=0}^{\infty} a_k z^k.$$

So, $a_0 = (-\alpha_1)(-\alpha_2) \ldots (-\alpha_m) = (-1)^m \alpha_1 \alpha_2 \ldots \alpha_m$ and

$$a_1 = a_0 \sum_{j=1}^{m} \left(\frac{1 - |\alpha_j|^2}{(-\alpha_j)} \right) = (-1)^{m+1} \alpha_1 \alpha_2 \ldots \alpha_m \sum_{j=1}^{m} \left(\frac{1 - |\alpha_j|^2}{(-\alpha_j)} \right).$$

Clearly, $|a_1|$ is maximal when all $1 - \epsilon < \alpha_j < 1$. So, we are led to maximize

$$\Psi(\alpha_1, \ldots, \alpha_m) = \alpha_1 \ldots \alpha_m \sum_{j=1}^{m} \left(\frac{1 - \alpha_j^2}{\alpha_j} \right), \quad 1 - \epsilon < \alpha_j < 1.$$

$$\frac{\partial \Psi}{\partial \alpha_1} = \alpha_2 \ldots \alpha_m \sum_{j=1}^{m} \left(\frac{1 - \alpha_j^2}{\alpha_j} \right) - \alpha_2 \ldots \alpha_m \left(\frac{1 + \alpha_1^2}{\alpha_1} \right) = 0.$$

Hence,

$$\sum_{j=1}^{m}\left(\frac{1-\alpha_j^2}{\alpha_j}\right) = \frac{1+\alpha_1^2}{\alpha_1}.$$

Similarly, by $\partial\Psi/\partial\alpha_2 = 0$, we get

$$\sum_{j=1}^{m}\left(\frac{1-\alpha_j^2}{\alpha_j}\right) = \frac{1+\alpha_2^2}{\alpha_2}.$$

Hence,

$$\frac{1+\alpha_1^2}{\alpha_1} = \frac{1+\alpha_2^2}{\alpha_2},$$

and so $(\alpha_2 - \alpha_1)(1 - \alpha_1\alpha_2) = 0$. Since $1 - \epsilon < \alpha_1, \alpha_2 < 1$ we deduce that $\alpha_2 = \alpha_1$ and working with at the indices $j = 3, \ldots, m$ we obtain that a necessary condition for an extremum is that $\alpha_1 = \cdots = \alpha_m = \alpha$, where $1 - \epsilon < \alpha < 1$. Thus, we need to maximize over $1 - \epsilon < \alpha < 1$ the following function of one variable:

$$\Psi(\alpha, \alpha, \ldots, \alpha) = m\alpha^{m-1}(1 - \alpha^2).$$

We find the zeros of its first derivative: $(m\alpha^{m-1}(1-\alpha^2))' = m\alpha^{m-2}((m-1) - (m+1)\alpha^2) = 0$. Since $\alpha = 0$ is outside the domain, we get

$$\alpha = \sqrt{\frac{m-1}{m+1}}.$$

Taking m large enough, we get $1 - \epsilon < \sqrt{(m-1)/(m+1)} < 1$. For that single value of α (per m), we have

$$m\alpha^{m-1}(1 - \alpha^2) = \left(\frac{2m}{m+1}\right)\left(1 - \frac{2}{m+1}\right)^{(m-1)/2} \to_{m\to\infty} \frac{2}{e}.$$

The last sequence (under the limit action) is an increasing sequence.

Already, **for** $n = 2$, the straightforward computation becomes involved:

$$B_m(z) = \sum_{k=0}^{\infty} a_k z^k = \prod_{j=1}^{m}\left(\frac{z - \alpha_j}{1 - \overline{\alpha_j}z}\right)$$

$$= \prod_{j=1}^{m}\left((-\alpha_j) + (1 - |\alpha_j|^2)z + (1 - |\alpha_j|^2)\overline{\alpha_j}z^2 + \cdots\right)$$

$$a_2 = (-\alpha_1)(-\alpha_2)\ldots(-\alpha_m) \sum_{1 \le j_1 < j_2 \le m} \left(\frac{1 - |\alpha_{j_1}|^2}{-\alpha_{j_1}} \right) \left(\frac{1 - |\alpha_{j_2}|^2}{-\alpha_{j_2}} \right)$$

$$+ (-\alpha_1)\ldots(-\alpha_m) \sum_{j=1}^{m} \left(\frac{(1 - |\alpha_j|^2)\overline{\alpha_j}}{-\alpha_j} \right)$$

$$= (-1)^m \alpha_1 \ldots \alpha_m \left\{ \sum_{1 \le j_1 < j_2 \le m} \left(\frac{1 - |\alpha_{j_1}|^2}{\alpha_{j_1}} \right) \left(\frac{1 - |\alpha_{j_2}|^2}{\alpha_{j_2}} \right) \right.$$

$$\left. - \sum_{j=1}^{m} \left(\frac{(1 - |\alpha_j|^2)\overline{\alpha_j}}{\alpha_j} \right) \right\}.$$

This time, the arguments of $(1/\alpha_{j_1})(1/\alpha_{j_2})$ and $(-\overline{\alpha_j}/\alpha_j)$ do not have to agree $\mathrm{mod}\,(2\pi)$ for all $1 \le j_1, j_2 \le m$, $1 \le j \le m$. We can simplify the expression by using a well-known result, see [30]. That will allow us to consider only those Blaschke products which are finite and all of whose zeros lie on a circle centered at the origin and with some radius r close to 1. Thus, we might assume that $|\alpha_1| = \cdots = |\alpha_m| = r$. So, $\alpha_j = re^{i\theta_j}$, $j = 1, \ldots, m$. Plugging that choice gives us

$$a_2 = (-1)^m \cdot r^m \exp\left(i \sum_{j=1}^{m} \theta_j \right) \left\{ \sum_{1 \le j_1 < j_2 \le m} \left(\frac{1 - r^2}{r} \right)^2 e^{-i(\theta_{j_1} + \theta_{j_2})} \right.$$

$$\left. - \sum_{j=1}^{m} (1 - r^2) e^{2i\theta_j} \right\}.$$

Using the identity

$$\left(\sum_{j=1}^{m} e^{-i\theta_j} \right)^2 = \left(\sum_{j=1}^{m} e^{-2i\theta_j} \right) + 2 \left(\sum_{1 \le j_1 < j_2 \le m} e^{-i(\theta_{j_1} + \theta_{j_2})} \right),$$

we can write a_2 in the following form:

$$a_2 = (-1)^m r^m e^{i\sum_{j=1}^{m} \theta_j} \left\{ \frac{1}{2} \left(\frac{1 - r^2}{r} \right)^2 \left(\sum_{j=1}^{m} e^{-i\theta_j} \right)^2 \right.$$

$$\left. - \left(\sum_{j=1}^{m} e^{-2i\theta_j} \right) \left(\frac{1 - r^4}{2r^2} \right) \right\}.$$

If we make the particular choice $r = 1 - (1/m)$ as we may by [30], then we get

$$a_2 = (-1)^m \left(1 - \frac{1}{m}\right)^m e^{i \sum_{j=1}^m \theta_j} \left\{ \frac{1}{2} \left(\frac{2m-1}{m^2-m}\right)^2 \left(\sum_{j=1}^m e^{-i\theta_j}\right)^2 \right.$$
$$\left. - \left(\sum_{j=1}^m e^{-2i\theta_j}\right) \frac{(2m-1)(2m^2-2m+1)}{2m^2(m^2-2m+1)} \right\}.$$

Let us play a little with the choice of the arguments θ_j, $1 \leq j \leq m$. If we take $\theta_j = 0$, $1 \leq j \leq m$ then $a_2 \to_{m\to\infty} 0$ which means that this choice is not a maximizing choice. If, however, we take $\theta_1 = \theta_3 = \theta_5 = \cdots = 0$, $\theta_2 = \theta_4 = \theta_6 = \cdots = \pi$, we get

$$a_2 = (-1)^m \left(1 - \frac{1}{m}\right)^m e^{im\pi/2} \left\{ -m \frac{(2m-1)(2m^2-2m+1)}{2m^2(m^2-2m+1)} \right\}.$$

So, $|a_2| \to_{m\to\infty} 2/e$. However, we did not justify the second choice of the arguments θ_j, as being a maximizing one of $|a_2|$. With some more computations, this can be done.

Remark 19.0.72. The family of Blaschke products of degree at most k and zeros in the closed ring $\{z \in \mathbb{C} \,|\, 1 - \epsilon \leq |z| \leq 1\}$ is a compact family. Thus, for fixed ϵ, n, k, we can employ variational methods.

Chapter 20

The Variational Method for the Blaschke Products Approach to the Krzyż Conjecture

Let us assume that

$$B(z) = \prod_{j=1}^{k} \left(\frac{z - \alpha_j}{1 - \overline{\alpha_j} z} \right),$$

maximizes $|a_n|$. The model we have in mind is that there are two parameters, $k \in \mathbb{Z}^+$ and $0 \leq \epsilon \leq 1$, and we maximize $|a_n|$ over the family of all the Blaschke products $B(z)$ as above, of order k or less for which the zero set $Z(B) \subset \{1 - \epsilon \leq |z| \leq 1\}$. Here, $B(z) = \sum_{m=0}^{\infty} a_m z^m$. We will call the set $\{1 - \epsilon \leq |z| \leq 1\}$ "the narrow ring" or "the ϵ narrow ring".

(I) Let $\delta \in \mathbb{R}$ and let us define the following:

$$B_\delta(z) = B(z) \cdot \left(\frac{z - e^{i\delta} \alpha_1}{1 - e^{-i\delta} \overline{\alpha_1} z} \right) \cdot \left(\frac{1 - \overline{\alpha_1} z}{z - \alpha_1} \right).$$

We have

$$\frac{z - e^{i\delta} \alpha_1}{1 - e^{-i\delta} \overline{\alpha_1} z} - \frac{z - \alpha_1}{1 - \overline{\alpha_1} z}$$

$$= \frac{\alpha_1(1 - e^{i\delta}) + |\alpha_1|^2 z (e^{i\delta} - e^{-i\delta}) - \overline{\alpha_1} z^2 (1 - e^{-i\delta})}{(1 - e^{-i\delta} \overline{\alpha_1} z)(1 - \overline{\alpha_1} z)}.$$

This proves that

$$B_\delta(z) = B(z) \left\{ 1 + \frac{\alpha_1(1 - e^{i\delta}) + |\alpha_1|^2 z (e^{i\delta} - e^{-i\delta}) - \overline{\alpha_1} z^2 (1 - e^{-i\delta})}{(1 - e^{-i\delta} \overline{\alpha_1} z)(z - \alpha_1)} \right\}.$$

115

Let us denote

$$\frac{\alpha_1(1 - e^{i\delta}) + |\alpha_1|^2 z(e^{i\delta} - e^{-i\delta}) - \overline{\alpha_1} z^2(1 - e^{-i\delta})}{(1 - e^{-i\delta}\overline{\alpha_1}z)(z - \alpha_1)} = \sum_{m=0}^{\infty} b_m(\alpha_1, \delta)z^m,$$

and $B_\delta(z) = \sum_{m=0}^{\infty} A_m z^m$. Then: $\sum_{m=0}^{\infty} A_m z^m = \sum_{m=0}^{\infty} a_m z^m \{1 + \sum_{m=0}^{\infty} b_m z^m\}$, and we obtain $A_n = a_n(1 + b_0) + a_{n-1}b_1 + \cdots + a_0 b_n = a_n + (a_n b_0 + a_{n-1}b_1 + \cdots + a_0 b_n)$. Using the extremality of $B(z)$, we have $|A_n| \leq |a_n|$ and hence,

$$|a_n b_0 + \cdots + a_0 b_n|^2 + 2\Re\{\overline{a_n}(a_n b_0 + \cdots + a_0 b_n)\} \leq 0, \qquad (20.0.1)$$

for every $\delta \in \mathbb{R}$.

Calculation of $b_m(\alpha, \delta)$: We will drop the index 1 and write $\alpha_1 = \alpha$. $b_0(\alpha, \delta) = -(1 - e^{i\delta})$. For $m \geq 1$ and assuming that $|z| < |\alpha|$,

$$\frac{\alpha(1 - e^{i\delta}) + |\alpha|^2 z(e^{i\delta} - e^{-i\delta}) - \overline{\alpha} z^2(1 - e^{-i\delta})}{(1 - e^{-i\delta}\overline{\alpha}z)(z - \alpha)}$$

$$= -(1 - e^{i\delta}) + \sum_{m=1}^{\infty} e^{i\delta}(1 - e^{i\delta})\left(\frac{|\alpha|^2 - 1}{|\alpha|^2}\right)\left\{\sum_{j=0}^{m-1}(e^{-i\delta}\overline{\alpha})^{m-j}\left(\frac{1}{\alpha}\right)^j\right\} z^m.$$

Hence,

$$b_0(\alpha, \delta) = -(1 - e^{i\delta}),$$

$$b_m(\alpha, \delta) = e^{i\delta}(1 - e^{i\delta})\left(\frac{|\alpha|^2 - 1}{|\alpha|^2}\right)\left\{\sum_{j=0}^{m-1}(e^{-i\delta}\overline{\alpha})^{m-j}\left(\frac{1}{\alpha}\right)^j\right\}, \quad m \geq 1.$$

$$(20.0.2)$$

We see from this that for $\delta \to 0$, we have the following asymptotics:

$$b_0(\alpha, \delta) \approx i\delta$$

$$b_m(\alpha, \delta) \approx \left(\frac{|\alpha|^{2m} - 1}{\alpha^m}\right)(-i\delta), \quad m \geq 1.$$

Also,

$$\lim_{\delta \to 0} \frac{|a_m b_0 + \cdots + a_0 b_m|^2}{\delta} = 0,$$

because the numerator is $\Omega(\delta^2)$. Hence, equations (20.0.2) and (20.0.1) give us after letting $\delta \to 0^+$, $\delta \to 0^-$:

$$\Im\left\{\overline{a_n}\left(\frac{|\alpha_j|^2-1}{\alpha_j}a_{n-1}+\frac{|\alpha_j|^4-1}{\alpha_j^2}a_{n-2}+\cdots+\frac{|\alpha_j|^{2n}-1}{\alpha_j^n}a_0\right)\right\}=0,$$

$$1\le j\le k. \tag{20.0.3}$$

(where we used the computation for each one of the zeros $\alpha_1,\ldots,\alpha_k$)
(II) Let $\delta \in \mathbb{R}$ and let us define

$$B_\delta(z)=B(z)\cdot\left(\frac{z-(1+\delta)\alpha_1}{1-(1+\delta)\overline{\alpha_1}z}\right)\cdot\left(\frac{1-\overline{\alpha_1}z}{z-\alpha_1}\right).$$

We have

$$\frac{z-(1+\delta)\alpha_1}{1-(1+\delta)\overline{\alpha_1}z}-\frac{z-\alpha_1}{1-\overline{\alpha_1}z}=-\delta\frac{\alpha_1-\overline{\alpha_1}z^2}{(1-(1+\delta)\overline{\alpha_1}z)(1-\overline{\alpha_1}z)}.$$

It follows that

$$B_\delta(z)=B(z)\left\{1-\delta\frac{\alpha_1-\overline{\alpha_1}z^2}{(1-(1+\delta)\overline{\alpha_1}z)(z-\alpha_1)}\right\}.$$

We denote

$$\frac{\alpha-\overline{\alpha}z^2}{(1-(1+\delta)\overline{\alpha}z)(z-\alpha)}=\sum_{m=0}^{\infty}c_m(\alpha,\delta)z^m.$$

By the extremality of $B(z)$,

$$|a_nc_0+\cdots+a_0c_n|^2+2\Re\left\{\overline{a_n}\left(a_nc_0+\cdots+a_0c_n\right)\right\}, \tag{20.0.4}$$

for all $\delta \in \mathbb{R}$.

Calculation of $c_m(\alpha,\delta)$: $c_0(\alpha,\delta)=-1$. For values $m\ge 1$, we use

$$\frac{\alpha-\overline{\alpha}z^2}{(1-(1+\delta)\overline{\alpha}z)(z-\alpha)}=-1+\sum_{m=1}^{\infty}\left\{\sum_{j=0}^{m-2}((1+\delta)\overline{\alpha})^{m-j}\left(\frac{1}{\alpha}\right)^j\right.$$

$$\left.\times\left(\frac{1-(1+\delta)^2|\alpha|^2}{(1+\delta)^2|\alpha|^2}\right)-(1+\delta)\overline{\alpha}\left(\frac{1}{\alpha}\right)^{m-1}-\left(\frac{1}{\alpha}\right)^m\right)\right\}z^m.$$

So, when $\delta \to 0$, we get the following asymptotics:

$$\begin{cases} c_0(\alpha, \delta) = -1 \\ c_m(\alpha, \delta) \approx -\dfrac{|\alpha|^{2m} + 1}{\alpha^m}, \quad m \geq 1. \end{cases} \qquad (20.0.5)$$

We let $\delta \to 0^{\pm}$ in equations (20.0.4) and (20.0.5) to get the following: If $1 - \epsilon < |\alpha_j| < 1$, then

$$|a_n|^2 = -\Re\left\{\overline{a_n}\left(\frac{|\alpha_j|^2 + 1}{\alpha_j}a_{n-1} + \cdots + \frac{|\alpha_j|^{2n} + 1}{\alpha_j^n}a_0\right)\right\}. \qquad (20.0.6)$$

If $|\alpha_j| = 1 - \epsilon$, then the equality sign "$=$" in equation (20.0.6) is replaced by "$\geq$". We now use equations (20.0.3) and (20.0.6):

$$|a_n|^2 = -\Re\left\{\overline{a_n}\left(\frac{|\alpha_j|^2 + 1}{\alpha_j}a_{n-1} + \cdots + \frac{|\alpha_j|^{2n} + 1}{\alpha_j^n}a_0\right)\right\}$$

$$= -\overline{a_n}\left(\frac{|\alpha_j|^2 - 1}{\alpha_j}a_{n-1} + \frac{|\alpha_j|^4 - 1}{\alpha_j^2}a_{n-2} + \cdots + \frac{|\alpha_j|^{2n} - 1}{\alpha_j^n}a_0\right)$$

$$- 2\Re\left\{\overline{a_n}\left(\frac{a_{n-1}}{\alpha_j} + \cdots + \frac{a_0}{\alpha_j^n}\right)\right\}.$$

We can derive other identities as well. Let us summarize our results in the following:

Theorem 20.0.73. *Let*

$$B(z) = \prod_{j=1}^{k}\left(\frac{z - \alpha_j}{1 - \overline{\alpha_j}z}\right) = \sum_{m=0}^{\infty} a_m z^m,$$

maximize $|a_n|$, *where* $1 - \epsilon \leq |\alpha_j| \leq 1$, $1 \leq j \leq k$. *Then*

$$\Im\left\{\overline{a_n}\left(\frac{|\alpha_j|^2 - 1}{\alpha_j}a_{n-1} + \frac{|\alpha_j|^4 - 1}{\alpha_j^2}a_{n-2} + \cdots + \frac{|\alpha_j|^{2n} - 1}{\alpha_j^n}a_0\right)\right\} = 0,$$

$$1 \leq j \leq k$$

and

$$-\Re\left\{\overline{a_n}\left(\frac{|\alpha_j|^2 + 1}{\alpha_j}a_{n-1} + \cdots + \frac{|\alpha_j|^{2n} + 1}{\alpha_j^n}a_0\right)\right\} = |a_n|^2, \quad 1 - \epsilon < |\alpha_j|,$$

and "$\leq$" *if* $1 - \epsilon = |\alpha_j|$.

Here is an interesting conclusion from Theorem 20.0.73:

Theorem 20.0.74. *Let*

$$B(z) = \prod_{j=1}^{k} \left(\frac{z - \alpha_j}{1 - \overline{\alpha_j} z} \right) = \sum_{m=0}^{\infty} a_m z^m,$$

maximize $|a_n|$, *where* $1 - \epsilon \leq |\alpha_j| \leq 1$, $1 \leq j \leq k$. *Then* B *has at most* n *different zeros in the open ring* $D(0,1) - \overline{D(0,1-\epsilon)}$.

Proof.   Since the problem $\max |a_n|$ has a symmetry with respect to rotations $(e^{i\beta} B(e^{i\gamma} z))$ we can assume that $a_n > 0$. True, this might multiply $B(z)$ by uni-modular numbers, but the number of zeros of $B(z)$ and that of $e^{i\beta} B(e^{i\gamma} z)$ in concentric rings of the type $U - \overline{D(0,1-\epsilon)}$ are identical. Using that extra assumption $(a_n > 0)$ and Theorem 20.0.73, we get the following: If $1 - \epsilon < |\alpha_j| < 1$, then

$$-a_n = (|\alpha_j|^2 + 1)\Re\left(\frac{a_{n-1}}{\alpha_j} \right) + \cdots + (|\alpha_j|^{2n} + 1)\Re\left(\frac{a_0}{\alpha_j^n} \right),$$

$$0 = (|\alpha_j|^2 - 1)\Im\left(\frac{a_{n-1}}{\alpha_j} \right) + \cdots + (|\alpha_j|^{2n} - 1)\Im\left(\frac{a_0}{\alpha_j^n} \right).$$

We multiply the second equation by the imaginary unit, i, and add the result to the first equation:

$$-a_n = \left(|\alpha_j|^2 \left(\frac{a_{n-1}}{\alpha_j} \right) + \overline{\left(\frac{a_{n-1}}{\alpha_j} \right)} \right) + \cdots + \left(|\alpha_j|^{2n} \left(\frac{a_0}{\alpha_j^n} \right) + \overline{\left(\frac{a_0}{\alpha_j^n} \right)} \right)$$

$$= \left(a_{n-1}\overline{\alpha_j} + \overline{a_{n-1}}\frac{1}{\alpha_j} \right) + \cdots + \left(a_0\overline{\alpha_j}^n + \overline{a_0}\frac{1}{\alpha_j^n} \right).$$

Let us define

$$R(\xi) = \sum_{m=0}^{n-1} \left(a_m \xi^{n-m} + \overline{a_m}\frac{1}{\xi^{n-m}} \right),$$

then we proved the following identity: $R(\overline{\alpha_j}) = -a_n$ for $1 \leq j \leq k$, $1 - \epsilon < |\alpha_j| < 1$. But the function we just defined has the form

$$R(\xi) = \frac{p_{2n}(\xi)}{\xi^n},$$

where p_{2n} is a polynomial of degree $2n$. This already proves that the number of zeros of $B(z)$ in the ring $1 - \epsilon < |\xi| < 1$ is less than or equals to $2n$. In order to reduce this number further to less than $n + 1$, we note that

$$R(\xi) = \overline{R\left(\frac{1}{\bar{\xi}}\right)}.$$

Hence, if $R(\alpha) = -a_n$, then also

$$R\left(\frac{1}{\bar{\alpha}}\right) = \overline{R(\alpha)} = -\overline{a_n} = -a_n.$$

Thus, corresponding to each $|\alpha| < 1$ that satisfies $R(\alpha) = -a_n$ we have $R(1/\bar{\alpha}) = -a_n$, where $|1/\bar{\alpha}| > 1$. This proves that the number of different zeros of $B(z)$ is less than or equal to n. $\qquad\square$

Remark 20.0.75. We can restate Theorem 20.0.74. In fact, we will give a statement that is more detailed: If

$$B(z) = \prod_{j=1}^{k} \left(\frac{z - \alpha_j}{1 - \overline{\alpha_j} z}\right),$$

is extremal for the parameters ϵ, n, k, then there are t complex numbers $\gamma_1, \ldots, \gamma_t$ such that

$$B(z) = \prod_{j=1}^{t} \left(\frac{z - \gamma_j}{1 - \overline{\gamma_j} z}\right)^{n_j} \times \prod_{j=1}^{s} \left(\frac{z - (1 - \epsilon)e^{i\beta_j}}{1 - (1 - \epsilon)e^{-i\beta_j} z}\right),$$

where $1 - \epsilon < |\gamma_j| < 1$, $t \le n$, $n_1 + \cdots + n_t = k - s$, where the s complex numbers $(1 - \epsilon)e^{i\beta}$, $1 \le j \le s$ are the total set of the zeros of $B(z)$ on the inner circle of the ring $1 - \epsilon \le |z| \le 1$. We recall that we can restrict the family of Blaschke products to be those finite Blaschke products whose zeros lie on a concentric circle $|z| = r$, where $r < 1$ is close to 1. This choice can enable us to drop altogether the following factor:

$$\prod_{j=1}^{s} \left(\frac{z - (1 - \epsilon)e^{i\beta_j}}{1 - (1 - \epsilon)e^{-i\beta_j} z}\right),$$

i.e., take $s = 0$ (no zeros of $B(z)$ on the inner circle $|z| = 1 - \epsilon$). This follows from the results in [30].

Remark 20.0.76. Remark 20.0.75 (i.e., Theorem 20.0.74) proves that for $n = 1$, the maximizing Blaschke product is of the form

$$\left(\frac{z - \alpha}{1 - \overline{\alpha}z} \right)^k$$

(assuming the zeros have modulus $> 1 - \epsilon$). That was the essential part in proving that $|a_1| < 2/e$ and so $2/e$ is the maximal value (in fact the supremum) of $|a_1|$. In the following two chapters, we will solve the cases $n = 1$ and $n = 2$, respectively, using the Blaschke products approximations. Of course, these two cases are well known and were solved by other methods. However, we will be able to solve the extremal Krzyż problem and also sort out the full set of local maxima. This will lead us to state an extension of the Krzyż conjecture that points at a beautiful geometric structure of all those local maxima and at the corresponding extremal functions.

Chapter 21

A Solution of max $|a_1|$ Using Finite Blaschke Products

We have the following elementary identity:

$$
\left(\frac{z-\alpha}{1-\overline{\alpha}z}\right)^m = (-\alpha)^m + m(1-|\alpha|^2)(-\alpha)^{m-1}z
$$

$$
+ \left\{ m(1-|\alpha|^2)\overline{\alpha}(-\alpha)^{m-1} + \binom{m}{2}(1-|\alpha|^2)^2(-\alpha)^{m-2} \right\} z^2
$$

$$
+ \left\{ m(1-|\alpha|^2)\overline{\alpha}^2(-\alpha)^{m-1} + m(m-1)(1-|\alpha|^2)^2\overline{\alpha}(-\alpha)^{m-2} \right.
$$

$$
\left. + \binom{m}{3}(1-|\alpha|^2)^3(-\alpha)^{m-3} \right\} z^3 + \cdots.
$$

According to Theorem 20.0.74, it is in this family of finite Blaschke products that we have to look for the maximum of $|a_1|$. Now, a_1 is the coefficient of z and so we look at the coefficient of z in $\left(\frac{z-\alpha}{1-\overline{\alpha}z}\right)^m$. This coefficient is $m(1-|\alpha|^2)(-\alpha)^{m-1}$. Let $r = |\alpha|$. Then $|m(1-|\alpha|^2)(-\alpha)^{m-1}| = m(1-r^2)r^{m-1}$. For a fixed m, we want to choose $r = r_m$ which maximizes $m(1-r^2)r^{m-1}$. We solve for the zeros of the r-derivative:

$$
-2r \cdot r^{m-1} + (1-r^2)(m-1) = 0 \Rightarrow -2r^2 + (1-r^2)(m-1) = 0
$$

$$
\Rightarrow (m-1) - (m+1)r^2 = 0.
$$

So,

$$r_m = \sqrt{\frac{m-1}{m+1}} = \sqrt{1 - \frac{2}{m+1}}, \quad 1 - r_m^2 = \frac{2}{m+1}.$$

$$\max|a_1| = m(1 - r_m^2)r_m^{m-1}$$

$$= m\left(\frac{2}{m+1}\right)\left(1 - \frac{2}{m+1}\right)^{(m-1)/2} \to_{m\to\infty} 2 \cdot \frac{1}{e} = \frac{2}{e}.$$

That solves the Krzyż problem for $n = 1$. But we will compute some other coefficients a_0, a_2 and a_3 for the choice $r = r_m$. We will discover that the arguments of the terms are very important. They are responsible for the cancellations in these Taylor coefficients. These coefficients are finite sums of terms. Each term is of the order of magnitude of $\max|a_1| = 2/e$. Only because these large terms have different arguments, we get cancellations among these terms so that the total sum is smaller than $2/e$ which appears only for $|a_1|$. Let us compute: $a_0 = (-\alpha)^m$, so $|a_0| = r^m$. If we plug into that $r = r_m$ we get

$$r_m^m = \left(1 - \frac{2}{m+1}\right)^{m/2} \to_{m\to\infty} \frac{1}{e}.$$

The coefficient a_2 is the sum of two terms which we will designate by $a_2(1)$ and by $a_2(2)$. Looking at the power series expansion of $\left(\frac{z-\alpha}{1-\bar{\alpha}z}\right)^m$ that appears at the beginning of this chapter, we get $a_2(1) = m(1 - |\alpha|^2)\bar{\alpha}(-\alpha)^{m-1}$. In absolute value $m(1 - r^2)r^m$. Substituting $r = r_m$, we obtain

$$m\left(\frac{2}{m+1}\right)\left(1 - \frac{2}{m+1}\right)^{m/2} \to_{m\to\infty} 2 \cdot \frac{1}{e} = \frac{2}{e}.$$

So, $|a_2(1)|$ is large at r_m. Now, for the second term, we get

$$a_2(2) = \binom{m}{2}(1 - |\alpha|^2)^2(-\alpha)^{m-2} = \binom{m}{2}(1 - r_m^2)^2 r_m^{m-2}.$$

Hence,

$$|a_2(2)| = \binom{m}{2}\left(\frac{2}{m+1}\right)^2\left(1 - \frac{2}{m+1}\right)^{(m-2)/2} \to_{m\to\infty} \frac{2 \cdot 2}{2} \cdot \frac{1}{e} = \frac{2}{e}.$$

So, $|a_2(2)|$ is large. However, when we compute the arguments of $a_2(1)$ and of $a_2(2)$, this is what we find:

$$\arg a_2(1) = \arg(\overline{\alpha}(-\alpha)^{m-1}) = \arg \overline{\alpha} + \arg(-\alpha)^{m-1}$$
$$= -\arg \alpha + (m-1)\arg(-\alpha) = -\arg \alpha + (m-1)(\arg \alpha + \pi)$$
$$= (m-2)\arg \alpha + (m-1)\pi.$$
$$\arg a_2(2) = \arg(-\alpha)^{m-2} = (m-2)\arg(-\alpha) = (m-2)(\arg \alpha + \pi)$$
$$= (m-2)\arg \alpha + (m-2)\pi.$$

This means that $\arg a_2(1) = \arg a_2(2) + \pi$ and so $a_2(1)$ and $a_2(2)$ differ (sign wise) by multiplication by (-1). So,

$$|a_2| = |a_2(1) + a_2(2)|_{r=r_m} \to_{m\to\infty} \frac{2}{e} - \frac{2}{e} = 0,$$

a small value because of cancellation. The next coefficient

$$a_3 = m(1 - |\alpha|^2)\overline{\alpha}^2(-\alpha)^{m-1} + m(m-1)(1 - |\alpha|^2)^2\overline{\alpha}(-\alpha)^{m-2}$$
$$+ \binom{m}{3}(1 - |\alpha|^2)^3(-\alpha)^{m-3},$$

is a sum of three terms. We noted above that

$$m(1 - r_m^2) \to_{m\to\infty} 2 \quad \text{and} \quad r_m^m \to_{m\to\infty} \frac{1}{e}.$$

Also, computations of the arguments give us the following:

$$\arg(\overline{\alpha}^2(-\alpha)^{m-1}) - \arg(\overline{\alpha}(-\alpha)^{m-2}) = \pi,$$
$$\arg(\overline{\alpha}(-\alpha)^{m-2}) - \arg(-\alpha)^{m-3} = \pi,$$

so,

$$|a_3|_{r-r_m} \to_{m\to\infty} \left| \frac{2}{e} - 2\cdot 2 \cdot \frac{1}{e} + \frac{2\cdot 2\cdot 2}{3!} \cdot \frac{1}{e} \right| = \frac{2}{3e}.$$

Once more, this is a sum of three large terms with alternating signs. Because of cancellations the result $(2/(3e))$ is smaller than $2/e$.

A Solution of $\max|a_2|$ Using Finite Blaschke Products, Local Maxima and an Extension of the Krzyż Conjecture

Let $B(z)$ be extremal for a_2. We fix its degree (the number of factors of $B(z)$, where by a factor we mean an automorphism of U and we count according to the multiplicities). We denote the degree of $B(z)$ by k. Only after the optimization (maximizing $|a_2|$), we will let $k \to \infty$. We also may assume the zeros of $B(z)$ to be in the narrow ring $1 - \epsilon < |z| < 1$ for some fixed and small $0 < \epsilon < 1$. We can also confine the family of our Blaschke products to have zeros of equal modulus $1 - \epsilon < r < 1$. The zeros may differ only in their arguments. By Theorem 20.0.74, we consider the following Blaschke products:

$$B(z) = \left(\frac{z - \alpha}{1 - \overline{\alpha}z}\right)^m \left(\frac{z - \beta}{1 - \overline{\beta}z}\right)^{k-m}, \quad 1 - \epsilon < |\alpha|, |\beta| < 1,$$

and denote the power series expansion of $B(z) = \sum_{j=0}^{\infty} a_j z^j$, $|z| < 1$. Using the power series of $\left(\frac{z-\alpha}{1-\overline{\alpha}z}\right)^m$ and of $\left(\frac{z-\beta}{1-\overline{\beta}z}\right)^{k-m}$ we obtain the following formula:

$$a_2 = (-\alpha)^m \left\{ (k - m)(1 - |\beta|^2)\overline{\beta}(-\beta)^{k-m-1} \right.$$

$$\left. + \binom{k - m}{2} (1 - |\beta|^2)^2 (-\beta)^{k-m-2} \right\}$$

$$+ m(1 - |\alpha|^2)(-\alpha)^{m-1}(k - m)(1 - |\beta|^2)(-\beta)^{k-m-1}$$

$$+ \left\{ m(1 - |\alpha|^2)\overline{\alpha}(-\alpha)^{m-1} + \binom{m}{2}(1 - |\alpha|^2)^2(-\alpha)^{m-2} \right\}(-\beta)^{k-m}.$$

We have $|\alpha| = |\beta| = r$ and we denote $\theta = \arg \alpha$, $\phi = \arg \beta$. The following formula is useful:

$$\arg(\overline{\alpha}^s(-\alpha)^t) = (t - s)\theta + t\pi.$$

Because: $\arg(\overline{\alpha}^s(-\alpha)^t) = \arg \overline{\alpha}^s + \arg(-\alpha)^t = s \arg \overline{\alpha} + t \arg(-\alpha) = -s \arg \alpha + t(\arg \alpha + \pi) = (t - s)\arg \alpha + t\pi$. We will write explicitly the five terms that add up to give a_2. We will denote these terms by $a_2(1)$, $a_2(2)$, $a_2(3)$, $a_2(4)$, $a_2(5)$. Here are the five formulas:

$$a_2(1) = (k - m)(1 - r^2)r^k \times \exp(i(m\theta + (k - m - 2)\phi + (k - 1)\pi)),$$

$$a_2(2) = \binom{k - m}{2}(1 - r^2)^2 r^{k-2} \times \exp(i(m\theta + (k - m - 2)\phi + (k - 2)\pi)),$$

$$a_2(3) = m(k - m)(1 - r^2)^2 r^{k-2} \times \exp(i((m - 1)\theta + (k - m - 1)\phi + (k - 2)\pi)),$$

$$a_2(4) = m(1 - r^2)r^k \times \exp(i((m - 2)\theta + (k - m)\phi + (k - 1)\pi)),$$

$$a_2(5) = \binom{m}{2}(1 - r^2)^2 r^{k-2} \times \exp(i((m - 2)\theta + (k - m)\phi + (k - 2)\pi)).$$

$$(22.0.1)$$

We immediately note that $\operatorname{sgn} a_2(1) = -\operatorname{sgn} a_2(2)$, $\operatorname{sgn} a_2(4) = -\operatorname{sgn} a_2(5)$ which hint on cancellations to come. This time, cancellations will occur within a_2 which we maximize. That is unlike the case of a_1 where there was just a single term. We now make the relations among the five arguments more transparent. If we denote $\Lambda = (m - 1)\theta + (k - m - 1)\phi + (k - 2)\pi$, and $\psi = \theta - \phi$, then

$$a_2(1) = |a_2(1)| \times (-e^{i\Lambda}e^{i\psi}),$$

$$a_2(2) = |a_2(2)| \times (e^{i\Lambda}e^{i\psi}),$$

$$a_2(3) = |a_2(3)| \times (e^{i\Lambda}), \qquad (22.0.2)$$

$$a_2(4) = |a_2(4)| \times (-e^{i\Lambda}e^{-i\psi}),$$

$$a_2(5) = |a_2(5)| \times (e^{i\Lambda}e^{-i\psi}).$$

Using appropriate rotation by x radians, the angles (θ, ϕ) become $(\theta+x, \phi+x)$ so that $(m-1)(\theta+x) + (k-m-1)(\phi+x) + (k-2)\pi \equiv 0 \pmod{2\pi}$. This allows us to assume that $\Lambda = \arg a_2(3) = 0$. After these normalizations, we eventually have the following terms that add up to a_2:

$$a_2(1) = (k-m)(1-r^2)r^k \times (-e^{i\psi}),$$

$$a_2(2) = \binom{k-m}{2}(1-r^2)^2 r^{k-2} \times (e^{i\psi}),$$

$$a_2(3) = m(k-m)(1-r^2)^2 r^{k-2}, \tag{22.0.3}$$

$$a_2(4) = m(1-r^2)r^k \times (-e^{-i\psi}),$$

$$a_2(5) = \binom{m}{2}(1-r^2)^2 r^{k-2} \times (e^{-i\psi}).$$

The first thing to note is that $\psi \not\equiv 0 \pmod{2\pi}$ for if $\psi = 0$ then $\theta = \phi$, i.e., $\arg \alpha = \arg \beta$ and since $|\alpha| = |\beta| = r$, it follows that $\alpha = \beta$ so that

$$B(z) = \left(\frac{z-\alpha}{1-\overline{\alpha}z}\right)^k.$$

This $B(z)$ maximizes $|a_1|$ in the sense that $|a_1| \to_{k \to \infty} (2/e)$ and $r = (1 - 2/(k+1))^{1/2}$. We also know that $|a_2| \to_{k \to \infty} 0$ (same r), so $|a_2|$ is certainly not maximized. The problem we face is the following optimization problem:

$$A_2(k) = \max\left\{|a_2(1) + a_2(2) + a_2(3) + a_2(4) + a_2(5)| \,|\, 0 \right.$$
$$\left. \leq \psi < 2\pi, m \in \{1, 2, \ldots, k-1\},\ 0 < r < 1\right\},$$

and then compute $\lim_{k \to \infty} A_2(k)$.

We will not make a complete and accurate computation but show the role of each of the independent variables, m, ψ, r. The special feature here is that m is a discrete variable being a natural number between 1 and $k-1$. The other two, ψ and r, are continuous variables and so we have at our disposal the arsenal of calculus to compute them. Our strategy will be to determine m at the beginning. It is here that our arguments are sketchy (but reasonable). After that, determining ψ and r is a routine. We recall that our functions from the admissible family of Blaschke products have the following form:

$$B(z) = \left(\frac{z-\alpha}{1-\overline{\alpha}z}\right)^m \left(\frac{z-\beta}{1-\overline{\beta}z}\right)^{k-m},$$

where $|\alpha| = |\beta| = r$, $\alpha \neq \beta$. It is a plausible argument to claim that the role of α and β in the extremal function is the same in the sense that their multiplicities should coincide (or almost coincide). A symmetry argument. So, we will make our computations under the assumption that this is the case. This assumption (identical multiplicities) means that $m = k - m$ if k is an even number. If it is odd, then they differ by 1. Let us take even k. In this case, $m = (k/2) = k - m$. Hence, using equations (22.0.3)

$$a_2(1) + a_2(4) = \left(\frac{k}{2}\right)(1 - r^2)r^k(-2\cos\psi),$$

and

$$a_2(2) + a_2(5) = \binom{k/2}{2}(1 - r^2)^2 r^{k-2}(2\cos\psi).$$

Hence,

$$a_2(\psi, r) = a_2(1) + a_2(2) + a_2(3) + a_2(4) + a_2(5)$$

$$= -k(1 - r^2)r^k \cos\psi + \left(\frac{k}{2}\right)\left(\frac{k}{2} - 1\right)(1 - r^2)^2 r^{k-2}\cos\psi$$

$$+ \left(\frac{k}{2}\right)^2 (1 - r^2)^2 r^{k-2}.$$

By calculus, $\frac{\partial a_2}{\partial \psi} = 0$, i.e.,

$$\left\{\left(\frac{k}{2}\right)\left(\frac{k}{2} - 1\right)(1 - r^2)^2 r^{k-2} - k(1 - r^2)r^k\right\}(-\sin\psi) = 0.$$

So,

$$\left\{\frac{1}{2}\left(\frac{k}{2} - 1\right) + \left(\frac{1}{2}\left(\frac{k}{2} - 1\right) + 1\right)r^2\right\}\sin\psi = 0.$$

One possibility is that

$$r^2 = \frac{k - 2}{k + 2}.$$

The other possibility is that $\sin \psi = 0$ and since $\psi \not\equiv 0 (\mathrm{mod}\ 2\pi)$, we get $\psi = \pi$. We check the second necessary condition: $\frac{\partial a_2}{\partial r} = 0$. We obtain

$$\left\{ \left(\frac{k}{2}\right) \left(\frac{k}{2} - 1\right) (-2r2(1 - r^2)r^{k-2} + (k - 2)(1 - r^2)^2 r^{k-3}) \right.$$

$$\left. -k(-2r \cdot r^k + k(1 - r^2)r^{k-1}) \right\} \cos \psi$$

$$+ \left(\frac{k}{2}\right)^2 (-2r2(1 - r^2)r^{k-2} + (k - 2)(1 - r^2)^2 r^{k-3}) = 0,$$

so

$$\left\{ (k - 2)\left(-4(1 - r^2)r^2 + (k - 2)\left(1 - r^2\right)^2\right) - 4\left(-2r^4 + k(1 - r^2)r^2\right) \right\} \cos \psi$$

$$+ k\left(-4(1 - r^2)r^2 + (k - 2)(1 - r^2)\right) = 0.$$

If we substitute the first solution that annihilates the $\partial \psi$ derivative

$$r^2 = \left(\frac{k - 2}{k + 2}\right) \quad \text{and} \quad 1 - r^2 = \left(\frac{4}{k + 2}\right),$$

we find out that $-4(1 - r^2)r^2 + (k - 2)(1 - r^2)^2 = 0$ because

$$-4\left(\frac{4}{k + 2}\right)\left(\frac{k - 2}{k + 2}\right) + (k - 2)\left(\frac{4}{k + 2}\right)^2 = 0.$$

So, we deduce (by $\partial r = 0$) that $-4(-2r^4 + k(1 - r^2)r^2)\cos \psi = 0$. For the substitution of r^2 from above, we get

$$\frac{(k - 2)(2k - 1)}{(k + 2)^2} \cos \psi = 0.$$

We conclude that solution (I) is

$$\psi = \pm\frac{\pi}{2}, \quad r^2 = \left(\frac{k - 2}{k + 2}\right).$$

Otherwise, the solution (II) is

$$\psi = \pi, \quad r^4 = \left(\frac{k - 2}{k + 2}\right).$$

To see the r^4 value in solution (II), we remember that $\psi = \pi$ annihilates $\partial \psi = 0$. Now, $\cos \psi = (-1)$ and if we plug that into $\partial r = 0$ we get since in this case $a_2 = 4(1 - r^2) + 2(1 - r^2)^2 r^{k-2}$, r-differentiation leads to the following equation on r: $2(-2r)r^k + 2(1 - r^2)kr^{k-1} + (-2r)2(1 - r^2)r^{k-2} + (1 - r^2)^2(k - 2)r^{k-3} = 0$. Hence, $-r^4(k + 2) + (k - 2) = 0$, which gives the value $r^4 = \left(\frac{k-2}{k+2}\right)$.

The two solutions we obtained are as follows:

Solution (I):

$$\psi = \pm\frac{\pi}{2}, \quad r^2 = \left(\frac{k-2}{k+2}\right), \quad 1 - r^2 = \frac{4}{k+2}.$$

So,

$$a_2 = \left(\frac{k}{2}\right)^2 \left(\frac{4}{k+2}\right)^2 \left(\frac{k-2}{k+2}\right)^{(k-2)/2} = \left(\frac{2k}{k+2}\right)^2 \left(1 - \frac{4}{k+2}\right)^{(k-2)/2}.$$

Finally,

$$a_2 \to_{k\to\infty} 4 \cdot \frac{1}{e^2} = \left(\frac{2}{e}\right)^2.$$

Solution (II):

$$\psi = \pi, \quad r^4 = \left(\frac{k-2}{k+2}\right), \quad 1 - r^2 = \frac{2}{k+2} + \cdots.$$

So,

$$a_2 = k(1 - r^2)r^k + \left(\frac{k}{2}\right)(1 - r^2)^2 r^{k-2}$$

$$= k\left(\frac{2}{k+2} + \cdots\right)\left(1 - \frac{4}{k+2}\right)^{(1/2)\cdot(k/2)}$$

$$+ \left(\frac{k}{2}\right)\left(\frac{2}{k+2} + \cdots\right)^2 \left(1 - \frac{4}{k+2}\right)^{(1/2)\cdot((k-2)/2)}$$

$$= \left(\frac{2k}{k+2} + \cdots\right)\left(1 - \frac{4}{k+2}\right)^{k/4}$$

$$+ \left(\frac{2k}{(k+2)^2} + \cdots\right)\left(1 - \frac{4}{k+2}\right)^{(k-2)/4}.$$

Finally,

$$a_2 \to_{k \to \infty} 2 \cdot \frac{1}{e} + 0 \cdot \frac{1}{e} = \left(\frac{2}{e}\right).$$

Thus, $\left(\frac{2}{e}\right)^2$ is a local solution and $\left(\frac{2}{e}\right)$ is the global maximum. We proved the following:

Theorem 22.0.77. *The following extremal problem (Krzyż problem for* $n = 2$*):*

$$A_2 = \max \left\{ |a_2| \,\Big|\, f(z) = \sum_{j=0}^{\infty} a_j z^j \in B \right\},$$

has the solution

$$A_2 = \frac{2}{e},$$

which is approximated by the following sequence of finite Blaschke products:

$$\left(\frac{z - \left(\frac{k-2}{k+2}\right)^{1/4}}{1 - \left(\frac{k-2}{k+2}\right)^{1/4} z}\right)^{k/2} \left(\frac{z + \left(\frac{k-2}{k+2}\right)^{1/4}}{1 + \left(\frac{k-2}{k+2}\right)^{1/4} z}\right)^{k/2} , \quad (k \to \infty), \ (k \in 2\mathbb{Z}^+).$$

It has one more local maximum,

$$\text{local maximum} = \left(\frac{2}{e}\right)^2,$$

which is approximated by the following sequence of Blaschke products:

$$\left(\frac{z - \left(\frac{k-2}{k+2}\right)^{1/2}}{1 - \left(\frac{k-2}{k+2}\right)^{1/2} z}\right)^{k/2} \left(\frac{z - i\left(\frac{k-2}{k+2}\right)^{1/2}}{1 + i\left(\frac{k-2}{k+2}\right)^{1/2} z}\right)^{k/2} , \quad (k \to \infty), \ (k \in 2\mathbb{Z}^+).$$

This theorem brings me to one of my favorite parts in mathematical research: speculations!

Conjecture 22.0.78. An extension of the Krzyż conjecture: *The extremal problem*

$$A_n = \max \left\{ |a_n| \,\Big|\, f(z) = \sum_{j=0}^{\infty} a_j z^j \in B \right\}, \quad n \in \mathbb{Z}^+,$$

has the following solution:

$$A_n = \frac{2}{e},$$

and it has $n - 1$ more local maxima:

$$\left(\frac{2}{e}\right)^2, \ldots, \left(\frac{2}{e}\right)^n.$$

Coming back to Theorem 22.0.77, we naturally ask what are the two limiting functions that correspond to the two solutions (I) and (II)? The answer is that solution (II), the one that produces the absolute maximum $A_2 = \frac{2}{e}$, approximates

$$\exp\left(-\frac{1 + z^2}{1 - z^2}\right),$$

and rotations of this function, as is well known.

Solution (I), that produces the local maximum $\left(\frac{2}{e}\right)^2$, approximates the function

$$\exp\left(-\frac{1 + z}{1 - z}\right)\exp\left(-\frac{i + z}{i - z}\right).$$

We note that

$$\exp\left(-\frac{1 + z}{1 - z}\right) = \frac{1}{e} - \frac{2}{e}z + 0 \cdot z^2 + \cdots,$$

and

$$\exp\left(-\frac{i + z}{i - z}\right) = \exp\left(-\frac{1 + (-iz)}{1 - (iz)}\right) = \frac{1}{e} + \frac{2i}{e}z + 0 \cdot z^2 + \cdots.$$

Hence,

$$\exp\left(-\frac{1 + z}{1 - z}\right)\exp\left(-\frac{i + z}{i - z}\right)$$

$$= \left(\frac{1}{e} - \frac{2}{e}z + 0 \cdot z^2 + \cdots\right)\left(\frac{1}{e} + \frac{2i}{e}z + 0 \cdot z^2 + \cdots\right)$$

$$= \left(\frac{1}{e}\right)^2 + \frac{2}{e^2}(i - 1)z + \left(-\frac{2}{e}\right)\left(\frac{2i}{e}\right)z^2 + \cdots,$$

and indeed,

$$\left| \left(-\frac{2}{e} \right) \left(\frac{2i}{e} \right) \right| = \left(\frac{2}{e} \right)^2.$$

The computations that support the above are elementary:

$$\frac{k-2}{k+2} = \frac{k+2-4}{k+2} = 1 - \frac{4}{k+2},$$

$$\left(\frac{k-2}{k+2} \right)^{1/4} = \left(1 - \frac{4}{k+2} \right)^{1/4} = 1 - \frac{1}{k+2} + \cdots,$$

$$\left(\frac{z - \left(\frac{k-2}{k+2} \right)^{1/4}}{1 - \left(\frac{k-2}{k+2} \right)^{1/4} z} \right)^{k/2} = \left(\frac{z - \left(1 - \frac{1}{k+2} \right)}{1 - \left(1 - \frac{1}{k+2} \right) z} \right)^{k/2}$$

$$= (-1)^{k/2} \left(1 - \frac{1}{2} \left(\frac{1+z}{1-z} \right) \frac{2}{k+2} \right)^{k/2} \to \pm \exp\left(-\frac{1}{2} \left(\frac{1+z}{1-z} \right) \right).$$

Similarly,

$$\left(\frac{z + \left(\frac{k-2}{k+2} \right)^{1/4}}{1 + \left(\frac{k-2}{k+2} \right)^{1/4} z} \right)^{k/2} \to \pm \exp\left(-\frac{1}{2} \left(\frac{1-z}{1+z} \right) \right).$$

The product of these two limiting functions is

$$\exp\left(-\frac{1}{2} \left(\frac{1+z}{1-z} \right) \right) \exp\left(-\frac{1}{2} \left(\frac{1-z}{1+z} \right) \right) = \exp\left(-\frac{1+z^2}{1-z^2} \right).$$

Similarly,

$$\left(\frac{k-2}{k+2} \right)^{1/2} = \left(1 - \frac{4}{k+2} \right)^{1/2} = 1 - \frac{2}{k+2} + \cdots,$$

so that

$$
\left(\frac{z - \left(\frac{k-2}{k+2} \right)^{1/2}}{1 - \left(\frac{k-2}{k+2} \right)^{1/2} z} \right)^{k/2}
= \left(\frac{z - \left(1 - \frac{2}{k+2} \right)}{1 - \left(1 - \frac{2}{k+2} \right) z} \right)^{k/2}
$$

$$
= (-1)^{k/2} \left(1 - \frac{1+z}{1-z} \frac{2}{k+2} \cdots \right)^{k/2}
$$

$$
\to \pm \exp\left(-\frac{1+z}{1-z} \right).
$$

Finally,

$$
\left(\frac{z - i\left(1 - \frac{2}{k+2} \right)}{1 - i\left(1 - \frac{2}{k+2} \right) z} \right)^{k/2}
= (-i)^{k/2} \left(1 - \frac{i+z}{i-z} \cdot \frac{2}{k+2} \right)^{k/2},
$$

which gives us

$$
\exp\left(-\frac{i+z}{i-z} \right).
$$

Chapter 23

On the Formulation of $A_n(2)$ Using a Variant of the Calculus of Variations[*]

In Chapter 16, we gave the formulation referred to in the title. In this chapter, we will take it a bit further and demonstrate how one can derive variational inequalities using this idea. If $f(z) \in H(U)$, $f(z) \neq 0 \ \forall \, |z| < 1$, then there is an holomorphic function $g(z) = u(z) + iv(z) \in H(U)$, where $u(z) = \Re\{g(z)\}$, and $v(z) = \Im\{g(z)\}$, such that $f(z) = \exp(g(z))$. We are dealing here with the problem $A_n(2)$, so we assume that $f(z)$ is an admissible function for that problem. Thus, in addition to the conditions above, $f(z)$ is assumed to satisfy $\|f\|_{H^2(U)} = 1$. This means that

$$1 = \lim_{r \to 1^-} \frac{1}{2\pi} \int_0^{2\pi} |f(re^{i\theta})|^2 d\theta = \lim_{r \to 1^-} \frac{1}{2\pi} \int_0^{2\pi} \exp(2u(re^{i\theta}))d\theta.$$

We can assume (by a multiplication by an appropriate $e^{i\alpha}$) that $a_n \in \mathbb{R}_{\geq 0}$, and then

$$|a_n| = a_n = \lim_{r \to 1^-} \frac{1}{2\pi} \int_0^{2\pi} f(re^{i\theta})e^{-in\theta} d\theta$$

$$= \lim_{r \to 1^-} \frac{1}{2\pi} \int_0^{2\pi} \exp(u(re^{i\theta})) \cos(v(re^{i\theta}) - n\theta)d\theta.$$

So, **a reformulation of $A_n(2)$ is**

$$\max \text{ or } \sup \lim_{r \to 1^-} \frac{1}{2\pi} \int_0^{2\pi} \exp(u(re^{i\theta})) \cos(v(re^{i\theta}) - n\theta)d\theta,$$

---

[*]A Continuation of Chapter 16.

137

where the max *(or the* sup) *is taken over all the harmonic conjugates* $u(z)$, $v(z)$ *in* $|z| < 1$, *we may assume that* $v(0) = 0$, *for which*

$$\lim_{r \to 1^-} \frac{1}{2\pi} \int_0^{2\pi} \exp(2u(re^{i\theta}))d\theta = 1.$$

More explicitly and in Cartesian coordinates,

$$\frac{\partial^2 u}{\partial x^2} + \frac{\partial^2 u}{\partial y^2} = \frac{\partial^2 v}{\partial x^2} + \frac{\partial^2 v}{\partial y^2} = 0,$$

and

$$\frac{\partial u}{\partial x} = \frac{\partial v}{\partial y}, \quad \frac{\partial u}{\partial y} = -\frac{\partial v}{\partial x}.$$

The (CR) system in polar coordinates, (r, θ),

$$u_r = \frac{1}{r}v_\theta, \quad \frac{1}{r}u_\theta = -v_r.$$

Now, we would like to point at an explicit construction of a variation for the $A_n(2)$ problem, as stated above. What distinguishes the above formulation from a standard problem in the calculus of variations is the fact that the admissible functions are restricted to be harmonic and conjugate in U. That comes on the top of the $H^2(U)$-norm, unit side condition $\| \exp(u+iv)\|_2^2 = 1$. It causes the difficulty in constructing the variation, because the nearby constructed pair of functions (u^*, v^*) has to be harmonic conjugates in U. Let $h(z) = s(z) + it(z) \in H(U)$, where $s(z) = \Re\{h(z)\}$ and $t(z) = \Im\{h(z)\}$. We would like to have

$$\lim_{r \to 1^-} \frac{1}{2\pi} \int_0^{2\pi} \exp(2(u(re^{i\theta}) + s(re^{i\theta})))d\theta = 1,$$

so if $f(z)$ is extremal for our functional, then

$$\lim_{r \to 1^-} \int_0^{2\pi} \exp(u(re^{i\theta}) + s(re^{i\theta})) \cos(v(re^{i\theta}) + t(re^{i\theta}) - n\theta)d\theta$$

$$\leq \lim_{r \to 1^-} \leq \int_0^{2\pi} \exp(u(re^{i\theta})) \cos(v(re^{i\theta}) - n\theta)d\theta.$$

We note that

$$\exp(u + s)\cos(v + y - n\theta) = e^u e^s \{\cos(v - n\theta)\cos t - \sin(v - n\theta)\sin t\}$$

$$= (e^u \cos(v - n\theta))$$

$$\times (e^s \cos t) - e^u e^s \sin(v - n\theta)\sin t.$$

Hence, the δ created the following:

$$\exp(u+s)\cos(v+t-n\theta) - \exp(u)\cos(v-n\theta)$$
$$= (e^u\cos(v-n\theta)) \times (e^s\cos t - 1) - e^u e^s\sin(v-n\theta)\sin t,$$

and we are especially interested in a choice of harmonic conjugates in U, (s,t) for which both $e^s\cos t - 1$ and $e^u e^s\sin(v-n\theta)\sin t$ will be small.

We recall that the boundary values of the harmonic function $s(z)$ can, to a large extent, be chosen at will. Things are classical and well known when the boundary function is a continuous function. This is the Dirichlet problem. In our case, the domain of definition of the functions is the unit disc, $U = \{z \in \mathbb{C}\,|\,|z| < 1\}$, so there is no geometric obstacle of a complicated boundary. But we will find it useful to have boundary values of s on ∂U, which are not continuous. Say allow finitely many (finite) jumps. Here, we base the flexibility to construct $s(z)$ with those non-continuous boundary values on the result of Norbert Wiener in [106]. For the unit circle, we have an especially simple and explicit solution via the Poisson integral. Integration according to Wiener, see [106], is Daniel's integration.

Example 23.0.79. Let $f(z) = \sum_{k=0}^{\infty} a_k z^k = \exp(u(z) + iv(z))$ be extremal for $A_n(2)$ and, in particular satisfies $a_n \in \mathbb{R}_{\geq 0}$, and, of course,

$$\lim_{r\to 1^-} \frac{1}{2\pi} \int_0^{2\pi} \exp(2u(re^{i\theta}))d\theta = 1.$$

Let us split the unit circle $\{e^{i\theta}\,|\,0 \leq \theta < 2\pi\}$ into two complementary arcs, $L = \{e^{i\theta}\,|\,0 \leq \theta < \alpha\}$ and $R = \{e^{i\theta}\,|\,\alpha \leq \theta < 2\pi\}$ so that

$$\lim_{r\to 1^-} \frac{1}{2\pi} \int_0^{\alpha} \exp(2u(re^{i\theta}))d\theta = \eta, \qquad \lim_{r\to 1^-} \frac{1}{2\pi} \int_\alpha^{2\pi} \exp(2u(re^{i\theta}))d\theta = 1 - \eta.$$

Here, η is a fixed real number such that $0 < \eta < 1$. Usually, we will be interested in the case where η is close to 1, and so α is close to 2π. Next, we pick two numbers λ_1 and λ_2, such that $0 < \lambda_1$, λ_2 and $\lambda_1\eta + \lambda_2(1 - \eta) = 1$, and define the boundary values of an harmonic function $s(z)$, $|z| < 1$ as follows:

$$\exp(2 \cdot s(e^{i\theta})) = \begin{cases} \lambda_1, & 0 < \theta < \alpha \\ \lambda_2, & \alpha < \theta < 2\pi \end{cases}.$$

Here and in the future, we shorten our notation as follows:

$$\exp(2 \cdot s(e^{i\theta})) = \lim_{r\to 1^-} \exp(2 \cdot s(re^{i\theta})),$$

and similarly for other functions in $|z| < 1$. Then

$$\frac{1}{2\pi} \int_0^{2\pi} \exp(2(u(e^{i\theta}) + s(e^{i\theta})))d\theta$$

$$= \frac{1}{2\pi} \int_0^{2\pi} \exp(2u(e^{i\theta})) \exp(2s(e^{i\theta}))d\theta$$

$$= \frac{1}{2\pi} \int_0^{\alpha} \exp(2u(e^{i\theta})) \exp(2s(e^{i\theta}))d\theta$$

$$+ \frac{1}{2\pi} \int_\alpha^{2\pi} \exp(2u(e^{i\theta})) \exp(2s(e^{i\theta}))d\theta$$

$$= \frac{\lambda_1}{2\pi} \int_0^{\alpha} \exp(2u(e^{i\theta}))d\theta + \frac{\lambda_2}{2\pi} \int_\alpha^{2\pi} \exp(2u(e^{i\theta}))d\theta$$

$$= \lambda_1 \eta + \lambda_2(1 - \eta) = 1.$$

If η is very close to 1 and also λ_1 is very close to 1, then $s(z)$ could be a reasonable generator of a variation of $u(z)$.

A computation of $s(z) = s(re^{i\theta})$, $0 \le r < 1$, $0 \le \theta < 2\pi$: Since by our definition, the boundary values of s are determined by

$$\exp(2 \cdot s(e^{i\theta})) = \begin{cases} \lambda_1, & 0 < \theta < \alpha \\ \lambda_2, & \alpha < \theta < 2\pi \end{cases},$$

we have

$$s(e^{i\theta}) = \begin{cases} \dfrac{1}{2} \log \lambda_1, & 0 < \theta < \alpha \\ \dfrac{1}{2} \log \lambda_2, & \alpha < \theta < 2\pi \end{cases}.$$

Following Norbert Wiener in [106], we use Poisson integrals:

$$s(re^{i\theta}) = \frac{(1/2) \log \lambda_1}{2\pi} \int_0^{\alpha} \frac{(1 - r^2)d\psi}{1 - 2r \cos(\theta - \psi) + r^2}$$

$$+ \frac{(1/2) \log \lambda_2}{2\pi} \int_\alpha^{2\pi} \frac{(1 - r^2)d\psi}{1 - 2r \cos(\theta - \psi) + r^2}.$$

We compute

$$\int_a^b \frac{(1-r^2)d\psi}{1-2r\cos(\theta-\psi)+r^2} = \int_{\theta-b}^{\theta-a} \frac{(1-r^2)d\phi}{1-2r\cos\phi+r^2}$$

$$= 2\tan^{-1}\left(\frac{1+r}{1-r}\tan\left(\frac{\phi}{2}\right)\right)\Bigg|_{\phi=\theta-b}^{\theta-a}$$

$$= 2\left\{\tan^{-1}\left(\frac{1+r}{1-r}\tan\left(\frac{\theta-a}{2}\right)\right)\right.$$

$$\left. -\tan^{-1}\left(\frac{1+r}{1-r}\tan\left(\frac{\theta-b}{2}\right)\right)\right\}.$$

Using this last identity, we compute $s(re^{i\theta})$:

$$s(re^{i\theta}) = \frac{(1/2)\log\lambda_1}{2\pi}\cdot 2\left\{\tan^{-1}\left(\frac{1+r}{1-r}\tan\left(\frac{\theta}{2}\right)\right)\right.$$

$$\left. -\tan^{-1}\left(\frac{1+r}{1-r}\tan\left(\frac{\theta-\alpha}{2}\right)\right)\right\}$$

$$+ \frac{(1/2)\log\lambda_2}{2\pi}\cdot 2\left\{\tan^{-1}\left(\frac{1+r}{1-r}\tan\left(\frac{\theta-\alpha}{2}\right)\right)\right.$$

$$\left. -\tan^{-1}\left(\frac{1+r}{1-r}\tan\left(\frac{\theta-2\pi}{2}\right)\right)\right\}$$

$$= \frac{\log\left(\frac{\lambda_1}{\lambda_2}\right)}{2\pi}\left\{\tan^{-1}\left(\frac{1+r}{1-r}\tan\left(\frac{\theta}{2}\right)\right)\right.$$

$$\left. -\tan^{-1}\left(\frac{1+r}{1-r}\tan\left(\frac{\theta-\alpha}{2}\right)\right)\right\}.$$

By $\lambda_1\eta+\lambda_2(1-\eta)=1$, we obtain $\lambda_2 = \frac{1-\lambda_1\eta}{1-\eta}$. So,

$$\frac{\lambda_1}{\lambda_2} = \frac{\lambda_1(1-\eta)}{1-\lambda_1\eta}.$$

We can drop the index 1 and write $\lambda = \lambda_1$. Now, we can conclude the example and obtain a variational inequality of a different kind than the inequalities obtained, for example, in [49]: Let $f(z) \in H(U)$, $f(z) \neq 0$ $\forall\, |z| < 1$, $\frac{1}{2\pi}\int_0^{2\pi}|f(e^{i\theta})|^2 d\theta = 1$. Suppose that $f(z)$ maximizes $a_n > 0$, where $f(z) = \sum_{j=0}^{\infty} a_j z^j$. Here, we fix $n \in \mathbb{Z}^+$. Let the pair η and α satisfy $0 < \eta < 1,\ 0 < \alpha < 2\pi$ and be related by the following equation:

$$\frac{1}{2\pi}\int_0^\alpha |f(e^{i\theta})|^2 d\theta = \eta.$$

Also, fix λ, $0 < \lambda < 1$. Then the function,

$$s(re^{i\theta}) = \frac{1}{2\pi} \log \left(\frac{\lambda(1-\eta)}{1-\lambda\eta} \right) \left\{ \tan^{-1} \left(\frac{1+r}{1-r} \tan \left(\frac{\theta}{2} \right) \right) \right.$$
$$\left. - \tan^{-1} \left(\frac{1+r}{1-r} \tan \left(\frac{\theta-\alpha}{2} \right) \right) \right\},$$

$0 \le r < 1$, $0 \le \theta < 2\pi$, is an harmonic function in U with the boundary values:

$$s(e^{i\theta}) = \begin{cases} \dfrac{1}{2} \log \lambda, & 0 < \theta < \alpha \\ \dfrac{1}{2} \log \left(\dfrac{1-\lambda\eta}{1-\eta} \right), & \alpha < \theta < 2\pi \end{cases}.$$

We have the following variational inequality:

$$\frac{1}{2\pi} \int_0^{2\pi} |f(e^{i\theta})| \cdot \exp(s(e^{i\theta})) e^{-in\theta} d\theta \le a_n = |a_n|.$$

$s(z)$ is the real part of the holomorphic function:

$$g_1(z) = \frac{1}{2\pi} \log \left(\frac{1-\lambda\eta}{\lambda(1-\eta)} \right) \tan^{-1} \left(\frac{1+z}{1-z} \tan \left(\frac{\alpha}{2} \right) \right),$$

and if $g(z) = \exp\{g_1(z)\} = \sum_{j=0}^{\infty} b_j(\lambda) z^j$, then $\forall\, 0 < \lambda < 1$,

$$|a_0 b_n(\lambda) + \cdots + a_n b_0(\lambda)| \le |a_n|,$$

which is equivalent to

$$\Re\{a_n (\overline{a_n}(1 - |b_0(\lambda)|^2) - 2(\overline{a_0}\overline{b_n(\lambda)} + \cdots + \overline{a_{n-1}}\overline{b_1(\lambda)}) b_0(\lambda))\}$$
$$\ge |a_0 b_n(\lambda) + \cdots + a_{n-1} b_1(\lambda)|^2.$$

The b coefficients are computable. For example,

$$b_0(\lambda) = g(0) = \left(\frac{1-\lambda\eta}{\lambda(1-\eta)} \right)^{\alpha/(4\pi)}.$$

Chapter 24

Other Forms of the Functional Used for $A_n(2)$ Using Calculus of Variations

Let us recall how we expressed the problem $A_n(2)$ as a variational problem.

$$\sup \lim_{r \to 1^-} \frac{1}{2\pi} \int_0^{2\pi} \exp(u(re^{i\theta})) \cos(v(re^{i\theta}) - n\theta)d\theta,$$

where the sup is taken over all the harmonic conjugates $u(z)$, $v(z)$ in $|z| < 1$, where $v(0) = 0$ and for which

$$\lim_{r \to 1^-} \frac{1}{2\pi} \int_0^{2\pi} \exp(2u(re^{i\theta}))d\theta = 1.$$

The disadvantage of the above formulation is that for estimating the functional, we need to go over all the harmonic functions $u(z)$ in $|z| < 1$. For a given such an harmonic function $u(z)$, its conjugate $v(z)$, for which $v(0) = 0$, is uniquely determined. Thus, our admissible space is not a space of pair (u, v). It is the more conservative space of the harmonic functions u. One purpose of this chapter is to offer few formulas that express $v(z)$ in terms of $u(z)$. Embedding such a formula into the functional makes it more apparent that we have a variational problem over the space of all the harmonic functions in $|z| < 1$ and over the space of all those conjugate pairs (u, v). All the following formulas are well known in the classical literature of complex analysis.

A first formula is taken from [21, pp. 16–18]. We almost quote from the book: In a simply connected domain D, every harmonic function u has an harmonic conjugate v, defined by the line integral

$$v(z) = \int_\Gamma -\frac{\partial u}{\partial y}dx + \frac{\partial u}{\partial x}dy,$$

143

where the path of integration Γ extends from a fixed base point $z_0 \in D$ to a general point $z \in D$. The integral is independent of the path, by Green's theorem, because u is harmonic and D is simply connected. Also, it is nice to remember that if u is harmonic and g is analytic, then the composition $u \circ g$ is harmonic. In our especially simple case $D = U = \{z \in \mathbb{C}\,|\,|z|,1\}$, $z_0 = 0$ and we take the ray $x(t) = t\cos\theta$, $y(t) = t\sin\theta$, $0 \le t \le r$, and θ is fixed. $\dot{x}(t) = \cos\theta$, $\dot{y}(t) = \sin\theta$, so

$$v(re^{i\theta}) = \int_0^r \left(-\frac{\partial u}{\partial y}(te^{i\theta})\cos\theta + \frac{\partial u}{\partial x}(te^{i\theta})\sin\theta \right) dt. \tag{24.0.1}$$

A second formula originates in the following:

Theorem. *Let $f(z)$ be holomorphic in a neighborhood of the point a. Then if $f(z) = u(z) + iv(z)$, we have*

$$f(z) = 2u\left(\frac{z+\bar{a}}{2}, \frac{z-\bar{a}}{2i}\right) - \overline{f(a)} = 2iv\left(\frac{z+\bar{a}}{2}, \frac{z-\bar{a}}{2i}\right) + \overline{f(a)}.$$

In particular, for $a = 0$, we get

$$f(z) = 2u\left(\frac{z}{2}, \frac{z}{2i}\right) - \overline{f(0)} = 2iv\left(\frac{z}{2}, \frac{z}{2i}\right) + \overline{f(0)}.$$

Hence,

$$v(z) = \frac{f(z) - \overline{f(z)}}{2i} = \frac{1}{i}\left\{ u\left(\frac{z}{2}, \frac{z}{2i}\right) - \overline{u\left(\frac{z}{2}, \frac{z}{2i}\right)} \right\} + \frac{f(0) - \overline{f(0)}}{2i}.$$

So,

$$v(z) = 2\Im u\left(\frac{z}{2}, \frac{z}{2i}\right) + \Im f(0). \tag{24.0.2}$$

There is a formula that makes a use of the Hilbert kernel:

$$\frac{\sin(\sigma - \tau)}{1 - \cos(\sigma - \tau)} = \cot\left(\frac{\sigma - \tau}{2}\right),$$

and the Hilbert operator:

$$Hf(e^{i\tau}) := \frac{1}{2\pi}\int_0^{2\pi} f(e^{i\sigma})\cot\left(\frac{\sigma - \tau}{2}\right) d\sigma.$$

The key relations used here are the relations between the boundary values of u and v given by

$$v = -Hu + v(0), \quad u = Hv + u(0). \tag{24.0.3}$$

With the aid of the identities (24.0.1), (24.0.2) or (24.0.3), we can rewrite the $A_n(2)$ functional in terms of $u(z)$ only. For example, using (24.0.2), we obtain

$$\sup \frac{1}{2\pi} \int_0^{2\pi} \exp\left(u(e^{i\theta})\right) \cos\left(2\left(\Im u\left(\frac{e^{i\theta}}{2}, \frac{e^{i\theta}}{2i}\right) - \Im u(0,0)\right) - n\theta\right) d\theta, \tag{24.0.4}$$

where the sup is taken over the family of all the harmonic functions $u(z)$ in $|z| < 1$ that satisfy

$$\frac{1}{2\pi} \int_0^{2\pi} \exp(2u(e^{i\theta}))d\theta = 1.$$

That is an extremal problem on the family of all the harmonic functions in the unit disc U that have boundary values on the unit circle ∂U. Thus, its solution will express a genuine property of harmonic functions in U. The conjecture in [49] is that the sup is a max and the maximal value is the following number:

$$\sqrt{\frac{2}{e}}.$$

This extremal problem is parametrized by the natural number $n \in \mathbb{Z}^+$. The conjecture in [49] goes on and predicts that an extremal harmonic for the parameter value n is given by

$$u_n(z) = \frac{1}{2} \frac{|z|^{2n} - 1}{|z^n + 1|^2} + \log|z^n + 1| - \frac{1}{2}\log 2. \tag{24.0.5}$$

To see why this is what the conjecture in [49] asserts, we recall that the extremal for $A_n(2)$ is, according to that paper, the following function:

$$\left(\frac{(1 + z^n)^2}{2}\right)^{1/2} \left(\exp\left(\frac{z^n - 1}{z^n + 1}\right)\right)^{1/2}, \quad |z| < 1.$$

Our harmonic extremal (according to our reformulation of $A_n(2)$), u_n is the natural logarithm of the absolute value of the [49] extremal for the coefficient problem $A_n(2)$. Thus, indeed, u_n is given by formula (24.0.5). We note from formula (24.0.5) that we have

$$\lim_{n\to\infty} u_n(z) = \frac{1}{2}\frac{0-1}{|0+1|^2} + \log|0+1| - \frac{1}{2}\log 2 = \log\left(\frac{1}{2}\sqrt{\frac{2}{e}}\right),$$

uniformly on compact subsets of $U = \{z \in \mathbb{C}\,|\,|z| < 1\}$. On the other hand, there is no limit on the unit circle itself, i.e., the limit:

$$\lim_{n\to\infty} u_n(e^{i\theta}),$$

does not exist for almost all real $\theta \in \mathbb{R}$ because of the term $\log|e^{in\theta} + 1|$. Thus, by

$$\exp\log\left(\frac{1}{2}\sqrt{\frac{2}{e}}\right) = \frac{1}{2}\sqrt{\frac{2}{e}},$$

we note that the conjectured extremal for $A_n(2)$ in [49] satisfies

$$\left(\frac{(1+z^n)^2}{2}\right)^{1/2}\left(\exp\left(\frac{z^n-1}{z^n+1}\right)\right)^{1/2} = \frac{1}{2}\sqrt{\frac{2}{e}} \pm \sqrt{\frac{2}{e}}z^n + \cdots,$$

and we discover once more that the doubling principle holds also for $A_n(2)$, i.e., the extremal function for a_n satisfies $|a_n| = 2|a_0|$. We point now at the following conclusion that is immediate from our reformulation of $A_n(2)$ as a variational problem. The value of our $A_n(2)$ functional on the conjectured extremal $u_n(z)$ is given by

$$\lim_{r\to1^-}\frac{1}{2\pi}\int_0^{2\pi}\exp\left(u_n(re^{i\theta})\right)\cos\left(2\left(\Im u_n\left(\frac{re^{i\theta}}{2},\frac{re^{i\theta}}{2i}\right) - \Im u_n(0,0)\right) - n\theta\right)d\theta$$

and this is bounded from above in absolute value by

$$\lim_{r\to1^-}\frac{1}{2\pi}\int_0^{2\pi}\exp(u_n(re^{i\theta}))d\theta.$$

We did see above that uniformly on compact subsets of $0 \leq r < 1$, we have

$$\lim_{n\to\infty} u_n(re^{i\theta}) = \log\left(\frac{1}{2}\sqrt{\frac{2}{e}}\right),$$

and hence the following repeated limit is calculated:

$$\lim_{r\to 1^-}\lim_{n\to\infty}\frac{1}{2\pi}\int_0^{2\pi}\exp(u_n(re^{i\theta}))d\theta = \frac{1}{2}\sqrt{\frac{2}{e}}.$$

A similar argument works for the same repeated limit of the H^2 norm of the conjectured extremal function, i.e.,

$$\lim_{r\to 1^-}\lim_{n\to\infty}\frac{1}{2\pi}\int_0^{2\pi}\exp(2u_n(re^{i\theta}))d\theta = \frac{1}{2e}.$$

This might seem to lead to the conclusion that the function

$$\left(\frac{(1+z^n)^2}{2}\right)^{1/2}\left(\exp\left(\frac{z^n-1}{z^n+1}\right)\right)^{1/2},$$

given in the last section of [49] cannot be extremal for $A_n(2)$ (at least for a large enough n)! But, of course, we cannot prove that for when z is close to any of the n roots of unity, $e^{2\pi i k/n}$, $k = 0, 1, \ldots, n-1$, the conjectured function is close to $\sqrt{2}$, which is large.

Chapter 25

$A_n(2)$ and Calculus of Variations for Harmonic and for Holomorphic Functions in U

The purpose of the following few chapters is to try and make it clear which of the principles of the calculus of variations hold true in the setting of $A_n(2)$. Also, if such a principle does not hold as in the classical calculus of variations, to what extent is it partially effective. So, we will make a few repetitions for the sake of clarity. We recall that if $f(z) \in H(U)$ satisfies $f(z) \neq 0 \ \forall \, |z| < 1$, then there is a $g(z) = u(z) + iv(z) \in H(U)$ $(u(z) = \Re\{g(z)\}$ and $v(z) = \Im\{g(z)\})$ such that $f(z) = \exp(g(z))$. If $f(z)$ further satisfies $\|f\|_{H^2(U)} = 1$, then

$$1 = \lim_{r \to 1^-} \frac{1}{2\pi} \int_0^{2\pi} |f(re^{i\theta})|^2 d\theta = \lim_{r \to 1^-} \frac{1}{2\pi} \int_0^{2\pi} \exp(2u(re^{i\theta}))d\theta.$$

We may assume (by multiplying $f(z) = \sum_{j=0}^{\infty} a_j z^j$ by an appropriate $e^{i\alpha}$) that $a_n \in \mathbb{R}_{\geq 0}$ in which case

$$|a_n| = a_n = \lim_{r \to 1^-} \frac{1}{2\pi} \int_0^{2\pi} \Re\left\{ f(re^{i\theta})e^{-in\theta} \right\} d\theta$$

$$= \lim_{r \to 1^-} \frac{1}{2\pi} \int_0^{2\pi} \exp(u(re^{i\theta})) \cos(v(re^{i\theta}) - n\theta)d\theta.$$

This gave us the variational form of $A_n(2)$, namely find

$$\sup \lim_{r \to 1^-} \frac{1}{2\pi} \int_0^{2\pi} \exp(u(re^{i\theta})) \cos(v(re^{i\theta}) - n\theta)d\theta,$$

149

where the supremum is taken over the family of all the pairs u, v of harmonic conjugates in U, that satisfy

$$\lim_{r \to 1^-} \frac{1}{2\pi} \int_0^{2\pi} \exp(2u(re^{i\theta}))d\theta = 1.$$

Remark 25.0.80. It could be proved that the supremum above is in fact a maximum. That allows us to use variational methods in order to try and solve $A_n(2)$. Moreover, given $u(z)$ which is harmonic in $|z| < 1$, one can compute the harmonic conjugate of $u(z)$, for which $v(0) = b$, using the well-known identity:

$$v(z) = 2 \left(\Im u \left(\frac{z}{2}, \frac{z}{2i} \right) - \Im u(0,0) \right) + b,$$

where $b \in [0, 2\pi)$ is a constant that maximizes our functional above, and where we substitute $x = z/2$, $y = z/(2i)$ into $u(x,y)$. If 0 is not a good point, then the two coordinates can be linearly shifted by a, $\bar{a}$ for some non-zero $a \in \mathbb{C}$. This leads us to the following formulation of $A_n(2)$: Find

$$\max_{} \lim_{r \to 1^-} \frac{1}{2\pi} \int_0^{2\pi} \exp(u(re^{i\theta}))$$
$$\times \cos \left(2 \left(\Im u \left(\frac{re^{i\theta}}{2}, \frac{re^{i\theta}}{2i} \right) - \Im u(0,0) \right) + b - n\theta \right) d\theta,$$

where the maximum is taken over the family of all of the harmonic functions $u(z)$ in $|z| < 1$ that satisfy

$$\lim_{r \to 1^-} \frac{1}{2\pi} \int_0^{2\pi} \exp(2u(re^{i\theta}))d\theta = 1.$$

Next, let (u, v) be harmonic conjugates in $|z| < 1$. We ask for the form of an increment that corresponds to our problem. Let (u^*, v^*) be a point near (u, v) (we did not yet specify the metric we use!). Let us denote $s = u^* - u$ and $t = v^* - v$. Since (u, v) and (u^*, v^*) are two pairs of harmonic conjugates in U, it follows that (s, t) is also a pair of harmonic conjugates in U (For the Laplacian $\Delta s = \Delta(u^* - u) = \Delta u^* - \Delta u = 0 - 0 = 0$, and $\frac{\partial s}{\partial x} = \frac{\partial}{\partial x}(u^* - u) = \frac{\partial u^*}{\partial x} - \frac{\partial u}{\partial x} = \frac{\partial v^*}{\partial y} - \frac{\partial v}{\partial y} = \frac{\partial}{\partial y}(v^* - v) = \frac{\partial t}{\partial y}$. Similarly $\frac{\partial s}{\partial y} = -\frac{\partial t}{\partial x}$). So, $s(z) + it(z) = h(z) \in H(U)$. Moreover, by

$$\lim_{r \to 1^-} \frac{1}{2\pi} \int_0^{2\pi} \exp(2u^*(re^{i\theta}))d\theta = 1,$$

it follows that

$$\lim_{r \to 1^-} \frac{1}{2\pi} \int_0^{2\pi} \exp(2(u(re^{i\theta}) + s(re^{i\theta})))d\theta = 1.$$

Hence, if $f(z) = \exp(u(z) + iv(z))$ is extremal for a_n, then

$$\lim_{r \to 1^-} \frac{1}{2\pi} \int_0^{2\pi} \exp(u(re^{i\theta}) + s(re^{i\theta})) \cos((v(re^{i\theta}) + t(re^{i\theta})) - n\theta)d\theta$$

$$\leq \lim_{r \to 1^-} \frac{1}{2\pi} \int_0^{2\pi} \exp(u(re^{i\theta})) \cos(v(re^{i\theta}) - n\theta)d\theta.$$

If we use holomorphic functions $g(z) \in H(U)$ instead of pairs of harmonic conjugates $(u(z), v(z))$, where these are related by $g(z) = u(z) + iv(z)$, and where $f(z) = \exp(g(z))$ is extremal for $|a_n|$, then the increment $h(z) = s(z) + it(z) \in H(U)$ is such that if

$$\lim_{r \to 1^-} \frac{1}{2\pi} \int_0^{2\pi} |f(re^{i\theta}) \exp(h(re^{i\theta}))|^2 d\theta = \lim_{r \to 1^-} \frac{1}{2\pi} \int_0^{2\pi} |f(re^{i\theta})|^2 d\theta = 1,$$

then

$$\left| \lim_{r \to 1^-} \frac{1}{2\pi} \int_0^{2\pi} f(re^{i\theta}) \exp(h(re^{i\theta}))e^{-in\theta} d\theta \right| \leq \left| \lim_{r \to 1^-} \frac{1}{2\pi} \int_0^{2\pi} f(re^{i\theta})e^{-in\theta} d\theta \right|.$$

We could replace the absolute value by $\Re$ and obtain

$$\lim_{r \to 1^-} \frac{1}{2\pi} \Re \left\{ \int_0^{2\pi} f(re^{i\theta}) \exp\left(h(re^{i\theta}) \right) e^{-in\theta} d\theta \right\}$$

$$\leq \lim_{r \to 1^-} \frac{1}{2\pi} \Re \left\{ \int_0^{2\pi} f(re^{i\theta})e^{-in\theta} d\theta \right\}.$$

In the former version, our functional is

$$J(f) = \lim_{r \to 1^-} \frac{1}{2\pi} \left| \int_0^{2\pi} f(re^{i\theta})e^{-in\theta} d\theta \right|,$$

and in the later, the functional is defined by

$$J(f) = \lim_{r \to 1^-} \frac{1}{2\pi} \Re \left\{ \int_0^{2\pi} f(re^{i\theta})e^{-in\theta} d\theta \right\}.$$

Moreover, due to the orthogonality properties of the sequence of the exponentials $\{e^{ik\theta}\}_{k=-\infty}^{\infty}$ we could altogether get rid of the limit in the defining formulas of the functional $J(\cdot)$, and fix an r_0, $0 < r_0 < 1$ and use either

$$J(f) = \frac{1}{2\pi r_0^n} \left| \int_0^{2\pi} f(r_0 e^{i\theta}) e^{-in\theta} d\theta \right|, \quad \text{or}$$

$$J(f) = \frac{1}{2\pi r_0^n} \Re \left\{ \int_0^{2\pi} f(r_0 e^{i\theta}) e^{-in\theta} d\theta \right\}.$$

Chapter 26

A Necessary Condition for an Extremum

We will follow the presentation in [35]. We start by following 3.2 in [35, p. 11]. The concept of the variation (or differential) of a functional is introduced. The background field in [35] is $\mathbb{R}$. In our context, the admissible functions form a part of $H(U)$, so they act $U \to \mathbb{C}$. The values of our functionals are real ($\mathbb{R}$) so that we can compare sizes using the natural order of $\mathbb{R}$. Let $J[y]$ ([35] notations) be a functional defined on some normed linear space and let $\Delta J[h] = J[y + h] - J[y]$ be its increment, corresponding to the increment $h = h(x)$ of the "independent variable" $y = y(x)$. If y is fixed, then $\Delta J[h]$ is a functional of h, in general a nonlinear functional. Suppose that $\Delta J[h] = \phi[h] + \epsilon \cdot \|h\|$, where $\phi[h]$ is a linear functional and where $\epsilon \to 0$ as $\|h\| \to 0$. Then the functional $J[y]$ is said to be differentiable, and the principle linear part of the increment $\Delta J[h]$, i.e., the linear functional $\phi[h]$ which differs from $\Delta J[h]$ by infinitesimal of order higher than 1 relative to $\|h\|$, is called the variation (or differential) of $J[y]$ and is denoted by $\delta J[h]$. Strictly speaking, the increments and the variation of $J[y]$ are functionals of two arguments, y and h, and to emphasize this fact, we might write $\Delta J[y; h] = \delta J[y; h] + \epsilon \cdot \|h\|$. In trying to put that structure on our space of functions, we have two options. The first (in some sense, it looks natural) is to think of the space of all the functions $f(z) \in H(U)$ such that $f(z) \neq 0 \; \forall \, |z| < 1$ and

$$\|f\|_{H^2(U)}^2 = \lim_{r \to 1^-} \frac{1}{2\pi} \int_0^{2\pi} |f(re^{i\theta})|^2 d\theta = 1.$$

In that case, we use the notation $J[f]$, where either

$$J[f] = \frac{1}{2\pi r_0^n} \left| \int_0^{2\pi} f(r_0 e^{i\theta}) e^{-in\theta} d\theta \right|, \quad \text{or}$$

$$J[f] = \frac{1}{2\pi r_0^n} \Re \left\{ \int_0^{2\pi} f(r_0 e^{i\theta}) e^{-in\theta} d\theta \right\}.$$

However, in this model, the notation $\Delta J[h] = J[f+h] - J[f]$ does not make sense because we do not increment f, but rather multiply it, i.e., pass from the point $f(z)$ to $f(z)\exp(h(z))$. The second model is closer to the classical calculus of variations theory, as described in [35]. In that model, our space of admissible functions is the space of all the functions $g(z) \in H(U)$ such that

$$\| \exp(g) \|_{H^2(U)}^2 = \lim_{r \to 1^-} \frac{1}{2\pi} \int_0^{2\pi} | \exp(g(re^{i\theta}))|^2 d\theta = 1.$$

In that case, we use the notation $J[g]$, where either

$$J[g] = \frac{1}{2\pi r_0^n} \left| \int_0^{2\pi} \exp(g(r_0 e^{i\theta})) e^{-in\theta} d\theta \right|, \quad \text{or}$$

$$J[g] = \frac{1}{2\pi r_0^n} \Re \left\{ \int_0^{2\pi} \exp(g(r_0 e^{i\theta})) e^{-in\theta} d\theta \right\}.$$

In this model, the notation $\Delta J[h] = J[g+h] - J[g]$ makes sense. We get the following two formulas (depending on the choice of the functional we make, i.e., use the absolute value operator, or use the real part operator):

$$\Delta J[h] = \frac{1}{2\pi r_0^n} \left(\left| \int_0^{2\pi} \exp\left(g(r_0 e^{i\theta}) + h(r_0 e^{i\theta}) \right) e^{-in\theta} d\theta \right| \right.$$

$$\left. - \left| \int_0^{2\pi} \exp(g(r_0 e^{i\theta})) e^{-in\theta} d\theta \right| \right),$$

or

$$\Delta J[h] = \frac{1}{2\pi r_0^n} \Re \left\{ \int_0^{2\pi} (\exp(g(r_0 e^{i\theta}) + h(r_0 e^{i\theta})) \right.$$

$$\left. - \exp(g(r_0 e^{i\theta}))) e^{-in\theta} d\theta \right\}.$$

The increment h in these two formulas is a function that must satisfy the following:

(a) $h \in H(U)$.

(b) $\|\exp(g+h)\|_{H^2(U)}^2 = \lim\limits_{r \to 1^-} \dfrac{1}{2\pi} \int_0^{2\pi} |\exp(g(re^{i\theta}) + h(re^{i\theta}))|^2 d\theta = 1.$

Note that the set of increments $h(z)$ depends on the incremented function $g(z)$. This is an aspect which is different from the typical problems in the calculus of variations where there is (typically) a universal set of increments that can be used to increment any of the admissible functions. To emphasize the dependency of the increments on the incremented admissible function $g(z)$, we can designate those g-increments as the set $\mathrm{Inc}(g)$. Thus, by definition,

$$\mathrm{Inc}(g) = \{h(z)\,|\,h(z) \text{ satisfies } (a) + (b)\}.$$

Of course, the main restriction on such an increment $h(z)$ is (b). It originates in the requirement that both $g(z)$ and $g(z) + h(z)$ must be admissible for the functional J. Otherwise, if g is extremal for J, we will not be able to compare the values $J[g + h]$ and $J[g]$. In our case (of a maximum), we deduce that $J[g+h] \leq J[g]\,\forall\,h \in \mathrm{Inc}(g)$. Computing the increments in $\mathrm{Inc}(g)$ might be a difficult problem. For example, finding a useful representation of the functions in $\mathrm{Inc}(g)$ (in terms of g and other natural parameters) might be hard. We make a few observations regarding functions that belong to $\mathrm{Inc}(g)$ for each admissible g (i.e., $g \in H(U)$ such that $\|\exp(g)\|_{H^2(U)} = 1$): If $h \in \mathrm{Inc}(g)$, then $\{h(z) + ik\,|\,k \in \mathbb{R}\} \subseteq \mathrm{Inc}(g)$. This follows by $|\exp(h(z) + ik)| = |\exp(h(z))|$. In particular, since always $0 \in \mathrm{Inc}(g)$, it follows that $\{ik\,|\,k \in \mathbb{R}\} \subseteq \mathrm{Inc}(g)$. Next,

$$\left\{\log \psi(z)\,|\,\psi(z) \in H(U),\ \psi(z) \neq 0\,\forall\,|z| < 1 \text{ and } \lim\limits_{r \to 1^-}|\psi(re^{i\theta})| = 1\,a.e.\text{ in }\theta\right\}$$

is a subset of $\mathrm{Inc}(g)$. This follows by $\log \psi(z) \in H(U)$ and by $|\exp(\log \psi(re^{i\theta}))| = |\psi(re^{i\theta})| = 1$ a.e. in θ. So, the logarithm of any non-vanishing inner function is in the increment set of any admissible g. We saw in an example, how, using the solution of the Dirichlet problem in U, i.e., integrating appropriate boundary values against the Poisson kernel, maybe using Daniel's integration, we can construct g-increments. We will call such increments, g-increments of Dirichlet type and of order 2. The order of a Dirichlet increment counts the number of arcs into which we partitioned ∂U. Soon, we will describe the general case. Let us briefly describe the $(m+1)$st-order Dirichlet increment construction. We want to construct an increment

$h(z) = s(z) + it(z) \in H(U)$, where $s(z) = \Re\{h(z)\}$ and $t(z) = \Im\{h(z)\}$. Our admissible function is $g(z) = u(z) + iv(z) \in H(U)$, where $u(z) = \Re\{g(z)\}$ and $v(z) = \Im\{g(z)\}$, and where

$$\lim_{r \to 1^-} \frac{1}{2\pi} \int_0^{2\pi} |\exp(g(re^{i\theta}))|^2 d\theta = \lim_{r \to 1^-} \frac{1}{2\pi} \int_0^{2\pi} \exp(2u(re^{i\theta})) d\theta = 1.$$

We first will construct the harmonic function $s(z)$ (in $|z| < 1$), after which we will arrive at $h(z)$, the desired increment. Let us fix a finite partition of the interval $[0, 2\pi)$ into $m + 1$ (at least 2) sub-intervals. Thus, we let m be the number of partition points between 0 and 2π. Thus, $m \in \mathbb{Z}^+$ and the partition itself is sorted in an increasing order and will be denoted as follows: $0 < \alpha_1 < \cdots < \alpha_m < 2\pi$. It will be convenient to agree on $\alpha_0 = 0$, $\alpha_{m+1} = 2\pi$. Next, we break the integral of $|\exp(g(z))|^2$ accordingly and write the following:

$$\lim_{r \to 1^-} \frac{1}{2\pi} \int_{\alpha_j}^{\alpha_{j+1}} \exp(2u(re^{i\theta})) d\theta = \eta_j, \quad j = 0, 1, \ldots, m.$$

So, $0 < \eta_j < 1$ and by the fact that $g(z)$ is admissible, we have sum of unity, $\sum_{j=0}^m \eta_j = 1$. We now choose $(m + 1)$ weights λ_j, $j = 0, 1, \ldots, m$ such that $0 < \lambda_j$ and $\sum_{j=0}^m \lambda_j \eta_j = 1$. We define a Dirichlet problem for the boundary values of the function $s(z)$, which is harmonic in U, using our weights as follows:

$$\exp\left(2 \cdot s(e^{i\theta})\right) = \lambda_j, \quad \alpha_j < \theta < \alpha_{j+1}, \ j = 0, 1, \ldots, m.$$

Then with this choice, we get the main requirement from a g-increment being satisfied:

$$\lim_{r \to 1^-} \frac{1}{2\pi} \int_0^{2\pi} \exp(2(u(re^{i\theta}) + s(re^{i\theta}))) d\theta = \sum_{j=0}^m \lambda_j \eta_j = 1.$$

The boundary values of $s(z)$ are given by

$$s(e^{i\theta}) = \frac{1}{2} \log \lambda_j, \quad \alpha_j < \theta < \alpha_{j+1}, \ j = 0, 1, \ldots, m.$$

We do not define the boundary values of $s(z)$ at the jumping points $\alpha_0, \alpha_1, \ldots, \alpha_m$ (our partition points of $[0, 2\pi)$). We solve the Dirichlet problem:

$$s(re^{i\theta}) = \sum_{j=0}^m \left(\frac{1}{2} \log \lambda_j\right) \left(\frac{1}{2\pi} \int_{\alpha_j}^{\alpha_{j+1}} \frac{(1 - r^2) d\psi}{1 - 2r \cos(\theta - \psi) + r^2}\right), \quad 0 \le r < 1.$$

We will explicitly compute $s(z)$ using the following well-known formula:

$$\int_a^b \frac{(1-r^2)d\psi}{1-2r\cos(\theta-\psi)+r^2} = 2\left(\tan^{-1}\left(\frac{1+r}{1-r}\tan\left(\frac{\theta-a}{2}\right)\right)\right.$$
$$\left. -\tan^{-1}\left(\frac{1+r}{1-r}\tan\left(\frac{\theta-b}{2}\right)\right)\right).$$

We arrive at the solution of our Dirichlet problem:

$$s(re^{i\theta}) = \sum_{j=0}^{m}\left(\frac{1}{2\pi}\log\lambda_j\right)\left(\tan^{-1}\left(\frac{1+r}{1-r}\tan\left(\frac{\theta-\alpha_j}{2}\right)\right)\right.$$
$$\left. -\tan^{-1}\left(\frac{1+r}{1-r}\tan\left(\frac{\theta-\alpha_{j+1}}{2}\right)\right)\right).$$

Remark 26.0.81. The following harmonic function

$$w(re^{i\theta}) = \frac{a}{2\pi}\left(\tan^{-1}\left(\frac{1+r}{1-r}\tan\left(\frac{\theta}{2}\right)\right) - \tan^{-1}\left(\frac{1+r}{1-r}\tan\left(\frac{\theta-\alpha}{2}\right)\right)\right),$$

is the real part of the following holomorphic function:

$$\phi(z) = \frac{-a}{2\pi}\tan^{-1}\left(\frac{1+z}{1-z}\tan\left(\frac{\alpha}{2}\right)\right),$$

hence, we obtain the following formula for our Dirichlet increment of order $(m+1)$st:

$$h(z) = -\sum_{j=0}^{m}\left(\frac{1}{2\pi}\log\lambda_j\right)\tan^{-1}\left(\frac{e^{i\alpha_j}+z}{e^{i\alpha_j}-z}\cdot\tan\left(\frac{\alpha_{j+1}-\alpha_j}{2}\right)\right),$$

where $0 = \alpha_0 < \alpha_1 < \cdots < \alpha_m < \alpha_{m+1} = 2\pi$, and where,

$$\lim_{r\to 1^-}\frac{1}{2\pi}\int_{\alpha_j}^{\alpha_{j+1}}|\exp(g(re^{i\theta}))|^2 d\theta = \eta_j, \quad j = 0,1,\ldots,m.$$

So, $\sum_{j=0}^{m}\eta_j = 1$, $0 < \eta_j$, and where $\sum_{j=0}^{m}\lambda_j\eta_j = 1$, $0 < \lambda_j$.

Chapter 27

The Vanishing of the Differential, $\delta J[h] = 0^*$

We recall that for an admissible function $g(z)$ and for an increment $h(z)$, we denote (as is standard in [35]) $\Delta J[h] = J[g + h] - J[g]$. If (as we agree in our case) $g(z)$ is fixed, then $\Delta J[h]$ is a functional of h, in general a nonlinear functional. In fact, in our case, the set of increments h ($h \in H(U)$, and most importantly $\| \exp(g + h) \|_{H^2(U)} = 1$), is not a linear space. For example, we will see below that if $h(z)$ is an increment, then $-h(z)$ is not always an increment (for the same admissible function $g(z)$). For if we take a Dirichlet increment of the order $(m + 1)$st

$$h(z) = -\sum_{j=0}^{m} \left(\frac{1}{2\pi} \log \lambda_j \right) \tan^{-1} \left(\frac{e^{i\alpha_j} + z}{e^{i\alpha_j} - z} \cdot \tan \left(\frac{\alpha_{j+1} - \alpha_j}{2} \right) \right),$$

then by construction, we have $\sum_{j=0}^{m} \lambda_j \eta_j = 1$, $\lambda_j > 0$, where

$$\lim_{r \to 1^-} \frac{1}{2\pi} \int_{\alpha_j}^{\alpha_{j+1}} |\exp(g(re^{i\theta}))|^2 d\theta = \eta_j, \quad j = 0, 1, \ldots, m.$$

Assuming that also $h_1(z) = -h(z)$ is an increment for $g(z)$, we get by construction and by $-\log \lambda_j = \log(1/\lambda_j)$ that $\sum_{j=0}^{m} \frac{\eta_j}{\lambda_j} = 1$. Clearly that is generically false. The property that if $h(z)$ is an increment then also $-h(z)$ is an increment is what lies behind the necessity of the condition $\delta J[h] = 0$ in calculus of variations. We point at Theorem 2 [35, p. 13] and its proof.

Theorem 2 ([35, p. 13]). *A necessary condition for the differentiable functional $J[y]$ to have an extremum for $y = \hat{y}$ is that its variation vanishes for $y = \hat{y}$, i.e., that $\delta J[h] = 0$ for $y = \hat{y}$ and all admissible h.*

---

*A Necessary Condition for Extremum in the Calculus of Variations.

Proof ([35, p. 13]). Suppose that $J[y]$ has a minimum for $y = \hat{y}$. According to the definition of the differential $\delta J[h]$, we have $\Delta J[h] = \delta J[h] + \epsilon \cdot \|h\|$, where $\epsilon \to 0$ as $\|h\| \to 0$. Thus, for sufficiently small $\|h\|$, then the sign of $\Delta J[h]$ will be the same as the sign of $\delta J[h]$, if the last one has a sign. Now, suppose that $\delta J[h_0] \neq 0$ for some admissible h_0. Then for any $\alpha > 0$, no matter how small, we have $\delta J[-\alpha h_0] = -\delta J[\alpha h_0]$. Hence, $\Delta J[h]$ can be made to have either signs for arbitrary small $\|h\|$. But this is impossible, since by hypothesis $J[y]$ has a minimum for $y = \hat{y}$, i.e., $\Delta J[h] = J[\hat{y}+h] - J[\hat{y}] \geq 0$ for sufficiently small $\|h\|$. This contradiction proves the theorem. $\qquad\square$

This proof makes an essential use of $\delta J[-\alpha h_0] = -\delta J[\alpha h_0]$. As we noted, that property is not true in our case because if h is an increment, then $-h$ might not be and so $\delta J[-h]$ does not (in such a case) even make sense. One might argue that maybe there is another proof of Theorem 2 that is valid also in our case. Before tackling that essential point, let us compute the differential $\delta J[h]$ for our functional. We recall that we have two candidates for J. Either

$$J[g] = \frac{1}{2\pi r_0^n} \left| \int_0^{2\pi} \exp(g(r_0 e^{i\theta})) e^{-in\theta} d\theta \right|, \quad \text{or}$$

$$J[g] = \frac{1}{2\pi r_0^n} \Re \left\{ \int_0^{2\pi} \exp(g(r_0 e^{i\theta})) e^{-in\theta} d\theta \right\}.$$

Let us take for the purpose of computations the second definition for J. It seems less technical than the first because $\Re\{\cdot\}$ is linear, while $|\cdot|$ is sub-additive only (satisfies the triangle inequality). Then

$$\Delta J[h] = \frac{1}{2\pi r_0^n} \Re \left\{ \int_0^{2\pi} (\exp(g(r_0 e^{i\theta}) + h(r_0 e^{i\theta})) - \exp(g(r_0 e^{i\theta}))) e^{-in\theta} d\theta \right\}$$

$$= \frac{1}{2\pi r_0^n} \Re \left\{ \int_0^{2\pi} \exp(g(r_0 e^{i\theta})) \cdot h(r_0 e^{i\theta}) e^{-in\theta} d\theta \right\} + o(h(r_0 e^{i\theta})),$$

when $\|h(r_0 e^{i\theta})\| \to 0$. So, for any "reasonable metric", we decide to use the principle linear part of $\Delta J[h]$, which is

$$\delta J[h] = \frac{1}{2\pi r_0^n} \Re \left\{ \int_0^{2\pi} \exp(g(r_0 e^{i\theta})) \cdot h(r_0 e^{i\theta}) e^{-in\theta} d\theta \right\}.$$

So, to say that $\delta J[h] = 0$ is the same thing as to say that if the holomorphic function $\exp(g(z))h(z)$ (in U) has the power series expansion: $\exp(g(z))h(z) = b_0 + b_1 z + \cdots + b_n z^n + \cdots$, then $\Re\{b_n\} = 0$ no matter what the increment h is, provided that h is sufficiently small (close to 0). We noted before that $\{ik \mid k \in \mathbb{R}\} \subseteq \operatorname{Inc}(g)$ and for any $k \in \mathbb{R}^*$ the holomorphic functions $\exp(g(z))$ and $\exp(g(z)) \cdot ik$ have non-zero Taylor coefficients for identical powers of z. When $\exp(g(z)) = a_0 + a_1 z + \cdots + a_n z^n + \cdots$ is extremal for J, i.e., it maximizes $\Re\{a_n\}$, then clearly $a_n \in \mathbb{R}^+$. But then, this implies that for any $k \in \mathbb{R}$ we have $\Re\{a_n \cdot ik\} = 0$ in agreement with Theorem 2. This is perhaps not surprising because the set of increments $\{ik \mid k \in \mathbb{R}\}$ is symmetric about the 0 (ik is an increment $\Leftrightarrow$ $(-ik)$ is an increment). So, the proof given in [35] to Theorem 2 works here too. It is much more interesting to inquire if $\delta J[h] = 0$ also for increments h for which $-h$ is not an increment, i.e., $\|\exp(g+h)\|_{H^2(U)} = 1$ but $\|\exp(g-h)\|_{H^2(U)} \neq 1$, for example, the Dirichlet increments. Is the following "Theorem" true?

Theorem 27.0.82. *Let $\exp(g(z)) = a_0 + a_1 z + \cdots + a_n z^n + \cdots$ be extremal for $A_n(2)$. We noted that $a_n \in \mathbb{R}^+$. Then for any finite partition of $[0, 2\pi)$, $0 = \alpha_0 < \alpha_1 < \cdots < \alpha_m < \alpha_{m+1} = 2\pi$, if*

$$\lim_{r \to 1^-} \frac{1}{2\pi} \int_{\alpha_j}^{\alpha_{j+1}} |\exp(g(re^{i\theta}))|^2 d\theta = \eta_j, \quad j = 0, 1, \ldots, m,$$

and for any $m + 1$ weights $0 < \lambda_j$, $j = 0, 1, \ldots, m$ such that $\sum_{j=0}^m \lambda_j \eta_j = 1$, if,

$$\exp\left\{ g(z) - \sum_{j=0}^m \left(\frac{1}{2\pi} \log \lambda_j \right) \tan^{-1}\left(\frac{e^{i\alpha_j} + z}{e^{i\alpha_j} - z} \cdot \tan\left(\frac{\alpha_{j\mid 1} - \alpha_j}{2} \right) \right) \right\}$$

$$= b_0 + b_1 z + \cdots + b_n z^n + \cdots$$

then we have $\Re\{b_n\} = 0$.

That "Theorem" (if true!) is interesting because generically if $h(z)$ is a Dirichlet increment, then $-h(z)$ is not an increment (and so the [35] argument in the proof of Theorem 2 there, fails). We can easily turn the inquiry about the validity of "Theorem" 27.0.82 into a purely concrete computational problem using the conjectured extremal $\exp(g(z))$ that appears in the last section of [49]. Namely, the conjecture is that

$$\exp(g(z)) = \frac{1}{\sqrt{2}}(1 + z^n) \exp\left(\frac{1}{2} \cdot \frac{z^n - 1}{z^n + 1} \right).$$

Thus, the computational task we face is the following: Let $0 = \alpha_0 < \alpha_1 < \cdots < \alpha_m < \alpha_{m+1} = 2\pi$ be a partition of $[0, 2\pi)$. Let

$$\eta_j = \lim_{r \to 1^-} \frac{1}{2\pi} \int_{\alpha_j}^{\alpha_{j+1}} \left| \frac{(1 + re^{i\theta})^2}{2} \exp\left(\frac{(re^{i\theta}) - 1}{(re^{i\theta}) + 1} \right) \right| d\theta, \quad j = 0, 1, \ldots, m,$$

(check $\sum_{j=0}^{m} \eta_j = 1$). Let $\lambda_j > 0$, $j = 0, 1, \ldots, m$ satisfy $\sum_{j=0}^{m} \lambda_j \eta_j = 1$. Then in the following power series expansion

$$\frac{1}{\sqrt{2}}(1 + z^n) \exp\left(\frac{1}{2} \cdot \frac{z^n - 1}{z^n + 1}\right) \cdot \left\{-\sum_{j=0}^{m}\left(\frac{1}{2\pi} \log \lambda_j\right)\right.$$

$$\left. \times \tan^{-1}\left(\frac{e^{i\alpha_j} + z}{e^{i\alpha_j} - z} \cdot \tan\left(\frac{\alpha_{j+1} - \alpha_j}{2}\right)\right)\right\} = b_0 + b_1 z + \cdots + b_n z^n + \cdots$$

we have $\Re\{b_n\} = 0$.

If we find a counterexample, then either the conjectured extremal in [49] is wrong, or the necessary condition of calculus of variations in Theorem 2 [35, p. 13] is false here. If, on the other hand, the computations indicate that the computational task is true, then it strengthens two interesting speculations:

(a) [49]: The function

$$\frac{1}{\sqrt{2}}(1 + z^n) \exp\left(\frac{1}{2} \cdot \frac{z^n - 1}{z^n + 1}\right),$$

     is extremal for $A_n(2)$.

(b) The classical necessary condition of calculus of variations in Theorem 2 [35, p. 13] is valid even though the proof given in [35] fails in our case.

Remark 27.0.83. We recall that a basic tool in classical calculus of variations is the Euler–Lagrange system of PDE's. These usually are derived for special kind of functionals from the fundamental necessary condition $\delta J[h] = 0$ for an extremal $y = \hat{y}$. Hence, the importance of that condition to our less standard setting (where the increments h are dependent on the incremented admissible function g, and where they do not satisfy: $h \in \mathrm{Inc}(g) \Leftrightarrow (-h) \in \mathrm{Inc}(g)$.). We mention here the pioneering work of Khavinson ([55], [56]).

Chapter 28

Using Lagrange Multipliers on the Dirichlet Increments

In this chapter, we point at the use of the elementary tool of Lagrange multipliers on the special set of Dirichlet increments.

(1) Let us fix an admissible function $g(z)$ for $A_n(2)$. This means that $g(z) \in H(U)$ and that $\|exp(g)\|_{H^2(U)} = 1$. We do not assume that g is an extremal function for $A_n(2)$.

(2) Let us fix a finite partition of $[0, 2\pi)$, $0 = \alpha_0 < \alpha_1 < \cdots < \alpha_m < \alpha_{m+1} = 2\pi$.

(3) We compute the corresponding $(m+1)$ arc integrals:

$$\eta_j = \lim_{r \to 1-} \frac{1}{2\pi} \int_{\alpha_j}^{\alpha_{j+1}} |g(re^{i\theta})|^2 d\theta, \quad j = 0, 1, \ldots, m.$$

Notation 28.0.84. The space of all the Dirichlet increments of order $(m+1)$ corresponding to the above data is defined and denoted by

$$Dir(g; (\alpha_1, \ldots, \alpha_m)) = \left\{ h(z) = -\sum_{j=0}^{m} \left(\frac{1}{2\pi} \log \lambda_j \right) \right.$$

$$\times \tan^{-1} \left(\frac{e^{i\alpha_j} + z}{e^{i\alpha_j} - z} \cdot \tan \left(\frac{\alpha_{j+1} - \alpha_j}{2} \right) \right) \bigg|,$$

$$\left. 0 < \lambda_j, \ j = 0, 1, \ldots, m \text{ and } \sum_{j=0}^{m} \lambda_j \eta_j = 1 \right\}.$$

We consider the following extremal problem:

$$\sup_{h \in Dir(g;(\alpha_1,\,\dots\,,\alpha_m))} \frac{1}{2\pi r_0^n} \Re \left\{ \int_0^{2\pi} \exp(g(r_0 e^{i\theta}) + h(r_0 e^{i\theta}) - in\theta)d\theta \right\}.$$

This means that we want to find the largest Taylor coefficient of z^n in the power series expansions of all those functions that deviate from g by a Dirichlet increment of order $(m+1)$ for a fixed partition of the unit circle. Of course, an answer to such a question is expected to be given as a set of $(m+1)$-vectors with non-negative components, i.e., the vectors of the extremal weights $(\lambda_0, \dots, \lambda_m) \in (\mathbb{R}_{\geq 0})^{m+1}$.

An observation. If g happens to be extremal for $A_n(2)$, then we expect to get the vector of 1's: $(\lambda_0, \dots, \lambda_m) = (1, 1, \dots, 1)$. So, maybe one can find a metric d on $(\mathbb{R}_{\geq 0})^{m+1}$ that can tell us the quality of a solution we found, i.e., how close we are to the actual solution of $A_n(2)$. The measure to the quality of the solution will (maybe) be based on the distance of a solution we found, $(\lambda_0, \lambda_1, \dots, \lambda_m)$ to the $(m+1)$-dimensional vector of 1's, $d((\lambda_0, \lambda_1, \dots, \lambda_m), (1, 1, \dots, 1))$. If this distance is 0, then its a perfect score. So, how can we try and solve the above extremal problem? We might try and use the method of Lagrange multipliers on this extremal problem subject to one (affine) side condition: $\sum_{j=0}^{m} \lambda_j \eta_j = 1$. So, there will be a single Lagrange multiplier which we will denote by λ. We construct the Lagrange auxiliary function:

$$F(\lambda_0, \lambda_1, \dots, \lambda_m; \lambda) = \frac{1}{2\pi r_0^n} \Re \left\{ \int_0^{2\pi} \exp\left(g(r_0 e^{i\theta}) - \sum_{j=0}^{m} \left(\frac{1}{2\pi} \log \lambda_j \right) \tan^{-1} \right. \right.$$

$$\left. \left. \times \left(\frac{e^{i\alpha_j} + r_0 e^{i\theta}}{e^{i\alpha_j} - r_0 e^{i\theta}} \cdot \tan\left(\frac{\alpha_{j+1} - \alpha_j}{2} \right) \right) \right) - in\theta \right) d\theta \right\}$$

$$+ \lambda \left(\sum_{j=0}^{m} \lambda_j \eta_j - 1 \right),$$

and get the necessary condition in the form of a system of $(m+2)$ equations:

$$\frac{\partial F}{\partial \lambda_j} = \frac{\partial F}{\partial \lambda} = 0, \quad j = 0, 1, \dots, m.$$

Let's make a use of the following notations:

$$\psi = \psi(r_0, \theta) = g(r_0 e^{i\theta}) - in\theta,$$

$$\phi_j = \phi_j(r_0, \theta) = \frac{-1}{2\pi} \tan^{-1}\left(\frac{e^{i\alpha_j} + r_0 e^{i\theta}}{e^{i\alpha_j} - r_0 e^{i\theta}} \cdot \tan\left(\frac{\alpha_{j+1} - \alpha_j}{2}\right)\right),$$

$$j = 0, 1, \ldots, m.$$

These functions are independent of $\lambda_0, \lambda_1, \ldots, \lambda_m$ and of λ. Using these, we have

$$F(\lambda_0, \lambda_1, \ldots, \lambda_m; \lambda) = \frac{1}{2\pi r_0^n} \Re\left\{\int_0^{2\pi} \exp\left(\sum_{j=0}^m (\log \lambda_j)\phi_j + \psi\right) d\theta\right\}$$

$$+ \lambda\left(\sum_{j=0}^m \lambda_j \eta_j - 1\right).$$

So,

$$F(\lambda_0, \lambda_1, \ldots, \lambda_m; \lambda) = \frac{1}{2\pi r_0^n} \int_0^{2\pi} \exp\left(\sum_{j=0}^m (\log \lambda_j)\Re\{\phi_j\} + \Re\{\psi\}\right)$$

$$\times \cos\left(\sum_{j=0}^m (\log \lambda_j)\Im\{\phi_j\} + \Im\{\psi\}\right) d\theta$$

$$+ \lambda\left(\sum_{j=0}^m \lambda_j \eta_j - 1\right).$$

Let $k \in \{0, 1, \ldots, m\}$ and let us compute the partial derivative of F relative to λ_k:

$$\frac{\partial F}{\partial \lambda_k} = \left(\frac{1}{\lambda_k}\right)\left(\frac{1}{2\pi r_0^n}\right) \Re\left\{\int_0^{2\pi} \phi_k \cdot \exp\left(\sum_{j=0}^m (\log \lambda_j)\phi_j + \psi\right) d\theta\right\} + \lambda\eta_k.$$

We recall that our $(m+1)$st-order Dirichlet increment is given by

$$h(z) = -\sum_{j=0}^m \left(\frac{1}{2\pi}\log \lambda_j\right) \tan^{-1}\left(\frac{e^{i\alpha_j} + z}{e^{i\alpha_j} - z} \cdot \tan\left(\frac{\alpha_{j+1} - \alpha_j}{2}\right)\right).$$

Thus, the main part of the Lagrange system of equations, $\frac{\partial F}{\partial \lambda_k} = 0$, is given by

$$\Re \left\{ \frac{1}{2\pi r_0^n} \int_0^{2\pi} \exp(g(r_0 e^{i\theta}) + h(r_0 e^{i\theta}) - in\theta) \right.$$

$$\left. \times \tan^{-1} \left(\frac{e^{i\alpha_k} + r_0 e^{i\theta}}{e^{i\alpha_k} - r_0 e^{i\theta}} \cdot \tan \left(\frac{\alpha_{k+1} - \alpha_k}{2} \right) \right) \right) d\theta \right\} = 2\pi \lambda \lambda_k \eta_k,$$

$$k = 0, 1, \ldots, m. \tag{28.0.1}$$

If we add those m equations, taking into an account that $\sum_{k=0}^m \lambda_k \eta_k = 1$, we obtain the value of the Lagrange multiplier, namely

$$\lambda = \Re \left\{ \frac{1}{2\pi r_0^n} \int_0^{2\pi} \exp \left(g(r_0 e^{i\theta}) + h(r_0 e^{i\theta}) \right) \right.$$

$$\left. \times \sum_{j=0}^m \left(\frac{1}{2\pi} \tan^{-1} \left(\frac{e^{i\alpha_j} + r_0 e^{i\theta}}{e^{i\alpha_j} - r_0 e^{i\alpha_j}} \cdot \tan \left(\frac{\alpha_{j+1} - \alpha_j}{2} \right) \right) \right) \cdot e^{-in\theta} d\theta \right\}. \tag{28.0.2}$$

We note that the function,

$$\sum_{j=0}^m \left(\frac{1}{2\pi} \tan^{-1} \left(\frac{e^{i\alpha_j} + r_0 e^{i\theta}}{e^{i\alpha_j} - r_0 e^{i\theta}} \cdot \tan \left(\frac{\alpha_{j+1} - \alpha_j}{2} \right) \right) \right),$$

the second factor inside the integral in equation (28.0.2) differs from the Dirichlet increment by that each element should be multiplied by

$$\frac{-1}{2\pi} \log \lambda_k = \frac{1}{2\pi} \log \left(\frac{1}{\lambda_k} \right).$$

So, if we multiply equation (28.0.1) by that factor and add for $k = 0, 1, \ldots, m$ we obtain

$$\Re \left\{ \frac{1}{2\pi r_0^n} \int_0^{2\pi} \exp(g(r_0 e^{i\theta}) + h(r_0 e^{i\theta})) \cdot h(r_0 e^{i\theta}) \cdot e^{-in\theta} d\theta \right\}$$

$$= \lambda \sum_{k=0}^m (-\lambda_k \log \lambda_k) \eta_k. \tag{28.0.3}$$

If we multiply equation (28.0.2) by the number $\sum_{k=0}^{m} \eta_k \lambda_k \log \lambda_k$ and add to equation (28.0.3), we get

$$\Re \left\{ \frac{1}{2\pi r_0^n} \int_0^{2\pi} \exp(g(r_0 e^{i\theta}) + h(r_0 e^{i\theta})) \cdot \left(h(r_0 e^{i\theta}) + \left(\sum_{k=0}^{m} \frac{1}{2\pi} \eta_k \lambda_k \log \lambda_k \right) \right) \right.$$

$$\left. \times \sum_{k=0}^{m} \tan^{-1} \left(\frac{e^{i\alpha_k} + r_0 e^{i\theta}}{e^{i\alpha_k} - r_0 e^{i\theta}} \cdot \tan \left(\frac{\alpha_{k+1} - \alpha_k}{2} \right) \right) \cdot e^{-in\theta} d\theta \right\} = 0 \quad (28.0.4)$$

Chapter 29

Continuous Dirichlet Increments

First, we recall the construction of Dirichlet increments of some finite order $(m+1)$. We call these discrete Dirichlet increments. The purpose of this chapter is to give the continuous version of Dirichlet increments. This will be achieved by refining *ad infinity* the underline $[0, 2\pi)$ partition of the discrete increments. Suppose that $g(z) \in H(U)$ satisfies $\exp(g(z)) \in H^2(U)$. Let us fix a partition of the interval $[0, 2\pi)$, say $0 = \alpha_0 < \alpha_1 < \cdots < \alpha_m < \alpha_{m+1} = 2\pi$. The mesh size of this partition is given by

$$\max\{\alpha_{j+1} - \alpha_j \mid j = 0, 1, \ldots, m\}.$$

A refinement of this partition is another partition that includes it. This means $0 = \beta_0 < \beta_1 < \cdots < \beta_M < \beta_{M+1} = 2\pi$ where $\{\alpha_0, \alpha_1, \ldots, \alpha_{m+1}\} \subset \{\beta_0, \beta_1, \ldots, \beta_{M+1}\}$. Clearly, a refinement of a partition has a mesh size not larger than that of the former partition. This means that

$$\max\{\beta_{j+1} - \beta_j \mid j = 0, 1, \ldots, M\} \leq \max\{\alpha_{j+1} - \alpha_j \mid j = 0, 1, \ldots, m\}.$$

Refining a partition *ad infinity* means constructing an infinite sequence of partitions, the first of which is the original partition and such that each partition starting from the second one is a refinement of the preceding partition (the one with an index one less in the sequence) and such that the sequence of the corresponding mesh sizes converges (decrease) to 0. Given the partition $0 = \alpha_0 < \alpha_1 < \cdots < \alpha_m < \alpha_{m+1} = 2\pi$, we compute the corresponding $(m+1)$, η numbers:

$$\eta_j = \lim_{r \to 1^-} \frac{1}{2\pi} \int_{\alpha_j}^{\alpha_{j+1}} |\exp(g(re^{i\theta}))|^2 d\theta, \quad j = 0, 1, \ldots, m.$$

169

We sometimes will shorten the notation by writing

$$\eta_j = \lim_{r \to 1^-} \frac{1}{2\pi} \int_{\alpha_j}^{\alpha_{j+1}} |\exp(g(re^{i\theta}))|^2 d\theta = \frac{1}{2\pi} \int_{\alpha_j}^{\alpha_{j+1}} |\exp(g(e^{i\theta}))|^2 d\theta.$$

Next, we assume that the $H^2(U)$-norm of $\exp(g)$ is a unity:

$$\|\exp(g)\|_{H^2(U)}^2 = \lim_{r \to 1^-} \frac{1}{2\pi} \int_0^{2\pi} |\exp(g(re^{i\theta}))|^2 d\theta = 1.$$

Hence, $\sum_{j=0}^m \eta_j = 1$. Let us have a finite sequence of λ's, $\{\lambda_0, \lambda_1, \ldots, \lambda_m\}$ which are positive $\lambda_j > 0$, $j = 0, 1, \ldots, m$ and satisfy the following equation:

$$\sum_{j=0}^m \lambda_j \eta_j = 1.$$

A continuous Dirichlet increment is a non-negative continuous function, defined on $|z| < 1$, say $0 \leq \lambda(re^{i\theta})$, $0 \leq r < 1$, $\theta \in [0, 2\pi)$ where we think of the connection with the finite partition via,

$$\eta_j = \frac{1}{2\pi} \int_{\alpha_j}^{\alpha_{j+1}} |\exp(g(e^{i\theta}))|^2 d\theta \approx \frac{1}{2\pi} |\exp(g(e^{i\alpha_j})|^2 \Delta\alpha_j,$$

$\Delta\alpha_j = \alpha_{j+1} - \alpha_j$ and $\lambda_j = \lambda(e^{i\alpha_j}) > 0$. So,

$$1 = \sum_{j=0}^m \lambda_j \eta_j \approx \frac{1}{2\pi} \sum_{j=0}^m \lambda(e^{i\alpha_j})|\exp(g(e^{i\alpha_j}))|^2 \Delta\alpha_j.$$

One recognizes on the right-hand side a Riemann sum of the integral,

$$\lim_{r \to 1^-} \frac{1}{2\pi} \int_0^{2\pi} \lambda(re^{i\theta})|\exp(g(re^{i\theta}))|^2 d\theta.$$

In fact, under mild conditions, these Riemann sums converge to the integral when we refine our partition *ad infinity*. This data induces the following Dirichlet increment:

$$h(z) = -\sum_{j=0}^m \left(\frac{1}{2\pi} \log \lambda_j\right) \tan^{-1}\left(\frac{e^{i\alpha_j} + z}{e^{i\alpha_j} - z} \cdot \tan\left(\frac{\alpha_{j+1} - \alpha_j}{2}\right)\right)$$

$$= -\sum_{j=0}^m \left(\frac{1}{2\pi} \log \lambda_j\right) \tan^{-1}\left(\frac{e^{i\alpha_j} + z}{e^{i\alpha_j} - z} \cdot \tan\left(\frac{\Delta\alpha_j}{2}\right)\right).$$

For $\max_{j=0,1,\ldots,m} \Delta\alpha_j \ll 1$ we have approximately,

$$\tan^{-1}\left(\frac{e^{i\alpha_j} + z}{e^{i\alpha_j} - z} \cdot \tan\left(\frac{\Delta\alpha_j}{2}\right)\right) \approx \frac{e^{i\alpha_j} + z}{e^{i\alpha_j} - z} \cdot \frac{\Delta\alpha_j}{2}.$$

Thus, when our partitions of $[0, 2\pi)$: $0 = \alpha_0 < \alpha_1 < \cdots < \alpha_m < \alpha_{m+1} = 2\pi$ are refined *ad infinity,* we obtain (when the mesh size is small enough),

$$h(z) \approx -\sum_{j=0}^{m}\left(\frac{1}{2\pi}\log\lambda_j\right)\cdot\left(\frac{e^{i\alpha_j} + z}{e^{i\alpha_j} - z}\right)\cdot\frac{\Delta\alpha_j}{2}$$

$$= -\sum_{j=0}^{m}\left(\frac{1}{2\pi}\log\lambda_j\right)\cdot\left(\frac{e^{i\alpha_j} + z}{e^{i\alpha_j} - z}\right)\cdot\frac{\Delta\alpha_j}{2}.$$

Once again, one can recognize that the last expression is a Riemann sum of the following integral,

$$\lim_{r\to 1^-}\left(\frac{1}{2\pi}\right)\int_0^{2\pi}\log\lambda(re^{i\phi})\cdot\left(\frac{re^{i\phi} + z}{re^{i\phi} - z}\right)\cdot\frac{d\phi}{2}.$$

In other words, the limiting Dirichlet increment (which will still be denoted by $h(z)$) is given by

$$h(z) = \left(-\frac{1}{2}\right)\cdot\lim_{r\to 1^-}\left(\frac{1}{2\pi}\right)\int_0^{2\pi}\log\lambda(re^{i\phi})\cdot\left(\frac{re^{i\phi} + z}{re^{i\phi} - z}\right)d\phi.$$

This result is pleasing because we can recognize a familiar important expression from the theory of $H^p(U)$ spaces, $0 < p < \infty$. The integral in the last formula resembles the logarithm of an outer function in the theory of $H^p(U)$ spaces. In fact, if $f(z) \in H(U)$, then in the theory of factorization of $H^p(U)$ functions, one defines the outer function of $f(z)$ by the following:

$$F(z) = \exp\left\{\frac{1}{2\pi}\int_0^{2\pi}\left(\frac{e^{i\phi} + z}{e^{i\phi} - z}\right)\cdot\log|f(e^{i\phi})|d\phi\right\}.$$

See [25, p. 74]. The $H^p(U)$ function $f(z)$ has a unique factorization $f(z) = \psi(z)\cdot F(z)$, where $\psi(z)$ is an inner function. In the terminology introduced by Beurling, [5], an inner function is any function $\psi \in H^\infty(U)$ whose boundary values satisfy $|\psi(e^{i\phi})| = 1$ a.e. Blaschke products are inner functions. Another type of inner function is

$$\nu(z) = \exp\left\{-\int_0^{2\pi}\frac{e^{i\phi} + z}{e^{i\phi} - z}d\mu(\phi)\right\},$$

where μ is a finite singular measure. In other words, $\mu(\phi)$ is a bounded non decreasing function with $\mu(\phi) = 0$ a.e. These functions $\nu(z)$ are called singular inner functions. An important example is the atomic singular inner function,

$$\nu(z) = \exp\left\{-\frac{1+z}{1-z}\right\},$$

with its measure μ concentrated at one point. The most general inner function is a product $\Psi(z) = e^{i\gamma}B(z)\nu(z)$, where $e^{i\gamma}$ is a unimodular constant, B is a Blaschke product, and ν is a singular inner function. Thus, in general, any $H^p(U)$ function $f(z)$ has the representation $f(z) = e^{i\gamma}B(z)\nu(z)F(z)$ which is unique and is known as the canonical factorization of an $H^p(U)$ function. Conversely, any such a function $f(z)$ belongs to $H^p(U)$, where B is an arbitrary Blaschke product, $\nu(z)$ is an arbitrary singular inner function and F is an arbitrary outer function for $H^p(U)$. Moreover, $|f(e^{i\phi})| = |F(e^{i\phi})|$ a.e. Returning to our case of the limiting Dirichlet increment,

$$h(z) = \left(-\frac{1}{2}\right) \cdot \lim_{r\to 1^-} \left(\frac{1}{2\pi}\right) \int_0^{2\pi} \log \lambda(re^{i\phi}) \cdot \left(\frac{re^{i\phi}+z}{re^{i\phi}-z}\right) d\phi. \qquad (29.0.1)$$

We inquire if we can find an holomorphic function (in $H^p(U)$), say $f(z)$ such that $|f(e^{i\phi})| = \lambda(e^{i\phi})$ a.e. Or (even better) $|f(e^{i\phi})| = \lambda(e^{i\phi})^{-1/2}$ a.e. For, in that case, we will obtain a logarithm of an outer function (that of $f(z)$). We refer to our computations in Chapter 26. We recall that the construction of our finite order Dirichlet increments went by solving the Dirichlet problem in U with finitely many boundary values of the harmonic function $s(z)$. Each boundary value belonged to a single open arc on ∂U, and so the unit circle was partitioned into finitely many arcs, corresponding to the arguments $0 = \alpha_0 < \alpha_1 < \cdots < \alpha_m < \alpha_{m+1} = 2\pi$. In fact, the harmonic function $s(z)$ was the real part of the Dirichlet increment $h(z)$ we were looking for. The imaginary part of $h(z)$ was denoted by $t(z)$. Thus $h(z) = s(z) + it(z)$ where $s(z) = \Re\{h(z)\}$ and $t(z) = \Im\{h(z)\}$. The relation to the continuous λ function was $\lambda(e^{i\phi}) = \exp(2 \cdot s(e^{i\phi}))$. This can be written as follows: $\lambda(e^{i\phi}) = |\exp(2 \cdot (s(e^{i\theta}) + it(e^{i\phi})))|$. We conclude by the two equations,

$$|f(e^{i\phi})| = \lambda(e^{i\phi}), \quad \text{and} \quad \lambda(e^{i\phi}) = |\exp(2h(e^{i\phi}))|,$$

such that there is indeed an holomorphic function $f(z)$. It is given by $f(z) = \exp(h(z))^2$ and since $h(z) \in H^2(U)$, $f(z)$ is an $H^p(U)$ function (say $p = 1$). We note that by equation (29.0.1), we obtain the following interesting representation:

$$h(z) = - \lim_{r \to 1^-} \frac{1}{2\pi} \int_0^{2\pi} \log |\exp(h(re^{i\phi}))| \cdot \left(\frac{re^{i\phi} + z}{re^{i\phi} - z} \right) d\phi.$$

Taking the exponent, we get

$$\exp(-h(z)) = \exp \left\{ \frac{1}{2\pi} \int_0^{2\pi} \log |\exp(h(e^{i\phi}))| \cdot \left(\frac{e^{i\phi} + z}{e^{i\phi} - z} \right) d\phi \right\}. \qquad (29.0.2)$$

This means that $\exp(-h(z))$ is the outer function of $\exp(h(z))$ and by the canonical factorization in $H^p(U)$, there is an inner function $\psi(z)$ such that

$$\exp(h(z)) = \psi(z) \exp(-h(z)).$$

Hence, $|\exp(h(e^{i\phi}))| \equiv 1$ a.e. So, $\exp(h(z))$ is a singular inner function. This, however, does not make sense for if indeed $|\exp(h(e^{i\phi}))| \equiv 1$ a.e., then by equation (29.0.2), $h(z) \equiv 0$. This clearly is not true (there are non-trivial Dirichlet increments). So, there is an error somewhere in our computations. If we assume that there is a sign change in equation (29.0.2), then the fixed equation is

$$\exp(h(z)) = \exp \left\{ \frac{1}{2\pi} \int_0^{2\pi} \log |\exp(h(e^{i\phi}))| \cdot \left(\frac{e^{i\phi} + z}{e^{i\phi} - z} \right) d\phi \right\}. \qquad (29.0.3)$$

This means that $\exp(h(z))$ is the outer function of (itself) $\exp(h(z))$. We proved the following (the sign change will be resolved in Chapter 30):

Theorem 29.0.85. *If* $g(z) \in H(U)$, $0 \le \lambda(z)$ *sub-harmonic in* U *and these functions satisfy the following:*

$$\frac{1}{2\pi} \int_0^{2\pi} |\exp(g(e^{i\phi}))|^2 d\phi = \frac{1}{2\pi} \int_0^{2\pi} \lambda(e^{i\phi}) |\exp(g(e^{i\phi}))|^2 d\phi = 1,$$

$(\| \exp(g) \|_{H^2(U)} = \sum \lambda_j \eta_j = 1)$ *then*

$$\frac{1}{2\pi} \int_0^{2\pi} \left| \exp \left(g(e^{i\phi}) + \frac{1}{2\pi} \int_0^{2\pi} \log \lambda(e^{it}) \left(\frac{e^{it} + e^{i\phi}}{e^{it} - e^{i\phi}} \right) \frac{dt}{2} \right) \right|^2 d\phi = 1.$$

The function,

$$h(z) = \frac{1}{2\pi} \int_0^{2\pi} \log \lambda(e^{it}) \left(\frac{e^{it} + e^{i\phi}}{e^{it} - e^{i\phi}} \right) \frac{dt}{2},$$

is called a Dirichlet increment of $g(z)$. *It satisfies the identity (29.0.3), i.e.,* $\exp(h(z))$ *is the outer function of itself.*

Definition 29.0.86. Let $f(z) \in H^2(U)$, $f(z) \neq 0 \ \forall |z| < 1$. An invariant multiplier of $f(z)$ is a function $\nu(z) \in H^2(U)$, $\nu(z) \neq 0 \ \forall |z| < 1$, which satisfies $\|\nu \cdot f\|_{H^2(U)} = \|f\|_{H^2(U)}$.

Using this terminology, we can rewrite a part of Theorem 29.0.85 as follows:

Theorem 29.0.87. *Let $f(z) \in H^2(U)$, $f(z) \neq 0 \ \forall |z| < 1$. If $\nu(z)$ is an invariant multiplier of $f(z)$, then*

$$\exp\left\{ \frac{1}{2\pi} \int_0^{2\pi} \log |\nu(e^{it})| \cdot \left(\frac{e^{it} + z}{e^{it} - z} \right) dt \right\},$$

is also an invariant multiplier of $f(z)$.

Definition 29.0.88. Let $\nu(z) \in H^2(U)$, $\nu(z) \neq 0 \ \forall |z| < 1$, be an invariant multiplier of $f(z)$ (and so $f(z) \in H^2(U)$, $f(z) \neq 0 \ \forall |z| < 1$). The Dirichlet operator, $\mathcal{D}$, has the following action:

$$\mathcal{D}(\nu) = \exp\left\{ \frac{1}{2\pi} \int_0^{2\pi} \log |\nu(e^{it})| \cdot \left(\frac{e^{it} + z}{e^{it} - z} \right) dt \right\}.$$

In other words, $\mathcal{D}$ assigns to ν its outer function.

Definition 29.0.89.

$$H^2_{\neq 0}(U) = \left\{ \psi(z) \in H^2(U) \mid \psi(z) \neq 0 \ \forall |z| < 1 \right\}.$$

Remark 29.0.90. We note that

$$\mathcal{D} : H^2_{\neq 0}(U) \to H^2_{\neq 0}(U).$$

Next, we have the following theorem.

Theorem 29.0.91. *Let $\nu(z) \in H^2_{\neq 0}(U)$ be an invariant multiplier of some function in $H^2_{\neq 0}(U)$. If the sequence of $\mathcal{D}$-iterates $\{\mathcal{D}^n \nu\}_{n=0}^{\infty}$ converges as $n \to \infty$ uniformly on compact subsets of U, to the limiting function $\nu_f(z)$, then $\nu_f(z) \in H^2_{\neq 0}(U)$ and $\mathcal{D}(\nu_f) = \nu_f$.*

Proof. By uniform convergence on compact subsets of U, $\nu_f \in H(U)$. By the theorem of Hurwitz, either $\nu_f(z) \neq 0 \ \forall |z| < 1$ or $\nu_f(z) \equiv 0$. The later possibility cannot occur because for each n, $\mathcal{D}^n(\nu)$ is an invariant multiplier of the same function in $H^2_{\neq 0}(U)$. Finally, the operator $\mathcal{D}$ is continuous on $H^2_{\neq 0}(U)$ and so the limiting function ν_f is a fixed-point of the operator $\mathcal{D}$. $\qquad \square$

Chapter 30

Computations of the Dirichlet Increment, that of a Finite Order and the Continuous Case

In this important but technical chapter, we will go through all the details of the computations of the Dirichlet increments — the discrete (finite order) ones and after that the continuous case. One reason is that we are convinced that there is some error in the expression we got before. At least a sign change.

The elements in the sum of the real part of $h(z)$
(the finite order case)

The Dirichlet increment $h(z) = s(z) + it(z) \in H(U)$, $s(z) = \Re\{h(z)\}$, $t(z) = Im\{h(z)\}$, was constructed by first solving for the harmonic function $s(z)$. That was done by finding the solution of a Dirichlet problem in U for piecewise constant boundary values. More precisely, the unit circle, ∂U, was partitioned into finitely many non-degenerate arcs and we were looking for the harmonic function in U that has these boundary values. The solution $s(z)$ was explicitly written in terms of constant multiples of the integrals of the Poisson kernel in U, between the end points of each one of the arcs in the partition. The function $s(z)$ was the finite sum of these integrals. The order of the Dirichlet increment $h(z)$ was the number of arcs in our partition of ∂U. We now carefully compute a typical element in this sum.

$$\left(\frac{1}{2} \log \lambda_j \right) \left(\frac{1}{2\pi} \int_{\alpha_j}^{\alpha_{j+1}} \frac{(1-r^2)d\psi}{1 - 2r\cos(\theta - \psi) + r^2} \right). \tag{30.0.1}$$

Thus, we need to resolve integrals of the following type:

$$\int_a^b \frac{(1-r^2)d\psi}{1-2r\cos(\theta-\psi)+r^2} = \int_{\theta-b}^{\theta-a} \frac{(1-r^2)d\phi}{1-2r\cos\phi+r^2}, \qquad (30.0.2)$$

$\phi = \theta - \psi$, $d\phi = -d\psi$. So, we will find

$$\int \frac{d\phi}{\alpha + \beta\cos\phi},$$

where $\alpha = 1 + r^2$, $\beta = -2r$. We make the standard substitution,

$$x = \tan\left(\frac{\phi}{2}\right), \quad dx = \frac{1}{2}\frac{d\phi}{\cos^2(\phi/2)} = \frac{1}{2}(1+x^2)d\phi, \quad d\phi = \frac{2dx}{1+x^2}.$$

Also,

$$\cos\phi = \cos^2\left(\frac{\phi}{2}\right) - \sin^2\left(\frac{\phi}{2}\right) = \frac{1}{1+x^2} - \frac{x^2}{1+x^2} = \frac{1-x^2}{1+x^2}.$$

Hence,

$$\int \frac{d\phi}{\alpha+\beta\cos\phi} = \int \frac{2dx/(1+x^2)}{\alpha+\beta(1-x^2)/(1+x^2)} = \int \frac{2dx}{\alpha(1+x^2)+\beta(1-x^2)}$$

$$= \int \frac{2dx}{(\alpha+\beta)+(\alpha-\beta)x^2}$$

$$= \left(\frac{2}{\alpha+\beta}\right)\int \frac{dx}{1+\left(\left(\frac{\alpha-\beta}{\alpha+\beta}\right)^{1/2}x\right)^2}.$$

We note that in our case $\alpha \pm \beta = 1 + r^2 \mp 2r = (1\mp r)^2 > 0$. We get

$$\int \frac{d\phi}{\alpha+\beta\cos\phi} = \left(\frac{2}{\alpha+\beta}\right)\cdot\left\{\tan^{-1}\left(\left(\frac{\alpha-\beta}{\alpha+\beta}\right)^{1/2}x\right)\right\}\cdot\left(\frac{\alpha+\beta}{\alpha-\beta}\right)^{1/2}$$

$$= \frac{2}{\sqrt{\alpha^2-\beta^2}}\tan^{-1}\left(\left(\frac{\alpha-\beta}{\alpha+\beta}\right)^{1/2}x\right)$$

$$= \left(\frac{2}{1-r^2}\right)\tan^{-1}\left(\frac{1+r}{1-r}\cdot x\right), \quad \alpha^2-\beta^2 - (1-r^2)^2.$$

$$\int \frac{d\phi}{\alpha+\beta\cos\phi} = \left(\frac{2}{1-r^2}\right)\tan^{-1}\left(\frac{1+r}{1-r}\cdot\tan\left(\frac{\phi}{2}\right)\right). \qquad (30.0.3)$$

We obtain by equation (30.0.2),

$$\int_a^b \frac{(1 - r^2)d\psi}{1 - 2r\cos(\theta - \psi) + r^2} = 2\tan^{-1}\left(\frac{1 + r}{1 - r} \cdot \tan\left(\frac{\phi}{2}\right)\right)\Bigg|_{\phi = \theta - b}^{\theta - a}$$

$$= 2\left\{\tan^{-1}\left(\frac{1 + r}{1 - r} \cdot \tan\left(\frac{\theta - a}{2}\right)\right)\right.$$

$$\left. - \tan^{-1}\left(\frac{1 + r}{1 - r} \cdot \tan\left(\frac{\theta - b}{2}\right)\right)\right\}. \quad (30.0.4)$$

So, our final identity is

$$\left(\frac{1}{2}\log \lambda_j\right)\left(\frac{1}{2\pi}\int_{\alpha_j}^{\alpha_{j+1}} \frac{(1 - r^2)d\psi}{1 - 2r\cos(\theta - \psi) + r^2}\right)$$

$$= \left(\frac{1}{2}\log \lambda_j\right)\left\{\tan^{-1}\left(\frac{1 + r}{1 - r} \cdot \tan\left(\frac{\theta - a}{2}\right)\right)\right.$$

$$\left. - \tan^{-1}\left(\frac{1 + r}{1 - r} \cdot \tan\left(\frac{\theta - b}{2}\right)\right)\right\}. \quad (30.0.5)$$

That indeed is what we used in Chapter 26. Now that we have an explicit formula for $s(z)$, we need to construct the analytic function $h(z)$ whose real part we found. In remark 26.0.81, we had the following claim. The harmonic function,

$$w(re^{i\theta}) = \frac{a}{2\pi}\left(\tan^{-1}\left(\frac{1 + r}{1 - r}\tan\left(\frac{\theta}{2}\right)\right) - \tan^{-1}\left(\frac{1 + r}{1 - r}\tan\left(\frac{\theta - \alpha}{2}\right)\right)\right),$$

is the real part of the holomorphic function:

$$\phi(z) = \frac{-a}{2\pi}\tan^{-1}\left(\frac{1 + z}{1 - z}\tan\left(\frac{\alpha}{2}\right)\right).$$

We are not sure that this is true. So, we compute independently.

The analytic function whose real part we found.

We start with equation (30.0.4),

$$2\left\{\tan^{-1}\left(\frac{1 + r}{1 - r} \cdot \tan\left(\frac{\theta - a}{2}\right)\right) - \tan^{-1}\left(\frac{1 + r}{1 - r} \cdot \tan\left(\frac{\theta - b}{2}\right)\right)\right\}$$

$$= \int_a^b \frac{(1 - r^2)d\psi}{1 - 2r\cos(\theta - \psi) + r^2} = \int_a^b \Re\left\{\frac{1 + re^{i(\theta - \psi)}}{1 - re^{i(\theta - \psi)}}\right\}d\psi$$

$$= \Re\left\{\int_a^b \frac{1 + ze^{-i\psi}}{1 - ze^{-i\psi}}d\psi\right\} = \Re\left\{\int_a^b \frac{e^{i\psi} + z}{e^{i\psi} - z}d\psi\right\}, \quad (z = re^{i\theta}).$$

That gives us our analytic function,

$$\int_a^b \frac{e^{i\psi} + z}{e^{i\psi} - z}\, d\psi = \int_{e^{ia}}^{e^{ib}} \frac{w + z}{w - z} \cdot \frac{dw}{iw} = \frac{1}{i} \int_{e^{ia}}^{e^{ib}} \left(\frac{2}{w - z} - \frac{1}{w} \right) dw$$

$$= \frac{2}{i} \log \left(\frac{e^{ib} - z}{e^{ia} - z} \right) + (a - b).$$

We used the following substitution:

$$w = e^{i\psi}, \quad dw = ie^{i\psi}\, d\psi = iw\, d\psi, \quad d\psi = \frac{dw}{iw} \quad \text{and} \quad \frac{w + z}{w - z}\frac{1}{w} = \frac{2}{w - z} - \frac{1}{w}.$$

We conclude the following formula:

$$h(z) = \sum_{j=0}^{m} \left(\frac{\log \lambda_j}{2\pi} \right) \left\{ \frac{1}{i} \log \left(\frac{e^{i\alpha_{j+1}} - z}{e^{i\alpha_j} - z} \right) + \left(\frac{e^{i\alpha_j} - e^{i\alpha_{j+1}}}{2} \right) \right\}$$

$$= \sum_{j=0}^{m} \left(\frac{\log \lambda_j}{2\pi i} \right) \log \left(\frac{e^{i\alpha_{j+1}} - z}{e^{i\alpha_j} - z} \right) - \sum_{j=0}^{m} \left(\frac{\log \lambda_j}{2\pi} \right) \left(\frac{e^{i\alpha_j} - e^{i\alpha_{j+1}}}{2} \right).$$

This changes the second formula in Theorem 27.0.82 to

$$\exp \left\{ g(z) + \sum_{j=0}^{m} \left(\frac{\log \lambda_j}{2\pi} \right) \left(\frac{1}{i} \log \left(\frac{e^{i\alpha_{j+1}} - z}{e^{i\alpha_j} - z} \right) + \left(\frac{e^{i\alpha_j} - e^{i\alpha_{j+1}}}{2} \right) \right) \right\}$$

$$= b_0 + b_1 z + \cdots + b_n z^n + \cdots .$$

The Dirichlet increment, the continuous case.

For a fixed partition $0 = \alpha_0 < \alpha_1 < \cdots < \alpha_m < \alpha_{m+1} = 2\pi$, we found the following formula for the Dirichlet increment of order $(m+1)$st corresponding to the partition,

$$h(z) = \sum_{j=0}^{m} \left(\frac{\log \lambda_j}{2\pi} \right) \left\{ \frac{1}{i} \log \left(\frac{e^{i\alpha_{j+1}} - z}{e^{i\alpha_j} - z} \right) + \left(\frac{e^{i\alpha_j} - e^{i\alpha_{j+1}}}{2} \right) \right\}.$$

We now take a sequence of refinements of this partition *ad infinity* so that the mesh size shrinks to zero. We have a continuous positive $\lambda(z)$ that corresponds to the sequences of positive weights λ_j, $j = 0, 1, \ldots, m$. We denote $\Delta \alpha_j = \alpha_{j+1} - \alpha_j$, $e^{i\alpha_{j+1}} - e^{i\alpha_j} = ie^{i\tilde{\alpha}_j} \Delta \alpha_j$ for a point $\tilde{\alpha}_j$ intermediate on the segment between α_j and α_{j+1}. If we write

$$\frac{e^{i\alpha_{j+1}} - z}{e^{i\alpha_j} - z} = 1 + X_j,$$

then we get

$$X_j = \frac{e^{i\alpha_{j+1}} - e^{i\alpha_j}}{e^{i\alpha_j} - z} = \frac{ie^{i\tilde{\alpha}_j}}{e^{i\alpha_j} - z}\Delta\alpha_j \to 0.$$

So, when the mesh size is small enough and we use first-order approximation, we get

$$\frac{1}{i}\log\left(\frac{e^{i\alpha_{j+1}} - z}{e^{i\alpha_j} - z}\right) = \frac{1}{i}\log\left(1 + X_j\right) \approx \frac{1}{i}X_j = \left(\frac{e^{i\tilde{\alpha}_j}}{e^{i\alpha_j} - z}\right)\Delta\alpha_j.$$

Hence,

$$\frac{1}{i}\log\left(\frac{e^{i\alpha_{j+1}} - z}{e^{i\alpha_j} - z}\right) + \left(\frac{e^{i\alpha_j} - e^{i\alpha_{j+1}}}{2}\right) \approx \left(\frac{e^{i\tilde{\alpha}_j}}{e^{i\alpha_j} - z} - \frac{1}{2}\right)\Delta\alpha_j$$

$$\approx \left(\frac{e^{i\alpha_j} + z}{e^{i\alpha_j} - z}\right)\cdot\frac{\Delta\alpha_j}{2}.$$

Thus, $h(z)$ is approximated by the following Riemann sum:

$$\frac{1}{2\pi}\sum_{j=0}^{m}\log\lambda(re^{i\alpha_j})\cdot\left(\frac{re^{i\alpha_j} + z}{re^{i\alpha_j} - z}\right)\cdot\frac{\Delta\alpha_j}{2},$$

that converges when the refinements go *ad infinity* with mesh size shrinking to zero, to

$$\frac{1}{2\pi}\int_0^{2\pi}\log\lambda(re^{i\phi})\cdot\left(\frac{re^{i\phi} + z}{re^{i\phi} - z}\right)\cdot\frac{d\phi}{2}.$$

That is exactly the logarithm of our outer function but now the sign is flipped as we expected, and as we needed in our theorems of the previous chapter. Thus, Theorem 29.0.85 is correct as is. In fact, the positive function $\lambda(e^{it})$ may be less regular than the assumptions we made there.

Chapter 31

Lagrange Multipliers, a Deeper Use

We suggest to use the method on the parameters $\alpha_1, \ldots, \alpha_m$ as well as on $\lambda_0, \ldots, \lambda_m$. Let us fix an admissible function $g(z)$ for $A_n(2)$. This means that $\exp(g(z)) \in H^2(U)$ and that $\|\exp(g)\|_{H^2(U)} = 1$. We consider the family of partitions, $0 = \alpha_0 < \alpha_1 < \cdots < \alpha_m < \alpha_{m+1} = 2\pi$. The first condition to be fulfilled is

$$\frac{1}{2\pi} \sum_{j=0}^{m} \int_{\alpha_j}^{\alpha_{j+1}} |g(e^{i\theta})|^2 d\theta = 1. \tag{31.0.1}$$

The weights $\lambda_0, \lambda_1, \ldots, \lambda_m > 0$ give us a second condition to be fulfilled:

$$\frac{1}{2\pi} \sum_{j=0}^{m} \lambda_j \int_{\alpha_j}^{\alpha_{j+1}} |g(e^{i\theta})|^2 d\theta = 1. \tag{31.0.2}$$

We want to maximize the following function:

$$\frac{1}{2\pi r_0^n} \Re \left\{ \int_0^{2\pi} \exp(g(r_0 e^{i\theta}) + h(r_0 e^{i\theta}) - in\theta) d\theta \right\}, \tag{31.0.3}$$

where

$$h(z) = \sum_{j=0}^{m} \left(\frac{\log \lambda_j}{2\pi} \right) \left\{ \frac{1}{i} \log \left(\frac{e^{i\alpha_{j+1}} - z}{e^{i\alpha_j} - z} \right) + \left(\frac{e^{i\alpha_j} - e^{i\alpha_{j+1}}}{2} \right) \right\}.$$

181

This time, we need two Lagrange multipliers. The first, x, corresponds to condition (31.0.1) and the second one, y, corresponds to condition (31.0.2). The Lagrange auxiliary function is

$$F(\alpha_1, \ldots, \alpha_m; \lambda_0, \ldots, \lambda_m; x; y)$$

$$= \frac{1}{2\pi r_0^n} \Re \left\{ \int_0^{2\pi} \exp(g(r_0 e^{i\theta}) + h(r_0 e^{i\theta}) - in\theta) d\theta \right\}$$

$$+ x \left(\frac{1}{2\pi} \sum_{j=0}^{m} \int_{\alpha_j}^{\alpha_{j+1}} |g(e^{i\theta})|^2 d\theta - 1 \right)$$

$$+ y \left(\frac{1}{2\pi} \sum_{j=0}^{m} \lambda_j \int_{\alpha_j}^{\alpha_{j+1}} |g(e^{i\theta})|^2 d\theta - 1 \right). \tag{31.0.4}$$

We use the following notations:

$$\psi = \psi(r_0, \theta) = g(r_0 e^{i\theta}) - in\theta,$$

$$\phi_j = \phi_j(r_0, \theta, \alpha_j, \alpha_{j+1}) = \frac{1}{2\pi} \left\{ \frac{1}{i} \log \left(\frac{e^{i\alpha_{j+1}} - z}{e^{i\alpha_j} - z} \right) + \left(\frac{e^{i\alpha_j} - e^{i\alpha_{j+1}}}{2} \right) \right\},$$

$j = 0, \ldots, m$. These functions are independent of the weights $\lambda_0, \ldots, \lambda_m$ and of the Lagrange multipliers x and y. We can write

$$F(\alpha_1, \ldots, \alpha_m; \lambda_0, \ldots, \lambda_m; x; y)$$

$$= \frac{1}{2\pi r_0^n} \int_0^{2\pi} \Re \left\{ \exp \left(\sum_{j=0}^{m} \phi_j \cdot (\log \lambda_j) + \psi \right) \right\} d\theta$$

$$+ x \left(\frac{1}{2\pi} \sum_{j=0}^{m} \int_{\alpha_j}^{\alpha_{j+1}} |g(e^{i\theta})|^2 d\theta - 1 \right)$$

$$+ y \left(\frac{1}{2\pi} \sum_{j=0}^{m} \lambda_j \int_{\alpha_j}^{\alpha_{j+1}} |g(e^{i\theta})|^2 d\theta - 1 \right).$$

So,

$$F(\alpha_1, \ldots, \alpha_m; \lambda_0, \ldots, \lambda_m; x; y)$$

$$= \frac{1}{2\pi r_0^n} \int_0^{2\pi} \exp\left(\sum_{j=0}^{m} (\log \lambda_j) \cdot \Re\{\phi_j\} + \Re\{\psi\}\right)$$

$$\times \cos\left(\sum_{j=0}^{m} (\log \lambda_j) \cdot \Im\{\phi_j\} + \Im\{\psi\}\right) d\theta$$

$$+ x \left(\frac{1}{2\pi} \sum_{j=0}^{m} \int_{\alpha_j}^{\alpha_{j+1}} \left|g(e^{i\theta})\right|^2 d\theta - 1\right)$$

$$+ y \left(\frac{1}{2\pi} \sum_{j=0}^{m} \lambda_j \int_{\alpha_j}^{\alpha_{j+1}} \left|g(e^{i\theta})\right|^2 d\theta - 1\right). \qquad (31.0.5)$$

$$\frac{\partial F}{\partial \alpha_k} = \frac{1}{2\pi r_0^n} \int_0^{2\pi} \left\{ \frac{\partial}{\partial \alpha_k}\left(\sum_{j=0}^{m}(\log \lambda_j) \cdot \Re\{\phi_j\} + \Re\{\psi\}\right) \cdot \exp(\ldots) \cdot \cos(\ldots)\right.$$

$$\left. + \frac{\partial}{\partial \alpha_k}\left(\sum_{j=0}^{m}(\log \lambda_j) \cdot \Im\{\phi_j\} + \Im\{\psi\}\right) \cdot \exp(\ldots) \cdot (-\sin(\ldots))\right\} d\theta$$

$$+ x \cdot \frac{1}{2\pi}(|g(e^{i\alpha_k})|^2 - |g(e^{i\alpha_k})|^2)$$

$$+ y \cdot \frac{1}{2\pi}(\lambda_{k-1}|g(e^{i\alpha_k})|^2 - \lambda_k|g(e^{i\alpha_k})|^2)$$

$$= \frac{1}{2\pi r_0^n} \int_0^{2\pi} \left\{ \left((\log \lambda_k) \cdot \frac{\partial}{\partial \alpha_k}\Re\{\phi_k\} + (\log \lambda_{k-1})\frac{\partial}{\partial \alpha_k}\Re\{\phi_{k-1}\}\right)\right.$$

$$\times \exp(\ldots) \cdot \cos(\ldots)$$

$$+ \left((\log \lambda_k) \cdot \frac{\partial}{\partial \alpha_k}\Im\{\phi_k\} + +(\log \lambda_{k-1})\frac{\partial}{\partial \alpha_k}\Im\{\phi_{k-1}\}\right)$$

$$\left. \times \exp(\ldots) \cdot (-\sin(\ldots))\right\} d\theta$$

$$+ \frac{y}{2\pi}|g(e^{i\alpha_k})|^2 \cdot (\lambda_{k-1} - \lambda_k).$$

By our definition,

$$\phi_k = \frac{1}{2\pi}\left\{\frac{1}{i}\log\left(\frac{e^{i\alpha_{k+1}}-z}{e^{i\alpha_k}-z}\right)+\left(\frac{e^{i\alpha_k}-e^{i\alpha_{k+1}}}{2}\right)\right\}$$

$$=\frac{1}{2\pi}\left\{\frac{1}{i}\log\left(e^{i\alpha_{k+1}}-r_0e^{i\theta}\right)+\left(\frac{r_0e^{i\theta}-e^{i\alpha_{k+1}}}{2}\right)\right.$$

$$\left.-\frac{1}{i}\log\left(e^{i\alpha_k}-r_0e^{i\theta}\right)+\left(\frac{e^{i\alpha_k}-r_0e^{i\theta}}{2}\right)\right\}.$$

We have $e^{i\alpha}-r_0e^{i\theta}=(\cos\alpha-r_0\cos\theta)+i(\sin\alpha-r_0\sin\theta)$. Since $\log z = \log|z|+i\arg\{z\}$, $z=c+id$, $|z|=\sqrt{c^2+d^2}$, $\arg\{z\}=\tan^{-1}\left(\frac{d}{c}\right)$ we need

$$|e^{i\alpha}-r_0e^{i\theta}|^2=(\cos\alpha-r_0\cos\theta)^2+(\sin\alpha-r_0\sin\theta)^2$$

$$=1+r_0^2-2r_0\cos(\alpha-\theta),$$

$$\arg(e^{i\alpha}-r_0e^{i\theta})=\tan^{-1}\left(\frac{\sin\alpha-r_0\sin\theta}{\cos\alpha-r_0\cos\theta}\right).$$

So,

$$\phi_k=\frac{1}{2\pi}\left\{\frac{1}{i}\cdot\frac{1}{2}\log(1+r_0^2-2r_0\cos(\alpha_{k+1}-\theta))\right.$$

$$+\frac{1}{i}\cdot i\cdot\tan^{-1}\left(\frac{\sin\alpha_{k+1}-r_0\sin\theta}{\cos\alpha_{k+1}-r_0\cos\theta}\right)-\frac{1}{2}(\cos\alpha_{k+1}-r_0\cos\theta)$$

$$-\frac{i}{2}(\sin\alpha_{k+1}-r_0\sin\theta)-\frac{1}{i}\cdot\frac{1}{2}\cdot\log(1+r_0^2-2r_0\cos(\alpha_k-\theta))$$

$$-\frac{1}{i}\cdot i\tan^{-1}\left(\frac{\sin\alpha_k-r_0\sin\theta}{\cos\alpha_k-r_0\cos\theta}\right)+\frac{1}{2}(\cos\alpha_k-r_0\cos\theta)$$

$$\left.+\frac{i}{2}(\sin\alpha_k-r_0\sin\theta)\right\}.$$

$$\Re\{\phi_k\}=\frac{1}{2\pi}\left\{\tan^{-1}\left(\frac{\sin\alpha_{k+1}-r_0\sin\theta}{\cos\alpha_{k+1}-r_0\cos\theta}\right)-\frac{1}{2}(\cos\alpha_{k+1}-r_0\cos\theta)\right.$$

$$\left.-\tan^{-1}\left(\frac{\sin\alpha_k-r_0\sin\theta}{\cos\alpha_k-r_0\cos\theta}\right)+\frac{1}{2}(\cos\alpha_k-r_0\cos\theta)\right\},$$

$$\Im\{\phi_k\} = \frac{1}{2\pi} \left\{ -\frac{1}{2} \log\left(1 + r_0^2 - 2r_0\cos(\alpha_{k+1} - \theta)\right) \right.$$

$$-\frac{1}{2}(\sin\alpha_{k+1} - r_0\sin\theta)$$

$$\left. +\frac{1}{2}\log\left(1 + r_0^2 - 2r_0\cos(\alpha_k - \theta)\right) + \frac{1}{2}(\sin\alpha_k - r_0\sin\theta) \right\},$$

$$\frac{\partial}{\partial\alpha_k}\Re\{\phi_k\} = -\frac{1}{2\pi}\left\{ \frac{1 - r_0\cos(\alpha_k - \theta)}{1 + r_0^2 - 2r_0\cos(\alpha_k - \theta)} + \frac{1}{2}\sin\alpha_k \right\},$$

$$\frac{\partial}{\partial\alpha_k}\Im\{\phi_k\} = \frac{1}{2\pi}\left\{ \frac{r_0\sin(\alpha_k - \theta)}{1 + r_0^2 - 2r_0\cos(\alpha_k - \theta)} + \frac{1}{2}\cos\alpha_k \right\}.$$

Also,

$$\frac{\partial}{\partial\alpha_k}\Re\{\phi_k\} = -\frac{\partial}{\partial\alpha_k}\Re\{\phi_{k-1}\}, \qquad \frac{\partial}{\partial\alpha_k}\Im\{\phi_k\} = -\frac{\partial}{\partial\alpha_k}\Im\{\phi_{k-1}\}.$$

So,

$$\frac{\partial F}{\partial\alpha_k} = \frac{1}{2\pi r_0^n}\int_0^{2\pi}\left\{ \frac{\partial}{\partial\alpha_k}\Re\{\phi_{k-1}\}\cdot(\log\lambda_{k-1} - \log\lambda_k)\cdot\exp(\ldots)\cdot\cos(\ldots) \right.$$

$$+\frac{\partial}{\partial\alpha_k}\Im\{\phi_k\}\cdot(\log\lambda_k - \log\lambda_{k-1})\cdot\exp(\ldots)$$

$$\left. \cdot(-\sin(\ldots)) \right\}d\theta$$

$$+\frac{y}{2\pi}|g(e^{i\alpha_k})|^2(\lambda_{k-1} - \lambda_k).$$

Finally, we get the first group of the Lagrange equation,

$$\frac{\partial F}{\partial\alpha_k} = \frac{1}{2\pi r_0^n}\int_0^{2\pi}(\log\lambda_{k-1} - \log\lambda_k)\cdot\exp(\ldots)$$

$$\times\left\{ \frac{1}{2\pi}\left(\frac{1 - r_0\cos(\alpha_k - \theta)}{1 + r_0^2 - 2r_0\cos(\alpha_k - \theta)} + \frac{1}{2}\sin\alpha_k \right) \right.$$

$$\times\frac{1}{2\pi}\left(\frac{r_0\sin(\alpha_k - \theta)}{1 + r_0^2 - 2r_0\cos(\alpha_k - \theta)} + \frac{1}{2}\cos\alpha_k \right)$$

$$\left. \times\sin\left(\sum_{j=0}^{m}(\log\lambda_j)\Im\{\phi_j\} + \Im\{\psi\} \right) \right\}d\theta$$

$$+\frac{y}{2\pi}|g(e^{i\alpha_k})|^2(\lambda_{k-1} - \lambda_k) = 0, \quad k = 0, 1, \ldots, m. \qquad (31.0.6)$$

The second group of the Lagrange equations is easier to derive as shown:

$$\frac{\partial F}{\partial \lambda_k} = \frac{1}{2\pi r_0^n} \int_0^{2\pi} \left\{ \frac{1}{\lambda_k} \Re\{\phi_k\} \exp(\ldots) \cos(\ldots) \right.$$

$$\left. + \frac{1}{\lambda_k} \Im\{\phi_k\} \exp(\ldots)(-\sin(\ldots)) \right\} d\theta$$

$$+ \frac{y}{2\pi} \int_{\alpha_k}^{\alpha_{k+1}} \left| g(e^{i\theta}) \right|^2 d\theta = 0.$$

$$\frac{\partial F}{\partial \lambda_k} = \frac{1}{2\pi r_0^n} \int_0^{2\pi} \frac{1}{\lambda_k} \cdot \exp(\ldots)$$

$$\times \left\{ \frac{1}{2\pi} \left(\tan^{-1} \left(\frac{\sin\alpha_{k+1} - r_0\sin\theta}{\cos\alpha_{k+1} - r_0\cos\theta} \right) - \frac{1}{2} \left(\cos\alpha_{k+1} - r_0\cos\theta \right) \right. \right.$$

$$\left. - \tan^{-1} \left(\frac{\sin\alpha_k - r_0\sin\theta}{\cos\alpha_k - r_0\cos\theta} \right) + \frac{1}{2} \left(\cos\alpha_k - r_0\cos\theta \right) \right) \cos(\ldots)$$

$$- \frac{1}{2\pi} \cdot \frac{1}{2} \left(-\log(1 + r_0^2 - 2r_0\cos(\alpha_{k+1} - \theta)) - (\sin\alpha_{k+1} - r_0\sin\theta) \right.$$

$$\left. + \log(1 + r_0^2 - 2r_0\cos(\alpha_k - \theta)) + (\sin\alpha_k - r_0\sin\theta)) \sin(\ldots) \right\} d\theta$$

$$+ \frac{y}{2\pi} \int_{\alpha_k}^{\alpha_{k+1}} \left| g(e^{i\theta}) \right|^2 d\theta = 0, \quad k = 0, 1, \ldots, m. \tag{31.0.7}$$

Chapter 32

Dirichlet Increments for p-norms

Let us fix a partition of $[0, 2\pi)$, $0 = \alpha_0 < \alpha_1 < \cdots < \alpha_m < \alpha_{m+1} = 2\pi$. Let us choose weights $\lambda_0, \lambda_1, \ldots, \lambda_m > 0$ such that $\sum_{j=0}^{m} \lambda_j(\alpha_{j+1} - \alpha_j) = 2\pi$. We want to construct a function $s(z)$, $|z| < 1$, harmonic in U that has the following boundary values:

$$\lim_{r \to 1^-} \exp(ps(re^{i\theta})) = \exp(ps(e^{i\theta})) = \lambda_j, \quad \alpha_j < \theta < \alpha_{j+1}, \quad j = 0, 1, \ldots, m.$$

Such a function satisfies

$$\lim_{r \to 1^-} \frac{1}{2\pi} \int_0^{2\pi} \exp(s(re^{i\theta}))^p d\theta = \lim_{r \to 1^-} \frac{1}{2\pi} \int_0^{2\pi} \exp(ps(re^{i\theta})) d\theta$$

$$= \lim_{r \to 1^-} \frac{1}{2\pi} \sum_{j=0}^{m} \int_{\alpha_j}^{\alpha_{j+1}} \exp(ps(re^{i\theta})) d\theta = \frac{1}{2\pi} \sum_{j=0}^{m} \lambda_j(\alpha_{j+1} - \alpha_j) = 1.$$

The solution is (see the equation prior to Remark 26.0.81),

$$s(re^{i\theta}) = \sum_{j=0}^{m} \left(\frac{1}{2\pi} \log \lambda_j \right) \left(\tan^{-1} \left(\frac{1+r}{1-r} \tan \left(\frac{\theta - \alpha_j}{2} \right) \right) \right.$$

$$\left. - \tan^{-1} \left(\frac{1+r}{1-r} \tan \left(\frac{\theta - \alpha_{j+1}}{2} \right) \right) \right).$$

$s(z)$ is the real part of the holomorphic function $h(z)$ in U, where

$$h(z) = \sum_{j=0}^{m} \left(\frac{\log \lambda_j}{p\pi} \right) \left\{ \frac{1}{i} \log \left(\frac{e^{i\alpha_{j+1}} - z}{e^{i\alpha_j} - z} \right) + \left(\frac{e^{i\alpha_j} - e^{i\alpha_{j+1}}}{2} \right) \right\}.$$

$h(z)$ satisfies

$$\| \exp(h) \|_{H^p(U)} = \lim_{r \to 1^-} \left(\frac{1}{2\pi} \int_0^{2\pi} | \exp(h(re^{i\theta})) |^p d\theta \right)^{1/p}$$

$$= \lim_{r \to 1^-} \left(\frac{1}{2\pi} \int_0^{2\pi} | \exp(ps(re^{i\theta})) |^p d\theta \right)^{1/p} = 1.$$

Chapter 33

The Nevanlinna Class
or the Class of Functions
of a Bounded Characteristic

This chapter is a brief summary of a well-known subject. That important class of holomorphic functions enters the list of topics that are related to the Krzyż conjecture in a natural way. The functions of this class are characterized by a growth condition, which is controlled by the Nevanlinna characteristic function. An important theorem implies that this growth condition is equivalent to a representation of the members of the Nevanlinna class as quotients of bounded analytic functions. Let us recall that $B = \{f \in H(U) \mid \forall |z| < 1, 0 < |f(z)| \leq 1\}$. The Krzyż conjecture is merely a conjecture on the solution of the coefficients problem in B. Let $\langle B \rangle$ be the group of functions generated by B with multiplication of functions as the binary operation. Then one easily observes that $\langle B \rangle = \{\frac{f_1}{f_2} \mid f_1, f_2 \in B\}$. In other words, the Abelian group $\langle B \rangle$ is precisely the subclass of the Nevanlinna class which is composed of those functions that never vanish at any of the points of U. This connection motivates a study of the Nevanlinna class related to the Krzyż conjecture. There are many sources and we will start with [22], and will mostly use Chapter 2 (pp. 15–31). The Nevanlinna class is denoted by N. It is also known as the class of functions of bounded characteristic. A function $f \in H(U)$ is said to be of class N if the integrals $\int_0^{2\pi} \log^+ |f(re^{i\theta})| d\theta$ are uniformly bounded for $r < 1$. Here, $\log^+ x = \max\{\log x, 0\}$ for $x \geq 0$. Since $\log^+ |f|$ is sub-harmonic whenever f is analytic, these integrals always increase with r. We have $H^p \subseteq N$ for every $p > 0$. For $\log^+ x < x$, $x \geq 0$ and hence $\log^+ |f(re^{i\theta})|^p < |f(re^{i\theta})|^p$. So, if $\int_0^{2\pi} |f(re^{i\theta})|^p d\theta$ is bounded for $0 \leq r < 1$, then so is $\int_0^{2\pi} \log^+ |f(re^{i\theta})|^p d\theta = p \int_0^{2\pi} \log^+ |f(re^{i\theta})| d\theta$. Since

$p > 0$, this implies that $\int_0^{2\pi} \log^+ |f(re^{i\theta})| d\theta$ are uniformly bounded for $r < 1$. Thus, indeed, $H^p \subseteq N$ for any $p > 0$.

Theorem 2.1 (F. and R. Nevanlinna) [22, p. 16]. *A function in $H(U)$ belongs to the class N if and only if it is a quotient of two bounded analytic functions.*

Theorem 2.2 [22, p. 17]. *For each function $f \in N$, the non-tangential limit $f(e^{i\theta})$ exists almost everywhere, and $\log|f(e^{i\theta})|$ is integrable unless $f(z) \equiv 0$. If $f \in H^p$ for some $p > 0$, then $f(e^{i\theta}) \in L^p$.*

So, if $f \in N$ and if $f(e^{i\theta}) = 0$ on a set of positive measure, then $f(z) \equiv 0$. In other words, a function of class N is uniquely determined by its boundary values on any set of a positive measure. It is evident from the representation $f = \phi/\psi$ (ϕ, ψ are bounded and $\phi, \psi \in H(U)$) that $\int \log^- |f(re^{i\theta})| d\theta$ is bounded if $f \in N$ ($\log^- x = \max\{-\log x, 0\}$). Hence, $f \in N$ if and only if $\int_0^{2\pi} |\log|f(re^{i\theta})|| d\theta$ are bounded for $0 \leq r < 1$.

Theorem 2.3 [22, p. 18]. *Let $f(z) \not\equiv 0$ be analytic in $|z| < 1$, and let $z_1, z_2, \ldots$ be its zeros, repeated according to multiplicity in $|z| < 1$. Then $\int_0^{2\pi} \log|f(re^{i\theta})| d\theta$ is bounded if and only if $\sum_{n=1}^\infty (1 - |z_n|) < \infty$.*

Corollary [22, p. 18]. *The zeros $\{z_n\}$ of a function $\not\equiv 0$ in N must satisfy $\sum_{n=1}^\infty (1 - |z_n|) < \infty$.*

Theorem 2.5 (F. Riesz) [22, p. 20]. *Each $f \in N$ has a factorization $f = Bg$ where $B(z)$ is a Blaschke product and g is a non-vanishing function of class N.*

An outer function for class H^p is a function of the form

$$F(z) = e^{i\gamma} \exp\left\{ \frac{1}{2\pi} \int_0^{2\pi} \frac{e^{it} + z}{e^{it} - z} \log \psi(t) dt \right\},$$

where $\gamma \in \mathbb{R}$, $\psi(t) \geq 0$, $\log \psi(t) \in L^1$, and $\psi(t) \in L^p$. An outer function for class N is a function $F(z)$ as above, where $\psi(t) \geq 0$ and $\log \psi(t) \in L^1$ (the condition $\psi \in L^p$ has been dropped). An inner function is any function $f \in H(U)$, having the properties $|f(z)| \leq 1$ and $|f(e^{i\theta})| = 1$ a.e. Every inner function has a factorization $e^{i\gamma} B(z)s(z)$, where $B(z)$ is a Blaschke product and $s(z)$ is a function of the form

$$s(z) = \exp\left\{ -\int_0^{2\pi} \frac{e^{it} + z}{e^{it} - z} d\mu(t) \right\},$$

$\mu(t)$ being a bounded non-decreasing singular function ($\mu'(t) = 0$ a.e.). Such a function $s(z)$ is called a singular inner function.

Theorem 2.8 (The canonical factorization theorem) [22, p. 24]. *Every function $f(z) \not\equiv 0$ of class $H^p, (p > 0)$ has a unique factorization of the form $f(z) = B(z)s(z)F(z)$, where $B(z)$ is a Blaschke product, $s(z)$ is a singular inner function, and $F(z)$ is an outer function for the class H^p (with $\psi(t) = |f(e^{it})|$). Conversely, every such product belongs to H^p.*

Theorem 2.9 [22, p. 25]. *Every function $f(z) \not\equiv 0$ of class N can be expressed in the form*

$$f(z) = B(z) \left(\frac{s_1(z)}{s_2(z)} \right) \cdot F(z), \tag{33.0.1}$$

where $B(z)$ is a Blaschke product, $s_1(z)$ and $s_2(z)$ are singular inner functions, and $F(z)$ is an outer function for the class N (with $\psi(t) = |f(e^{it})|$). Conversely, every such function $f(z)$ of the form in equation (33.0.1) belongs to N.

The two preceding theorems point out the sharp structural difference between functions in the classes H^p and N. In factoring functions of class N, it is necessary not only to enlarge the class of admissible outer functions, but also to replace the singular factor by a quotient of two singular inner functions. This allows, for instance, the "pathological" example $\exp\left\{ \frac{1+z}{1-z} \right\}$, which we now recognize as the reciprocal of a singular inner function. It is useful to distinguish the class N^+ of all functions in N for which $s_2(z) \equiv 1$. That is, $f \in N^+$ if it has the form $f = B \cdot s \cdot F$, where B is a Blaschke product, s is a singular inner function, and F is an outer function for the class N. In a sense, N^+ is the natural limit of H^p as $p \to 0^+$. We have the proper inclusions $H^p \subset N^+ \subset N$.

Theorem 2.10 [22, p. 26]. *A function $f \in N$ belongs to the class N^+ if and only if*

$$\lim_{r \to 1^-} \int_0^{2\pi} \log^+ |f(re^{i\theta})|d\theta = \int_0^{2\pi} \log^+ |f(e^{i\theta})|d\theta.$$

Theorem 2.11 [22, p. 28]. *If $f \in N^+$ and $f(e^{i\theta}) \in L^p$ for some $p > 0$, then $f \in H^p$.*

The *a priori* assumption $f \in N^+$ cannot be relaxed. The reciprocal of any (non-trivial) singular inner function is bounded on $|z| = 1$, but is not of class N^+.

Chapter 34

On the Critical Points of Functions in the Unit Ball of $H^\infty(U)$

Let $f \in H^\infty(U) - \mathbb{C}$. A critical point of $f(z)$ is a zero of $f'(z)$ in U. Let us assume that $\|f\|_{H^\infty}(U) \le 1$ and let $\{z_j\}_j$ be the zeros of f within U. In general, if a zero z_j appears in this sequence m times ($m \ge 1$), then f has a zero at z_j of precise order (or multiplicity) m. Let $\{w_j\}_j$ be that sub-sequence of $\{z_j\}_j$ that includes each z_j exactly once. The Blaschke product with the zero set $\{w_j\}_j$ is called the reduced Blaschke product of $f(z)$. We will denote this Blaschke product by $B_f(z)$. Thus, by definition it follows that

Proposition 34.0.92. (i) *We have the following representation:*

$$B_f(z) = z^\epsilon \prod_j \left(\frac{|w_j|}{w_j} \frac{w_j - z}{1 - \overline{w}_j z} \right)$$

where $\epsilon = 1$ if $f(0) = 0$ and $\epsilon = 0$ otherwise, in which case we interpret z^0 as the constant 1.

(ii) Let $g(z) = f(z)/B_f(z)$. Then $g \in H^\infty(U)$, $\|g\|_{H^\infty(U)} = \|f\|_{H^\infty(U)}$ and the zero set of g, $Z(g)$ in U is precisely the set of all those zeros of $f(z)$ of multiplicity $m \ge 2$.

(iii) $Z(g)$ is a subset of the set of the critical points of f in U.

Corollary 34.0.93. *Let $f \in H^\infty(U) - \mathbb{C}$, $\|f\|_{H^\infty(U)} \le 1$. For each $a \in f(U)$, let us denote*

$$f_a(z) = \left(\frac{|a|}{a} \right) \left(\frac{a - f(z)}{1 - \overline{a}f(z)} \right) \bigg/ B_{\left(\frac{f(z') - a}{1 - \overline{a}f(z')} \right)}(z).$$

193

Then the critical set (i.e., the set of all the critical points) of f is precisely $\bigcup_{a \in f(U)} Z(f_a)$. In fact, this union contains only a countable number of non-empty zero sets $Z(f_a)$. We have $Z(f_a) \neq \emptyset$ if and only if $\exists\, b \in f^{-1}(a)$ such that $f'(b) = 0$. In other words, $Z(f_a) \neq \emptyset$ if and only if the fiber of f at a (i.e., $f^{-1}(a)$) contains a zero of $f'(z)$. So, we can modify the union representation of $Z(f')$ (the critical set of f in U) and write the following:

$$Z(f') = \bigcup_{\{a \,|\, Z(f') \cap f^{-1}(a) \neq \emptyset\}} Z(f_a).$$

Let $\{a_n\}_n$ be the set of all the numbers in $f(U)$ for which $Z(f_{a_n}) \neq \emptyset$. This means that $Z(f') = \bigcup_n Z(f_{a_n})$ and in fact this is a partition of $Z(f')$. Let us consider the following product:

$$\prod_n f_{a_n}(z) = f(z)^\epsilon \prod_n \left(\frac{|a_n|}{a_n} \right) \left(\frac{a_n - f(z)}{1 - \overline{a_n} f(z)} \right) \Big/ \prod_n B_{\left(\frac{f(z') - a_n}{1 - \overline{a_n} f(z')} \right)}(z).$$

The product in the numerator is the composition $B \circ f$, where

$$B(z) = z^\epsilon \prod_n \left(\frac{|a_n|}{a_n} \right) \left(\frac{a_n - z}{1 - \overline{a_n} z} \right)$$

is a Blaschke product provided that $\{a_n\}_n$ is a Blaschke sequence, i.e., $\sum_n (1 - |a_n|) < \infty$. The product in the denominator is also a Blaschke product provided that the union of the sets

$$\bigcup_n Z\left(B_{\left(\frac{f - a_n}{1 - \overline{a_n} f} \right)} \right)$$

forms a Blaschke sequence. This sequence is the sequence of zeros of the $\frac{f - a_n}{1 - \overline{a_n} f}$ for all n, each zero appears exactly once. We note that if $\prod_n f_{a_n}(z)$ converges uniformly on compact subsets of U, then in fact $\prod_n f_{a_n} \in H^\infty(U)$, $\| \prod_n f_{a_n} \|_{H^\infty(U)} \leq 1$. In this case, $Z(f')$ is a Blaschke sequence. So, the interesting case is the complimentary case. In general, we need to use Weierstrass products as in [102].

A Survey on Inner Functions

The Krzyż conjecture was reduced in [49] to a problem on the coefficients of singular inner functions in U whose measure is supported on finitely many atoms. We recall that proving $|a_n| \le \frac{2}{e}$, $n \ge 2$ with equality only for rotations of $\exp\left(-\frac{1+z^n}{1-z^n}\right)$ was reduced in [49] to finding the maximum of the absolute value of the coefficient of z^n in the family of functions of the following form:

$$\exp\left(-\sum_{j=1}^{m} \lambda_j \frac{1+k_j z}{1-k_j z}\right), \quad m \le n, \quad \lambda_j > 0, \ |k_j| = 1, \ 1 \le j \le m.$$

Hence, the relevance of studying inner functions and in particular singular inner functions to the Krzyż conjecture. We will closely follow the first two chapters of [71]. Infinite Blaschke products were introduced by Blaschke in 1915. In 1929, Rolf Nevanlinna introduced the class of bounded analytic functions with almost everywhere unimodular boundary values. However, the term inner function was coined by Beurling in his seminal work on the invariant subspaces of the shift operator, [5]. The first extensive studies of the properties of inner functions were made by Frostman (1935), Siedel (1934), and Riesz (1923). Their efforts to understand the zeros and boundary behavior of bounded analytic functions led to the celebrated canonical factorization theorem. The special factorization for inner functions says that any inner function is the product of a Blaschke product and a zero free inner function, the so-called singular inner function, which is generated by a singular measure residing on the unit circle. Roughly speaking, we can say that the Blaschke product is formed with the zeros of an inner function inside the open unit disk, and the singular part stems from its zeros on the boundary.

Chapter 35

The Poisson Integral
of a Measure [71, p. 1]

Let μ be a complex Borel measure on the unit circle $\mathbb{T}$. Then the Poisson integral of μ on the open unit disk U is defined by the formula

$$P_\mu(z) = \int_\mathbb{T} \frac{1 - |z|^2}{|z - \xi|^2} d\mu(\xi), \quad (z \in U).$$

If $d\mu(e^{i\theta}) = u(e^{i\theta})\frac{d\theta}{2\pi}$, where $u \in L^1(\mathbb{T})$, instead of P_μ we write P_u. It can be verified that $h = P_\mu$ is an harmonic function on U. Using Fubini's theorem and the identity

$$\frac{1}{2\pi} \int_0^{2\pi} \frac{1 - |z|^2}{|z - e^{i\theta}|^2} d\theta = 1, \quad (z \in U), \tag{35.0.1}$$

we see that

$$\frac{1}{2\pi} \int_0^{2\pi} |h(re^{i\theta})| d\theta \leq \int_\mathbb{T} \left(\frac{1}{2\pi} \int_0^{2\pi} \frac{(1 - r^2)d\theta}{|re^{i\theta} - \xi|^2} \right) d|\mu|(\xi) = \|\mu\|,$$

where $\|\mu\|$ is the total variation of the measure μ on $\mathbb{T}$. Hence, h fulfills the growth restriction

$$\sup_{0 \leq r < 1} \int_0^{2\pi} |h(re^{i\theta})| d\theta < \infty. \tag{35.0.2}$$

Hence, the Poisson integral of a Borel measure on $\mathbb{T}$ is an harmonic function on U which satisfies equation (35.0.2). The converse of this assertion is also true and we have the following complete characterization.

Theorem 1.1 (Plessner). *Let h be a function defined on U. Then the following assertions are equivalent.*

(i) *h is an harmonic function on U which satisfies the condition (35.0.2).*

(ii) *There exists a (unique) Borel measure μ on $\mathbb{T}$ such that $h = P_\mu$.*

As a special case, if μ is positive, then $h = P_\mu$ is a positive harmonic function on U which satisfies (35.0.2), and if h is a given positive harmonic function, then, by the mean value property, it satisfies

$$\int_0^{2\pi} |h(re^{i\theta})|d\theta = \int_0^{2\pi} h(re^{i\theta})d\theta = 2\pi \cdot h(0), \quad (0 \le r < 1).$$

Therefore, in this case, Theorem 1.1 can be rewritten as follows.

Corollary 1.2 (Herglotz Theorem). *Let h be a function defined on U. Then the following assertions are equivalent.*

(i) *h is a positive harmonic function on U.*

(ii) *There exists a (unique) positive Borel measure on $\mathbb{T}$ such that $h = P_\mu$.*

The following celebrated result of Fatou provides a sufficient condition for the existence of radial limits of P_μ.

Theorem 1.3 (Fatou). *Let μ be a complex Borel measure on $\mathbb{T}$. Suppose that at $e^{i\theta} \in \mathbb{T}$ the symmetric derivative,*

$$\mu'(e^{i\theta}) = \lim_{t \to 0^+} \frac{\mu(\{e^{is} \mid \theta - t < s < \theta + t\})}{2t}$$

exists. Then $\lim_{r \to 1^-} P_\mu(re^{i\theta}) = 2\pi\mu'(e^{i\theta})$.

Proof. Without loss of generality, we assume $\theta = 0$. We define the following function of x, $\mathcal{U}(x) = \mu(\{e^{is} \mid -\pi \le s < x\})$, $x \in [-\pi, \pi)$. Then integration by parts gives

$$P_\mu(r) = \int_{\mathbb{T}} \frac{1 - r^2}{1 + r^2 - 2r\cos t} d\mu(e^{it})$$

$$= \left\{ \frac{1 - r^2}{1 + r^2 - 2r\cos t} \mathcal{U}(t) \right\}_{-\pi}^{\pi} - \int_{-\pi}^{\pi} \frac{\partial}{\partial t} \left\{ \frac{1 - r^2}{1 + r^2 - 2r\cos t} \right\} \mathcal{U}(t)dt$$

$$= \frac{1 - r}{1 + r}\mathcal{U}(\pi) + \frac{2r}{1 + r} \int_{-\pi}^{\pi} \frac{(1 + r)^2(1 - r)t\sin t}{(1 + r^2 - 2r\cos t)^2} \cdot \frac{\mathcal{U}(t) - \mathcal{U}(-t)}{2t} dt.$$

Let us define

$$\Phi(t) = \frac{\mathcal{U}(t) - \mathcal{U}(-t)}{2t} - \mu'(1) = \frac{1}{2t}\int_{-t}^{t} d\mu(e^{is}) - \mu'(1), \quad (-\pi \le t < \pi),$$

and let us note, by the assumption (the symmetric derivative of μ exists), that

$$\lim_{t\to 0^+} \Phi(t) = 0. \tag{35.0.3}$$

Let us define further

$$F_r(t) = \frac{(1+r)^2(1-r)t\sin t}{(1+r^2 - 2r\cos t)^2}, \quad (0 \le r < 1, \ -\pi \le t \le \pi).$$

This function satisfies the following properties:

(i) $F_r \ge 0$.

(ii)

$$\frac{1}{2\pi}\int_{-\pi}^{\pi} F_r(t)dt = 1.$$

(iii) For each fixed $0 < \delta < \pi$, we have, $\lim_{r\to 0^+}\left(\sup_{\delta < |t| \le \pi} F_r(t)\right) = 0$.

Thus F_r is a positive approximate identity on $[-\pi, \pi]$. Using the above notations, we have

$$\lim_{r\to 1^-} P_\mu(r) = \lim_{r\to 1^-}\int_{-\pi}^{\pi} F_r(t)(|Phi(t) + \mu'(1)|)dt$$

$$= 2\pi\mu'(1) + \lim_{r\to 1^-}\int_{-\pi}^{\pi} F_r(t)\Phi(t)dt.$$

Using equation (35.0.3), given $\epsilon > 0$, there is a $\delta > 0$ such that $|\Phi(t)| < \epsilon$ for $|t| < \delta$. We assume without losing the generality that $\delta < \pi$. Then we have

$$\left|\int_{-\pi}^{\pi} F_r(t)\Phi(t)dt\right| \le \int_{-\delta}^{\delta} F_r(t)|\Phi(t)|dt + \int_{\delta < |t| \le \pi} F_r(t)|\Phi(t)|dt$$

$$\le \epsilon\int_{-\delta}^{\delta} F_r(t)dt + \left(\max_{-\pi \le t \le \pi} |\Phi(t)|\right)\cdot\int_{\delta < |t| \le \pi} F_r(t)dt$$

$$\text{(by property (ii))}$$

$$\le 2\pi\epsilon + 2\pi\left(\max_{-\pi \le t \le \pi}|\Phi(t)|\right)\left(\sup_{\delta < |t| \le \pi} F_r(t)\right).$$

Therefore, for each $\epsilon > 0$, we get (by property (iii))

$$\limsup_{r \to 1^+} \left| \int_{-\pi}^{\pi} F_r(t)\Phi(t)dt \right| \leq 2\pi\epsilon.$$

This implies that $\lim_{r \to 1^-} P_\mu(r) = 2\pi\mu'(1)$ (using $\lim_{r \to 1^-} P_\mu(r) = 2\pi\mu'(1) + \lim_{r \to 1^-} \int_{-\pi}^{\pi} F_r(t)\Phi(t)dt$). $\square$

By Lebesgue's decomposition theorem, for each complex Borel measure μ, there is a function $u \in L^1(\mathbb{T})$ and a complex singular Borel measure σ such that $d\mu(e^{i\theta}) = u(e^{i\theta})\frac{d\theta}{2\pi} + d\sigma(e^{i\theta})$. Moreover, for almost all $e^{i\theta} \in \mathbb{T}$,

$$\mu'(e^{i\theta}) = \lim_{t \to 0^+} \frac{\mu(\{e^{is} \mid \theta - t < s < \theta + t\})}{2t} = \frac{u(e^{i\theta})}{2\pi}.$$

Hence, we obtain the following two results. First, if $\mu = \sigma$ is a complex singular Borel measure on $\mathbb{T}$, then

$$\lim_{r \to 1^-} P_\sigma(re^{i\theta}) = 0, \quad \text{for almost all } e^{i\theta} \in \mathbb{T}. \tag{35.0.4}$$

Second, if $d\mu = u(e^{i\theta})\frac{d\theta}{2\pi}$ is absolutely continuous, then

$$\lim_{r \to 1^-} P_\mu(re^{i\theta}) = u(e^{i\theta}), \quad \text{for almost all } e^{i\theta} \in \mathbb{T}. \tag{35.0.5}$$

The following variant of Fatou's theorem is useful. Since F_r is a positive approximate identity, the proof of Theorem 1.3 with some modification works also in the following case.

Theorem 1.4. *Let μ be a finite positive Borel measure on $\mathbb{T}$, and let $e^{i\theta} \in \mathbb{T}$ be such that*

$$\mu'(e^{i\theta}) = \lim_{t \to 0^+} \frac{\mu(\{e^{is} \mid \theta - t < s < \theta + t\})}{2t} = \infty.$$

Then $\lim_{r \to 1^-} P_\mu(re^{i\theta}) = \infty$.

Proof. We assume, without losing the generality that $\theta = 0$. We use the same notations as in the proof of Theorem 1.3. We have $\lim_{r \to 1^-} P_\mu(r) = \lim_{r \to 1^-} \int_{-\pi}^{\pi} F_r(t)\Phi(t)dt$, except that in this case

$$\Phi(t) = \frac{\mathcal{U}(t) - \mathcal{U}(-t)}{2t} = \frac{1}{2t}\int_{-t}^{t} d\mu(e^{is}) \geq 0, \quad (-\pi \leq t < \pi),$$

and $\lim_{t\to 0+} \Phi(t) = \infty$. Hence, given $M > 0$, there is a $\delta > 0$ such that $\Phi(t) \geq M$, whenever $|t| < \delta$. Without losing the generality, we assume that $\delta < \pi$. Then we have

$$\int_{-\pi}^{\pi} F_r(t)\Phi(t)dt \geq \int_{-\delta}^{\delta} F_r(t)\Phi(t)dt \geq M \int_{-\delta}^{\delta} F_r(t)dt$$

$$= M \cdot \left(2\pi - \int_{\delta < |t| \leq \pi} F_r(t)dt \right)$$

$$\geq 2\pi M - M \left(\sup_{\delta < |t| \leq \pi} F_r(t) \right).$$

This implies that $\lim_{r\to 1-} P_\mu(r) = \infty$. $\qquad\square$

The set $S_C(e^{i\theta}) = \{z \in U \,|\, |z - e^{i\theta}| \leq C(1 - |z|)\}$, where $C > 1$ is a constant, is called a Stolz domain. This domain looks like an angle with vertex at $e^{i\theta}$, and the bigger the C, the wider the angle. A function f defined on U has a non-tangential limit at $e^{i\theta}$ if for each fixed constant C, the limit $\lim_{z\to e^{i\theta}, z\in S_C(e^{i\theta})} f(z)$ exists and its value is independent of C. In this case, we write $\lim_{z\to e^{i\theta}, \lhd} f(z)$. In all the preceding results, the radial limit can be replaced by a non-tangential limit. The proofs though become more complicated. However, the same technique works. In some applications, we will also need the following approach region: $S_{C,\delta}(e^{i\theta}) = \{z \in U \,|\, |e^{i\theta} - z| \leq C(1 - |z|)^\delta\}$, where $C \geq 1$ and $0 < \delta \leq 1$. This is referred to as a generalized Stolz domain anchored at $e^{i\theta} \in \mathbb{T}$. The smaller the δ, the more tangential $S_{C,\delta}$ is to the boundary ($\mathbb{T}$) at $e^{i\theta}$.

Chapter 36

The Hardy Space $H^p(U)$ [71, p. 6]

Let f be an analytic function on the open unit disk U. Define

$$\|f\|_p = \sup_{0 \le r < 1} \|f_r\|_p = \sup_{0 \le r < 1} \left(\frac{1}{2\pi} \int_0^{2\pi} |f(re^{i\theta})|^p d\theta \right)^{1/p},$$

if $p \in (0, \infty)$, and $\|f\|_\infty = \sup_{z \in U} |f(z)|$. Then the Hardy space $H^p(U)$ is the family of all the analytic functions f which satisfy the growth restriction $\|f\|_p < \infty$. The Hardy spaces $H^1(U)$, $H^2(U)$ and $H^\infty(U)$ will frequently appear in our discussion. A simple application of Hölder's inequality shows that $H^\infty(U) \subset H^q(U) \subset H^p(U)$ for $0 < p < q < \infty$. In particular, $H^\infty(U) \subset H^2(U) \subset H^1(U)$. By equation (35.0.5), for each $f \in H^p(U)$, $0 < p \le \infty$, the non-tangential limit,

$$f^*(e^{i\theta}) = \lim_{z \to e^{i\theta}, \vartriangleleft} f(z) \tag{36.0.1}$$

exists for almost all $e^{i\theta} \in \mathbb{T}$. Moreover, $f^* \in L^p(\mathbb{T})$ and $\|f^*\|_{L^p(\mathbb{T})} = \|f\|_{H^p(U)}$, $0 < p \le \infty$, and

$$\lim_{r \to 1^-} \|f_r - f^*\|_p = 0, \quad (0 < p < \infty). \tag{36.0.2}$$

This establishes a norm preserving correspondence between $H^p(U)$ and a closed subspace of $L^p(\mathbb{T})$, which we denote by $H^p(\mathbb{T})$. Hence, we will write f for f^*. For any non-zero $f \in H^p(\mathbb{T})$, we have

$$\log |f| \in L^1(\mathbb{T}), \quad \text{and} \tag{36.0.3}$$

$$\log |f(re^{i\theta})| \le \frac{1}{2\pi} \int_0^{2\pi} \frac{1 - r^2}{1 + r^2 - 2r \cos(\theta - t)} \log |f(e^{it})| dt. \tag{36.0.4}$$

If $1 \leq p \leq \infty$, $H^p(\mathbb{T})$ has the equivalent characterization, $H^p(\mathbb{T}) = \{f \in L^p(\mathbb{T}) \mid \hat{f}(n) = 0, n \leq -1\}$, where $\hat{f}(n) = \frac{1}{2\pi} \int_0^{2\pi} f(e^{it})e^{-int}dt$, $(n \in \mathbb{Z})$, is the n-th Fourier coefficient of f. We also have

$$f(z) = \frac{1}{2\pi} \int_0^{2\pi} \frac{1 - |z|^2}{|z - e^{it}|^2} f(e^{it})dt, \quad (z \in U), \tag{36.0.5}$$

or equivalently

$$f(re^{i\theta}) = \frac{1}{2\pi} \int_0^{2\pi} \frac{1 - r^2}{1 + r^2 - 2r\cos(\theta - t)} f(e^{it})dt, \quad (re^{i\theta} \in U). \tag{36.0.6}$$

In other words, f is the Poisson integral of its boundary values. A function $f \in L^\infty(\mathbb{T})$ is called unimodular if $|f(e^{it})| = 1$ for almost all $e^{it} \in \mathbb{T}$. A function $f \in H^\infty(U)$ is called inner if it is unimodular on the boundary. Note that if an analytic function on U has unimodular boundary values almost everywhere on $\mathbb{T}$, we cannot deduce that it is an inner function. For example, the function

$$f(z) = \exp\left(\frac{1 + z}{1 - z}\right), \quad (z \in U),$$

satisfies $f(e^{i\theta}) = \exp(i\cot(\frac{\theta}{2}))$ for all $e^{i\theta} \in \mathbb{T} - \{1\}$. However, f is not an inner function since $f(r) = \exp(\frac{1+r}{1-r}) \to \infty$ as $r \to 1^-$.

Chapter 37

Two Classes of Inner Functions [71, p. 8]

We introduce two classes of inner functions, Blaschke products and singular inner functions. Then we will see that any arbitrary inner function has a unique decomposition as the product of two such inner functions. Let $\{z_n\}_{n\geq 1}$ be a sequence of non-zero complex numbers inside the open unit disk U with $\lim_{n\to\infty} |z_n| = 1$. The sequence is indexed so that $0 < |z_1| \leq |z_2| \leq \cdots$. z_n can repeat a finite number of times. Let

$$\Omega = \Omega_{\{z_n\}} = \mathbb{C} - \overline{\left\{ \frac{1}{\overline{z_n}} \right\}}.$$

The notation $\overline{A}$ for $A \subseteq \mathbb{C}$ represents the topological closure of A in the complex plane. Since $\lim_{n\to\infty} |z_n| = 1$, all the accumulation points of $\left\{ \frac{1}{\overline{z_n}} \right\}$ lie on the unit circle $\mathbb{T}$. Clearly, this set has at least one accumulation point. It might have just one accumulation point, for example, if all the z_n belong to a single radius. On the other hand, there are sequences $\{z_n\}_{n\geq 1}$ that accumulate at every point of $\mathbb{T}$, like $z_n = (1 - e^{-n})e^{i\theta_n}$ where $\{\theta_n \mid n \geq 1\}$ is an enumeration of the rationals, $\mathbb{Q}$. So, the set Ω contains the open unit disk, U, and

$$\{z \in \mathbb{C} \mid |z| > 1\} - \left\{ \frac{1}{\overline{z_n}} \,\Big|\, n \geq 1 \right\}.$$

It might also contain some open arcs on $\mathbb{T}$. Let us fix $m \in \mathbb{Z}_{\geq 0}$, $\beta \in \mathbb{R}$ and define

$$B_k(z) = e^{i\beta} z^m \prod_{n=1}^{k} \frac{|z_n|}{z_n} \cdot \frac{z_n - z}{1 - \overline{z_n}z}. \tag{37.0.1}$$

Under a certain growth condition on $\{z_n\}_{n\geq 1}$, the partial products $B_k(z)$, $k \geq 1$, converge uniformly on compact subsets of Ω to an analytic function, which we denote by

$$B(z) = e^{i\beta} z^m \prod_{n=1}^{\infty} \frac{|z_n|}{z_n} \cdot \frac{z_n - z}{1 - \overline{z_n}z}. \tag{37.0.2}$$

The function $B(z)$ is called an infinite Blaschke product for the open unit disk, U.

Theorem 1.5 (Blaschke). *Let the points $z_n \in U - \{0\}$, $n \geq 1$, be such that $\lim_{n\to\infty} |z_n| = 1$. Then a necessary and sufficient condition for the uniform convergence of the $B_k(z)$, given by equation (37.0.1), on compact subsets of*

$$\Omega = \Omega_{\{z_n\}} = \mathbb{C} - \left\{ \overline{\frac{1}{\overline{z_n}}} \right\}.$$

to a non-zero analytic function, is that $\sum_{n=1}^{\infty}(1 - |z_n|) < \infty$.

Remark. There is no restriction on $\arg z_n$.

Proof. Suppose that $B_k(z)$ is uniformly convergent to $B(z)$ on compact subsets of Ω. Without losing the generality, we assume that $B(0) \neq 0$. Since otherwise we can consider the partial products $B_k(z)/z^m$, which uniformly converge to $B(z)/z^m$ on compact subsets of Ω. The inequality $t \leq e^{t-1}$, $0 \leq t \leq 1$, implies that $|B(0)| = \prod_{n=1}^{\infty} |z_n| \leq \exp\left(-\sum_{n=1}^{\infty}(1 - |z_n|)\right)$. Since $B(0) \neq 0$, we deduce $\sum_{n=1}^{\infty}(1 - |z_n|) < \infty$. The identity

$$\frac{|z_n|}{z_n} \cdot \frac{z_n - z}{1 - \overline{z_n}z} = 1 - (1 - |z_n|)\frac{z_n + |z_n|z}{z_n(1 - \overline{z_n}z)} \tag{37.0.3}$$

implies that

$$\left| 1 - \frac{|z_n|}{z_n} \cdot \frac{z_n - z}{1 - \overline{z_n}z} \right| \leq (1 - |z_n|)\frac{1 + |z_n|}{|1 - \overline{z_n}z|}.$$

On compact subsets of Ω, we have

$$\frac{1 + |z_n|}{|1 - \overline{z_n}z|} \leq \frac{1 + |z|}{|z_n||z - 1/\overline{z_n}|} \leq \frac{1 + |z|}{|z_1|\,\mathrm{dist}(z, \partial\Omega)} \leq C,$$

where C is a constant independent of n. However, C depends on the choice of the compact set. This dependence is harmless. Therefore, for all z in a compact subset of Ω,

$$\left| 1 - \frac{|z_n|}{z_n} \cdot \frac{z_n - z}{1 - \overline{z_n}z} \right| \leq C \cdot (1 - |z_n|), \quad (n \geq 1).$$

This inequality establishes the uniform and absolute convergence of the partial products $B_k(z)$ on compact subsets of Ω. $\qquad\square$

A sequence $\{z_n\}$ of complex numbers in the open unit disk satisfying the condition

$$\sum_{n=1}^{\infty}{}'(1 - |z_n|) < \infty, \qquad (37.0.4)$$

is called a Blaschke sequence. The growth restriction (37.0.4) is called the Blaschke condition. Under this condition, the partial products $B_k(z)$ converge uniformly on any compact subset of U to the infinite Blaschke product $B(z)$. Since $|B_k(z)| \le 1$ on U, we necessarily have

$$|B(z)| \le 1, \quad (z \in U). \qquad (37.0.5)$$

Hence, a Blaschke product $B(z)$, restricted to U, belongs to $H^\infty(U)$. Thus, by equation (37.0.5), for almost all $e^{i\theta} \in \mathbb{T}$, $B(e^{i\theta}) = \lim_{z \to e^{i\theta}, \triangleleft} B(z)$ exists and satisfies the norm identity $\|B\|_{H^\infty(U)} = \|B\|_{L^\infty(\mathbb{T})}$. Moreover, by equation (36.0.6),

$$B(re^{i\theta}) = \frac{1}{2\pi} \int_0^{2\pi} \frac{1 - r^2}{1 + r^2 - 2r\cos(\theta - t)} \cdot B(e^{it})dt, \quad (re^{i\theta} \in U).$$

If $B(z)$ is a finite Blaschke product, then $B(e^{i\theta})$ is a well-defined analytic function on $\mathbb{T}$. But, if $B(z)$ is an infinite Blaschke product, then $B(e^{i\theta})$ defined almost everywhere on $\mathbb{T}$. According to Theorem 1.5, an infinite Blaschke product is analytic and unimodular at $e^{i\theta} \in \mathbb{T}$ if and only if $e^{i\theta}$ is not an accumulation point of its zeros. But, if $e^{i\theta} \in \mathbb{T}$ is an accumulation point of the zeros of $B(z)$, then $B(z)$ is not even continuous there. Nevertheless, $B(e^{i\theta})$ is still unimodular almost everywhere on $\mathbb{T}$. This result is of special interest when the zeros of $B(z)$ accumulate on a subset of $\mathbb{T}$ with a positive measure. In particular, the zeros of $B(z)$ may accumulate at all points of $\mathbb{T}$.

Lemma 1.6. *Let $B(z)$ be a Blaschke product for U. Then*

$$\lim_{r \to 1^-} \frac{1}{2\pi} \int_0^{2\pi} \log|B(re^{i\theta})|d\theta = 0.$$

Proof. By equation (37.0.5), we have

$$\frac{1}{2\pi} \int_0^{2\pi} \log|B(re^{i\theta})|d\theta \le 0, \qquad (37.0.6)$$

for all r, $0 \leq r < 1$. Without losing the generality, we assume that $B(0) \neq 0$. Since otherwise we can divide $B(z)$ by z^m, where m is the order of the zero of $B(z)$ at $z = 0$, and this modification does not change the limit that we want to evaluate. By Jensen's formula,

$$\log |B(0)| = \sum_{|z_n|<r} \log \left(\frac{|z_n|}{r} \right) + \frac{1}{2\pi} \int_0^{2\pi} \log |B(re^{i\theta})| d\theta,$$

for all r, $0 < r < 1$. Since $|B(0)| = \prod_{n=1}^{\infty} |z_n|$, we obtain

$$\frac{1}{2\pi} \int_0^{2\pi} \log |B(re^{i\theta})| d\theta = \sum_{|z_n|<r} \log \left(\frac{r}{|z_n|} \right) - \sum_{n=1}^{\infty} \log \left(\frac{1}{|z_n|} \right).$$

At this point, we can use the monotone convergence theorem and finish the proof. However, a direct proof is also possible. Given $\epsilon > 0$, choose N so large that $\sum_{n=N+1}^{\infty} \log(1/|z_n|) < \epsilon$. Then for $r > |z_N|$, we have

$$\frac{1}{2\pi} \int_0^{2\pi} \log |B(re^{i\theta})| d\theta \geq \sum_{n=1}^{N} \log \left(\frac{r}{|z_n|} \right) - \sum_{n=1}^{N} \log \left(\frac{1}{|z_n|} \right) - \epsilon.$$

Therefore, $\liminf_{r \to 1-} \frac{1}{2\pi} \int_0^{2\pi} \log |B(re^{i\theta})| d\theta \geq -\epsilon$. Since ϵ is an arbitrary positive number $\liminf_{r \to 1-} \frac{1}{2\pi} \int_0^{2\pi} \log |B(re^{i\theta})| d\theta \geq 0$. Finally, equation (37.0.6) and the last inequality imply that the limit exists and equals to zero. $\square$

We will see that the property described in Lemma 1.6 is a characterization of Blaschke products among the inner functions. Firstly, let us see that the property implies that $B(z)$ is an inner function.

Theorem 1.7 (Riesz). *Let $B(z)$ be a Blaschke product for U. Then $|B(e^{i\theta})| = 1$ for almost all $e^{i\theta} \in \mathbb{T}$.*

Proof. By equation (37.0.5), we have $|B(e^{i\theta})| \leq 1$ for almost all $e^{i\theta} \in \mathbb{T}$. Moreover, by Fatou's Lemma and Lemma 1.6,

$$\frac{1}{2\pi} \int_0^{2\pi} \log \left| \frac{1}{B(e^{i\theta})} \right| d\theta \leq 0. \tag{37.0.7}$$

Since $\log |1/B(e^{i\theta})| \geq 0$, and (37.0.7) holds, we must have $|B(e^{i\theta})| = 1$ for almost all $e^{i\theta} \in \mathbb{T}$. $\square$

Remark. Fatou's Lemma: If $\{f_n\}$ is a sequence of non-negative measurable functions, then

$$\int \liminf_{n\to\infty} f_n d\mu \le \liminf_{n\to\infty} \int f_n d\mu.$$

We now introduce the singular inner functions. Let σ be a positive singular Borel measure on $\mathbb{T}$, and let

$$S(z) = S_\sigma(z) = \exp\left(-\int_\mathbb{T} \frac{e^{it}+z}{e^{it}-z} d\sigma(t)\right). \tag{37.0.8}$$

Clearly, $S(z)$ is analytic in U. Since

$$|S_\sigma(re^{i\theta})| = \exp\left(-\int_\mathbb{T} \frac{1-r^2}{1+r^2-2r\cos(\theta-t)} d\sigma(t)\right) \tag{37.0.9}$$

and since σ is positive, it follows that $|S(re^{i\theta})| \le 1$ for all $re^{i\theta} \in U$, i.e., $S \in H^\infty(U)$. Moreover, by (35.0.4), $\lim_{r\to 1^-}|S(re^{i\theta})| = 1$ for almost all $e^{i\theta} \in \mathbb{T}$. In other words, $S(z)$ is an inner function. The function $S(z)$ is called a singular inner function. We end this chapter by showing a property which always holds for singular inner functions. Finite Blaschke products never satisfy such a property. Later, we will construct an infinite Blaschke product which does not have that property.

Lemma 1.8. *Let σ be a positive singular Borel measure on $\mathbb{T}$, with $\sigma \ne 0$. Then there is at least one point $e^{i\theta} \in \mathbb{T}$ such that $\lim_{r\to 1^-} S_\sigma(re^{i\theta}) = 0$.*

Proof. Since σ is singular with respect to Lebesgue measure

$$\mu'(e^{i\theta}) = \lim_{t\to 0^+} \frac{\mu(\{e^{is} \,|\, \theta-t < s < \theta+t\})}{2t} = \infty,$$

for all $e^{i\theta} \in E$, where $E \subset \mathbb{T}$ is such that $\sigma(\mathbb{T} - E) = 0$. Since $\sigma \ne 0$, we have $E \ne \emptyset$. Hence, by Theorem 1.4, at all points of E, we have

$$\lim_{r\to 1^-} \int_\mathbb{T} \frac{1-r^2}{1+r^2-2r\cos(\theta-t)} d\sigma(t) = \infty,$$

which by equation (37.0.9), implies that $S_\sigma(re^{i\theta}) \to_{r\to 1^-} 0$. $\qquad\square$

Chapter 38

The Canonical Factorization [71, p. 12]

By a celebrated result of Weierstrass, if $\{z_n\}_{n\geq 1}$ is a sequence in U with no accumulation point in U, or equivalently $\lim_{n\to\infty} |z_n| = 1$, then there is an analytic function f on U, $f \not\equiv 0$, such that $f(z_n) = 0$, $(n \geq 1)$. We can even choose f such that it has no other zeros, i.e., $Z(f) = \{z_n\}_{n\geq 1}$. However, if there are restrictions on the rate of growth of $|f|$ as z approaches the boundary, then there will be restriction on the rate of growth of the zeros of $f(z)$. Here is a typical example.

Lemma 1.9. *Let f be an analytic function in U, $f \not\equiv 0$, and let $\{z_n\}_{n\geq 1}$ denote the sequence of zeros of $f(z)$ in U. Then $\sum_{n=1}^{\infty}(1 - |z_n|) < \infty$ if and only if $\sup_{0\leq r<1} \int_0^{2\pi} \log |f(re^{i\theta})|d\theta < \infty$.*

Remark. If $f(z)$ has a finite number of zeros in U, then we replace $\sum_{n=1}^{\infty}$ by $\sum_{n=1}^{N}$.

Proof. Let us denote $M = \sup_{0\leq r<1} \int_0^{2\pi} \log |f(re^{i\theta})|d\theta < \infty$. Without losing the generality, we assume that $f(0) \neq 0$. Then by Jensen's formula,

$$\frac{1}{2\pi} \int_0^{2\pi} \log |f(re^{i\theta})|d\theta = \log |f(0)| + \sum_{|z_n|<r} \log \left(\frac{r}{|z_n|}\right), \quad (0 < r < 1).$$

Hence, either by the monotone convergence theorem or by direct verification as was done in the proof of Lemma 1.6, $M = \log |f(0)| + \sum_{n=1}^{\infty} \log(1/|z_n|)$. Therefore, $M < \infty$ if and only if $\sum_{n=1}^{\infty} \log |z_n| > -\infty$, which is equivalent to the Blaschke condition. $\square$

If $f \in H^p(U)$, $0 < p < \infty$ and if $f \not\equiv 0$, then

$$\exp \left(\frac{1}{2\pi} \int_0^{2\pi} \log |f(re^{i\theta})|^p d\theta\right) \leq \frac{1}{2\pi} \int_0^{2\pi} |f(re^{i\theta})|^p d\theta \leq \|f\|^p_{H^p(U)}.$$

Thus, $\frac{1}{2\pi}\int_0^{2\pi}\log|f(re^{i\theta})|d\theta \le \log\|f\|_{H^p(U)}$. This inequality is also true for $p = \infty$. By Lemma 1.9, it follows that we can form the Blaschke product with zeros of $f(z)$. Riesz discovered that we can extract those zeros in a special way.

Theorem 1.10 (Riesz). *Let $f \in H^p(U)$, $f \not\equiv 0$, and let B be the Blaschke product with the zeros of f in U. Then there is a zero free $g \in H^p(U)$ such that $f = B \cdot g$, and $\|f\|_{H^p(U)} = \|g\|_{H^p(U)}$.*

Proof. Let $g(z) = f(z)/B(z)$. Clearly, $g(z)$ is a zero free analytic function on U, and moreover $\|f\|_{H^p(U)} \le \|g\|_{H^p(U)}$. First, we suppose that $0 < p < \infty$. Fix $N \ge 1$ and $0 < r < 1$ such that there are no zeros of f on the circle $|z| = r$. Let B_N be the finite Blaschke product formed with the first N zeros of f. Then $g_N(z) = f(z)/B_N(z)$ is an analytic function on U, and for all ρ with $r < \rho < 1$, we have

$$\frac{1}{2\pi}\int_0^{2\pi}|g_N(re^{i\theta})|^p d\theta \le \frac{1}{2\pi}\int_0^{2\pi}|g_N(\rho e^{i\theta})|^p d\theta$$

$$\le \frac{1}{\inf_\theta |B_N(\rho e^{i\theta})|^p}\left(\frac{1}{2\pi}\int_0^{2\pi}|f(\rho e^{i\theta})|^p d\theta\right)$$

$$\le \frac{\|f\|_{H^p(U)}^p}{\inf_\theta |B_N(\rho e^{i\theta})|^p}.$$

Let $\rho \to 1^-$. Since $|B_N|$ tends uniformly to 1, we obtain $\frac{1}{2\pi}\int_0^{2\pi}|g_N(re^{i\theta})|^p d\theta \le \|f\|_{H^p(U)}^p$. Now let $N \to \infty$. On the circle $|z| = r$, B_N tends uniformly to B and hence, $\frac{1}{2\pi}\int_0^{2\pi}|g(re^{i\theta})|^p d\theta \le \|f\|_{H^p(U)}^p$. Finally, let $r \to 1^-$ through a sequence r_n such that there are no zeros of $f(z)$ on the circles $|z| = r_n$ and we obtain $\|g\|_{H^p(U)} \le \|f\|_{H^p(U)}$. With some modifications, the proof works also for $p = \infty$. $\square$

Corollary 1.11. *Let $0 < p < \infty$, and let $f \in H^p(U)$, $f \not\equiv 0$. Let B be the Blaschke product formed with the zeros of f in U. Then there is $g \in H^2(U)$, with no zeros in U, such that $f = B \cdot g^{2/p}$. Moreover, $\|f\|_{H^p(U)}^p = \|g\|_{H^2(U)}^2$.*

Proof. By Theorem 1.10, f has the decomposition $f = B \cdot h$, where h is a zero free function in $H^p(U)$ with $\|f\|_{H^p(U)} = \|h\|_{H^p(U)}$. Since h has no zeros in U, there is an analytic function Φ such that $h = e^\Phi$, i.e., $\log(h)$ is well defined in U. We define $g = \exp(p\Phi/2)$ and note that $\|g\|_{H^2(U)}^2 = \|h\|_{H^p(U)}^p = \|f\|_{H^p(U)}^p$. $\square$

We can now prove the main result of this chapter.

Theorem 1.12. *Let $f \in H^p(U)$, $0 < p \le \infty$, $f \not\equiv 0$. Then we have the unique factorization $f = B \cdot S \cdot h$, where B is the Blaschke product formed with the zeros of f,*

$$S(z) = \exp\left(-\int_{\mathbb{T}} \frac{e^{it} + z}{e^{it} - z} d\sigma(e^{it})\right), \quad (z \in U),$$

is a singular inner factor formed with a finite, positive and singular Borel measure σ on $\mathbb{T}$, and h is given by the formula

$$h(z) = \exp\left(-\int_{\mathbb{T}} \frac{e^{it} + z}{e^{it} - z} \log|f(e^{it})| dt\right), \quad (z \in U).$$

Moreover, $h \in H^p(U)$ and $\|f\|_{H^p(U)} = \|h\|_{H^p(U)}$.

Remark. Theorem 1.12 is known as the canonical factorization theorem for $H^p(U)$.

Proof. The property (36.0.3) enables us to define $h(z)$ as in the statement of the theorem. The function h is analytic in U and satisfies the crucial identity

$$\log|h(re^{i\theta})| = \frac{1}{2\pi} \int_0^{2\pi} \frac{1 - r^2}{1 + r^2 - 2r\cos(\theta - t)} \cdot \log|f(e^{it})| dt. \qquad (38.0.1)$$

therefore, by (36.0.4), we have

$$|f(z)| \le |h(z)| \quad (z \in U), \qquad (38.0.2)$$

and by (35.0.5),

$$\lim_{r \to 1^-} \log|h(re^{i\theta})| = \log|f(e^{i\theta})| \qquad (38.0.3)$$

for almost all $e^{i\theta} \in \mathbb{T}$. The identity (38.0.1) implies that

$$|h(re^{i\theta})|^p \le \frac{1}{2\pi} \int_0^{2\pi} \frac{1 - r^2}{1 + r^2 - 2r\cos(\theta - t)} |f(e^{it})|^p dt,$$

and thus $h \in H^p(U)$ with $\|h\|_{H^p(U)} \le \|f\|_{H^p(U)}$, while (38.0.2) shows that $\|f\|_{H^p(U)} \le \|h\|_{H^p(U)}$. Hence, $\|f\|_{H^p(U)} = \|h\|_{H^p(U)}$. We define $\Phi(z) = f(z)/h(z)$. Since $h(z)$ has no zeros in U, $\Phi(z)$ is an analytic function on U. Moreover, by (38.0.2), $\Phi \in H^\infty(U)$, and by (38.0.3), $\lim_{r \to 1^-} |\Phi(re^{i\theta})| = 1$, for almost all $e^{i\theta} \in \mathbb{T}$. In other words, $\Phi(z)$ is an inner function. By Theorem 1.10, $\Phi = B \cdot S$, where $S \in H^\infty(U)$ and is zero free in U. For almost all $e^{i\theta} \in \mathbb{T}$, the radial limits of Φ and of B are unimodular. Hence, $S(z)$ is also an inner function. Since $S(z)$ is inner and free of zeros in U, $-\log|S(z)|$ is a positive harmonic function in U. By Corollary 1.2, there is a positive Borel measure σ on $\mathbb{T}$ such that

$$\log|S(re^{i\theta})| = -\int_{\mathbb{T}} \frac{1-r^2}{1+r^2-2r\cos(\theta-t)}d\sigma(t). \tag{38.0.4}$$

On the one hand, by Theorem 1.3, $\lim_{r\to 1^-} \log|S(re^{i\theta})| = -2\pi\sigma'(e^{i\theta})$, and on the other hand, since $S(z)$ is inner, $\lim_{r\to 1^-} \log|S(re^{i\theta})| = 0$, for almost all $e^{i\theta} \in \mathbb{T}$. Hence, σ is singular with respect to Lebesgue measure. Finally, the identity (38.0.4) can be re-written as follows:

$$\Re\log(S(z)) = \Re\left(-\int_{\mathbb{T}} \frac{e^{it}+z}{e^{it}-z}d\sigma(t)\right).$$

Therefore, there is a real constant β such that

$$S(z) = e^{i\beta}\exp\left(-\int_{\mathbb{T}} \frac{e^{it}+z}{e^{it}-z}d\sigma(t)\right).$$

The constant $e^{i\beta}$ can be absorbed into the Blaschke product. $\square$

In the canonical factorization described in Theorem 1.12, the function $h(z)$ is called the outer part of $f(z)$ and $B \cdot S$ is called the inner part of $f(z)$. Sometimes, instead of h and $B \cdot S$, we will write O_f and I_f, respectively. Hence, a function $f \in H^p(U)$, $0 < p \le \infty$, is called outer if its inner part is a unimodular constant. Therefore, $f(z)$ is outer if and only if it has no zeros in U and the harmonic function $\log|f|$ is given by the Poisson integral of its boundary values, i.e.,

$$\log|f(re^{i\theta})| = \frac{1}{2\pi}\int_0^{2\pi} \frac{1-r^2}{1+r^2-2r\cos(\theta-t)}\log|f(e^t)|dt, \quad (re^{i\theta} \in U).$$

We note that if f, g are in some Hardy spaces (not necessarily the same) and $g \not\equiv 0$, then O_f/O_g is also an outer function and

$$\frac{O_f}{O_g} = O_{f/g}. \tag{38.0.5}$$

Up to now, we have seen two classes of inner functions: Blaschke products and singular inner functions. Clearly, any product of the form $B \cdot S$ is also an inner function. More importantly, by the canonical factorization theorem any inner function has this form. This is implicit in Theorem 1.12. Because of its importance, we explicitly repeat it in the following.

Corollary 1.13. *Let Φ be an inner function for U. Let B be the Blaschke product formed with the zeros of Φ. Then there is a finite, positive and singular Borel measure σ on $\mathbb{T}$ such that $\Phi = B \cdot S_\sigma$.*

Chapter 39

A Characterization of Blaschke Products [71, p. 17]

In Corollary 1.13, we saw that an inner function Φ decomposes as $\Phi = B \cdot S_\sigma$, where B is a Blaschke product and S_σ is a singular inner function. The factor B reduces to a unimodular constant if and only if Φ has no zeros in U. We naturally may propose a similar question for the singular factor S_σ. In other words, under which extra conditions, an inner function has to be a Blaschke product? In this chapter, we study two conditions of this type.

Theorem 1.14. *Let Φ be an inner function for U. Then the limit*

$$\lim_{r \to 1^-} \int_0^{2\pi} \log |\Phi(re^{i\theta})| d\theta$$

exists. Moreover, Φ is a Blaschke product if and only if

$$\lim_{r \to 1^-} \int_0^{2\pi} \log |\Phi(re^{i\theta})| d\theta = 0.$$

Proof. By Corollary 1.13, we have the decomposition $\Phi = B \cdot S_\sigma$. Hence, by (37.0.9),

$$\log |\Phi(re^{i\theta})| = \log |B(re^{i\theta})| - \frac{1}{2\pi} \int_{\mathbb{T}} \frac{1 - r^2}{1 + r^2 - 2r \cos(\theta - t)} d\sigma(t),$$

which by (35.0.1) implies that $\log |\Phi(re^{i\theta})| = \log |B(re^{i\theta})| - \int_{\mathbb{T}} d\sigma(t)$. Therefore by Lemma 1.6, the required limit exists and is equal to

$$\lim_{r \to 1^-} \int_0^{2\pi} \log |\Phi(re^{i\theta})| d\theta = \lim_{r \to 1^-} \int_0^{2\pi} |B(re^{i\theta})| d\theta - \int_{\mathbb{T}} d\sigma(t) = -\sigma(\mathbb{T}).$$

The above identities show that $\lim_{r\to1^-}\int_0^{2\pi}\log|\Phi(re^{i\theta})|d\theta = 0$, holds if and only if $\sigma(\mathbb{T}) = 0$. But since σ is positive, $\sigma(\mathbb{T}) = 0$ is equivalent to $\sigma \equiv 0$. $\square$

In Theorem 1.14, we used integral means to detect the Blaschke products. In the following theorem, we will consider the radial limits.

Theorem 1.15. *Let Φ be an inner function for U. Suppose that $\lim_{r\to1^-}|\Phi(re^{i\theta})| \neq 0$ for all $e^{i\theta} \in \mathbb{T}$. Then Φ is a Blaschke product.*

Remark. The assumption $\lim_{r\to1^-}|\Phi(re^{i\theta})| \neq 0$ means that either the radial limit $\lim_{r\to1^-}|\Phi(re^{i\theta})|$ does not exist, or it exists but its value is not zero.

Proof. By Corollary 1.13, we have the decomposition $\Phi = B \cdot S_\sigma$, and thus $|\Phi(re^{i\theta})| \leq |S_\sigma(re^{i\theta})|$, $(re^{i\theta} \in U)$. Suppose that $\sigma \neq 0$. Then by Lemma 1.8, there is at least one point $e^{i\theta} \in \mathbb{T}$ such that $\lim_{r\to1^-}S_\sigma(re^{i\theta}) = 0$. Thus, at this point, we have $\lim_{r\to1^-}|\Phi(re^{i\theta})| = 0$, which contradicts our assumption. $\square$

Example 1.16. Theorem 1.15 enables us to detect some non trivial Blaschke products. Let σ be the Dirac measure with unit mass at the point 1. Then equation (37.0.8) gives us the following atomic inner function:

$$S(z) = S_\sigma(z) = \exp\left(-\frac{1+z}{1-z}\right).$$

We compute

$$\lim_{r\to1^-} S(r(e^{i\theta}) = S(e^{i\theta}) = \exp\left(-i\cot\left(\frac{\theta}{2}\right)\right)$$

for every $e^{i\theta} \in \mathbb{T} - \{1\}$, and $\lim_{r\to1^-} S(r) = 0$. Fix any $\alpha \in U$, $\alpha \neq 0$ and define

$$\Phi(z) = \frac{\alpha - S(z)}{1 - \overline{\alpha}S(z)}, \quad (z \in U).$$

Then $|\Phi(z)| < 1$, $(z \in U)$, and for every $e^{i\theta} \neq 1$,

$$\lim_{r\to1^-} \Phi(re^{i\theta}) = \frac{\alpha - S(e^{i\theta})}{1 - \overline{\alpha}S(e^{i\theta})} \in \mathbb{T}$$

and also $\lim_{r\to1^-} \Phi(r) = \alpha \neq 0$. Hence, $\Phi(z)$ is an inner function and by Theorem 1.15, $\Phi(z)$ is a Blaschke product. The zeros of $\Phi(z)$ are the solutions of the equation

$$S(z) = \exp\left(-\frac{1+z}{1-z}\right) = \alpha.$$

The solutions are

$$z_n = \frac{\log|\alpha| + i(\arg\alpha + 2n\pi) + 1}{\log|\alpha| + i(\arg\alpha + 2n\pi) - 1}, \quad (n \in \mathbb{Z}).$$

In particular, for $\alpha = 1/e$, these zeros are

$$z_n = \frac{i \cdot n\pi}{i \cdot n\pi - 1}, \quad (n \in \mathbb{Z}).$$

We note that without using Theorem 1.15, it is not clear that

$$\Phi(z) = \frac{\alpha - \exp\left(-\frac{1+z}{1-z}\right)}{1 - \overline{\alpha} \cdot \exp\left(-\frac{1+z}{1-z}\right)}, \quad (\alpha \in U - \{0\}),$$

is a Blaschke product.

Chapter 40

The Nevanlinna Class $\mathcal{N}$ and its Subclass $\mathcal{N}^+$ [71, p. 20]

The Nevanlinna class $\mathcal{N}$ is the family of all the analytic functions f on U which satisfy the following growth restriction:

$$\sup_{0 \le r < 1} \int_{-\pi}^{\pi} \log^+ |f(re^{i\theta})| d\theta < \infty. \tag{40.0.1}$$

It is well known that $f \in \mathcal{N}$ if and only if f is the quotient of two bounded analytic functions on U. By the canonical factorization theorem, this characterization implies the following properties of an element $f \in \mathcal{N}$.

(i) The zeros of $f(z)$ satisfy the Blaschke condition. This also follows by Lemma 1.9.

(ii) For almost all $e^{i\theta} \in \mathbb{T}$, the radial limits $f(e^{i\theta}) = \lim_{r \to 1^-} f(re^{i\theta})$ exist and are finite.

(iii) If $f \not\equiv 0$, then $\log |f| \in L^1(\mathbb{T})$.

(iv) There is a finite signed measure, singular and Borel on $\mathbb{T}$ which we denote by σ, such that $f = B \cdot S_\sigma \cdot O_{|f|}$, where B is the Blaschke product formed with the zeros of $f(z)$.

On the other hand, if $B(z)$ is any Blaschke product, σ is any finite signed singular Borel measure on $\mathbb{T}$, h is any positive measurable function with $\log f \in L^1(\mathbb{T})$, then $B \cdot S_\sigma \cdot O_h \in \mathcal{N}$. We emphasize that the measure σ appearing in the canonical decomposition of functions in the Nevanlinna class is not necessarily positive. Hence, we define the proper subclass $\mathcal{N}^+ = \{f \in \mathcal{N} \mid f = BS_\sigma O_{|f|} \text{ with} \sigma \ge 0\}$. Based on the canonical factorization theorem, we have $\bigcup_{p>0} H^p(U) \subset \mathcal{N}^+$, and the inclusion is proper. The main advantage of $\mathcal{N}^+$ over $\mathcal{N}$ is the following.

Theorem 1.17. *Let $f \in \mathcal{N}^+$. Then $f \in H^p(U)$, $0 < p < \infty$, if and only if $f \in L^p(\mathbb{T})$. In particular, if $f \in H^s(U)$ for some $s \in (0, \infty]$ and $f \in L^p(\mathbb{T})$, for some $0 < p \leq \infty$, then $f \in H^p(U)$.*

Proof. Since $f \in \mathcal{N}^+$, it has the canonical factorization $f = BS_\sigma O_{|f|}$, where B is the Blaschke product formed with the zeros of f and σ is a finite positive singular Borel measure on $\mathbb{T}$. Therefore, $\Phi = B \cdot S_\sigma$ is an inner function. Hence, it is enough to show that $O_{|f|} \in H^p(U)$. But, $O_{|f|} \in H^p(U)$ if and only if $f \in L^p(\mathbb{T})$. $\qquad\square$

The function

$$f(z) = \exp\left(\frac{1+z}{1-z}\right), \quad (z \in U),$$

is such that $f \in \mathcal{N}$ with the unimodular boundary values $f(e^{i\theta}) = \exp(i \cot(\theta/2))$, $(e^{i\theta} \in \mathbb{T})$. However, $f \notin H^\infty(U)$. This example shows that in Theorem 1.17 the assumption $f \in \mathcal{N}^+$ cannot be weakened to $f \in \mathcal{N}$. The following result provides a characterization of the elements of $\mathcal{N}^+$ in the Nevanlinna class $\mathcal{N}$.

Theorem 1.18. *Let $f \in \mathcal{N}$. Then $f \in \mathcal{N}^+$ if and only if*

$$\lim_{r \to 1^-} \int_{-\pi}^{\pi} \log^+ \left| f(re^{i\theta}) \right| d\theta = \int_{-\pi}^{\pi} \log^+ \left| f(e^{i\theta}) \right| d\theta. \tag{40.0.2}$$

Proof. Suppose that $f \in \mathcal{N}^+$. Then $f = B \cdot S_\sigma \cdot O_{|f|}$, where σ is a positive singular Borel measure on $\mathbb{T}$. Therefore, $|f| \leq |O_{|f|}|$, which implies that

$$\log^+ |f(re^{i\theta})| \leq \frac{1}{2\pi} \int_{-\pi}^{\pi} \frac{1-r^2}{1+r^2 - 2r\cos(\theta - t)} \log^+ |f(e^{it})| dt.$$

By Fubini's theorem and (35.0.1), $\int_{-\pi}^{\pi} \log^+ |f(re^{i\theta})| d\theta \leq \int_{-\pi}^{\pi} \log^+ |f(e^{it})| dt$, $(0 \leq r < 1)$. The left-hand side is an increasing function of r, and by Fatou's Lemma

$$\int_{-\pi}^{\pi} \log^+ |f(e^{i\theta})| d\theta \leq \liminf_{r \to 1^-} \int_{-\pi}^{\pi} \log^+ |f(re^{i\theta})| d\theta.$$

The last two inequalities show that equation (40.0.2) holds. The other direction is more delicate. Suppose that equation (40.0.2) holds. Let $f = B \cdot S_\sigma \cdot O_{|f|}$ be the canonical factorization of $f \in \mathcal{N}$. We denote $g = S_\sigma \cdot O_{|f|}$.

Since $f = B \cdot g$, it is enough to prove that $g \in \mathcal{N}^+$. We have $|f| = |g|$ on $\mathbb{T}$ and

$$\log |B(z)| + \log^+ |g(z)| \leq \log^+ |f(z)| \leq \log^+ |g(z)|,$$

for $\log |B(z)| \leq 0$ and $|f(z)| \leq |g(z)|$, $z \in U$. So, by Theorem 1.14 and by our assumption (40.0.2),

$$\lim_{r \to 1^-} \int_{-\pi}^{\pi} \log^+ \left|g(re^{i\theta})\right| d\theta = \int_{-\pi}^{\pi} \log^+ \left|g(e^{i\theta})\right| d\theta. \tag{40.0.3}$$

The function g has the integral representation

$$\log |g(re^{i\theta})| = \frac{1}{2\pi} \int_{\mathbb{T}} \frac{1 - r^2}{1 + r^2 - 2r\cos(\theta - t)} d\lambda(e^{it}), \quad (re^{i\theta} \in U),$$

where $d\lambda(e^{it}) = \log |g(e^{it})|dt - d\sigma(e^{it})$. Hence, the measures $\log |g(re^{it})|dt$ converge in the weak* topology to $d\lambda(e^{it})$. Take any sequence $r_n > 0$, $n \geq 1$ such that $r_n \to 1^-$. Since the sequences $\{\log^+ |g(r_n e^{it})|dt\}_{n \geq 1}$ and $\{\log^- |g(r_n e^{it})|dt\}_{n \geq 1}$ are uniformly bounded in the Banach space $\mathcal{M}(\mathbb{T})$, there is a subsequence $\{n_k\}_{k \geq 1}$ and two positive measures $\lambda_1, \lambda_2 \in \mathcal{M}(\mathbb{T})$ such that

$$\log^+ \left|g(r_{n_k} e^{it})\right| dt \to d\lambda_1(e^{it}) \quad \text{and} \quad \log^- \left|g(r_{n_k} e^{it})\right| dt \to d\lambda_2(e^{it})$$

in the weak* topology. Hence, the measure λ is given by $\lambda = \lambda_1 - \lambda_2$. Our main task is to show that λ_1 is absolutely continuous with respect to the Lebesgue measure. This is equivalent to $\sigma \geq 0$, and thus it will conclude the proof of the theorem. Let E be any Borel subset of $\mathbb{T}$. Then, by the Fatou's Lemma,

$$\int_E \log^+ |g(e^{it})|dt \leq \liminf_{r \to 1^-} \int_E \log^+ |g(re^{it})|dt, \quad \text{and}$$

$$\int_{\mathbb{T}-E} \log^+ |g(e^{it})|dt \leq \liminf_{r \to 1^-} \int_{\mathbb{T}-E} \log^+ |g(re^{it})|dt.$$

If, in any one of the last two inequalities, the inequality is strict, then we add them up and obtain a strict inequality with integrals on $\mathbb{T}$, which contradicts (40.0.3). Hence,

$$\int_E \log^+ |g(e^{it})|dt = \liminf_{r \to 1^-} \int_E \log^+ |g(re^{it})|dt$$

for all the Borel subsets E of $\mathbb{T}$. In particular, if we take $r = r_{n_k}$, $k \to \infty$, then we see that the measures $\log^+ |g(r_{n_k} e^{it})|dt$ converge to $\log^+ |g(e^{it})|dt$ in the weak* topology. Therefore, we conclude that $d\lambda_1(e^{it}) \log^+ |g(e^{it})|dt \geq 0$. $\qquad \square$

Chapter 41

Bergman Space [71, p. 23]

We fix $0 < p < 1$. The Banach space $B^p(U)$ consists of all the analytic functions on U for which

$$\|f\|_{B^p(U)} = \int_0^1 \int_0^{2\pi} |f(re^{i\theta})|(1-r)^{\frac{1}{p}-2} dr d\theta < \infty.$$

Some authors prefer to replace $dr d\theta$ by $r dr d\theta$ or to multiply the norm by a normalization constant. This comment applies also to other norms which we define in what follows. We have

$$0 < p < q < 1 \Rightarrow B^q(U) \subset B^p(U).$$

It is also well known that $B^p(U) \subset H^p(U)$, $(0 < p < 1)$. The classical Bergman space $A^p(U)$, $0 < p < \infty$, consists of all the analytic functions f on U such that

$$\|f\|_{A^p(U)} = \left(\frac{1}{\pi} \int_0^1 \int_0^{2\pi} |f(re^{i\theta})|^p r dr d\theta\right)^{1/p} < \infty.$$

The constant in the definition is adjusted so that the norm of the constant function $f(z) \equiv 1$ is 1. We note that $A^1(U) = B^{1/2}(U)$. The weighted Bergman spaces $A^p_\gamma(U)$ are defined via the growth restriction

$$\|f\|_{A^p_\gamma(U)} = \left(\frac{1+\gamma}{\pi} \int_0^1 \int_0^{2\pi} |f(re^{i\theta})|^p r(1-r^2)^\gamma dr d\theta\right)^{1/p} < \infty,$$

where $\gamma > -1$.

Hence, we have $A^p(U) = A^p_0(U)$. We have $0 < p < q < \infty \Rightarrow A^q(U) \subset A^p(U)$. More generally, for a fixed γ, $0 < p < q < \infty \Rightarrow A^q_\gamma(U) \subset A^p_\gamma(U)$.

If $0 < p < 1$, then $d(f,g) = \|f - g\|^p_{A^p_\gamma(U)}$ defines a complete translation-invariant metric on $A^p_\gamma(U)$, and thus $A^p_\gamma(U)$ is a Fréchet space. For $1 \leq p < \infty$, $\| \cdot \|_{A^p_\gamma(U)}$ is a well-defined complete norm on $A^p_\gamma(U)$ and so $A^p_\gamma(U)$ is a Banach space. In particular, for $p = 2$, the norm is generated by the inner product

$$\langle f, g \rangle_{A^2_\gamma(U)} = \frac{1}{\pi(1 + \gamma)} \int_0^1 \int_0^{2\pi} f(re^{i\theta})\overline{g(re^{i\theta})}r(1 - r^2)^\gamma dr d\theta.$$

Thus, $A^2_\gamma(U)$ is an Hilbert space. If we have the Taylor series expansions, $f(z) = \sum_{n=0}^\infty a_n z^n$ and $g(z) = \sum_{n=0}^\infty b_n z^n$ on U, then either by direct verification or by applying the Parseval identity, we obtain

$$\langle f, g \rangle_{A^2_\gamma(U)} = \sum_{n=0}^\infty \frac{n!\Gamma(\gamma + 2)}{\Gamma(\gamma + n + 2)} a_n \overline{b_n}. \tag{41.0.1}$$

Hence,

$$\|f\|^2_{A^2_\gamma(U)} = \sum_{n=0}^\infty \frac{n!\Gamma(\gamma + 2)}{\Gamma(\gamma + n + 2)} |a_n|^2. \tag{41.0.2}$$

In particular, in the classical case $A^2(U)$, we have

$$\langle f, g \rangle_{A^2(U)} = \sum_{n=0}^\infty \frac{a_n \overline{b_n}}{n + 1}, \tag{41.0.3}$$

$$\|f\|^2_{A^2(U)} = \sum_{n=0}^\infty \frac{|a_n|^2}{n + 1}. \tag{41.0.4}$$

Theorem 1.19. (Hardy–Littlewood). *Let $0 < p < \infty$. Then every function $f \in H^p(U)$ also belongs to $A^{2p}(U)$ and satisfies $\|f\|_{A^{2p}(U)} \leq \|f\|_{H^p(U)}$, with equality if and only if*

$$f(z) = \frac{c}{(1 - \lambda z)^{2/p}},$$

where $\lambda \in U$ and $c \in \mathbb{C}$ are arbitrary constants.

Proof. We first consider the case $p = 2$. If $f(z) = \sum_{n=0}^\infty a_n z^n$, then $f^2(z) = \sum_{n=0}^\infty (\sum_{k=0}^n a_k a_{n-k}) z^n$. Therefore, by (41.0.4),

$$\|f\|^4_{A^4(U)} = \|f^2\|_{A^2(U)^2} = \sum_{n=0}^\infty \frac{1}{n + 1} \left| \sum_{k=0}^n a_k a_{n-k} \right|^2.$$

By Cauchy–Schwarz inequality

$$\left| \sum_{k=0}^{n} a_k a_{n-k} \right|^2 \le (n+1) \sum_{k=0}^{n} |a_k|^2 |a_{n-k}|^2. \tag{41.0.5}$$

Hence,

$$\|f\|_{A^4(u)}^4 \le \sum_{n=0}^{\infty} \sum_{k=0}^{n} |a_k|^2 |a_{n-k}|^2 = \left(\sum_{m=0}^{\infty} |a_m|^2 \right)^2 = \|f\|_{H^4(U)}^4.$$

The equality holds if and only if we have equality in (41.0.5) for all $n \ge 0$, and this happens if and only if $a_k a_{n-k} = c_n$, $(0 \le k \le n)$, where c_n does not depend on k. In particular, we must have $a_n^2 = a_0 a_{2n} = c_{2n}$. This observation shows that $a_0 = 0$ forces all other coefficients to be zero and thus $f(z) \equiv 0$. If $a_0 \ne 0$, we use $a_n a_0 = a_{n-1} a_1 = c_n$ to deduce that $a_n = \lambda a_{n-1}$, $(n \ge 1)$, where $\lambda = \frac{a_1}{a_0}$. Hence, $a_n = \lambda^n a_0$, $(n \ge 0)$. With this choice of coefficients, we obtain

$$f(z) = \frac{a_0}{1 - \lambda z}, \tag{41.0.6}$$

and, for this function, equality holds in (41.0.5). For other values of p, we use the Riesz technique (Corollary 1.11). Write $f(z) = B(z) g(z)^{2/p}$, with $g \in H^2(U)$ and $\|f\|_p^2 = \|g\|_2^2$. Then, by the above $\|f\|_{A^{2p}(U)}^{2p} \le \|g\|_{A^4(U)}^4 \le \|g\|_{H^2(U)}^4 = \|f\|_{H^p(U)}^{2p}$. The first inequality becomes an equality if and only if there is no Blaschke factor. The second is an equality if and only if $g(z)$ has the form (41.0.6). Hence, $\|f\|_{A^{2p}(U)} = \|f\|_{H^p(U)}$ if and only if $f(z) = c/(1 - \lambda z)^{2/p}$. $\qquad\square$

The proof above shows that when $1 \le p < \infty$, the injection $H^p(U) \to A^{2p}(U)$, $f \to f$ has norm one. Moreover, this result is sharp in the sense that the exponent $2p$ is the best possible choice. More precisely, $H^p(U) \not\subset H^q(U)$ for any $q > 2p$. See Example 4.5.

We pause with the survey on inner functions and return to our current research theme.

Chapter 42

More on Increments. Not Just Dirichlet Increments!

We refer to Chapters 16, 23, 29, 30 and alike, in which we introduced the Dirichlet increments, both the finite and the continuous ones. A much simpler and a straightforward construction is possible. Let $g(z) \in H(U)$ satisfy

$$\lim_{r \to 1^-} \frac{1}{2\pi} \int_0^{2\pi} |\exp(g(re^{i\theta}))|^2 d\theta = 1.$$

We would like to construct increments, i.e., functions $h(z) \in H(U)$ such that

$$\lim_{r \to 1^-} \frac{1}{2\pi} \int_0^{2\pi} |\exp(g(re^{i\theta}) + h(re^{i\theta}))|^2 d\theta = 1.$$

For that purpose, we take any $H(z) \in H(U)$ that satisfies the following condition:

$$M_H = \lim_{r \to 1^-} \frac{1}{2\pi} \int_0^{2\pi} |\exp(g(re^{i\theta}) + H(re^{i\theta}))|^2 d\theta < \infty.$$

We let $M_H = \exp(2m_H)$ and define $h(z) = H(z) - m_H$. Then $h(z) \in H(U)$ and we have

$$\lim_{r \to 1^-} \frac{1}{2\pi} \int_0^{2\pi} |\exp(g(re^{i\theta}) + h(re^{i\theta}))|^2 d\theta$$

$$= \lim_{r \to 1^-} \frac{1}{2\pi} \int_0^{2\pi} |\exp(g(re^{i\theta}) + H(re^{i\theta}) - m_H)|^2 d\theta$$

$$= \lim_{r \to 1^-} \frac{1}{2\pi} \int_0^{2\pi} |\exp(g(re^{i\theta}) + H(re^{i\theta}))|^2 d\theta \exp(-2m_H)$$

$$= \lim_{r \to 1^-} \frac{1}{2\pi} \int_0^{2\pi} |\exp(g(re^{i\theta}) + H(re^{i\theta}))|^2 d\theta \cdot M_H^{-1} = 1.$$

The same simple idea works directly for the Krzyż conjecture, instead of $A_n(2)$. We use the notation $B_0(U) = \{f(z) \in H(U) \,|\, \forall z \in U,\, 0 < |f(z)| \leq 1\}$. The Krzyż conjecture is that for $n \geq 1$, we have

$$\max\left\{ \Re\{a_n\} \,\middle|\, f(z) = \sum_{k=0}^{\infty} a_k z^k \in B_0(U) \right\} = \frac{2}{e},$$

where the extremal functions are precisely the rotations of $\exp\left(-\frac{1+z^n}{1-z^n}\right)$. By [49], an extremal function has the following form:

$$\exp\left(-\sum_{j=1}^{N} \lambda_j \frac{1 + k_j z}{1 - k_j z} \right), \quad \text{where } N \leq n,\ \lambda_j > 0 \ \text{ and } |k_j| = 1,\ j = 1, \ldots, N,$$

and k_j are different from one another. If $S_i(U)$ is the family of all the singular inner functions, then $S_i(U) \subset B_0(U)$ and we have the following reduction:

$$\max\left\{ \Re\{a_n\} \,\middle|\, f(z) = \sum_{k=0}^{\infty} a_k z^k \in B_0(U) \right\}$$
$$= \max\left\{ \Re\{a_n\} \,\middle|\, f(z) = \sum_{k=0}^{\infty} a_k z^k \in S_i(U) \right\}.$$

The closed convex hull of $S_i(U)$, which is denoted by $\overline{C_0(S_i(U))}$, is the closure with respect to the topology of uniform convergence on compact subsets of U of the family of all the functions of the form $\lambda_1 S_{\sigma_1} + \cdots + \lambda_j S_{\sigma_j}$ where $\lambda_1, \ldots, \lambda_j > 0$, $\sum_{m=1}^{j} \lambda_j = 1$, and the S_{σ_m} are singular inner functions, where the σ_m are singular Borel measures on $\mathbb{T}$.

Theorem 42.0.94.

$$\max\left\{ \Re\{a_n\} \,\middle|\, f(z) = \sum_{k=0}^{\infty} a_k z^k \in S_i(U) \right\}$$
$$= \max\left\{ \Re\{a_n\} \,\middle|\, f_0(z) \cdot T(z) = \sum_{k=0}^{\infty} a_k z^k,\ T \in \overline{C_0(S_i(U))} \right\},$$

where $f_0(z)$ is an extremal for the Krzyż conjecture.

Proof. Let $S \in S_i(U)$. Then $f_0(z) \cdot S(z) \in S_i(U)$ and by the assumption that $f_0(z)$ is an extremal for the n'th Krzyż problem, we deduce that the real part of the n'th coefficient of $f_0(z) \cdot S(z)$ is less than or equal to the real part of the n'th coefficient of $f_0(z)$. So,

$$\Re\left\{\int_0^{2\pi} f_0(re^{i\theta})(1 - S(re^{i\theta}))e^{-in\theta}d\theta\right\} \geq 0.$$

Taking the convex combination with the coefficients $\lambda_1, \ldots, \lambda_j > 0$, $\sum_{m=1}^j \lambda_m = 1$ of these inequalities with the choices $S = S_{\sigma_1}, \ldots, S_{\sigma_j}$ we obtain

$$\Re\left\{\int_0^{2\pi} f_0(re^{i\theta})\left(1 - \sum_{m=1}^j \lambda_m S_{\sigma_m}(re^{i\theta})\right)e^{-in\theta}d\theta\right\} \geq 0,$$

i.e., $\forall T \in C_0(S_i(U))$, we have

$$\Re\left\{\int_0^{2\pi} f_0(re^{i\theta})(1 - T(re^{i\theta}))e^{-in\theta}d\theta\right\} \geq 0.$$

Finally, by uniform limits on compact subsets of U, we obtain our result, i.e., the last inequality holds true for any $T \in \overline{C_0(S_i(U))}$. $\square$

Remark 42.0.95. If $\lambda_1, \ldots, \lambda_j > 0$, $\sum_{m=1}^j \lambda_m = 1$, $S_{\sigma_1}, \ldots, S_{\sigma_j} \in S_i(U)$, then clearly $\|\lambda_1 S_{\sigma_1} + \cdots + \lambda_j S_{\sigma_j}\|_{H^\infty(U)} \leq 1$. However, the functions $\lambda_1 S_{\sigma_1} + \cdots + \lambda_j S_{\sigma_j}$ can have zeros in U. For example,

$$\frac{1}{2}\left(\exp\left(-\frac{1+z}{1-z}\right) + \exp\left(-\frac{1-z}{1+z}\right)\right)$$

$$= \frac{1}{2}\exp\left(-\frac{1+z}{1-z}\right)\left(1 + \exp\left(\frac{-4z}{1-z^2}\right)\right).$$

The zeros are the solutions of

$$\frac{-4z}{1-z^2} = (2k+1)\pi i, \quad (k \in \mathbb{Z}).$$

So,

$$z = \frac{2 \pm \sqrt{4 - (2k+1)^2\pi^2}}{(2k+1)\pi i} \quad \text{and} \quad \left|\frac{2 \pm \sqrt{4 - (2k+1)^2\pi^2}}{(2k+1)\pi i}\right| < 1$$

for large k. We also note that the above function is not an inner function, because on $\mathbb{T}$ it equals

$$\frac{1}{2}\left(\exp\left(i\cot\left(\frac{\theta}{2}\right)\right) + \exp\left(-i\tan\left(\frac{\theta}{2}\right)\right)\right)$$

and this is not unimodular.

We deduce an interesting approximation theorem.

Theorem 42.0.96. *Let $B(U) = \{f(z) \in H(U) \mid \|f\|_{H^\infty(U)} \leq 1\}$ be the closed unit ball of $H^\infty(U)$. Then $\overline{C_0(S_i(U))} \subsetneq B(U)$. In other words, there are functions in the closed unit ball of $H^\infty(U)$ that are not uniform limits on compact subset of U of sequences of convex combinations of singular inner functions.*

Proof. The following extremal problem $\max\{\Re\{a_n\} \mid f(z) = \sum_{k=0}^\infty a_k z^k \in B(U)\}$, for $n \geq 1$, has the solution 1 and it is attained for z^n. On the other hand, by a result of Charles Horowitz [48], there is an absolute constant c, $0 < c < 1$, such that

$$\max\left\{ \Re\{a_n\} \,\middle|\, f(z) = \sum_{k=0}^\infty a_k z^k \in B_0(U) \right\} \leq c < 1.$$

In particular, we have (recall [49]),

$$\max\left\{ \Re\{a_n\} \,\middle|\, f(z) = \sum_{k=0}^\infty a_k z^k \in S_i(U) \right\} \leq c < 1.$$

Hence, $\Re\{a_n\} \leq c$ for any convex combination of singular inner functions, i.e., for all the functions in $C_0(S_i(U))$. Hence, by taking the closure in the topology of uniform convergence on compact subsets of U, the same bound c, for $\Re\{a_n\}$ holds for all the functions in $\overline{C_0(S_i(U))}$. $\qquad\square$

Remark 42.0.97. The above proof of Theorem 42.0.3 shows, in fact, the following.

Theorem 42.0.98. $\forall\, n \in \mathbb{Z}^+,\ z^n \notin \overline{C_0(S_i(U))}$

It might be interesting to evaluate the H^∞ distances:

$$\mathrm{dist}_{H^\infty(U)}(z^n, \overline{C_0(S_i(U))}) = \min\left\{ \|z^n - f(z)\|_{H^\infty(U)} \,\middle|\, f \in \overline{C_0(S_i(U))} \right\},$$

and also $\mathrm{dist}_{H^\infty(U)}(z^n, B_0(U))$, $(n \geq 1)$. We note that also the last distances are positive numbers depending only on $n \in \mathbb{Z}^+$, by the theorem of Horowitz, [48], and by the reduction we mentioned before that is implied in [49]. See the reduction mentioned before the statement of Theorem 42.0.1. We also note that $B_0(U)$ is not a closed subspace of $B(U)$, but by the theorem of Hurwitz (NOT Horowitz!) $\overline{B_0(U)} = \{0\} \cup B_0(U)$, so indeed the second distance is a positive number depending only on $n \in \mathbb{Z}^+$. We recall that the family of Blaschke products exhibits a very different behavior with respect to approximation. Once more, we give the following result from [30]:

Corollary 2.3. ([30, p. 1107]). *Suppose that g is analytic and non-vanishing in a neighborhood of $|z| \leq r < 1$. For $\epsilon > 0$ and $\delta > 0$, there exists a constant C_B and a Blaschke product B having all zeros on $|z| = r$ such that $|g(z) - C_B \cdot B(z)| < \epsilon$ for $|z| < r \cdot (1 - \delta)$.*

We remark that if g is analytic and $g \not\equiv 0$, but we do not assume that it is non-vanishing, then we can form the Blaschke product B_g with the zeros of g in a neighborhood of $|z| \leq r < 1$. Then the function g/B_g is analytic and non-vanishing in a neighborhood of $|z| \leq r < 1$. Using Corollary 2.3 quoted above from [30], for all $\epsilon > 0$ and $\delta > 0$, there exists a constant C_B and a Blaschke product B having all of its zeros on $|z| = r$ such that for $|z| < r \cdot (1 - \delta)$, we have

$$\left| \frac{g(z)}{B_g(z)} - C_B \cdot B(z) \right| < \epsilon$$

and hence $|g(z) - C_B B(z) B_g(z)| < \epsilon |B_g(z)| \leq \epsilon$. Thus, all of the functions in the closed unit ball of $H^\infty(U)$ can be approximated uniformly on compact subsets of U by constant multiples of Blaschke products. In fact, a theorem of Frostman asserts the following:

Let ϕ be an inner function for the open unit disk U. Given $\epsilon > 0$, there is a Blaschke product B such that $\|\phi - B\|_{H^\infty(U)} < \epsilon$.

See in [71], Corollary 2.2 (p. 29).

A Survey on the Exceptional Set of an Inner Function

We continue our survey on inner functions taken from [71]; we now go through the second chapter of [71] (pp. 27–37).

Chapter 43

Frostman Shifts
and the Exceptional Set ξ_ϕ

For a fixed $w \in U$, the mapping, $\tau_w(z) = \frac{w-z}{1-\overline{w}z}$, $(z \in U)$, is an automorphism of U. Hence, for an inner function ϕ, the following function

$$\phi_w(z) = \frac{w - \phi(z)}{1 - \overline{w}\phi(z)}, \quad (z \in U)$$

maps U into itself. Hence, ϕ_w is an element of the closed unit ball of $H^\infty(U)$. Moreover, for almost all $e^{i\theta} \in \mathbb{T}$,

$$\lim_{r \to 1^-} \phi_w(re^{i\theta}) = \frac{w - \phi(e^{i\theta})}{1 - \overline{w}\phi(e^{i\theta})} \in \mathbb{T}.$$

This follows because $|\phi(e^{i\theta})| = 1$ a.e. Therefore, for each $w \in U$, $\phi_w(z)$ is an inner function. What is not obvious is that $\phi_w(z)$ has a good chance to be a Blaschke product. Thus, the exceptional set

$$\xi_\phi = \{w \in U \mid \phi_w \text{ is not a Blaschke product}\}$$

is small. In this chapter, we will show that the Lebesgue measure of ξ_ϕ is 0. In Chapter 46, this result will be refined, showing that the logarithmic capacity of ξ_ϕ is 0. The functions, ϕ_w, $w \in U$, are called the Frostman shifts of ϕ. The following version of Frostman's theorem says that the two-dimensional Lebesgue measure of ξ_ϕ is 0.

Theorem 2.1 (Frostman). *Let ϕ be an inner function for U. Fix $0 < \rho < 1$ and define*

$$\xi_\rho(\phi) = \{e^{i\theta} \in \mathbb{T} \mid \phi_{\rho e^{i\theta}} \text{ is not a Blaschke product}\}.$$

Then $\xi_\rho(\phi)$ has one-dimensional Lebesgue measure 0.

Proof. For each $\alpha \in U$ we have the following identity,

$$\frac{1}{2\pi} \int_0^{2\pi} \log \left| \frac{\rho e^{i\theta} - \alpha}{1 - \rho e^{-i\theta} \alpha} \right| d\theta = \max \left\{ \log \rho, \log |\alpha| \right\}. \tag{43.0.1}$$

Since ϕ is inner, we can substitute $\phi(re^{it})$ for α and then integrate with respect to t. We obtain

$$\frac{1}{2\pi} \int_0^{2\pi} \left(\int_0^{2\pi} \log |\phi_{\rho e^{i\theta}}(re^{it})| d\theta \right) dt = \int_0^{2\pi} \max \left\{ \log \rho, \log |\phi(re^{it})| \right\} dt,$$

where, as usual,

$$\phi_{\rho e^{i\theta}}(z) = \frac{\rho e^{i\theta} - \phi(z)}{1 - \rho e^{-i\theta} \phi(z)}, \quad (z \in U).$$

Since ρ is fixed and $|\phi| \leq 1$, the family $f_r(e^{it}) = \max\{\log \rho, \log |\phi(re^{it})|\}$, $(e^{it} \in \mathbb{T})$, where the parameter $r \in [0, 1)$, is uniformly bounded. In fact we have $\log \rho \leq f_r(e^{it}) \leq 0$, $(e^{it} \in \mathbb{T})$. Moreover,

$$\lim_{r \to 1^-} f_r(e^{it}) = \max \left\{ \log \rho, \lim_{r \to 1^-} \log |\phi(re^{it})| \right\} = \max\{\log \rho, 0\} = 0,$$

for almost all $e^{it} \in \mathbb{T}$. Hence by the dominated convergence theorem, $\lim_{r \to 1^-} \int_0^{2\pi} f_r(e^{it}) dt = 0$. Therefore

$$\lim_{r \to 1^-} \int_0^{2\pi} \left(\int_0^{2\pi} \log |\phi_{\rho e^{i\theta}}(re^{it})| d\theta \right) dt = 0.$$

The integrand $\log |\phi_{\rho e^{i\theta}}(re^{it})|$ is negative. Thus, using Fubini's theorem, we can write

$$\lim_{r \to 1^-} \int_0^{2\pi} \left(\int_0^{2\pi} \log |\phi_{\rho e^{i\theta}}(re^{it})| dt \right) d\theta = 0. \tag{43.0.2}$$

We now use Fatou's Lemma. First, we denote $M(r, w) = \int_0^{2\pi} \{- \log |\phi_w(re^{it})|\} dt$. By Theorem 1.14, we have for each fixed point $w \in U$, the limit $\lim_{r \to 1^-} M(r, w)$ exists and is a positive number. By Fatou's Lemma,

$$\int_0^{2\pi} \left(\liminf_{r \to 1^-} M(r, \rho e^{i\theta}) \right) d\theta \leq \liminf_{r \to 1^-} \int_0^{2\pi} M(r, \rho e^{i\theta}) d\theta.$$

Hence by equation (43.0.2) and by $M \geq 0$, $\int_0^{2\pi} \left(\lim_{r \to 1^-} M(r, \rho e^{i\theta}) \right) d\theta = 0$, which implies that $\lim_{r \to 1^-} M(r, \rho e^{i\theta}) = 0$ for almost all $\theta \in [0, 2\pi)$. Therefore by Theorem 1.14, $\phi_{\rho e^{i\theta}}$ is a Blaschke product for almost all $\theta \in [0, 2\pi)$. This implies that the Lebesgue measure of $\xi_\rho(\phi)$ is 0. $\qquad \square$

This result implies an interesting approximation theorem. It shows that the set of Blaschke products is uniformly dense in the set of all the inner functions.

Corollary 2.2 (Frostman). *Let ϕ be an inner function for U. Then, given $\epsilon > 0$, there is a Blaschke product B such that $\|\phi - B\|_{H^\infty(U)} < \epsilon$.*

Proof. Take $\rho \in (0,1)$ small enough that $\frac{2\rho}{1-\rho} < \epsilon$. By Theorem 2.1, on the circle $\{z \in \mathbb{C} \mid |z| = \rho\}$, there are many points $\rho e^{i\theta}$ such that $\phi_{\rho e^{i\theta}}$ is a Blaschke product. We pick one such a point $\rho e^{i\theta}$. Then we have

$$|\phi(z) + \phi_{\rho e^{i\theta}}(z)| = \left| \frac{\rho e^{i\theta} - \rho e^{-i\theta} \phi^2(z)}{1 - \rho e^{-i\theta} \phi(z)} \right| \leq \frac{2\rho}{1-\rho} < \epsilon$$

for all $z \in U$. This means that $\|\phi + \phi_{\rho e^{i\theta}}\|_{H^\infty(U)} < \epsilon$. Hence, the Blaschke product $B = -\phi_{\rho e^{i\theta}}$ satisfies the required inequality $\|\phi - B\|_{H^\infty(U)} < \epsilon$. $\qquad\square$

Chapter 44

Capacity

Let us define the following family of functions of r,

$$\Phi_\alpha = \Phi_\alpha(r) = \begin{cases} 1/r^\alpha & \text{if } \alpha > 0 \\ \log(1/r) & \text{if } \alpha = 0 \end{cases}.$$

Let μ be a positive Borel measure with a compact support in $\mathbb{C}$. Then the potential function created by μ and the kernel Φ_α is defined by

$$P_{\alpha,\mu} = \int_{\mathbb{C}} \Phi_\alpha(|z - w|) d\mu(w). \tag{44.0.1}$$

In particular, $P_\mu = P_{0,\mu}$ is the logarithmic potential. Since the kernel Φ_0 assumes both positive and negative values, the treatment of the logarithmic potential is slightly different from others. $P_{\alpha,\mu}$ is a superharmonic function with values in $(-\infty, \infty]$ on $\mathbb{C}$. We have

$$P_{\alpha,\mu}(z) = \begin{cases} \|\mu\|(1 + o(1))/|z|^\alpha & \text{if } \alpha > 0 \\ \|\mu\| \log(1/|z|)(1 + o(1)) & \text{if } \alpha = 0 \end{cases},$$

as $z \to \infty$. In some applications, especially, when we treat subsets of U, it is easier to work with the Green potential,

$$G_{\alpha,\mu}(z) = \int_{\mathbb{C}} \Phi_\alpha\left(\left|\frac{w - z}{1 - \overline{w}z}\right|\right) d\mu(w). \tag{44.0.2}$$

The advantage of this potential is that, as w ranges on a compact subset of U, the quantity $\left|\frac{w-z}{1-\overline{w}z}\right|$ uniformly tends to 1 as $|z| \to 1$. Hence, in this situation, $G_{\alpha,\mu}(z) \to 0$. In other words, $G_{\alpha,\mu}(z)$ is continuous on an annulus around $\mathbb{T}$ and, $G_{\alpha,\mu}(z) \equiv 0$ on $\mathbb{T}$. The following quantity,

$$\xi_{\alpha,\mu} = \int_{\mathbb{C}} P_{\alpha,\mu}(z)d\mu(z) = \int_{\mathbb{C}} \int_{\mathbb{C}} \Phi_\alpha(|z - w|)d\mu(w)d\mu(z) \qquad (44.0.3)$$

is called the energy of μ. In equations (44.0.1) and (44.0.3), we can replace $\int_{\mathbb{C}}$ by $\int_{\operatorname{supp}\mu}$ where $\operatorname{supp}\mu$ is the support of μ. For a given compact set K, the quantities

$$\inf_{\mu}\left(\sup_{z\in K} P_{\alpha,\mu}(z)\right) \quad \text{and} \quad \inf_{\mu} \xi_{\alpha,\mu}, \qquad (44.0.4)$$

where in both cases the infimum is taken over is taken over all the probability measures μ whose supports are in K, are important for us. The maximum principle says that $\sup_{z\in K} P_{\alpha,\mu}(z) = \sup_{z\in\mathbb{C}} P_{\alpha,\mu}(z)$, so in equation (44.0.4) we can replace K by $\mathbb{C}$. By a theorem of Frostman, the two quantities in (44.0.4) are equal, hence we define $\Delta_\alpha(K) = \inf_\mu (\sup_{z\in K} P_{\alpha,\mu}(z)) = \inf_\mu \xi_{\alpha,\mu}$. Using this fact, the C_α-capacity of K, denoted by $C_\alpha(K)$, is defined by the equation $\Phi_\alpha(C_\alpha(K)) = \Delta_\alpha(K)$. Since Φ_α is one-to-one, $C_\alpha(K)$ is well defined.

$$C_\alpha(K) = \begin{cases} \Delta_\alpha(K)^{-1/\alpha} & \text{if } \alpha > 0 \\ \exp(-\Delta_\alpha(K)) & \text{if } \alpha = 0 \end{cases}. \qquad (44.0.5)$$

The C_0-capacity is also referred to as the logarithmic capacity. It is worthwhile mentioning that the above Frostman's theorem goes further and says that there is a particular measure, called the equilibrium measure, whose choice depends on K and α, such that $\Delta_\alpha(K) = \sup_{z\in K} P_{\alpha,\mu}(z) = \xi_{\alpha,\mu}$. Usually it is hard to evaluate $C_\alpha(K)$. In most applications, we want to know if $C_\alpha(K) = 0$ or $C_\alpha(K) > 0$. The following result gives a characterization for $C_\alpha(K) > 0$.

Theorem 2.3. *Let K be a compact subset of U, and let $\alpha \geq 0$. Then the following assertions are equivalent.*

(i) $C_\alpha(K) > 0$.

(ii) *There is a positive Borel measure μ, with $\mu \neq 0$, whose support is in K and the potential $P_{\alpha,\mu}$ is bounded above on $\mathbb{C}$.*

(iii) *There is a positive Borel measure μ, with $\mu \neq 0$, whose support is in K and the Green potential $G_{\alpha,\mu}$ is bounded above on $\mathbb{C}$.*

Proof. (i) ⇔ (ii): This is a direct consequence of (44.0.5).

(ii) ⇔ (iii): It follows from the estimations $0 < 1 - d \leq |1 - \overline{w}z| < 2$, where $d = \max_{z \in K} |z|$. $\qquad\qquad\square$

An arbitrary set $E \subseteq \mathbb{C}$ is said to have positive α-capacity if there is a compact subset of E whose α-capacity is positive. The next result shows that a Borel set of logarithmic capacity 0 has 0, two-dimensional Lebesgue measure. Knowing that the logarithmic capacity of an interval is positive, we deduce that the family of Borel subsets of $\mathbb{C}$ with logarithmic capacity 0 is a proper subclass of the family of all the Borel subsets of $\mathbb{C}$ of two-dimensional Lebesgue measure 0.

Corollary 2.4. *Let E be a Borel subset of $\mathbb{C}$ whose two-dimensional Lebesgue measure is positive. Then, for each $0 \leq \alpha \leq 2$, we have $C_\alpha(E) > 0$.*

Proof. The result essentially follows from the fact that the functions $\log(1/|z|)$ and $1/|z|^\alpha$, for $0 < \alpha < 2$, are locally integrable. By regularity, there is a compact subset K is E whose two-dimensional Lebesgue measure is positive. Without loss of generality, we may assume that $d = \text{diam}(K) < 1$. Denoting by m the Lebesgue measure, we define μ by $\mu(A) = m(A \cap K)$, where A is a Borel subset of $\mathbb{C}$. Then μ is a positive Borel measure, with $\mu \neq 0$ and $\text{supp}\,\mu = K$, and

$$P_{\alpha,\mu}(z) = \int_{\mathbb{C}} \Phi_\alpha(|z - w|)d\mu(w) \leq 2\pi \int_0^1 \Phi_\alpha(r)dr < \infty.$$

Theorem 2.3 implies that $C_\alpha(K) > 0$. So, by definition, $C_\alpha(E) > 0$. $\qquad\square$

If E is a Borel set of diameter at most 1, we have $C_\beta(E) \leq kC_\alpha(E)$, $(\beta \geq \alpha \geq 0)$, where the constant $k > 0$ depends on the parameters α and β. Hence, for any Borel set, the condition $C_\alpha(E) = 0$ implies $C_\beta(E) = 0$ for $\beta \geq \alpha$. Based on this observation, the capacity dimension of E is defined by

$$\dim_C(E) = \inf\{\alpha \,|\, C_\alpha(E) = 0\}.$$

Chapter 45

Hausdorff Dimension

To define a Hausdorff measure, we need a positive increasing and continuous function h which is defined on the interval $[0, \infty)$ and $h(0) = 0$. In particular, we will consider the following family of functions:

$$h_\alpha(t) = \begin{cases} t^\alpha & \text{if } \alpha > 0 \\ 1/(\log \frac{1}{t}) & \text{if } \alpha = 0 \end{cases}. \tag{45.0.1}$$

Since, for our purposes, the behavior of $h_\alpha(t)$ for small values of t is important, we sometimes give the definition of $h_\alpha(t)$ only on intervals $[0, \delta]$, where $\delta > 0$ is small. The extension to (δ, ∞) is rather arbitrary. Let E be any subset of $\mathbb{C}$. If there are open disks $B(z_n, r_n)$, $n \geq 1$, such that $E \subseteq \bigcup_n B(z_n, r_n)$, we say that $\{B(z_n, r_n)\}_n$ is a covering of E. Fixing a function h as above, define

$$\mu_{h,\delta}(E) = \inf\left\{ \sum_n h(r_n) \mid E \subseteq \bigcup_n B(z_n, r_n), \text{ and } r_n < \delta \right\},$$

where the infimum is taken over all possible coverings of E. The quantity $\mu_{h,\delta}(E)$ is a decreasing function of δ. Hence, we can define $\mu_h(E) = \mu_{h,0}(E) = \lim_{\delta \to 0^+} \mu_{h,\delta}(E)$ and $\mu_{h,\infty}(E) = \lim_{\delta \to \infty}(E) = \inf\{\sum_n h(r_n) \mid E \subseteq \bigcup_n B(z_n, r_n)\}$. Therefore, in evaluating $\mu_h(E)$, just the fine coverings of E are important, while in $\mu_{h,\infty}(E)$, all coverings are considered. The functions $\mu_{h,\delta}$, where $0 \leq \delta \leq \infty$, are outer regular measures such that all Borel subsets of $\mathbb{C}$ are measurable with respect to them. These measures become more interesting when we exploit the family h_α. In this situation, we write $\mu_{\alpha,\delta}$ and μ_α, respectively, for $\mu_{h_\alpha,\delta}$ and μ_{h_α}. By the above definition, $\mu_{h,\infty}(E) \leq \mu_{h,\delta} \leq \mu_h(E)$. Moreover, if $\mu_{h,\infty}(E) = 0$, then given any $\epsilon > 0$,

243

     The Krzyż Conjecture: Theory and Methods

there is a covering such that $E \subseteq \bigcup_n B(z_n, r_n)$ and $\sum_n h(r_n) < \epsilon$. Therefore, $h(r_n) < \epsilon$ which implies

$$r_n < h^{-1}(\epsilon), \quad (n \geq 1). \tag{45.0.2}$$

We can now say that $\mu_{h,h^{-1}(\epsilon)}(E) < \epsilon$. Let $\epsilon \to 0^+$ to get $\mu_h(E) = 0$. Therefore, we have

$$\mu_{h,\infty}(E) = 0 \Leftrightarrow \mu_h(E) = 0. \tag{45.0.3}$$

The above definition also implies that $\mu_{h_1}(E) \leq c\mu_{h_2}(E)$ provided that the inequality $h_1(t) \leq ch_2(t)$ holds at least in a right neighborhood of 0. This property gives $0 \leq \mu_\alpha(E) < \infty \Rightarrow \mu_\beta(E) = 0$, $(\beta > \alpha)$, and $0 < \mu_\alpha(E) \leq \infty \Rightarrow \mu_\beta(E) = \infty$, $(\beta < \alpha)$. Hence, for a fixed Borel set E, the graph of $\alpha \to \mu_\alpha(E)$ is a step function with two possible values, ∞ and 0, which breaks at some point in the interval $[0, \infty)$. At the breaking point, it can take a positive finite value. Based on this observation, the Hausdorff dimension of E is defined by $\dim_H(E) = \inf\{\alpha \,|\, \mu_\alpha(E) = 0\}$. If $\{\alpha \,|\, \mu_\alpha = \infty\} \neq \emptyset$, we can also say that $\dim_H(E) = \sup\{\alpha \,|\, \mu_\alpha(E) = \infty\}$. It is striking to know that

$$\dim_H(E) = \dim_C(E). \tag{45.0.4}$$

That is why we will simply write $\dim(E)$ to refer to the above common quantity. For integers j, k and n, the set

$$Q = \left[\frac{j}{2^n}, \frac{j+1}{2^n}\right) \times \left[\frac{k}{2^n}, \frac{k+1}{2^n}\right)$$

is called a dyadic square the plane $\mathbb{C}$. It has the side length $l(Q) = 2^{-n}$. If Q_1 and Q_2 are any two dyadic squares, it is important to observe that if $Q_1 \cap Q_2 \neq \emptyset$, then either $Q_1 \subset Q_2$ or $Q_2 \subseteq Q_1$. We define

$$m_h(E) = \inf\left\{\sum_n h(l(Q_n)) \,\middle|\, E \subseteq \bigcup_n Q_n\right\},$$

where the infimum is taken over all the dyadic covers of E. Since every dyadic square of side r is contained in a disk of radius r and each disk of radius r is contained in at most 25 dyadic squares Q with $\frac{r}{2} < l(Q) \leq Q$, we have

$$\mu_{h,\infty}(E) \leq m_h(E) \leq 25\mu_{h,\infty}(E). \tag{45.0.5}$$

Hence, by (45.0.3) and (45.0.5), we deduce that

$$\mu_{h,\infty}(E) = 0 \Leftrightarrow \mu_h(E) = 0 \Leftrightarrow m_h(E) = 0, \tag{45.0.6}$$

and this fact implies

$$\dim(E) = \inf\{\alpha \mid \mu_\alpha(E)\} = \inf\{\alpha \mid \mu_{\alpha,\infty}(E) = 0\} = \inf\{\alpha \mid m_\alpha(E) = 0\}.$$

We now discuss the construction of a generalized Cantor set on $[0,1]$. With small modifications, this method can be applied to $\mathbb{T}$. Let $F_0 = [0,1]$. Remove an open interval from the middle of E_0 to obtain E_1 which is the union of two closed intervals of length l_1. In the next step, we remove an open interval from the middle of each interval in E_1 to obtain E_2 which is the union of four closed intervals of length l_2. Hence, continuing this procedure, E_n would be the union of 2^n closed intervals each of length l_n. Put $E = \bigcap_{n=0}^{\infty} E_n$. The generalized Cantor set E is a perfect set and

$$\mu_\alpha(E) = C \lim_{n\to\infty} 2^n l_n^\alpha, \tag{45.0.7}$$

where the constant C depends only on α. Whenever needed, we can enumerate all the intervals that appear in this procedure, say I_n, $n \geq 1$, such that $I_{2^n}, I_{2^n+1}, \ldots, I_{2^{n+1}-1}$ represent the 2^n intervals that make E_n.

Chapter 46

ξ_ϕ has Logarithmic Capacity Zero

In Chapter 43, the exceptional set ξ_ϕ for an inner function ϕ was defined and we saw that this set has Lebesgue measure zero. Now, we refine this result by showing that its logarithmic capacity is zero. Hence, in light of Corollary 2.4, this result is a generalization of Theorem 2.1.

Theorem 2.5 (Frostman). *Let ϕ be a non-constant inner function for U. Then the logarithmic capacity of ξ_ϕ is zero.*

Proof. Suppose, to the contrary, that the logarithmic capacity of ξ_ϕ is strictly positive, i.e., $C_0(\xi_\phi) > 0$. By Theorem 2.3, there is a positive Borel measure μ, $\mu \neq 0$, whose support K lies in ξ_ϕ and the Green potential

$$G_{0,\mu}(z) = \int_K \log \left| \frac{1 - \overline{w}z}{w - z} \right| d\mu(w), \quad (z \in \mathbb{C}),$$

is bounded from above on $\mathbb{C}$. Clearly, we have $G_{0,\mu} \geq 0$ on U. The trick is to consider the function

$$\Phi(z) = (G_{0,\mu} \circ \phi)(z) = \int_K \log \left| \frac{1 - \overline{w}\phi(z)}{w - \phi(z)} \right| d\mu(w), \quad (z \in U).$$

Since $K \subseteq U$, the potential $G_{0,\mu}$ is continuous at all points of $\mathbb{T}$ (indeed, locally harmonic at such points) and, moreover, $G_{0,\mu} \equiv 0$ on $\mathbb{T}$. Hence, remembering that ϕ is inner, we deduce that

$$\lim_{r \to 1^-} \Phi(re^{i\theta}) = \lim_{r \to 1^-} G_{0,\mu}(\phi(re^{i\theta})) = G_{0,\mu}\left(\lim_{r \to 1^-} \phi(re^{i\theta}) \right) = 0$$

for almost all $e^{i\theta} \in \mathbb{T}$. Thus, by the dominated convergence theorem,

$$\lim_{r \to 1^-} \int_0^{2\pi} \Phi(re^{i\theta}) d\theta = 0. \tag{46.0.1}$$

Note that we used the assumption that Φ is bounded on U. By Fubini's theorem

$$\int_0^{2\pi} \Phi(re^{i\theta})d\theta = \int_K \left(\int_0^{2\pi} \log \left| \frac{1 - \overline{w}\phi(re^{i\theta})}{w - \phi(re^{i\theta})} \right| d\theta \right) d\mu(w),$$

and thus, using the notation $M(r,w)$ used in the proof of Theorem 2.1, the relation (46.0.1) becomes

$$\lim_{r \to 1^-} \int_K M(r,w)d\mu(w) = 0. \tag{46.0.2}$$

According to Theorem 1.14, we know that, for each fixed $w \in U$, the following limit exists:

$$\lim_{r \to 1^-} M(r,w) > 0. \tag{46.0.3}$$

Moreover, by Fatou's Lemma

$$\int_0^{2\pi} \left(\liminf_{r \to 1^-} M(r,w) \right) d\mu w \le \liminf_{r \to 1^-} \int_0^{2\pi} M(r,w)d\mu(w).$$

Therefore, by (46.0.2) and (46.0.3), we must have

$$\int_0^{2\pi} \left(\lim_{r \to 1^-} M(r,w) \right) d\mu(w) = 0.$$

Hence, $\lim_{r \to 1^-} M(r,w) = 0$, ($\mu$-a.e. $w \in K$). Hence, by Theorem 1.14, ϕ_w is a Blaschke product for all $w \in K$, except possibly on a set of μ-measure zero. But the inclusion of $K \subseteq \xi_\phi$ means that for no point of K, this assertion holds. We arrived at a contradiction which proves the theorem. $\qquad\square$

Chapter 47

The Cluster Set at a Boundary Point

Let $f \in H^\infty(U)$ and $\xi \in \mathbb{T}$. The Euclidean disk of radius r and center ξ are denoted by $D(\xi, r)$. The cluster set of f at ξ is $C(f, \xi) = \bigcap_{r>0} \overline{f(U \cap D(\xi, r))}$, and its range set is $R(f, \xi) = \bigcap_{r>0} f(U \cap D(\xi, r))$. From these definitions, it follows that $C(f, \xi)$ is a compact connected subset of $\mathbb{C}$ and $R(f, \xi) \subseteq C(f, \xi)$. By the finite intersection property of compact sets, it follows that $C(f, \xi) \neq \emptyset$. Moreover, f is continuous at ξ if and only if $C(f, \xi)$ is a singleton. $R(f, \xi)$ is a connected G_δ subset of $\mathbb{C}$ which might be empty. For example, if f is not constant and has analytic continuation across ξ, then, by the uniqueness theorem for analytic functions, we have $C(f, \xi) = \{f(\xi)\}$ and $R(f, \xi) = \emptyset$. As a matter of fact, we can use the following topological interpretation to verify the preceding claim. A point w belongs to $C(f, \xi)$ if and only if there is a sequence $\{z_n\}_{n \geq 1} \subset U$ such that $\lim_{n \to \infty} z_n = \xi$ and $\lim_{n \to \infty} f(z_n) = w$. But $w \in R(f, \xi)$ if and only if there is a sequence $\{z_n\}_{n \geq 1} \subset U$ such that $\lim_{n \to \infty} z_n = \xi$ and $f(z_n) = w$ for all $n \geq 1$. In other words, $R(f, \xi)$ is the set of values assumed by f infinitely many times in every neighborhood of ξ in U, while, roughly speaking $C(f, \xi)$ is the set of values that can be approximated by f in every neighborhood of ξ in U.

Theorem 2.6. *Let ϕ be a non-constant inner function, and let $\xi \in \mathbb{T}$ be a singular point of ϕ. Then $C(\phi, \xi) = \overline{U}$ and $U - \xi_\phi \subseteq R(\phi, \xi) \subseteq U$.*

Proof. By Theorem 2.5, ξ_ϕ has no interior. Also, for an inner function, $C(\phi, \xi)$ is necessarily a compact subset of $\overline{U}$. Hence, the second assertion implies the first one. We recall that the singular points of a Blaschke product are precisely the accumulation points of the zero set on $\mathbb{T}$. Now, if $w \in U - \xi_\phi$, then, by the definition of ξ_ϕ,

$$\phi_w(z) = (\tau_w \circ \phi)(z) = \frac{w - \phi(z)}{1 - \overline{w}\phi(z)}, \quad (z \in U),$$

249

is a Blaschke product. Moreover, since ϕ cannot be analytically extended across ξ, then neither can ϕ_w. Otherwise, $\phi = \tau_w \circ \phi_w$ would have an analytic extension across ξ, which is a contradiction. In other words, ξ is also a singular point of the Blaschke product ϕ_w. Hence, it is an accumulation point of the zeros of ϕ_w, which means that there is a sequence $\{z_n\}_{n \geq 1} \subset U$ such that

$$\phi_w(z_n) = \frac{w - \phi(z_n)}{1 - \overline{w}\phi(z_n)} = 0 \quad \text{and} \quad \lim_{n \to \infty} z_n = \xi.$$

Therefore, $w \in R(\phi, \xi)$. $\qquad\qquad\qquad\qquad\qquad\qquad\qquad\qquad\qquad\qquad$ $\square$

This result shows that a non-constant inner function assumes all points in U, except possibly a subset of logarithmic capacity zero, infinitely many times in each neighborhood of any of its singular points. Consider a Blaschke product whose zeros accumulate at all the points of $\mathbb{T}$, or a singular inner function constructed by a measure whose support is $\mathbb{T}$. Then such a function exhibits this rough behavior at all the points of $\mathbb{T}$ and at the same time, according to Fatou's theorem, it has non-tangential limits almost everywhere. This is a function of a wild boundary behavior. Despite the wild behavior, still, we can choose zeros such that the non-tangential limits exist at all (!!!) points of $\mathbb{T}$.

Chapter 48

Some More Advanced Facts on Bounded Analytic Functions

We gather facts on bounded analytic functions. Those are collected from papers according to the author's (of this book) taste. We sometimes append the quoted results with results that are noted by the author and apparently are new.

1. [50]: Suppose S is a set of functions defined on U. A sequence $\{z_j\} \subset U$ is said to separate S if $\sum_{j=1}^{\infty} |f(z_j)| = \infty$ for every $f \in S$. In this paper, a sequence is constructed that separates the Blaschke products and a sequence that separates the non-vanishing bounded analytic functions, both of which do NOT separate $H^{\infty}(U)$.

Theorem 1 ([50]). *There is a sequence $\{z_j \mid j = 1, 2, \ldots\} \subset U$ such that*

(i) $\sum_{j=1}^{\infty} |f(z_j)| < \infty$ *for some* $f(\not\equiv 0) \in H^{\infty}(U)$.
(ii) $\sum_{j=1}^{\infty} |B(z_j)| = \infty$ *for every Blaschke product B.*

Theorem 2 ([50]). *There is a Blaschke sequence $\{z_j \mid j = 1, 2, \ldots\}$ such that $\sum_{j=1}^{\infty} |f(z_j)| = \infty$ for every $f(\not\equiv 0) \in H^{\infty}(U)$ that does not vanish on U.*

Corollary ([50]). *Suppose $0 < p < 1$ and*

$$|z_n| = 1 - \frac{1}{(\log n)^p}, \quad n = 2, 3, \ldots.$$

Then for every $f(\not\equiv 0) \in H^{\infty}(U)$, we have $\sum_{n=2}^{\infty} |f(z_n)| = \infty$.

Lemma 3 ([50]). *Suppose $\sum_{j=2}^{\infty} |f(z_j)| < \infty$ for a function $f \in H^{\infty}(\{z \mid \Re z > 0\})$ that does not vanish on the right half plane. We denote*

$z_n = x_n + iy_n$, $n = 2, 3, \ldots$. *Then there is a finite positive measure ν supported on $\{y_n \mid n = 2, 3, \ldots\}$ such that*

$$\sum_{j=2}^{\infty} \exp\left(\int_{-\infty}^{\infty} \frac{x_j}{(y_j - t)^2 + x_j^2} d\nu(t) \right) < \infty.$$

2. [18]: The paper shows that for any G_δ set F of Lebesgue measure zero on the unit circle $\mathbb{T}$, there exists a function $f \in H^\infty(U)$ such that the radial limits of f exist at each point of $\mathbb{T}$ and vanish precisely on F. This solves a problem proposed by Rubel in 1973. It is well known that every $g \in H^\infty(U)$ has radial limits $g(e^{i\theta})$ a.e. on $\mathbb{T}$. A point $e^{i\theta} \in \mathbb{T}$ is called a Fatou point for $g \in H^\infty(U)$ if $g(e^{i\theta})$ exists. The paper assumes that any $g \in H^\infty(U)$ is defined also a.e. on $\mathbb{T}$ by its radial limits $g(e^{i\theta})$. The paper gives an affirmative solution to Rubel's Problem 5.29 published in the well-known research problem collection of Hayman on the materials of "Symposium on complex analysis" held in 1973 at the University of Kent, Canterbury. The formulation of the problem is the following.

Problem 5.29. Let F be a G_δ of measure zero on $\mathbb{T}$. Then, does there exist an $f \in H^\infty(U)$, $f \not\equiv 0$, such that $f = 0$ on F and every point of $\mathbb{T}$ is a Fatou point of f?

The following is a more precise question.

Modified Problem 5.29. Let F be a G_δ of measure zero on $\mathbb{T}$. Then does there exist an $f \in H^\infty(U)$ such that $f = 0$ precisely on F and every point of $\mathbb{T}$ is a Fatou point of f?

Theorem 1 ([18]). *Let F be a G_δ of measure zero on $\mathbb{T}$. Then there exists a non-vanishing $f \in H^\infty(U)$ (even $\Re f > 0$ on U) such that $f = 0$ precisely on F and every point of $\mathbb{T}$ is a Fatou point of f.*

In a sense, Theorem 1 is an extension of Fatou's following classical interpolation theorem of 1906: If F is closed and of measure zero on $\mathbb{T}$, then there exists an element in the disk algebra that vanishes precisely on F (see [45, p. 80]). In fact, Theorem 1 can be formulated as the following "if and only if" result.

Corollary 1 ([18]). *Let $F \subseteq \mathbb{T}$. There exists an $f \in H^\infty(U)$ such that $f = 0$ precisely on F and every point of $\mathbb{T}$ is a Fatou point of f if and only if F is a G_δ of measure zero on $\mathbb{T}$.*

As a corollary of the proof of Theorem 1, we have the following description of the peak sets for those elements of $H^\infty(U)$ for which all the points of $\mathbb{T}$ are Fatou points.

Corollary 2 ([18]). *Let F be a G_δ of a measure zero on $\mathbb{T}$. Then there exists a $\lambda \in H^\infty(U)$ such that*

(a) *all the points of $\mathbb{T}$ are Fatou points of λ.*
(b) *$\lambda = 1$ on F. And*
(c) *$|\lambda| < 1$ on $\overline{U} - F$.*

Also, here, the converse implication is obvious and Corollary 2 is in fact a complete description of peak sets for those elements of $H^\infty(U)$ for which all the points of $\mathbb{T}$ are Fatou points.

The paper uses and discusses the results from [58].

Lemma 2 ([58]). *Let G be an open subset on $\mathbb{T}$ and let $F \subset G$ be a set of measure zero on $\mathbb{T}$. For any $\epsilon > 0$, there exists an open set O, $F \subset O \subsetneq G$, and a function $g \in H^\infty(U)$ such that:*

(1) *$|g(z)| < 2$, $0 < \Re g(z) < 1$ for $z \in U$*
(2) *the function g has a finite radial limit $g(\xi)$ at each point $\xi \in \mathbb{T}$*
(3) *at points $\xi \in O$, the function g is analytic and $\Re g(\xi) = 1$*
(4) *$|g(z)| \leq \epsilon$ on every radius R_{ξ_0} with end-point at $\xi_0 \in \mathbb{T} - G$.*

The main result in Kolesnikov's paper is a solution of the classical problem on the description of the sets of non-existence of radial limits of bounded analytic functions.

Theorem (Kolesnikov). *Let $E \subset \mathbb{T}$. There exists an $f \in H^\infty(U)$ such that the radial limits of f exist exactly on the set $\mathbb{T} - E$ if and only if E is a $G_{\delta\sigma}$ of measure zero.*

Relevant theorems:
Let μ be a complex Borel measure on the unit circle $\mathbb{T}$. Then the Poisson integral of μ on the open unit disk U is defined by the formula

$$P_\mu(z) = \int_{\mathbb{T}} \frac{1 - |z|^2}{|z - \xi|^2} d\mu(\xi), \quad (z \in U).$$

If $d\mu(e^{i\theta}) = u(e^{i\theta})\frac{d\theta}{2\pi}$, where $u \in L^1(\mathbb{T})$, instead of P_μ we write P_u. It can be verified that $h = P_\mu$ is an harmonic function on U. Using Fubini's theorem and the identity

$$\frac{1}{2\pi} \int_0^{2\pi} \frac{1 - |z|^2}{|z - e^{i\theta}|^2} d\theta = 1, \quad (z \in U), \tag{48.0.1}$$

we see that

$$\frac{1}{2\pi}\int_0^{2\pi}|h(re^{i\theta})|\,d\theta \leq \int_{\mathbb{T}}\left(\frac{(1-r^2)\,d\theta}{|re^{i\theta}-\xi|^2}\right)d|\mu|(\xi) = \|\mu\|,$$

where $\|\mu\|$ is the total variation of the measure μ on $\mathbb{T}$. Hence, h fulfills the following growth condition

$$\sup_{0\leq r<1}\int_0^{2\pi}\left|h(re^{i\theta})\right|\,d\theta < \infty. \tag{48.0.2}$$

Hence, the Poisson integral of a Borel measure on $\mathbb{T}$ is an harmonic function on U which satisfies condition (48.0.2). The converse of this assertion is also true and we have the following complete characterization.

Theorem (Plessner). *Let h be a function defined on U. Then the following assertions are equivalent.*

(i) *h is an harmonic function on U which satisfies the condition (48.0.2).*
(ii) *There exists a (unique) Borel measure μ on $\mathbb{T}$ such that $h = P_\mu$.*

In the special case in which μ is positive, $h = P_\mu$ is a positive harmonic function on U which satisfies (48.0.2). If h is a given positive harmonic function, it follows by the mean value property that

$$\int_0^{2\pi}|h(re^{i\theta})|\,d\theta = \int_0^{2\pi}h(re^{i\theta})\,d\theta = 2\pi h(0), \quad (0\leq r<1).$$

Therefore, in this case, the theorem of Plessner can be rewritten as follows.

Corollary (Herglotz). *Let h be a function defined on U. Then the following assertions are equivalent.*

(i) *h is a positive harmonic function on U.*
(ii) *There exists a (unique) positive Borel measure on $\mathbb{T}$ such that $h = P_\mu$.*

Here is a celebrated result of Fatou that provides a sufficient condition for the existence of radial limits of P_μ.

Theorem (Fatou). *Let μ be a complex Borel measure on $\mathbb{T}$. Suppose that at $e^{i\theta}\in\mathbb{T}$, the following symmetric derivative exists. Then*

$$\mu'(e^{i\theta}) = \lim_{t\to 0^+}\frac{\mu(\{e^{is}\,|\,\theta-t<s<\theta+t\}).}{2t}$$

$$\times \lim_{r\to 1^-}P_\mu(re^{i\theta}) = 2\pi\mu'(e^{i\theta}). \tag{48.0.3}$$

By Lebesgue's decomposition theorem, for each complex Borel measure μ, there is a function $u \in L^1(\mathbb{T})$ and a complex singular Borel measure σ, such that $d\mu(e^{i\theta}) = u(e^{i\theta})\frac{d\theta}{2\pi} + d\sigma(e^{i\theta})$. Moreover, for almost all $e^{i\theta} \in \mathbb{T}$,

$$\mu'(e^{i\theta}) = \lim_{t \to 0^+} \frac{\mu(\{e^{is} \mid \theta - t < s < \theta + t\})}{2t} = \frac{u(e^{i\theta})}{2\pi}.$$

Hence, we obtain the following two results. First, if $\mu = \sigma$ is a complex singular Borel measure on $\mathbb{T}$, then

$$\lim_{r \to 1^-} P_\sigma(re^{i\theta}) = 0, \tag{48.0.4}$$

for almost all $e^{i\theta} \in \mathbb{T}$. Second, if $d\mu = u(e^{i\theta})\frac{d\theta}{2\pi}$ is absolutely continuous, then

$$\lim_{r \to 1^-} P_\mu(re^{i\theta}) = u(e^{i\theta}) \tag{48.0.5}$$

for almost all $e^{i\theta} \in \mathbb{T}$. The following variant of Fatou's theorem is useful.

Theorem (Fatou). *Let μ be a finite positive Borel measure on $\mathbb{T}$, and let $e^{i\theta} \in \mathbb{T}$ be such that*

$$\mu'(e^{i\theta}) = \lim_{t \to 0^+} \frac{\mu(\{e^{is} \mid \theta - t < s < \theta + t\})}{2t} = \infty.$$

Then, $\lim_{r \to 1^-} P_\mu(re^{i\theta}) = \infty$.

Taking the original version of Fatou's theorem, we arrive at the following question: Let μ be a complex Borel measure on $\mathbb{T}$. What is the set of all the points at which the μ-symmetric derivative exists, i.e.,

$$\mathcal{D}(\mu) = \left\{ e^{i\theta} \in \mathbb{T} \,\middle|\, \lim_{t \to 0^+} \frac{\mu(\{e^{is} \mid \theta - t < s < \theta + t\})}{2t} \right\}?$$

This seems to have some relation to the Kolesnikov theorem. In particular, is $\mathcal{D}(\mu)^c$ a $G_{\delta\sigma}$ of measure zero? If $\mu = \sigma$ is singular, is $\mathcal{D}(\mu)^c$ a G_δ of measure zero? If the answers to these questions are known, it might give alternative proofs to the results of Kolesnikov and of Danielyan. If the answers are not known, we might be able to make a progress on them using the results of Kolesnikov and of Danielyan. It seems to be important, in this respect, to understand for which θ the following difference

$$\left| \frac{\mu(\{e^{is} \mid \theta - t_1 < s < \theta + t_1\})}{2t_1} - \frac{\mu(\{e^{is} \mid \theta - t_2 < s < \theta + t_2\})}{2t_2} \right|$$

tends to zero when $|t_1 - t_2| \to 0$, $t_1, t_2 > 0$. This difference is

$$(t_1 t_2)^{-1} |t_2 \mu(\{e^{is} \mid \theta - t_1 < s < \theta + t_1\}) - t_1 \mu(\{e^{is} \mid \theta - t_2 < s < \theta + t_2\})|.$$

We might want to distinguish the two cases according to the Lebesgue decomposition theorem.

3. [13]: We next discuss to what extent Fatou's theorem has a valid converse. This time, we quote from [13]. The converse of Fatou's theorem is true for positive measures but not for arbitrary measures. The paper includes the proof that the converse holds for Zygmund (smooth) measures, being a sharp result in some sense. In this paper, μ denotes a complex Borel measure on $\mathbb{R}$ or a positive Borel measure such that $\int_{\mathbb{R}} d\mu(t)/(1 + t^2) < \infty$. For each μ,

$$u(x, y) = P_\mu(x, y) = \frac{1}{\pi} \int_{\mathbb{R}} \frac{y}{y^2 + (x - t)^2} d\mu(t)$$

denotes its Poisson integral defined on the upper half plane $\Pi^+ = \{(x, y) \mid y > 0\}$. We note that $P_\mu(x, y) = P_y * \mu(x)$, where $P_y(x) = \frac{1}{\pi} \frac{y}{y^2 + x^2}$. Analogously, if $f(x)$ is a bounded function, we write $P_f(x, y) = P_y * f(x)$. A classical theorem of Fatou (see [90, p. 257]) relates some differentiability properties of μ at $x \in \mathbb{R}$ to the asymptotic behavior of $u(z)$ when z tends to x. To state this, we recall the definitions of the symmetric derivative of μ at x and the derivative of μ at x:

$$D_{sym}\mu(x) = \lim_{h \to 0+} \frac{\mu((x - h, x + h))}{2h}, \quad D\mu(x) = \lim_{t-s \to 0,\, s<x<t} \frac{\mu((s, t))}{t - s},$$

provided that both limits exist. The Stolz angle will be denoted by $\Delta_\alpha(x)$, $0 < \alpha < \infty$, where $\Delta_\alpha(x) = \{(t, y) \mid |t - x| < \alpha y\}$.

Theorem (Fatou). *Let u be the Poisson integral of a measure μ and let $L \in \mathbb{C}$.*

(a) *If $D\mu(x_0) = L$, then*

$$\lim_{z \to x_0,\, z \in \Delta_\alpha(x_0)} u(z) = L, \quad \forall \alpha > 0. \tag{48.0.6}$$

(b) *If $D_{sym}\mu(x_0) = L$, then $\lim_{y \to 0+} u(x_0, y) = L$.*

The converses of these two results are not true, see [67]. However, the positivity of μ is a Tauberian condition which makes true the converses of (a) (b) in Fatou's theorem.

Theorem (Loomis). *Suppose that μ is positive, $0 \le L < \infty$, and let $u = P_\mu$.*

(i) *If u has non-tangential limit L (defined by equation (48.0.6)) at x_0, then*
$$D\mu(x_0) = L.$$

(ii) *If $\lim_{y\to 0+} u(x_0, y) = L$, then $D_{sym}\mu(x_0) = L$.*

Loomis obtained several proofs of these results. The most direct one uses an integral representation of positive harmonic functions on Π^+ [91]. By applying a version of Wiener's Tauberian theorem, generalized to higher dimensions the statement (ii), in this way, another different proof is obtained. Moreover, he gave an example of a positive measure ν such that $\lim_{y\to 0+} P_\nu(0, y) = \infty$ but $D_{sym}\nu(0)$ does not exist. Therefore, the hypothesis $L < \infty$ is necessary in (ii). In [12], the authors obtain another condition for which the converse of (a) and (b) holds.

Definition 48.0.99. A complex measure μ on $\mathbb{R}$ is a Zygmund measure if there is a positive constant C such that $|\mu(I) - \mu(I')| \le C|I|$ for any two adjacent intervals of the same length. Here, $|I|$ is the Lebesgue measure of I.

Remark 48.0.100. If μ is a Zygmund measure, then its distribution function, $f_\mu(x) = \mu((-\infty, x))$, $x \in \mathbb{R}$, belongs to the Zygmund class Λ_*, i.e., f_μ is bounded and
$$|f_\mu(x + h) + f_\mu(x - h) - 2f_\mu(x)| \le C|h|, \tag{48.0.7}$$
for all $x, h \in \mathbb{R}$. The Zygmund norm of $f \in \Lambda_*(\mathbb{R})$ is $\|f\|_* = \|f\|_\infty + A$, where A is the infimum of the constants C for which equation (48.0.7) holds. The measures $f\,dx$, $f \in \mathrm{BMO}(\mathbb{R})$, are Zygmund measures.

Theorem 1 ([13]). *Let μ be a Zygmund measure on $\mathbb{R}$. If $u = P_\mu$ has non-tangential limit L at x_0, then $D\mu(x_0) = L$.*

Theorem 2 ([13]). *Let μ be a Zygmund measure on $\mathbb{R}$. If $\lim_{y\to 0+} u(x_0, y) = L$, then $D_{sym}\mu(x_0) = L$.*

These extend previous results by Brossard and Chevalier.

Definition 48.0.101. (a) The Poisson integral of μ, $u = P_\mu$, verifies the hypothesis $(\mathcal{H})$ if and only if the function $P_{|\mu|} - |P_\mu|$ is bounded in $V \cap \Pi^+$, where V is a neighborhood of $(0,0)$.

(b) If μ is a measure, the radial part of μ is defined by
$$\mu_{ra}(A) = \frac{\mu(A) + \mu(-A)}{2},$$
for each measurable set A.

Theorem A (Brossard–Chevalier). *Let μ be a measure. If $u = P_\mu$ has non-tangential limit at the point $(0,0)$ and u satisfies the hypothesis $(\mathcal{H})$, then μ is derivable at 0.*

Theorem B (Brossard–Chevalier). *Let μ be a measure and $u = P_\mu$. If the radial part μ_{ra} of μ satisfies the hypothesis $(\mathcal{H})$ and if the function u has radial limit at the point $(0,0)$, then the measure μ has symmetric derivative at the point 0.*

Example 48.0.102. If μ is positive or $\mu = f\,dx$, where $f \in \mathrm{BMO}(\mathbb{R})$, then μ and μ_{ra} satisfy $(\mathcal{H})$.

However, there are radial Zygmund measures that do not satisfy the hypothesis $(\mathcal{H})$. Therefore, it seems that the results of [13] are not consequences of Theorems A and B.

A question. Does Theorem 2 remain true when $L = \infty$?

In contrast to Rudin's example of the measure ν, the following is proved in [12].

Proposition 1 ([13]). *Let μ be a real Zygmund measure. Then the following statements are equivalent.*

(i) $D_{sym}\mu(x_0) = +\infty$.
(ii) $D\mu(x_0) = +\infty$.
(iii) $\lim_{y\to 0+} u(x_0, y) = +\infty$.
(iv) *The function u has non-tangential limit $+\infty$ at x_0.*

A question. To what extent can we weaken the hypothesis of μ being a Zygmund measure?

The modulus of continuity of a Zygmund function is $O(\delta \log(1/\delta))$ as $\delta \to 0^+$, [107, p. 44]. So, $|\mu(I)| \le C|I|\log(1/|I|)$, if $|I| \le \frac{1}{2}$. The following proposition shows a certain sharpness of the main results of [12].

Proposition 2 ([13]). *Let $\phi(x) = x\log(1/x)$ if $0 < x < 1/2$. Then there exists a real measure μ such that*

(i) $|\mu(I)| \le \phi(|I|)$, *for any interval I with $|I| \le 1/2$.*
(ii) P_μ *has non-tangential limit 0 at the origin.*
(iii) $D_{sym}\mu(0)$ *does not exist.*

The measure in Proposition 2 is a modification of Loomis's above example and shows (once more) that in general the converse of Fatou's theorem is not true. We say that μ is a little Zygmund measure ($\mu \in \lambda_*$) if $f_\mu \in \lambda_*$, i.e., $f_\mu \in \Lambda_*$ and

$$\sup_x \frac{|f_\mu(x+h) + f_\mu(x-h) - 2f_\mu(x)|}{|h|} \to 0, \quad \text{as } h \to 0.$$

Proposition 3 ([13]). *Let μ be a little Zygmund measure. If $\lim_{y \to 0+} u(x_0, y) = L$, then $D\mu(x_0) = L$.*

Proposition 4 ([13]). *Let $\mu \in \lambda_*$ be a positive measure, $\mu \neq 0$. Assume that μ is singular. Then for each $0 \leq \alpha \leq \infty$ the set $E_\alpha = \{x \in \mathbb{R} \mid D_\mu(x) = 0\}$ has Hausdorff dimension 1.*

Results that are used in [13] include the following.

Theorem (Zygmund). *Let u be an harmonic function on Π^+. Then $u = P_f$ with $f \in \Lambda_*(\mathbb{R})$ if and only if*

$$\sup_x \left| \frac{\partial^2 u}{\partial y^2}(x, y) \right| \leq \frac{C}{y} \quad \forall\, y > 0.$$

Lemma 1. *Let $f \in \Lambda_*$ and $u = P_f$. Then*

$$|u(x+t, y+s) + u(x-t, y-s) - 2u(x,y)| \leq C|h|, \tag{48.0.8}$$

for any $h = (t, s)$ and (x, y) with $y \geq |s| \geq 0$.

Lemma 2. *Let μ be a Zygmund measure. There exists a positive constant $C > 0$ so that*

$$\left| \int_s^t P_\mu(x, y)dx - \mu((s, t)) \right| \leq Cy \log\left(\frac{t-s}{y} \right),$$

for any $s < t$ and $0 < y \leq (t-s)/2$.

Theorem 3 (Wiener). *Let $k \in L^1(d\tau)$ be such that*

$$\hat{k}(y) = \int_0^\infty k(r)r^{-iy} \frac{dr}{r} \neq 0$$

for all $y \in \mathbb{R}$ ($d\tau = \frac{dr}{r}$). Assume that there exists $M \in L^\infty(d\tau)$ such that $\lim_{r \to 0}(M*k)(r) = L\hat{k}(0)$, then $\lim_{r \to 0}(M*f)(r) = L\hat{f}(0)$ for all $f \in L^1(d\tau)$. Here, the convolution is defined by

$$(f*g)(r) = \int_0^\infty f\left(\frac{r}{s}\right) g(s) \frac{ds}{s}.$$

Lemma 3. *Let μ be a Zygmund measure. Then $P_\mu(x_0, \cdot)$ is bounded if and only if $M \in L^\infty$. Here,*

$$M(t) = \frac{\mu((x_0 - t, x_0 + t))}{2t}.$$

If $k(t) = \frac{2t}{\pi(1+t^2)}$ then a straightforward computation gives

$$\hat{k}(y) = \frac{2}{\pi} \int_0^\infty \frac{t^{-iy}}{1+t^2} dt = \frac{1}{\pi} \Gamma\left(\frac{1}{2} + i\frac{y}{2}\right) \Gamma\left(\frac{1}{2} - i\frac{y}{2}\right) = \frac{1}{\cosh(\pi y/2)} \neq 0,$$

for all $y \in \mathbb{R}$, and $\hat{k}(0) = 1$. Using Fubini's theorem twice and the change of variables $y = s|x_0 - t|$, we get $(M * k)(r) = \frac{1}{r} \int_0^r u(x_0, y) dy$.

Lemma 4. *Let μ be as usual and consider $M(t)$ defined above. For any $c > 0$ there exists a positive function f_c defined on $\mathbb{R}^+$ and depending only on c, such that:*

(i) $\hat{f}_c(0) = \int_0^\infty f_c(x) \frac{dx}{x} = 1.$

(ii) $(M * f_c)(r) = \frac{1}{2r} \int_{x_0-r}^{x_0+r} u(x, cr) dx, \; r > 0.$

Finally, from Chapter 4 in [86], we derive the following theorem.

Theorem (Pommerenke). *Let f be an inner function in the little Bloch space which is not a finite Blaschke product. Then for each $|w| < 1$, there exists a set $E_w \subseteq \partial U = \mathbb{T}$ of Hausdorff dimension one, such that $\lim_{r\to 1} f(re^{i\theta}) = w$ for all $e^{i\theta} \in E_w$.*

Remark 48.0.103. We recall that $f \in \mathcal{B}_0$ (the little Bloch space) if $f \in H(U)$ and $\lim_{\delta\to 0+} \sup_{|z|\geq 1-\delta} (1 - |z|^2)|f'(z)| = 0$.

Chapter 49

Conclusions on the Topology of the Set of Points at Which a Measure has a Derivative

The results of this chapter are original results of the author.

Theorem 49.0.104. *Let μ be a positive Borel measure on $\mathbb{T}$ and let*

$$\mathcal{D}(\mu) = \{e^{i\theta} \in \mathbb{T} | D\mu(e^{i\theta}) \text{ exists and is finite, } 0 \leq D\mu(e^{i\theta}) < \infty\}.$$

Then $\mathbb{T} - \mathcal{D}(\mu)$ is a $G_{\delta\sigma}$ of Lebesgue measure zero.

Proof. By the theorem of Loomis, [67], P_μ has non-tangential finite limits at every point $e^{i\theta} \in \mathcal{D}(\mu)$. By Fatou's theorem, it follows that in this case the set on $\mathbb{T}$ at which P_μ has non-tangential limits is exactly $\mathcal{D}(\mu)$. By the theorem of Kolesnikov, [58], $\mathbb{T} - \mathcal{D}(\mu)$ is a $G_{\delta\sigma}$ of measure zero. $\square$

Remark 49.0.105. We used the fact that P_μ has a non-tangential limit at $e^{i\theta} \in \mathbb{T}$ if and only if it has a radial limit there.

Theorem 49.0.106. *Let μ be a positive Borel measure on $\mathbb{T}$ and let*

$$\mathcal{D}_\infty(\mu) = \{e^{i\theta} \in \mathbb{T} \mid D\mu(e^{i\theta}) = +\infty\}.$$

Assume that $\mathbb{T} = \mathcal{D}(\mu) \cup \mathcal{D}_\infty(\mu)$. Then $\mathcal{D}_\infty(\mu)$ is a G_δ of measure zero.

Proof. This time, the function $\exp(-P_\mu)$ is in $H^\infty(U)$ and every point of $\mathbb{T}$ is a Fatou point of it. It vanishes on $\mathbb{T}$ exactly on $\mathcal{D}_\infty(\mu)$. Now, the conclusion follows by Danielyan theorem. $\square$

We note that for positive Borel measures on $\mathbb{T}$, we used the following dictionary:

$\mathcal{D}(\mu)$,	the set where non-tangential limit exist.	$\mathbb{T} - \mathcal{D}(\mu)$ is $G_{\delta\sigma}$ of measure 0.
$\mathcal{D}_{sym}(\mu)$,	the set where radial limits exist.	$\mathbb{T} - \mathcal{D}_{sym}(\mu)$ is $G_{\delta\sigma}$ of measure 0.
$\mathcal{D}_{\infty}(\mu)$,	the set of zeros on $\mathbb{T}$.	$\mathcal{D}_{\infty}(\mu)$ is G_{δ} of measure 0.

Similar theorems are true for Zygmund measures on $\mathbb{T}$ (by Theorem 1, Theorem 2, Proposition 1 in [13], and from Fatou's theorem, [18, 58]). Similarly, we can use the theorems of Brossard and Chevalier (mentioned in [13]) for Borel measures on $\mathbb{T}$ that satisfy the $(\mathcal{H})$ hypothesis.

Chapter 50

Some Advanced Facts on Bounded Analytic Functions: A Continuation

We continue to gather facts on bounded analytic functions. These are collected from papers according to the author's (of this book) taste. We sometimes append the quoted results with results that are noted by the author and apparently are new.

5. [36]: As usual, U will denote the open unit disk in the complex plane $\mathbb{C}$, and $H^\infty(U)$ will denote the algebra of bounded analytic functions on U.

Theorem 1 ([36]). *Let $f \in H^\infty(U)$, and let E be a subset of $\partial U = \mathbb{T}$ such that f extends continuously to each point of E. Then there is a sequence $f_n \in H^\infty(U)$ such that each f_n extends to be analytic on some neighborhood of E, and f_n converges uniformly to f on U.*

For $E \subseteq \mathbb{T}$, let H^∞_E denote the subalgebra of $H^\infty(U)$ of functions which extend continuously to each point of E. The theorem asserts that the functions in $H^\infty(U)$ which extend analytically to a neighborhood of E are dense in H^∞_E. Combining the theorem with Carleson's Corona theorem, we obtain the following result of Détraz, [19].

Corollary ([19, 36]). *The open unit disk U is dense in the maximal ideal space of H^∞_E.*

The proof of Theorem 1 (above) relies on the following lemma.

Lemma 1 ([36]). *Let Q be a closed subset of $\mathbb{T}$, let W be an open subset of $\mathbb{C}$ at a positive distance from Q, and let $\epsilon > 0$. Let f be a Borel bounded function on $\mathbb{C}$, such that $f \in H(U)$. Suppose there is a continuous function* *u in a neighborhood of Q such that $|f(z) - u(z)| < d$, $\forall\, z \in U$ which are near Q. Then there is a bounded Borel function h such that:*

(i) *h is analytic on an open set containing $U \cup Q$.*
(ii) *h extends analytically across any arc on $\mathbb{T}$ across which f extends analytically.*
(iii) *$f - h$ is analytic on W and satisfies $|f - h| < \epsilon$ there.*
(iv) *$|f(z) - h(z)| < c_1 \cdot d$, $\forall\, z \in U$.*

The proof of Lemma 1 uses Vitushkin's scheme for approximation as developed in [104, 105]. The proof of Theorem 1 also uses the following lemma.

Lemma 2 ([36]). *Let $f \in H^\infty(U)$, and let $E \subseteq \mathbb{T}$. Suppose there is an open set V containing E, and a function u defined and continuous on V, such that $|f(z) - u(z)| < d$, $\forall\, z \in V \cap U$. Then there is an $h \in H^\infty(U)$ such that h extends to be analytic in a neighborhood of E, and $\sup_{z \in U} |f(z) - h(z)| \le c_0 \cdot d$.*

Corollary ([36]). *Let $f \in H^\infty(U)$, let $E \subseteq \mathbb{T}$, and let $d > 0$. Suppose that $\forall\, z \in E$, the diameter of the cluster set of f at z is less than d. Then there is $h \in H^\infty(U)$ such that h extends to be analytic in a neighborhood of E, and $\sup_{z \in U} |f(z) - h(z)| \le c_0 \cdot d$.*

Lemma 2 can be restated as follows.

Theorem 2 ([36]). *There is a universal constant c_0 such that for all $E \subseteq \mathbb{T}$ and all $f \in H^\infty(U)$, $d(f, L_E^\infty) \le d(f, H_E^\infty) \le c_0 \cdot d(f, L_E^\infty)$.*

In Theorem 2, L_E^∞ denotes the uniform closure of the functions in $L^\infty(d\theta)$ which extend continuously to an open set containing E. Then L_E^∞ consists of the functions in L^∞ which are constant on each "fiber" of the maximal ideal space of L^∞ lying over points of E. If we identify functions in $H^\infty(U)$ with their radial boundary values, we can regard $H^\infty(U)$ as a subalgebra of $L^\infty(d\theta)$. Under this identification, H_E^∞ becomes a subalgebra of L_E^∞. In fact, $H_E^\infty = H^\infty \cap L_E^\infty$ and H_E^∞ is a logmodular subalgebra of L_E^∞. For $f \in L^\infty(d\theta)$, we define as usual the distance from f to L_E^∞ by

$$d(f, L_E^\infty) = \inf\{\|f - g\|_\infty \,|\, g \in L_E^\infty\},$$

and we define $d(f, H_E^\infty)$ similarly.

6. **[74]:** For $f \in H^\infty(U)$, let $E(f) = \{\alpha \in \mathbb{T} \,|\, \lim_{r \to 1^-} f(r\alpha) = 0\}$. Any Blaschke product B can be factored as $B = CD$, where C and D are two Blaschke products such that $E(B) = E(C) = E(D)$. This type of factorization theorem is used in the description of the ideal structure of H^∞. Suppose that $f \in H^\infty(U)$, $\|f\|_{H^\infty} = \sup_{|z|<1} |f(z)| \leq 1$ and $\lim_{r \to 1-} f(r) = 0$. Are there g and h in $H^\infty(U)$ such that $f = g \cdot h$ and $\limsup_{r \to 1-} |h(r)| = \limsup_{r \to 1-} |g(r)| = 1$? According to the author, the inner function $S(z) = \exp\left(-\frac{1+z}{1-z}\right)$ shows that in general the answer is no. By the Riesz–Smirnov factorization theorem, the only factors of S are S^α ($0 \leq \alpha \leq 1$) but those all tend to 0 as $r \to 1^-$. On the other hand, a result of Oyma, [80], shows that every Blaschke product with radial limit 0 at 1 admits such a factorization. What about general singular inner functions or outer functions? Let f be a function in the unit ball $B = \{f \in H^\infty(U) \mid \|f\|_{H^\infty(U)} \leq 1\}$ of $H^\infty(U)$. A sequence $\{z_n\}_n \subset U$, $|z_n| \to 1$ is called a spectral sequence for f if $\lim_{n \to \infty} f(z_n) = 0$. We say that f in B is $\{z_n\}$-factorizable if there exist $g, h \in B$ such that $f = g \cdot h$ and $\limsup |h(z_n)| = \limsup |g(z_n)| = 1$, whenever $\{z_n\}$ is a spectral sequence for f. The aim of this book is to give a characterization of all factorizable functions in $H^\infty(U)$. Besides the known case of Blaschke products, it will be shown that outer functions as well as singular inner functions with associated continuous measures are factorizable. Concerning the discrete singular inner functions the result depends on the sequence $\{z_n\}$ itself. If $f \in B$ then either $f \equiv 0$ or $f = e^{i\theta} B F_\mu$, where $\theta \in [0, 2\pi]$, B is a Blaschke product and the zero-free function F_μ is given by

$$F_\mu(z) = \exp\left(\int_{|\xi|=1} \frac{z+\xi}{z-\xi} d\mu(\xi)\right), \quad \text{where } \mu = \mu_s + \log\left(\frac{1}{|f(\xi)|}\right),$$

for some positive, finite Borel measure μ_s, singular to the normalized Lebesgue measure σ on $\mathbb{T}$. The support of μ is the smallest compact subset of $\mathbb{T}$ for which $\mu(K) = \mu(\mathbb{T})$. We note that if f is not inner, then the support of μ is $\mathbb{T}$.

Theorem 1.1 ([74]). *Let $f = BF_\mu \in B$, $f \not\equiv 0$. Suppose that $\{z_n\}_n \subset U$ that converges to a point $\alpha \in \mathbb{T}$ with $\mu_s(\{\alpha\}) = 0$. Then there exists $g, h \in B$ with $f = gh$ and $\limsup |h(z_n)| = \limsup |g(z_n)| = 1$.*

Corollary 1.2 ([74]). (a) *Let f be an outer function of norm 1, or a singular inner function associated to a continuous measure, or a Blaschke product satisfying $\lim_{r \to 1-} f(re^{i\theta}) = 0$ for some $\theta \in [0, 2\pi)$. Then f admits a factorization $f = gh$, where $\limsup_{r \to 1-} |g(re^{i\theta})| = \limsup_{r \to 1-} |h(re^{i\theta})| = 1$.* (b) *The same holds if $f = S_\nu$ is a discrete singular inner function for which $e^{i\theta}$ is a point of the support of ν with $\nu(\{e^{i\theta}\}) = 0$.*

Now, a characterization of functions that are not $\{z_n\}$-factorizable.

Theorem 1.3 ([74]). *Let $f = BF_\mu \in B$, $f \not\equiv 0$. Suppose that $\{z_n\}_n$ is a spectral sequence for f, or more generally, any sequence in U converging to the boundary of U. Then f is not $\{z_n\}$-factorizable if and only if $\{z_n\}$ converges to a point $\alpha \in \mathbb{T}$ with $\mu_s(\{\alpha\}) > 0$ and*

$$\liminf \frac{1 - |z_n|^2}{|\alpha - z_n|^2} > 0.$$

Theorem 2.1 ([74]). *A non-zero prime ideal I in $H^\infty(U)$ is countably generated if and only if $I = M(z_0) := \{f \in H^\infty(U) | f(z_0) = 0\}$ for some $z_0 \in U$ or $I = I(S_\sigma, S_\sigma^{1/2}, S_\sigma^{1/3}, \dots)$ for some σ with $|\sigma| = 1$.*

A result communicated to the author of [74] by Gamelin is described in what follows.

Theorem 2.2 (Gamelin). *The only non-zero prime ideals in $H^\infty(U)$, whose hull (i.e., the zero set) are G_δ-sets, are the ideals $p = M(z_0) := \{f \in H^\infty(U) | f(z_0) = 0\}$ for some $z_0 \in U$ or $p = I(S_\sigma, S_\sigma^{1/2}, S_\sigma^{1/3}, \dots)$ for some σ with $|\sigma| = 1$.*

7. [52]: $H^\infty(U)$ functions are identified with their boundary functions. Thus, H^∞ is considered as an essentially supremum norm closed subalgebra of $L^\infty(\mathbb{T})$, the space of bounded measurable functions on $\mathbb{T}$ with respect to the Lebesgue measure. Sarason, [92], proved that $H^\infty(U) + C(\mathbb{T})$ is a closed subalgebra of $L^\infty(\mathbb{T})$, where $C(\mathbb{T})$ is the set of continuous functions on $\mathbb{T}$. We denote by $M(H^\infty + C)$ the maximal ideal space of $H^\infty(U) + C(\mathbb{T})$. Guillory and Sarason proved, [38], that there is a positive integer N such that if $f \in H^\infty + C$ and b is an inner function with $|f| \leq |b|$ on $M(H^\infty + C)$, then $f^N/b = f^n\bar{b} \in H^\infty + C$, and we cannot take $N = 1$. K. Izuchi and Y. Izuchi proved, [53], that we can take $N = 2$. In this book, the author assumes that b is a Blaschke product and studies the case $|f| \leq |b|$ on $M(H^\infty + C)$ and $f\bar{b} \notin H^\infty + C$. His aim is to investigate the kind of small changes of f or b, say g and ψ, respectively, that make $g\bar{b} \in H^\infty + C$ or $f\bar{\psi} \in H^\infty + C$. An important role in the proofs of several of the theorems is played by Hoffman's factorization theorem for Blaschke products, Theorem 5.2 in [46]. The author gives in Chapter 3 an additional property in Hoffman's factorization theorem, that zero sets of its factors having zeros of infinite order coincide with each other. In Chapter 4, the author proves that if $f \in H^\infty + C$ and b is a Blaschke product with $|f| \leq |b|$ on $M(H^\infty + C)$, then there is a subproduct ψ of b such that $f\bar{\psi} \in H^\infty + C$ and $Z(\psi) = Z(b)$, and there is a function g in $H^\infty + C$ such that $|g| = |f|$ on $M(H^\infty + C)$ and $g\bar{b} \in H^\infty + C$. In Chapter 5, the author gives a sufficient condition for which the absolute moduli of two Blaschke products coincide on $M(H^\infty + C)$. As usual, the function

$$b(z) = \prod_{n=1}^{\infty} \frac{-\overline{z_n}}{|z_n|} \frac{z - z_n}{1 - \overline{z_n}z}, \quad z \in U,$$

where the sequence $\{z_n\}_n \subset U$ satisfies the Blaschke condition $\sum_{n=1}^{\infty}(1 - |z_n|) < \infty$, is called a Blaschke product with zero sequence $\{z_n\}_n$. If $\{z_n\}_n$ satisfies in addition the condition

$$\inf_k \prod_{\{n|n \neq k\}} \left| \frac{z_k - z_n}{1 - \overline{z_n}z_k} \right| > 0,$$

then $\{z_n\}_n$ and $b(z)$ are called interpolating. In this case, for every bounded sequence $\{a_n\}_n$, there is a function $f \in H^\infty$ such that $f(z_n) = a_n$ for every $n \in \mathbb{Z}^+$. A function $I \in H^\infty(U)$ is called inner if $|I| = 1$ in $L^\infty(\mathbb{T})$. A Blaschke product is a typical inner function. An essentially supremum norm closed algebra between $H^\infty(U)$ and $L^\infty(\mathbb{T})$ is called a Douglas algebra. Every Douglas algebra is generated by $H^\infty(U)$ and the complex conjugates of some interpolating Blaschke products. For $\Lambda \subseteq L^\infty(\mathbb{T})$, we denote by $[H^\infty, \Lambda]$ the Douglas algebra generated by $H^\infty(U)$ and Λ. For a Douglas algebra B, $M(B)$ denotes the maximal ideal space of B. $M(B)$ can be considered as a closed subset of $M(H^\infty)$ and $M(H^\infty + C) = M(H^\infty) - U$. Also, $M(L^\infty)$ can be considered to be the Shilov boundary for every B. We identify a function with its Gelfand transform. For $E \subseteq M(H^\infty)$, $\operatorname{cl} E$ or $\overline{E}$ will denote the closure of E in $M(H^\infty)$ (provided that $\overline{E}$ cannot be confused with complex conjugation). For points $x, y \in M(H^\infty)$, we put

$$\rho(x, y) = \sup\left\{ |f(y)| \mid f \in H^\infty(U), \|f\|_{H^\infty(U)} \le 1, f(x) = 0 \right\}.$$

It is well known that $\rho(z, w) = |z - w|/|1 - \overline{z}w|$ for $z, w \in U$. A set $P(x) = \{y \in M(H^\infty)|\rho(x, y) < 1\}$ is called a Gleason part. In the paper of Hoffman [46], he studies those Gleason parts. He showed that if $P(x) \neq \{x\}$, there is a 1–1 continuous map L_x from U onto $P(x)$ such that $f \circ L_x \in H^\infty(U)$, $\forall f \in H^\infty(U)$ and $(f \circ L_x)(0) = f(x)$. For $f \in H^\infty + C$ we put, $Z(f) = \{\xi \in M(H^\infty + C) \mid f(\xi) = 0\}$. For $x \in Z(f)$, we put

$$\operatorname{ord}(f, x) = \begin{cases} \operatorname{ord}(f \circ L_x, 0) & \text{if} \quad P(x) \neq \{x\}, \\ \infty & \text{if} \quad P(x) = \{x\}, \end{cases}$$

where $\operatorname{ord}(f \circ L_x, 0)$ is the usual order of the zero of the analytic function $f \circ L_x$ at 0. We note that $\operatorname{ord}(f, x) = \infty$ if and only if $f = 0$ on $P(x)$.

We denote by $Z_\infty(f)$ the set of $x \in Z(f)$ with $\operatorname{ord}(f, x) = \infty$. If b is an interpolating Blaschke product with zeros $\{z_n\}_n$, then $Z(b) = \overline{\{z_n\}_n} - \{z_n\}_n$, and $\operatorname{ord}(b, x) = 1$ for $x \in Z(b)$. The following theorem gives an additional property of Hoffman's factorization theorem. For a Blaschke product b with zeros $\{z_n\}_n$, let

$$K_\sigma(b) = \bigcap_{n=1}^{\infty} \{z \mid \rho(z, z_n) \geq \sigma\} \quad \text{for } \sigma > 0.$$

Theorem 3.1 ([52]). *Let b be a Blaschke product with $Z_\infty(b) \neq \emptyset$. Then b admits a factorization $b = b_1 b_2$, b_1, b_2 are Blaschke products such that $Z_\infty(b_1) = Z_\infty(b_2) = Z_\infty(b)$.*

Corollary 3.1 ([52]). *Let b be a Blaschke product. Then the set $Z_\infty(b)$ is a closed G_δ-subset of $M(H^\infty)$.*

Proof. For $f \in H^\infty(U)$, put $Z_0(f) = \{x \in M(H^\infty) \mid f(x) = 0\}$. Then $Z_0(f)$ is a closed G_δ-subset of $M(H^\infty)$. Let $\Lambda = \{(i_1, i_2, \ldots, i_k) \mid i_j = 0 \text{ or } 1, k = 1, 2, \ldots\}$. Then Λ is a countable set. Using Theorem 3.1, we can define a sequence of Blaschke products $\{b_\alpha \mid \alpha \in \Lambda\}$ as follows: $b = b_0 b_1$ and $b_\alpha = b_{\alpha_1} b_{\alpha_2}$. Then $Z_0(b_\alpha) \supseteq Z_\infty(b_\alpha) = Z_\infty(b)$, so $Z_\infty(b) \subseteq \bigcap \{Z_0(b_\alpha) \mid \alpha \in \Lambda\}$. Let $x \in Z_0(b) - Z_\infty(b)$. Then $b_\alpha(x) \neq 0$ for some $\alpha \in \Lambda$. Hence, $\bigcap \{Z_0(b_\alpha) \mid \alpha \in \Lambda\} \subseteq Z_\infty(b)$. Therefore, we have the equality $Z_\infty(b) = \bigcap \{Z_0(b_\alpha) \mid \alpha \in \Lambda\}$ which is a closed G_δ-subset of $M(H^\infty)$. $\square$

Division problems in $H^\infty + C$.

Lemma 4.1 ([52]). *Let B be a Douglas algebra and let I be an inner function. Then $IB \subseteq H^\infty + C$ if and only if $I = 0$ on $M(H^\infty + C) - M(B)$.*

Lemma 4.2 ([52]). *Let b be a Blaschke product. If $f \in H^\infty + C$ and $|f| \leq |b|$ on $M(H^\infty + C)$, then $\operatorname{ord}(b, x) = \infty \ \forall x \in M(H^\infty + C) - M([H^\infty, f\bar{b}])$.*

Corollary 4.1 ([52]). *Let b be a Blaschke product, and let $b = b_1 b_2$ be a factorization given in Theorem 3.1. If $f \in H^\infty + C$ and $|f| \leq |b|$ on $M(H^\infty + C)$, then $f\bar{b_i} \in H^\infty + C$ for $i = 1, 2$.*

Lemma 4.3 ([52]). *Let b be a Blaschke product with distinct zeros $\{z_n\}_n$. Then b admits a factorization $b = b_0 b_1$ such that*

(i) *If $b_0(z_n) = 0$ then $(1 - |z_n|^2)|b_0'(z_n)| \geq |b_1(z_n)|$.*
(ii) *If $b_1(z_n) = 0$ then $(1 - |z_n|^2)|b_1'(z_n)| \geq |b_0(z_n)|$.*

Moreover, if x is a point in $Z(b)$ with $2 \leq \operatorname{ord}(b,x) < \infty$, then $b_0(x) = b_1(x) = 0$.

Lemma 4.4 ([52]). *Let b be a Blaschke product with zeros $\{z_n\}_n$. If $\operatorname{ord}(b,x) = 1 \ \forall\, x \in \operatorname{cl}\{z_n\}_n$, then b is an interpolating Blaschke product.*

Lemma 4.5 ([52]). *Let b be a Blaschke product with zeros $\{z_n\}_n$. If $\operatorname{ord}(b,x) < \infty \ \forall\, x \in \operatorname{cl}\{z_n\}_n$, then b is a product of finitely many interpolating Blaschke products.*

Lemma 4.6 ([52]). *Let b be a Blaschke product. Then there is a sequence of interpolating Blaschke products $\{b_n\}_n$ such that*

(i) *$b = \prod_{n=1}^{\infty} b_n$, and*
(ii) *If $x \in Z(b) - Z_{\infty}(b)$, then $\operatorname{ord}(\prod_{n=1}^{k} b_n, x) = \operatorname{ord}(b,x)$ for some k depending on x.*

Lemma 4.7 ([52]). *Let $\{V_{s,n}\}_{s,n=1}^{\infty}$ be a family of compact subsets of U. Let $\{I_j\}_j$ be a sequence of Blaschke products. Moreover, we assume that $b = \prod_{j=1}^{\infty} I_j$ is a Blaschke product. Then we have*
(i) *If*

$$\sup_{\xi \in V_{s,n}} \left| \left(b \prod_{j=1}^{k} \overline{I_j} \right)(\xi) \right| \to 0 \text{ as } n \to \infty \ \forall\, s,\, k,$$

then there is a sequence of tails J_j of I_j such that

$$\sup_{\xi \in V_{s,n}} \left| \left(b \prod_{j=1}^{\infty} \overline{J_j} \right)(\xi) \right| \to 0 \text{ as } n \to \infty \ \forall\, s.$$

(ii) (the dual version) *If*

$$\inf_{\xi \in V_{s,n}} \left| \left(\prod_{j=1}^{k} I_j \right)(\xi) \right| \to 1 \text{ as } n \to \infty \ \forall\, s,\, k,$$

then there is a sequence of tails J_j of I_j such that

$$\inf_{\xi \in V_{s,n}} \left| \left(\prod_{j=1}^{\infty} J_j \right)(\xi) \right| = 1 \text{ as } n \to \infty \ \forall\, s.$$

Theorem 4.1 ([52]). *Let b be a Blaschke product. If $f \in H^\infty + C$ and $|f| \le |b|$ on $M(H^\infty + C)$, then there is a subproduct ψ of b such that $f\overline{\psi} \in H^\infty + C$, $Z(\psi) = Z(b)$.*

Theorem 4.2 ([52]). *Let b be a Blaschke product. If $f \in H^\infty + C$ and $|f| \le |b|$ on $M(H^\infty + C)$, then there is a function $g \in H^\infty + C$ such that $|g| = |f|$ on $M(H^\infty + C)$ and $g\overline{b} \in H^\infty + C$.*

Problem 4.1. In Theorem 4.1 is there a subproduct Φ of b such that $f\overline{\Phi} \in H^\infty + C$ and $|\Phi| = |b|$ on $M(H^\infty + C)$?

Theorem 4.3 ([52]). *Let b be the Blaschke product*

$$b(z) = \prod_{n=1}^{\infty} \left(\frac{-\overline{z_n}}{|z_n|} \frac{z - z_n}{1 - \overline{z_n}z} \right)^{k_n},$$

where $k_n \to \infty$ as $n \to \infty$. If $f \in H^\infty + C$ and $|f| \le |b|$ on $M(H^\infty + C)$, then there is a subproduct ψ of b such that $f\overline{\psi} \in H^\infty + C$ and $|\psi| = |b|$ on $M(H^\infty + C)$.

Absolute moduli of Blaschke products on $M(H^\infty + C)$.

Theorem 5.1 ([52]). *Let b_1 and b_2 be Blaschke products with zeros $\{z_n\}_n$ and $\{w_n\}_n$ respectively. If $\rho(z_n, w_n) \to 0$ as $n \to \infty$, then $|b_1| = |b_2|$ on $M(H^\infty + C)$.*

Lemma 5.1 ([34], pp. 310, 404). *Let $\{\xi_n\}_n$ be an interpolating sequence with*

$$\inf_{k} \prod_{\{n \,|\, n \ne k\}} \rho(\xi_n, \xi_k) \ge \delta > 0.$$

Then there is $\lambda = \lambda(\delta)$, $0 < \delta < 1$, such that $V_n = \{z \in U \,|\, \rho(z, \xi_n) < \lambda\}$ are pairwise disjoint domains, and such that if $w_n \in V_n$, then $\{w_n\}_n$ is an interpolating sequence. Moreover, $\lambda(\delta) \to 1$ as $\delta \to 1^+$.

Theorem 5.2 ([52]). *Let b be a Blaschke product and let q be a product of finitely many interpolating Blaschke products with $|b| \le |q|$ on $M(H^\infty + C)$. Then there is a subproduct b_0 of b such that $|b| = |b_0 q|$ on $M(H^\infty + C)$.*

Lemma 5.2 ([34], p. 439). *Let E and F be subsets of U such that $(\mathrm{cl}\, E) \cap (\mathrm{cl}\, F)$ contains a point x with $P(x) \ne \{x\}$. Then $\inf\{\rho(z, w) \,|\, z \in E, w \in F\} = 0$.*

Lemma 5.3 ([52]). *Let $f \in H^\infty(U)$ and let q be an interpolating Blaschke product with $Z(q) \subseteq Z(f)$ and $Z(q) \not\subseteq Z_\infty(f)$. If B is the Blaschke factor of f, then there is an interpolating subproduct b_0 of B such that $Z(b_0) \subseteq Z(q)$ and $Z(b_0) \supseteq Z(q) - Z_\infty(f)$.*

Problem 5.1. Is the assertion of Theorem 5.2 true when q is a general Blaschke product?

Theorem 5.3 ([52]). *Let b be a product of finitely many interpolating Blaschke products. Let b_n be the n'th tail of b. Then for every $f \in H^\infty(U)$, $\lim_{n\to\infty} \|f + b_n H^\infty\| = \|f + b(H^\infty + C)\|$.*

Lemma 5.4 ([52]). *Let $f \in H^\infty(U)$ and let $q = \prod_{j=1}^n q_j$, where q_j is an interpolating Blaschke product. If $\mathrm{ord}(f, x) \geq \mathrm{ord}(q, x)\ \forall\, x \in Z(q)$, then there is a factorization $f = \prod_{j=1}^n f_j$ such that $f_j \in H^\infty(U)$, $\|f_j\|_{H^\infty(U)} = \|f\|_{H^\infty(U)}^{1/n}$, and $Z(f_j) \supset Z(q_j)$ for $j = 1, 2, \ldots, n$.*

Problem 5.2. Is the assertion of Theorem 1.1 true when b is a general Blaschke product?

8. [62]: Critical points of bounded analytic functions.

A sequence of points $\{z_j\}_j$ in a subdomain Ω of the complex plane $\mathbb{C}$ is called the zero set of an analytic function $f : \Omega \to \mathbb{C}$, if f vanishes precisely on this set. If the point $\xi \in \Omega$ occurs m times in the sequence, then f has a zero at ξ of precise order m, and $f(z) \neq 0$ for $z \in \Omega - \{z_j\}_j$.

Theorem A (Jensen, Blaschke, Nevanlinna and Nevanlinna). *Let $\{z_j\}_j$ be a sequence in U. Then the following statements are equivalent.*

(a) *There is an analytic self-map of U with zero set $\{z_j\}_j$.*
(b) *There is a Blaschke product with zero set $\{z_j\}_j$.*
(c) *The sequence $\{z_j\}_j$ is a Blaschke sequence, i.e., $\sum_j(1 - |z_j|) < \infty$.*

A sequence of points $\{z_j\}_j$ in a domain $\Omega \subseteq \mathbb{C}$ is the critical set of an analytic function $f : \Omega \to \mathbb{C}$, if $\{z_j\}_j$ is the zero set of $f'(z)$. One aim of this survey paper [62] is to point out at the following analogue of Theorem A, for the critical sets of bounded analytic functions (instead of their zero sets).

Theorem 1.1 ([62]). *Let $\{z_j\}_j$ be a sequence in U. Then the following statements are equivalent.*

(a) *There is an analytic self-map of U with the critical set $\{z_j\}_j$.*
(b) *There is a Blaschke product with the critical set $\{z_j\}_j$.*
(c) *There is a function in the weighted Bergman space,*

$$\mathcal{A}_1^2 = \left\{ g : U \to \mathbb{C} \ \middle| \ \iint_U \left(1 - |z|^2\right) |g(z)|^2 d\sigma_z < \infty \right\},$$

with zero set $\{z_j\}_j$.

We denote by σ_z the two-dimensional Lebesgue measure with respect to z. The special case of finite sequences, a result related to Theorem 1.1 can be found in [42, Section 29]. It was proved there that for every finite $\{z_j\}_j$ in U there is always a finite Blaschke product whose critical set is $\{z_j\}_j$. A generalization by Kraus and Roth is that every Blaschke sequence $\{z_j\}_j$ is the critical set of some Blaschke product. The converse though, known as the Bloch–Nevanlinna conjecture is false. According to a result of Frostman, there exist Blaschke products whose critical sets fail to satisfy the Blaschke condition. Hence, the critical sets of bounded analytic functions are not just the Blaschke sequences. Theorem 1.1 identifies the critical sets of bounded analytic functions with the zero sets of functions in the Bergman space $\mathcal{A}_1^2$. There is no simple geometric characterization of these zero sets yet. The implications (b)$\Rightarrow$(a) and (a)$\Rightarrow$(c) of Theorem 1.1 are not hard to prove. The statement (b)$\Rightarrow$(a) is clear. (a)$\Rightarrow$(c) follows directly from the Littlewood–Paley identity, which says that for any holomorphic function $f : U \to U$ the derivative, f', belongs to $\mathcal{A}_1^2$. It is, however, not true that any $\mathcal{A}_1^2$ function is the derivative of a bounded analytic function. So, (c)$\Rightarrow$(a) is more subtle.

Conformal metrics and maximal Blaschke products.

An essential tool of the proof of the implication (a)$\Rightarrow$(b) in Theorem 1.1 is the use of negatively curved conformal pseudo-metrics. The authors review this topic. A non-negative upper semi-continuous function λ on G, $\lambda : G \to [0, +\infty)$, $\lambda \not\equiv 0$, is called a conformal pseudo-metric on G. If $\lambda(z) > 0 \ \forall\, z \in G$, we say that $\lambda(z)|dz|$ is a conformal metric on G. We call a conformal pseudo-metric $\lambda(z)|dz|$ regular on G, if λ is of class C^2 in $\{z \in G \mid \lambda(z) > 0\}$.

Example 2.1 (The hyperbolic metric). This is the Poincaré metric

$$\lambda_U(z)|dz| = \frac{|dz|}{1 - |z|^2}, \quad z \in U.$$

Clearly, $\lambda_U(z)|dz|$ is a regular conformal metric on U.

The Gauss curvature $\mathcal{K}_\lambda$ of a regular conformal pseudo-metric $\lambda(z)|dz|$ on G is defined by

$$\mathcal{K}_\lambda(z) = -\frac{\Delta(\log \lambda)(z)}{\lambda(z)^2} \quad \forall z \in G \text{ where } \lambda(z) > 0.$$

Thus, if $\lambda(z)|dz|$ is a regular conformal metric with curvature $\mathcal{K}_\lambda = \mathcal{K}$ on G, then the function $u = \log \lambda$ satisfies the following PDE, the so-called Gauss curvature equation,

$$\Delta u = -\mathcal{K}(z)e^{2u} \text{ on } G. \tag{50.0.1}$$

Conversely, if a real-valued C^2-function u is a solution of the Gauss curvature equation (50.0.1) on G, then $\lambda(z) = e^{u(z)}$ induces a regular conformal metric $e^{u(z)}|dz|$ on G with curvature $\mathcal{K}$.

Example 2.2. The hyperbolic metric $\lambda_U(z)|dz|$ on U has a constant negative curvature -4.

Let $\lambda(w)|dw|$ be a conformal pseudo-metric on a domain D and let $f : G \to D$, $w = f(z)$ be a non-constant analytic map from another domain G to D. Then the pullback of $\lambda(w)|dw|$ under the map f is the following conformal pseudo-metric $(f^*\lambda)(z)|dz| = \lambda(f(z))|f'(z)||dz|$ defined on G. Gauss curvature is important because it is a conformal invariant.

Theorem B (Theorema Egregium). For every analytic map $w = f(z)$ and every regular conformal pseudo-metric $\lambda(w)|dw|$, the relation $\mathcal{K}_{f^*\lambda}(z) = \mathcal{K}_\lambda(f(z))$ holds, provided $\lambda(f(z))|f'(z)| > 0$.

If $\lambda(w)|dw|$ is a conformal metric on D (without zeros) and if $w = f(z)$ is a non-constant map from G to D, then $(f^*\lambda)(z)|dz|$ is a pseudo-metric on G with zeros exactly at the critical points of f. In order to take the multiplicity into account, it is convenient to introduce the following.

Definition (Zero set of a pseudo-metric). Let $\lambda|dz|$ be a conformal pseudo-metric on G. We say that $\lambda(z)|dz|$ has a zero of order $m_0 > 0$ at $z_0 \in G$ if

$$\lim_{z \to z_0} \frac{\lambda(z)}{|z - z_0|^{m_0}},$$

exists and is $\neq 0$.

We call a sequence $\mathcal{C} = \{\xi_j\}_j \subseteq G$, $\{\xi_j\}_j = \{\underbrace{z_1, \ldots, z_1}_{m_1}, \underbrace{z_2, \ldots, z_2}_{m_2}, \ldots\}$, $z_k \neq z_n$ if $k \neq n$, the zero set of a conformal pseudo-metric $\lambda(z)|dz|$, if $\lambda(z) > 0 \ \forall z \in G - \mathcal{C}$ and if $\lambda(z)|dz|$ has a zero of order $m_k \in \mathbb{Z}^+$ at z_k for all

k. Every conformal pseudo-metric of a constant curvature -4 can locally be represented by an holomorphic function. This is the content of the following important result.

Theorem C (Liouville's Theorem). *Let $\mathcal{C}$ be a sequence of points in a simply connected domain G and let $\lambda(z)|dz|$ be a regular conformal pseudo-metric on G with a constant curvature -4 on G and the zero set $\mathcal{C}$. Then $\lambda(z)|dz|$ is a pull-back of the hyperbolic metric $\lambda_U|dz|$ under some analytic map $f : G \to U$, i.e.,*

$$\lambda(z) = \frac{|f'(z)|}{1 - |f(z)|^2}, \quad z \in G. \tag{50.0.2}$$

If $g : G \to U$ is another analytic function, then $\lambda(z) = (g^*\lambda_U)(z)$ for $z \in G$ if and only if $g = T \circ f$ for some conformal automorphism T of U.

It is not so clear what is the role of mentioning $\mathcal{C}$ in the theorem except saying implicitly that this is the critical set of f (in U).

Definition. A holomorphic function f with the property (50.0.2) will be called a developing map for $\lambda(z)|dz|$. We note that the critical set of each developing map coincides with the zero set of the corresponding conformal pseudo-metric.

Example 2.3. The developing maps of the hyperbolic metric $\lambda_U(z)|dz|$ are precisely the conformal automorphisms of U, i.e., the finite Blaschke products of degree 1.

Here is a variant of Liouville's Theorem.

Theorem 2.5. *Let G be a simply connected domain and let $h : G \to \mathbb{C}$ be an analytic map. If $\lambda(z)|dz|$ is a regular conformal metric with curvature $-4|h(z)|^2$, then there exists a holomorphic function $f : G \to U$ such that*

$$\lambda(z) = \frac{1}{|h(z)|} \cdot \frac{|f'(z)|}{1 - |f(z)|^2}, \quad z \in G.$$

Moreover, f is uniquely determined up to post composition with a unit disk automorphism.

Maximal conformal pseudo-metrics and maximal Blaschke products.

Apart from having constant negative curvature, the hyperbolic metric on U has another important property: It is maximal among all regular conformal pseudo-metrics on U with curvature bounded above by -4. This is a result of Ahlfors which is usually referred to as Ahlfors' Lemma.

Theorem D (The Fundamental Theorem). *Let $\lambda(z)|dz|$ be a regular conformal pseudo-metric on U with curvature bounded above by -4. Then $\lambda(z) \leq \lambda_U(z)$ for every $z \in U$.*

Beardon and Minda proposed to call Ahlfors' Lemma the Fundamental Theorem. As a result of the Fundamental Theorem, we have

$$\lambda_U(z) = \max\left\{\lambda(z) \mid \lambda|dz| \text{ is a regular conformal pseudo-metric on U}\right.$$
$$\left.\text{with curvature} \leq (-4)\right\} \quad \forall\, z \in U.$$

Remark (Developing maps and universal coverings). Let G be a hyperbolic sub-domain of $\mathbb{C}$, i.e., the complement $\mathbb{C}$–G consists of more than one point. A regular conformal metric $\lambda_G(z)|dz|$ of constant curvature -4 on G is said to be the hyperbolic metric for G, if it is the maximal regular conformal pseudo-metric with curvature $\leq (-4)$ on G, i.e., $\lambda(z) \leq \lambda_G(z)$, $z \in G$, for all regular conformal pseudo-metrics $\lambda(z)|dz|$ on G with curvature bounded above by -4. Then $\lambda_G(z)|dz|$ and $\lambda_U(z)|dz|$ are connected via the universal covering $\pi : U \to G$ of G by the formula $(\pi^*\lambda_G)(z)|dz| = \lambda_U(z)|dz|$. Hence every branch of the inverse of a universal covering map $\pi : U \to G$ is locally the developing map of the hyperbolic metric $\lambda_G(z)|dz|$ of G. In particular, if G is a hyperbolic simply connected domain, then the developing maps for $\lambda_G(z)|dz|$ are precisely the conformal mappings from G onto U. We now consider prescribed zeros.

Theorem E. *Let $\mathcal{C} = \{\xi_j\}_j$ be a sequence of points in G and*

$$\Phi_{\mathcal{C}} = \{\lambda \mid \lambda|dz| \text{ is a regular conformal pseudo-metric in } G \text{ with}$$
$$\text{curvature} \leq -4 \text{ and zero set, } C^* \supseteq \mathcal{C}\}.$$

If $\Phi_{\mathcal{C}} \neq \emptyset$, then $\lambda_{\max}(z) = \sup\{\lambda(z) \mid \lambda \in \Phi_{\mathcal{C}}\}$, $z \in G$, induces the unique maximal regular conformal pseudo-metric $\lambda_{\max}(z)|dz|$ on G with constant curvature -4 and zero set $\mathcal{C}$.

Thus, if $\Phi_{\mathcal{C}} \neq \emptyset$, i.e., if there is at least one regular conformal pseudo-metric $\lambda(z)|dz|$ on G with curvature ≤ -4 whose zero set contains the given sequence $\mathcal{C}$, then there exists a (maximal) regular conformal pseudo-metric $\lambda(z)|dz|$ on G with constant curvature -4, whose zero set is exactly the sequence $\mathcal{C}$. In particular, Theorem E can be applied, if there exists a non-constant holomorphic function $f : G \to U$ with critical set $C^* \supseteq \mathcal{C}$ since then the pseudo-metric $(f^*\lambda_U)(z)|dz|$ belongs to $\Phi_{\mathcal{C}}$. The proof of Theorem E relies on a modification of Perron's method and can be found in [42, Sections 12 and 13].

Example 2.4. If G is a hyperbolic domain and $\mathcal{C} = \emptyset$, then the maximal regular conformal pseudo-metric $\lambda_{\max}(z)|dz|$ on G with constant curvature -4 and empty zero set is exactly the hyperbolic metric $\lambda_G(z)|dz|$ for G.

Thus, the maximal pseudo-metrics are generalizations of the hyperbolic metric and their developing maps are of special interest.

Definition (Maximal functions). Let $\mathcal{C}$ be a sequence of points in U such that there exists a maximal regular conformal pseudo-metric $\lambda_{\max}(z)|dz|$ on U with constant curvature -4 and zero set $\mathcal{C}$. Then every developing map for $\lambda_{\max}(z)|dz|$ is called a maximal function for $\mathcal{C}$.

Remark. By Theorem C, every maximal function is uniquely determined by its critical set $\mathcal{C}$ up to post composition with a unit disk automorphism and, conversely, the post composition of any maximal function with a unit disk automorphism is again a maximal function. Next, if $\mathcal{C} = \emptyset$, then the maximal functions for $\mathcal{C}$ are precisely the unit disk automorphisms, i.e., the finite Blaschke product of degree 1.

Theorem 2.9. *Every maximal function is a Blaschke product.*

Since the post composition of any maximal function with an automorphism of the unit disk is again a maximal function, the maximal function is an indestructible Blaschke product (we will talk about these later). Theorem E combined with Theorem 2.9 gives the implication "(a)$\Rightarrow$(b)" in Theorem 1.1. In fact, if a non-constant analytic self-map of U is with critical set $\mathcal{C}$, then $\lambda(z)|dz| = (f^*\lambda_U)(z)|dz|$ is a regular conformal pseudo-metric on U with zero set $\mathcal{C}$. So, by Theorem E, there exists a maximal conformal pseudo-metric on U with zero set $\mathcal{C}$. Now, Theorem 2.9 implies that the corresponding maximal function for $\mathcal{C}$ is a Blaschke product.

Remark. The case that the maximal function has finitely many critical points follows (i.e., Theorem 2.9 follows in this case) from the following result due to Heinz.

Theorem F. *Let $\mathcal{C} = \{z_1, \ldots, z_n\}$ be a finite sequence in U and $f : U \to U$ analytic. Then the following statements are equivalent.*

(a) *f is a maximal function for $\mathcal{C}$.*
(b) *f is a Blaschke product of degree $n + 1$ with the critical set $\mathcal{C}$.*

Definition (Maximal Blaschke products). A non-constant Blaschke product is called a maximal Blaschke product, if it is a maximal function for its critical set.

We mentioned earlier that every maximal Blaschke product is indestructible and every finite Blaschke product is a maximal Blaschke product. Moreover, if $\mathcal{C}$ is the critical set of any non-constant analytic function $f : U \to U$, then there is a maximal Blaschke product with the critical set $\mathcal{C}$. A maximal Blaschke product is uniquely determined by its critical set $\mathcal{C}$ up to post composition with a unit disk automorphism.

Properties of maximal Blaschke products.

Theorem 3.1 (Maximal Blaschke products as extremal functions). *Suppose that $\mathcal{C}$ is a sequence of points in U such that $\mathcal{F}_{\mathcal{C}} = \{f \in H^\infty(U) \mid f'(z) = 0 \,\forall\, z \in \mathcal{C}\}$ contains at least some non-constant function and let*

$$m = \min\{n \in \mathbb{Z}^+ \mid f^{(n)}(0) \neq 0 \text{ for some } f \in \mathcal{F}_{\mathcal{C}} \geq 1\}.$$

Then the unique extremal function to the extremal problem

$$\max\{\Re\, f^{(m)}(0) \mid f \in \mathcal{F}_{\mathcal{C}}, \|f\|_{H^\infty} \leq 1\}, \tag{50.0.3}$$

is the maximal Blaschke product F for $\mathcal{C}$ normalized by $F(0) = 0$ and $F^{(m)}(0) > 0$.

Note that if $m = 1$, then the extremal problem (50.0.3) is exactly the problem of maximizing the derivative at a point, i.e., exactly the character of Schwarz' lemma.

Remark (The Nehari–Schwarz lemma). If $\mathcal{C} = \emptyset$, then $\mathcal{F}_{\mathcal{C}} = H^\infty(U)$, i.e., the set of all the bounded analytic functions in U. In this case, Theorem 3.1 is just the statement of Schwarz' lemma. If $\mathcal{C}$ is a finite sequence, then Theorem 3.1 is Nehari's 1947 generalization of Schwarz' lemma. Just like the Schwarz–Pick lemma (says that

$$\frac{|f'(z)|}{1 - |f(z)|^2} \leq \frac{1}{1 - |z|^2}, \quad z \in U, \tag{50.0.4}$$

for any analytic map $f : U \to U$, with equality for some point $z \in U$ if and only if f is a conformal disk automorphism) is an invariant form of Schwarz' lemma, we have an invariant formulation of Theorem 3.1.

Theorem 3.4 (Sharpened Schwarz–Pick inequality). *Let $f : U \to U$ be a non-constant analytic function with critical set $\mathcal{C}$ and let $\mathcal{C}^*$ be a* *subsequence of* $\mathcal{C}$*. Then there exists a maximal Blaschke product* F *with critical set* $\mathcal{C}^*$ *such that*

$$\frac{|f'(z)|}{1 - |f(z)|^2} \leq \frac{|F'(z)|}{1 - |F(z)|^2}, \quad z \in U.$$

If $\mathcal{C}^*$ is finite then F is a finite Blaschke product. Furthermore, $f = T \circ F$ for some automorphism T of U if and only if

$$\lim_{z \to w} \frac{|f'(z)|}{1 - |f(z)|^2} \cdot \frac{1 - |F(z)|^2}{|F'(z)|} = 1,$$

for some $w \in U$. If $\mathcal{C}^* = \mathcal{C} \neq \emptyset$, this gives a sharp form of Schwarz–Pick inequality

$$\frac{|f'(z)|}{1 - |f(z)|^2} \leq \frac{|F'(z)|}{1 - |F(z)|^2} \leq \frac{1}{1 - |z|^2}, \tag{50.0.5}$$

for all $f \in \mathcal{F}_\mathcal{C}$ of the Schwarz–Pick inequality (50.0.4).

Related extremal problems in Hardy and Bergman spaces.

Let $\mathcal{C}$ be a sequence in U. Assume that $\mathcal{C}$ is the critical set of a bounded analytic function $f : U \to U$ and let N denote the multiplicity of the point 0 in $\mathcal{C}$. According to Theorem 3.1, the maximal Blaschke product F for $\mathcal{C}$ normalized by $F(0) = 0$ and $F^{(N+1)}(0) > 0$ is the unique solution of the extremal problem

$$\max\{\Re\, f^{(N+1)}(0) \,|\, f \in H^\infty(U),\ \|f\|_{H^\infty} \leq 1 \text{ and } f'(z) = 0 \text{ for } z \in \mathcal{C}\}.$$

This extremal property of a maximal Blaschke product is reminiscent of the well-known extremal property of

(i) Blaschke products in Hardy spaces $H^\infty(U)$ and $H^p(U)$ where $1 \leq p < +\infty$, and

(ii) Canonical divisors in the weighted Bergman spaces

$$\mathcal{A}_\alpha^p = \{f : U \to \mathbb{C} \text{ analytic } |\, \|f\|_{p,\alpha} < +\infty\},$$

where $-1 < \alpha < +\infty$ and $1 \leq p < +\infty$ and

$$\|f\|_{p,\alpha} \left(\frac{1}{\pi} \iint_U (1 - |z|^2)^\alpha |f(z)|^p d\sigma_z \right)^{1/p}.$$

We note that in (i) and (ii), the prescribed data are not the critical points, but the zeros. More precisely, let the sequence $\mathcal{C} = \{z_j\}_j$ in U be the zero set of an $H^p(U)$ function and let N be the multiplicity of the point 0 in $\mathcal{C}$.

Then the unique solution to the extremal problem

$$\max\{\Re f^{(N)}(0) \mid f \in H^p(U), \|f\|_{H^p} \leq 1 \text{ and } f(z) = 0 \text{ for } z \in \mathcal{C}\}.$$

is a Blaschke product B with zero set $\mathcal{C}$ which is normalized by $B^{(N)}(0) > 0$. In searching for an analogue of Blaschke products for Bergman spaces, Hedenmalm had the idea of posing an appropriate counterpart of the later extremal problem for Bergman spaces. As before, let $\mathcal{C} = \{z_j\}_j$ be a sequence in U where the point 0 occurs N times and assume that $\mathcal{C}$ is the zero set of a function in $\mathcal{A}^p_\alpha$. Then the extremal problem

$$\max\{\Re f^{(N)}(0) \mid f \in \mathcal{A}^p_\alpha, \|f\|_{p,\alpha} \leq 1 \text{ and } f(z) = 0 \text{ for } z \in \mathcal{C}\},$$

has a unique extremal function $\mathcal{G} \in \mathcal{A}^p_\alpha$, which vanishes precisely on $\mathcal{C}$ and is normalized by $\mathcal{G}^{(N)}(0) > 0$. The function $\mathcal{G}$ is called the canonical divisor for $\mathcal{C}$. It plays an important role in the theory of Bergman spaces. Because of the strong analogy, one expects that maximal Blaschke products have similar properties to finite Blaschke products and to canonical divisors. An example is the property of analytic continuation. It is well known that a Blaschke product can be analytically extended across every open arc of ∂U that contains no limit point of its zero set. The same is true for a canonical divisor in the Bergman spaces $\mathcal{A}^p_0$. This was proved by Sundberg in 1997. Moreover, for maximal Blaschke products, we have the following theorem.

Theorem 3.5. (Analytic continuability). *Let $F : U \to U$ be a maximal Blaschke product with the critical set $\mathcal{C}$. Then F can be analytically continued across each arc of ∂U which is free of limit points of $\mathcal{C}$. In particular, the limit points of the critical set coincide with the limit points of the zero points of F.*

Another strong property of finite Blaschke products is their semigroup property with respect to composition. In contrast, the composition of two infinite Blaschke products need not be a Blaschke product (consider destructible Blaschke products). However, for maximal Blaschke products we have the following theorem.

Theorem 3.6. (Semigroup property). *The set of maximal Blaschke products is closed under composition.*

Boundary behavior of maximal Blaschke products.

Ideally, one should be able to determine whether a bounded analytic function $F : U \to U$ is a maximal Blaschke product either from the behavior of

$$\frac{|F'(z)|}{1 - |F(z)|^2}, \quad \text{as } |z| \to 1^-,$$

or from the behavior of

$$\int_0^{2\pi} \log \frac{|F'(re^{it})|}{1 - |F(re^{it})|^2}\, dt, \quad \text{as } r \to 1^-.$$

We have only some partial results on these. First, we have for finite products.

Theorem 3.7. (Boundary behavior of finite Blaschke products). *Let $I \subseteq \partial U$ be some open arc and let $f : U \to U$ be an analytic function. Then the following statements are equivalent.*

(a)

$$\lim_{z \to \xi} (1 - |z|^2) \frac{|f''z)|}{1 - |f(z)|^2} = 1 \quad \forall \xi \in I.$$

(b)

$$\lim_{z \to \xi} \frac{|f'(z)|}{1 - |f(z)|^2} = +\infty \quad \forall \xi \in I.$$

(c) *f has a holomorphic extension across the arc I with $f(I) \subseteq \partial U$. In particular, if $I = \partial U$, then f is in either case a finite Blaschke product.*

The equivalence of conditions (a) and (b) in Theorem 3.7 for the special case $I = \partial U$ is due to Heinz. We now extend Theorem 3.7 beyond the class of finite Blaschke products. We start with the following proposition.

Proposition 3.8. *Let $f : U \to U$ be an analytic function and I some subset of ∂U.*

(1) *If*

$$\angle \lim_{z \to \xi} (1 - |z|^2) \frac{|f'(z)|}{1 - |f(z)|^2} = 1 \quad \forall\, \xi \in I,$$

then f has a finite angular derivative at a.e. $\xi \in I$. In particular, $\angle \lim_{z \to \xi} |f(z)| = 1$ a.e. $\xi \in I$.

(2) *If f has a finite angular derivative (and $\angle \lim_{z \to \xi} |f(z)| = 1$) at some $\xi \in I$, then*

$$\angle \lim_{z \to \xi} (1 - |z|^2) \frac{|f'(z)|}{1 - |f(z)|^2} = 1.$$

In particular, when $I = \partial U$, we obtain the following corollary.

Corollary 3.9. *Let $f : U \to U$ be analytic function. Then the following statements are equivalent.*

(a)
$$\angle \lim_{z \to \xi}(1 - |z|^2)\frac{|f'(z)|}{1 - |f(z)|^2} = 1 \quad \text{for a.e. } \xi \in \partial U.$$

(b) f *is an inner function with finite angular derivative at almost every point of ∂U.*

We further note that conditions (1) and (2) in Proposition 3.8 do not complement each other. Therefore, we ask if an analytic self-map of U which satisfies

$$\angle \lim_{z \to \xi}(1 - |z|^2)\frac{|f'(z)|}{1 - |f(z)|^2} = 1$$

has an angular limit or even a finite angular derivative at $z = 1$. This might be viewed as a converse of the Julia–Wolff–Caratheodory theorem. For maximal Blaschke products whose critical sets satisfy the Blaschke condition one can show that condition (a) of Corollary 3.9 holds true.

Theorem 3.10. *Let $\mathcal{C} = \{z_j\}_j$ be a Blaschke sequence in U.*

(a) *The maximal conformal pseudo-metric $\lambda_{\max}(z)|dz|$ on U with a constant curvature -4 and a zero set $\mathcal{C}$ satisfies*

$$\angle \lim_{z \to \xi}\frac{\lambda_{\max}(z)}{\lambda_U(z)} = 1 \quad \text{for a.e. } \xi \in \partial U.$$

(b) *Every maximal function for $\mathcal{C}$ has a finite angular derivative at almost every point of ∂U.*

The next result gives a sufficient condition for maximality of a Blaschke product F in terms of the boundary behavior of the integral means of the quantity

$$(1 - |z|^2)\frac{|F'(z)|}{1 - |F(z)|^2} = \frac{\lambda(z)}{\lambda_U(z)}.$$

Theorem 3.12. *Let $\lambda(z)|dz|$ be a conformal pseudo-metric on U with a constant curvature -4 and zero set $\mathcal{C}$ such that*

$$\lim_{r \to 1^-} \int_0^{2\pi} \log \frac{\lambda(re^{it})}{\lambda_U(re^{it})} dt = 0. \tag{50.0.6}$$

Then $\lambda(z)|dz|$ is the maximal conformal pseudo-metric $\lambda_{\max}(z)|dz|$ on U with the constant curvature -4 and zero set $\mathcal{C}$.

Theorem 3.13. *Let $\mathcal{C}$ be a Blaschke sequence in U. A conformal pseudo-metric $\lambda(z)|dz|$ on U with a constant curvature -4 and zero set $\mathcal{C}$ is the maximal conformal pseudo-metric $\lambda_{\max}(z)|dz|$ on U with a constant curvature -4 and zero set $\mathcal{C}$ if and only if (50.0.6) holds.*

Heinz' results on maximal functions.

Definition 3.14. Let $f : U \to U$ be an analytic function. A point $q \in U$ is called locally omitted by f provided that either $q \in U - f(U)$ or else $q \in f(U)$ and there exists a domain Ω, $q \in \Omega$ such that for some component V of $f^{-1}(\Omega)$ the restriction $f|_V$ omits q, i.e., $q \notin f(V)$.

Theorem G (Heinz). *A maximal function has no locally omitted point.*

A question. Given a function in H^∞, how can we recognize if it is a maximal function?

Definition 3.15. A holomorphic function $f : U \to U$ is called locally of island type if f is onto and if $\forall\, w \in U$ there is an open disk $K(w)$ about w such that each component of $f^{-1}(K(w))$ is compactly contained in U.

Every finite Blaschke product is locally of island type. This follows by the fact that every surjective analytic self-map of U with constant finite valence is locally of island type.

A question. Is every surjective analytic self-map of U with uniformly bounded (finite) valence is locally of island type?

Theorem H (Heinz). *Every function locally of island type is a maximal function.*

The Gauss curvature PDE and the Berger–Nirenberg problem.

We discuss the implication "(c)$\Rightarrow$(a)" in Theorem 1.1.

Theorem 4.1. *Let $h : U \to \mathbb{C}$ be a non-constant holomorphic function with zero set $\mathcal{C}$. Then the following statements are equivalent.*
(a) *There exists a holomorphic function $f : U \to U$ with critical set $\mathcal{C}$.*
(b) *There exists a C^2-solution $u : U \to \mathbb{R}$ to the Gauss curvature equation*

$$\Delta u = 4\,|h(z)|^2\, e^{2u}. \qquad (50.0.7)$$

If there exists a holomorphic function $f : U \to U$ with critical set $\mathcal{C}$, then

$$u(z) = \log\left(\frac{1}{|h(z)|} \cdot \frac{|f'(z)|}{1 - |f(z)|^2} \right)$$

is a C^2-solution of (50.0.7). This proves "(a)$\Rightarrow$(b)". Conversely, if there is a C^2-solution $u : U \to \mathbb{R}$ to the curvature equation (50.0.7), then $\lambda(z)|dz| = e^{u(z)}|dz|$ is a regular conformal metric with curvature $-4|h(z)|^2$ on U. By Theorem 2.5,

$$u(z) = \log\left(\frac{1}{|h(z)|} \cdot \frac{|f'(z)|}{1 - |f(z)|^2}\right)$$

with some analytic self-map of U. Hence, the zero set of h agrees with the critical set of f. By Theorem 4.1, we are motivated to characterize those holomorphic functions $h : U \to \mathbb{C}$ for which the PDE (50.0.7) has a solution. This problem is a special case of the Berger–Nirenberg problem in differential geometry.

The Berger–Nirenberg problem.

Given a function $K : R \to \mathbb{R}$ on a Riemann surface R, is there a conformal metric on R with Gauss curvature K?

The Berger–Nirenberg problem is well understood for compact Riemannian surfaces, as well as for the complex plane. However, much less is known for proper domains D of the complex plane. In this situation, the Berger–Nirenberg problem reduces to the question if for a given function $k : D \to \mathbb{R}$ the Gauss curvature equation

$$\Delta u = k(z)e^{2u}, \tag{50.0.8}$$

has a solution in D. We note that k is the negative of the curvature K of the conformal metric $e^{u(z)}|dz|$.

Theorem 4.2. *Let D be a bounded and regular domain (i.e., there exists Green's function g_D for D which vanishes continuously on ∂D) and let k be a non-negative locally Hölder continuous function on D.*

(1) *If for some (and therefore for every) $z_0 \in D$, $\iint_D g_D(z_0, \xi)k(\xi)d\sigma_\xi < +\infty$, then (50.0.8) has a C^2-solution $u : D \to \mathbb{R}$, which is bounded from above.*

(2) *If (50.0.8) has a C^2-solution $u : D \to \mathbb{R}$, which is bounded from below and if this solution has a harmonic majorant on D, then $\iint_D g_D(z, \xi)k(\xi)d\sigma_\xi < +\infty$ for all $z \in D$.*

(3) *There exists a bounded C^2-solution $u : D \to \mathbb{R}$ of (50.0.8) if and only if*

$$\sup_{z \in D} \iint_D g_D(z, \xi)k(\xi)d\sigma_\xi < +\infty.$$

If we choose $D = U$ and $z_0 = 0$, then $g_D(0, \xi) = -\log|\xi|$. As a consequence of the inequality

$$\frac{1 - |\xi|^2}{2} \leq \log \frac{1}{|\xi|} \leq \frac{1 - |\xi|^2}{|\xi|}, \quad 0 < |\xi| < 1,$$

we obtain the following equivalent formulation of Theorem 4.2.

Corollary 4.3. *Let k be a non-negative locally Hölder continuous function on U.*

(1) *If*

$$\iint_U (1 - |\xi|^2) k(\xi) d\sigma_\xi < +\infty, \tag{50.0.9}$$

Then (50.0.8) has a C^2-solution $u : U \to \mathbb{R}$ which is bounded from above.
(2) *If (50.0.8) has a C^2-solution $u : U \to \mathbb{R}$ which is bounded from below and if this solution has an harmonic majorant on U, then (50.0.9) holds true.*
(3) *There exists a bounded C^2-solution $u : U \to \mathbb{R}$ of (50.0.8) if and only if*

$$\sup_{z \in U} \iint_U \log \left| \frac{1 - \overline{\xi}z}{z - \xi} \right| k(\xi) d\sigma_\xi < +\infty.$$

Both, Theorem 4.2 and Corollary 4.3 are not best possible, because (50.0.8) can have solutions, even if $\iint_D g_D(z, \xi) \cdot k(\xi) d\sigma_\xi = +\infty$, for some (and therefore for all) $z \in D$. Although the condition (50.0.9) is not necessary for the existence of a solution of (50.0.8), it is strong enough to deduce a necessary and sufficient condition for the solvability of the Gauss curvature equation of the particular form (50.0.7).

Theorem 4.6. *Let $h : U \to \mathbb{C}$ be a holomorphic function. Then the Gauss curvature equation (50.0.7) has a solution if and only if h has a representation as a product of an $\mathcal{A}_1^2$ function and a non-vanishing analytic function.*

Note that Theorems 4.1 and 4.6 combined show that the class of all the holomorphic functions $h : U \to \mathbb{C}$ whose zero sets coincide with the critical sets of the class of bounded analytic functions is exactly the Bergman space $\mathcal{A}_1^2$. This proves the implication "(c)$\Rightarrow$(a)" in Theorem 1.1. We remark that Theorem 4.2(3) characterizes those curvature functions k for which (50.0.8) has at least one bounded solution. For the case of the unit disk, this result can be stated as follows.

Theorem 4.7. *Let $\phi : U \to \mathbb{C}$ be analytic and $k(z) = 4|\phi'(z)|^2$. Then there exists a bounded solution to the Gauss curvature equation* (50.0.8) *if and only if $\phi \in$ BMOA, where*

$$\text{BMOA} = \left\{ \phi : U \to \mathbb{C} \text{ analytic} \ \Big| \ \sup_{z \in U} \iint_U g_U(z, \xi) |\phi'(\xi)|^2 d\sigma_\xi < +\infty \right\}$$

is the space of analytic functions of bounded mean oscillation on U.

Finally, we note that in Theorem 4.2 and Corollary 4.3, condition (1) does not imply condition (3). The Gauss curvature equation (50.0.8) can have solutions none of which is bounded. For example, choose $\phi \in H^2(U) - \text{BMOA}$ and set $k(z) = 4|\phi'(z)|^2$. Then by Theorem 4.7, every solution of (50.0.8) must be unbounded. The following result of Heinz adds another item to the list of equivalent statements in Theorem 1.1.

Theorem I (Heinz). *Let $\mathcal{C} = \{z_j\}_j$ be a sequence in U. Then the following conditions are equivalent.*

(a) *There is an analytic function $f : U \to U$ with critical set $\{z_j\}_j$.*
(b) *There is a function in the Nevanlinna class $\mathcal{N}$ with critical set $\{z_j\}_j$.*

Here, a function f analytic in U is said to belong to the Nevanlinna class $\mathcal{N}$ if the integrals $\int_0^{2\pi} \log^+ |f(re^{it})| dt$ remain bounded as $r \to 1^-$.

Proof (using results from the current survey). (b)$\Rightarrow$(a). Let $\phi \in \mathcal{N}$. Then $\phi = \frac{\phi_1}{\phi_2}$, the quotient of two analytic self-maps of U. We may assume that ϕ_2 is zero free. We have

$$\phi'(z) = \frac{1}{\phi_2(z)}^2 \left(\phi_1'(z)\phi_2(z) - \phi_1(z)\phi_2'(z) \right).$$

Since $\phi_1', \phi_2' \in \mathcal{A}_1^2$ and $\mathcal{A}_1^2$ is a vector space, it follows that $\phi_1'\phi_2 - \phi_1\phi_2' \in \mathcal{A}_1^2$. Thus, by Theorem 4.6, there exists a solution $u : U \to \mathbb{R}$ to $\Delta u = |\phi'(z)|^2 e^{2u}$. Hence Theorem 4.1 gives the result. $\qquad\square$

Here is a question of Heinz. Jensen's formula connects the rate of growth of an analytic function to the density of its zeros. Thus, the restriction on the growth of the derivative of an analytic self-map of U imposed by the Schwarz–Pick lemma (3.2) forces an upper bound for the number of critical points of non-constant analytic self-maps of U. More precisely,

$$\sup_{f \in H^\infty(U), \|f\|_\infty \leq 1, f \not\equiv 1} \left(\limsup_{r \to 1^-} \frac{N(r, f')}{-\log(1 - r^2)} \right) \leq 1, \qquad (50.0.10)$$

where

$$N(r, f') = \int_0^r \frac{n(t, f')}{t} dt,$$

and $n(r, f')$ denotes the number of zeros of f' counted with multiplicity in the disk $D_r = \{z \in \mathbb{C} \,|\, |z| < r\}$, $0 < r < 1$. Heinz showed that equality holds in (50.0.10). More precisely, using his solution of the Schwarz–Picard problem, he proved that there exists for every $p = 2, 3, \ldots$, a non-constant analytic function $f_p : U \to U$ such that

$$\limsup_{r \to 1^-} \frac{N(r, f_p')}{\log\left(\frac{1}{1-r^2}\right)} = \frac{2p - 3}{2p - 2}.$$

Letting $p \to +\infty$ shows that (50.0.10) is best possible. Heinz was led to ask whether there is a bounded analytic function that realizes supremum in (50.0.10).

Theorem 4.8. *Let $\beta \in [0, 1]$. Then there exists a non-constant analytic function (and even a maximal Blaschke product) $f : U \to U$ such that*

$$\limsup_{r \to 1^-} \frac{N(r, f)}{\log\left(\frac{1}{1-r^2}\right)} = \beta.$$

Chapter 51

On the Critical Points of Functions in the Unit Ball of $H^\infty(U)$

We recall the essentials of the chapter having the same name in the two last pages of Research Notes XII', October 10–November 24. The approach is a straightforward one. For $f \in H^\infty(U) - \mathbb{C}$, $\|f\|_{H^\infty(U)} \leq 1$, we denote by $\{z_j\}_j$ the sequence of all the zeros of f in U, where multiplicity is taken into account. We denote by $\{w_j\}_j$ the subsequence of $\{z_j\}_j$ that includes each value of the z_j exactly once. This means that as sets $\{z_j\}_j$ and $\{w_j\}_j$ are equal but the sequence $\{w_j\}_j$ is composed of different numbers ($i \neq j \Rightarrow w_i \neq w_j$). We denote:

$$B_f(z) = z^\epsilon \prod_j \left(-\frac{\overline{w_j}}{|w_j|} \cdot \frac{z - w_j}{1 - \overline{w_j} z} \right),$$

where $\epsilon = 1$ if $f(0) = 0$ and $\epsilon = 0$ ($z^0 = 1$) if $f(0) \neq 0$. The convergence of the product is clear (because $\{w_j\}_j$ is a Blaschke sequence). We denote $f_0(z) = f(z)/B_f(z)$. It satisfies $f_0 \in H^\infty(U)$, $\|f_0\|_{H^\infty(U)} = \|f\|_{H^\infty(U)}$, and the zero set of $f_0(z)$, $Z(f_0)$, in U, is precisely the set of all the zeros of f of multiplicity $m \geq 2$, where each such a zero is of multiplicity $m - 1$ as a zero of f_0. $Z(f_0)$ is a set of critical points of f in U. Usually, not all of the critical points just those that lie in the fiber $f^{-1}(0)$. For each $a \in f(U) - \{0\}$, we define

$$f(z, a) = -\frac{\overline{a}}{|a|} \cdot \frac{f(z) - a}{1 - \overline{a} f(z)}, \text{ and } f_a(z) = \frac{f(z, a)}{B_{f(z,a)}(z)}.$$

Then $\bigcup_{a \in f(U)} Z(f_a)$ is precisely (as a set) the set of all the critical points of f. This union contains only a countable number of non-empty zero sets $Z(f_a)$.

287

It is a partition of the set of critical points of f, $Z(f')$. In fact, $Z(f_a) = z(f') \cap f^{-1}(a)$, $\forall\, a \in f(U)$. Thus, we have a countable set $\{a_n\}_n$ such that,

$$Z(f') = \bigcup_{a \in f(U)} Z(f_a) = \bigcup_n Z(f_{a_n}), \qquad \forall\, n,\ Z(f_{a_n}) \neq \emptyset.$$

We cannot assume that $\{a_n\}_n$ is a Blaschke sequence. It is a discrete subset of U (being $Z(f') \neq U$). The partition $\bigcup_n Z(f_{a_n})$ of $Z(f')$, (and here we count, taking the multiplicity into an account) is such that $|Z(f_{a_n})|$ equals $\sum(m-1)$, where $m \geq 2$ is the multiplicity of $f(z) - a_n = 0$ at each point of $f^{-1}(a_n)$ that has multiplicity $m \geq 2$.

Chapter 52

Back to the Krzyż Conjecture

A problem. Let us denote the $H^\infty(U)$ unit ball by

$$B(H^\infty(U)) = \{f \in H^\infty(U) \mid \|f\|_\infty \le 1\}.$$

We know that $f(z) = \sum_{n=0}^\infty a_n z^n \in B(H^\infty(U)) \Rightarrow \forall\, n \in \mathbb{Z}^+ \cup \{0\},\ |a_n| \le 1$. (for

$$|a_n| = \left| \frac{1}{2\pi} \int_0^{2\pi} f(e^{i\theta}) e^{-in\theta}\, d\theta \right| \le \frac{1}{2\pi} \int_0^{2\pi} \left| f(e^{i\theta}) \right| d\theta \le 1).$$

This bound is sharp, as shown by $f(z) = z^n$.

(1) Do multipliers reduce the size of the coefficients? That is, let $F(z) \in B(H^\infty(U))$. Can you find necessary and sufficient conditions on F, so that there exists a constant c_F depending only on F, such that $0 \le c_F < 1$ and such that

$$\forall\, f(z) = \sum_{n=0}^\infty a_n z^n \in F(z) \cdot B(H^\infty(U)),$$

$$\forall\, n \in \mathbb{Z}^+ \cup \{0\} \text{ we have } |a_n| \le c_F\,?$$

(2) If such a $c_F < 1$ exists, what is the smallest value it could have? That is, we look for min c_F where the minimum is taken over all the c_F's that satisfy the condition specified in (1).

Any function of the form $F(z) = \alpha z^N$, where $\alpha \in \mathbb{C}$, $|\alpha| < 1$, $N \in \mathbb{Z}^+$ satisfies (1) and the minimal c_F is $|\alpha|$.

Application. Suppose that

$$F(z) = \exp\left(-\lambda \frac{1+kz}{1-kz} \right),$$

where $\lambda > 0$ and $|k| = 1$ are constants. Let us imagine that F has the property (1) with some c_F, $0 \leq c_F < 1$. Let us denote the family of bounded (bounded by 1) non-vanishing functions by

$$B_0(U) = \{f(z) \in H(U) | \forall\, z \in U,\, 0 < |f(z)| \leq 1\}.$$

The Krzyż conjecture asserts that

$$\forall\, f(z) = \sum_{n=0}^{\infty} a_n z^n \in B_0(U) \quad \forall\, n \in \mathbb{Z}^+ \text{ we have } |a_n| \leq \frac{2}{e}.$$

Moreover, this inequality is sharp and it reduces to an equality exactly for rotations of

$$f(z) = \exp\left(-\frac{1 + z^n}{1 - z^n}\right).$$

By a well-known result in [49], we know that for every such $n \in \mathbb{Z}^+$, there are extremal functions and they all have the following form:

$$\exp\left(-\sum_{j=1}^{N} \lambda_j \frac{1 + k_j z}{1 - k_j z}\right), \text{ where } N \leq n,\, \lambda_j > 0 \text{ and}$$

$$|k_j| = 1 \text{ for } j = 1, \ldots, N.$$

We may assume that $1 \leq i < j \leq N \Rightarrow k_i \not\equiv k_j \pmod{2\pi}$. Clearly, we have

$$\exp\left(-\sum_{j=1}^{N} \lambda_j \frac{1 + k_j z}{1 - k_j z}\right) \in \exp\left(-\lambda_1 \frac{1 + k_1 z}{1 - k_1 z}\right) \cdot B(H^\infty(U)).$$

Hence, if indeed we can show that given a $\lambda > 0$ and a $|k| = 1$, the function

$$F(z) = \exp\left(-\lambda \frac{1 + kz}{1 - kz}\right)$$

has the property (1) it would be of interest. Here, $c_F = c_F(\lambda, k)$. It might enable us reproduce the celebrated result of Charles Horowitz from 1978, [48].

Remark 52.0.107. We note that we can include in our coefficient estimates also the free term, i.e., $n = 0$. For if $f(z) \in B(H^\infty(U))$, then

$$F(z) \cdot f(z) = \exp\left(-\lambda \frac{1 + kz}{1 - kz}\right) \cdot f(z)$$

and hence the free term of this function is bounded from above by

$$|F(0) \cdot f(0)| = e^{-\lambda} |f(0)| \leq e^{-\lambda} < 1.$$

Here, though we cannot expect $\sup c_F(\lambda, k) < 1$ also for $n = 0$ ($e^{-\lambda} \to_{\lambda \to 0^+} 1$).

The simplest and most direct approach to that question goes as follows: Let

$$\exp\left(-\lambda \frac{1 + kz}{1 - kz}\right) = \sum_{n=0}^{\infty} \alpha_n z^n$$

and let $f(z) = \sum_{n=0}^{\infty} a_n z^n \in B(H^{\infty}(U))$. Then

$$\exp\left(-\lambda \frac{1 + kz}{1 - kz}\right) \cdot f(z) = \sum_{n=0}^{\infty} \left(\sum_{j=0}^{n} \alpha_{n-j} a_j\right) z^n,$$

and so we search for good upper bounds of $\left|\sum_{j=0}^{n} \alpha_{n-j} a_j\right|$. We might want to try and use results of Goluzin, [37]. We refer to [37, Chapter XI, Section 8, p. 514]. The section is titled "Some inequalities for bounded functions". We quote the following.

Theorem 1 ([37], p. 515). *Let m denote a non-negative integer and let γ_n $(n = m, m + 1, \ldots)$ denote complex numbers such that $\gamma_m \neq 0$ and $\sum_{n=m}^{\infty} |\gamma_n| < \infty$. Then*

(1) If $m = 0$, a necessary and sufficient condition for the inequality

$$\left|\sum_{n=0}^{\infty} \gamma_n c_n\right| \leq |\gamma_0| \tag{52.0.1}$$

to hold for all functions $f(z) = \sum_{n=0}^{\infty} c_n z^n \in B(H^{\infty}(U))$ is that

$$\Re\left(\frac{1}{\gamma_0} \sum_{n=0}^{\infty} \gamma_n \xi^n\right) \geq \frac{1}{2} \quad \text{on } |\xi| = 1. \tag{52.0.2}$$

Equality holds in (52.0.1) only for $f(z) = c_0$, where $|c_0| = 1$.
(2) If $m > 0$, then a necessary and sufficient condition for the inequality

$$\left|\frac{1}{\gamma_m} \sum_{n=m}^{\infty} \gamma_n c_n + \frac{1}{\overline{\gamma}_m} \sum_{n=0}^{m-1} \overline{\gamma}_{2m-n} c_n\right| \leq 1 \tag{52.0.3}$$

to hold for all the functions $f(z) = \sum_{n=0}^{\infty} c_n z^n \in B(H^{\infty}(U))$ is that

$$\Re\left(\frac{1}{\gamma_m} \sum_{n=m}^{\infty} \gamma_n \xi^{n-m}\right) \geq \frac{1}{2} \quad \text{on } |\xi| = 1. \tag{52.0.4}$$

Equality holds in (52.0.3) only for $f(z) = c_m z^m$, where $|c_m| = 1$.

We note that, by virtue of the assumptions of the theorem, the series in (52.0.2) and (52.0.4) converge uniformly on the circle $|\xi| = 1$ and the series in (52.0.1) and (52.0.3) converge for all $f(z) \in B(H^\infty(U))$ since we have $|c_n| \leq 1$ $(n = 0, 1, \ldots)$.

To try and use that in our situation, we will write explicitly the estimates we need and compare them to inequality (52.0.1) say, in Theorem 1, of Goluzin. We search for upper bounds on $|\sum_{j=0}^{n} \alpha_{n-j} a_j|$, where

$$\exp\left(-\lambda\frac{1 + kz}{1 - kz}\right) = \sum_{m=0}^{\infty} \alpha_m z^m$$

and for all $f(z) = \sum_{m=0}^{\infty} a_m z^m \in B(H^\infty(U))$. We clearly can assume $k = 1$ (by rotation of z to $k^{-1}z = \bar{k}z$). Goluzin's inequality is

$$\left|\sum_{n=0}^{\infty} \gamma_n c_n\right| \leq |\gamma_0| \quad \text{for all} \quad f(z) = \sum_{m=0}^{\infty} c_m z^m \text{ in } B(H^\infty(U)).$$

Thus, a straightforward assignment of what we need versus what Goluzin offers gives us

$$\alpha_j = \gamma_{n-j}, \quad a_j = c_j \quad \text{for } j = 0, \ldots, n.$$

So, in Goluzin's notations,

$$\exp\left(-\lambda\frac{1 + kz}{1 - kz}\right) = \gamma_n + \gamma_{n-1}z + \gamma_{n-2}z^2 + \cdots + \gamma_0 z^n + \alpha_{n+1}z^{n+1} + \cdots$$

and we take $\gamma_{n+1} = \gamma_{n+2} = \cdots = 0$. Goluzin's necessary and sufficient condition is (52.0.2)

$$\Re\left(\frac{1}{\gamma_0} \sum_{m=0}^{\infty} \gamma_m \xi^m\right) \geq \frac{1}{2} \quad \text{on } |\xi| = 1$$

so in our circumstances it becomes

$$\Re\left(\frac{1}{\gamma_0}(\gamma_0 + \gamma_1 \xi + \cdots + \gamma_n \xi^n)\right) \geq \frac{1}{2}$$

and since our multiplier

$$\exp\left(-\lambda\frac{1 + kz}{1 - kz}\right) = \gamma_n + \gamma_{n-1}z + \gamma_{n-2}z^2 + \cdots + \gamma_0 z^n + \cdots \quad (\lambda > 0)$$

is a real function (real coefficients) if $k = 1$, we have $\gamma_0, \ldots, \gamma_n \in \mathbb{R}$, so Goluzin's condition is

$$\Re(\gamma_0 + \gamma_1 \xi + \cdots + \gamma_n \xi^n) \begin{cases} \geq \frac{1}{2}\gamma_0 & \text{if} \quad \gamma_0 > 0 \\ \leq \frac{1}{2}\gamma_0 & \text{if} \quad \gamma_0 < 0 \end{cases}.$$

We are assuming, of course, that $\gamma_0 \neq 0$. We now will perform some elementary computations in order to get some experience with these ideas. For $|\xi| = 1$, $\overline{\xi} = \xi^{-1}$ and so

$$\gamma_0 + \gamma_1 \xi + \cdots + \gamma_n \xi^n = \xi^n \left(\gamma_n + \cdots + \gamma_0 \xi^{-n} \right) = \xi^n \left(\gamma_n + \cdots + \gamma_0 \overline{\xi}^n \right).$$

We note that $\gamma_n + \gamma_{n-1}\overline{\xi} + \cdots + \gamma_0 \overline{\xi}^n = S_{n+1}(\overline{\xi})$ the $(n+1)$-st partial sum of

$$\exp\left(-\lambda \frac{1 + \overline{\xi}}{1 - \overline{\xi}} \right).$$

As mentioned above we have $S_{n+1}(\overline{\xi}) = S_{n+1}\left(\frac{1}{\xi} \right)$. Thus, Goluzin's condition can be written as follows:

$$\Re\left(\xi^n S_{n+1}\left(\frac{1}{\xi} \right) \right) \begin{cases} \geq \frac{1}{2}\gamma_0 & \text{if} \quad \gamma_0 > 0 \\ \leq \frac{1}{2}\gamma_0 & \text{if} \quad \gamma_0 < 0 \end{cases}.$$

Let us examine the power series of

$$\exp\left(-\lambda \frac{1 + \overline{\xi}}{1 - \overline{\xi}} \right) = \sum_{m=0}^{\infty} \alpha_m z^m.$$

$$-\lambda \frac{1 + z}{1 - z} = -\lambda \left(1 + 2 \sum_{n=1}^{\infty} z^n \right), \qquad \left(-\lambda \frac{1 + z}{1 - z} \right)' = -2\lambda \sum_{m=0}^{\infty} (m + 1)z^m.$$

By differentiation, we get

$$\sum_{m=1}^{\infty} m\alpha_m z^{m-1} = \exp\left(-\lambda \frac{1 + \overline{\xi}}{1 - \overline{\xi}} \right) \cdot \left(-\lambda \frac{1 + z}{1 - z} \right)',$$

$$\sum_{m=0}^{\infty} (m + 1)\alpha_{m+1} z^m = \left(\sum_{m=0}^{\infty} \alpha_m z^m \right) \left(\sum_{m=0}^{\infty} (m + 1)z^m \right) (-2\lambda).$$

$$\left(\sum_{m=0}^{\infty} \alpha_m z^m \right) \left(\sum_{m=0}^{\infty} (m + 1)z^m \right) = \sum_{m=0}^{\infty} \left(\sum_{j=0}^{m} (m + 1 - j)\alpha_j \right) z^m.$$

By comparing coefficients, we obtain a recursion for the coefficients α_m.

$$(m + 1)\alpha_{m+1} = -2\lambda \cdot \sum_{j=0}^{m} (m + 1 - j)\alpha_j, \qquad \alpha_0 = e^{-\lambda}.$$

Let us compute the first four coefficients, using the recursion.

$$\alpha_0 = e^{-\lambda},\ \alpha_1 = -2\lambda e^{-\lambda},\ 2\alpha_2 = 4\lambda(\lambda-1)e^{-\lambda}, 3\alpha_3 = (-2\lambda)(2\lambda^2-6\lambda+3)e^{-\lambda}.$$

We see that for certain values of λ certain α_j's vanish. But those "bad" λ's are very few.

Proposition 52.0.108. $\forall\, m \in \mathbb{Z}^+ \cup \{0\}$, $e^\lambda \alpha_m$ *is a polynomial in* λ, *with rational coefficients and of degree* m. *In particular, if* λ *is not an algebraic number, then* $\forall\, m \in \mathbb{Z}^+ \cup \{0\}$, $\alpha_m \neq 0$. *Also, for* $\lambda > 0$, *small enough* $\alpha_0 > 0$ *and* $\alpha_m < 0$ *for all* $m \in \mathbb{Z}^+$, *as large as we please.*

Proof. Using inductive argument and the recursion we found above for the coefficients α_m, we deduce all the assertions except for the last one, regarding $\alpha_0 > 0$ and $\alpha_m < 0$, $m \in \mathbb{Z}^+$ for $\lambda > 0$ close enough to zero. For this last assertion, we use the recursion as follows:

$$(m+1)\alpha_{m+1} = -2\lambda \cdot \sum_{j=0}^{m}(m+1-j)\alpha_j$$

$$= -2\lambda \cdot \left(O(\lambda) + (m+1)\right)e^{-\lambda}. \qquad \square$$

Instead of concentrating further on the singular inner function above, in particular regarding the Goluzin conditions, we will make a general discussion. That will be done in the following chapter.

Chapter 53

$B(H^\infty(U))$ Multipliers that Reduce the Size of the Coefficients

As usual, $B(H^\infty(U)) = \{f \in H^\infty(U) | \|f\|_\infty \le 1\}$ will denote the closed unit ball of $H^\infty(U)$. If $f(z) = \sum_{n=0}^\infty a_n z^n \in B(H^\infty(U))$ then $\forall n \in \mathbb{Z}^+ \cup \{0\}$, $|a_n| \le 1$.

Definition 53.0.109. A multiplier that reduces the size of the coefficients in $B(H^\infty(U))$, or simply (in this chapter), a $B(H^\infty(U))$-multiplier is a function $F \in H(U)$ that has the following properties:

(1) $F(z) \cdot B(H^\infty(U)) := \{F(z) \cdot f(z) | f \in B(H^\infty(U))\} \subseteq B(H^\infty(U))$.
(2) There exists a constant c_F depending only on $F(z)$ such that $0 \le c_F < 1$ and for any $g(z) = \sum_{n=0}^\infty b_n z^n \in F(z) \cdot B(H^\infty(U))$ and for any $n \in \mathbb{Z}^+ \cup \{0\}$, $|b_n| \le c_F$.

Remark 53.0.110. If F is a function that satisfies (1) in Definition 53.0.109, then $F \in B(H^\infty(U))$ (because $1 \in B(H^\infty(U))$ and by (1) $F = F \cdot 1 \in F \cdot B(H^\infty(U)) \subseteq B(H^\infty(U))$). Also, the converse is true ($F \in B(H^\infty(U)) \Rightarrow F \cdot B(H^\infty(U)) \subseteq B(H^\infty(U))$).

Example 53.0.111. For each $\alpha \in \mathbb{C}$, $|\alpha| < 1$ and for each $N \in \mathbb{Z}^+ \cup \{0\}$, the function $F(z) = \alpha z^N$ is a $B(H^\infty(U))$-multiplier.

Proof. By $|\alpha| < 1$, it follows that $F \in B(H^\infty(U))$ so by Remark 53.0.110, F satisfies (1). Let $c_F = |\alpha|$. Then $0 \le c_F < 1$. Let $f(z) = \sum_{n=0}^\infty a_z z^n \in B(H^\infty(U))$. Then $g(z) = F(z) \cdot f(z) = \sum_{n=0}^\infty \alpha a_n z^{n+N} = \sum_{n=0}^\infty b_n z^n$. So $b_n = 0$ for $n < N$, and $b_n = \alpha a_{n-N}$ for $N \le n$. Hence, for any $n \in \mathbb{Z}^+ \cup \{0\}$, $|b_n| \le \alpha$ (because $|a_{n-N}| \le 1$ for $N \le n$). Thus, F satisfies (2) with $c_F = |\alpha|$. $\qquad\square$

Example 53.0.112. What pairs of numbers α, $\beta \in \mathbb{C}$ qualify to generate a linear $B(H^\infty(U))$-multiplier? Thus, we look for multipliers of the form $F(z) = \alpha + \beta z$. By $F \in H^\infty(U))$, it follows that $|\alpha| + |\beta| \leq 1$, and conversely, $|\alpha| + |\beta| \leq 1 \Rightarrow \alpha + \beta z \in B(H^\infty(U))$. Let $f(z) = a_0 + a_1 z + \cdots \in B(H^\infty(U))$. Then

$$(\alpha + \beta z)f(z) = (\alpha + \beta z)(a_0 + a_1 z + a_2 z^2 + \cdots)$$

$$= (\alpha a_0) + (\alpha a_1 + \beta a_0)z + (\alpha a_2 + \beta a_1)z^2 + \cdots,$$

and so we look for pairs α, $\beta \in \mathbb{C}$, such that $|\alpha| + |\beta| \leq 1$ and such that there is a constant $c_{\alpha,\beta}$, such that $0 \leq c_{\alpha,\beta} < 1$ and such that $|\alpha a_0| \leq c_{\alpha,\beta}$ and $|\alpha a_{n+1} + \beta a_n| \leq c_{\alpha,\beta}$ for every $f(z) = \sum_{n=0}^\infty a_n z^n \in B(H^\infty(U))$ and every $n \in \mathbb{Z}^+ \cup \{0\}$. An easy sufficient condition is $|\alpha| + |\beta| < 1$ in which case we may take $c_{\alpha,\beta} = |\alpha| + |\beta|$. Then, $|\alpha a_0| \leq |\alpha| \leq c_{\alpha,\beta}$ by $|a_0| \leq 1$. Also, $|\alpha a_{n+1} + \beta a_n| \leq |\alpha| + |\beta|$ by $|a_{n+1}|$, $|a_n| \leq 1$ and by the triangle inequality. However, we can do better.

Proposition 53.0.113. $\alpha + \beta z$ is a $B(H^\infty(U))$-multiplier if and only if $|\alpha| + |\beta| \leq 1$ and $\max\{|\alpha|, |\beta|\} < 1$. In that case, we can take $c_{\alpha,\beta} = \max\{|\alpha|, |\beta|\}$.

Proof. Let us assume that $|\alpha| \geq |\beta|$. We can choose $f(z) = a_0 + a_1 z$ in $B(H^\infty(U))$ such that $\alpha a_1 + \beta a_0 = |\alpha|x + |\beta|(1 - x)$ for a given $0 \leq x \leq 1$. Hence, $\alpha a_1 + \beta a_0 = (|\alpha| - |\beta|)x + |\beta|$ That function is an increasing function for $0 \leq x \leq 1$ from $|\beta|$ ($x = 0$) to $|\alpha| = c_{\alpha,\beta}$ ($x = 1$). $\qquad\square$

In the general case, let $F(z) = \sum_{n=0}^\infty c_n z^n$ be a $B(H^\infty(U))$-multiplier. Let $0 \leq c_F < 1$ be a corresponding constant. Then for any function $f(z) = \sum_{n=0}^\infty a_n z^n \in B(H^\infty(U))$, we have

$$F(z) \cdot f(z) = \sum_{n=0}^\infty c_n z^n \cdot \sum_{n=0}^\infty a_n z^n = \sum_{n=0}^\infty \left(\sum_{j=0}^n a_j c_{n-j} \right) z^n$$

and $\forall n \in \mathbb{Z}^+\{0\}$, $|\sum_{j=0}^n a_j c_{n-j}| \leq c_F$.

Proposition 53.0.114. If $F(z) \in B(H^\infty(U))$ satisfies $\|F\|_2 < 1$, then F is a $B(H^\infty(U))$-multiplier. We can take $c_F = \|F\|_2$.

Proof. If we use the notations prior to the statement of Proposition 53.0.114, we obtain using the Cauchy–Schwarz inequality.

$$\left| \sum_{j=0}^n a_j c_{n-j} \right| \leq \left(\sum_{j=0}^n |a_j|^2 \right)^{1/2} \left(\sum_{j=0}^n |c_{n-j}|^2 \right)^{1/2}$$

$$\leq \|f\|_2 \cdot \|F\|_2 \leq \|f\|_\infty \cdot c_F \leq c_F. \qquad\square$$

We recall that $1 \le p, q \le \infty$ are Hölder conjugates if $\frac{1}{p} + \frac{1}{q} = 1$ (where we agree on $\frac{1}{\infty} = 0$). Hölder inequality implies that if $f, g \in H(U)$, have radial limits $f(e^{i\theta})$, $g(e^{i\theta})$ a.e. θ, then

$$\frac{1}{2\pi} \int_0^{2\pi} |(fg)(e^{i\theta})| \, d\theta \le \left(\frac{1}{2\pi} \int_0^{2\pi} |f(e^{i\theta})|^p d\theta \right)^{1/p} \left(\frac{1}{2\pi} \int_0^{2\pi} |g(e^{i\theta})|^q d\theta \right)^{1/q}$$

and there is equality if and only if $|f(e^{i\theta})|^p$ and $|g(e^{i\theta})|^q$ are effectively proportional [44, Theorem 189, p. 140]. In particular, if $1 < p_1 < p_2 < \infty$ and $f \in H^{p_2}(U)$, then

$$\frac{1}{2\pi} \int_0^{2\pi} |f(e^{i\theta})|^{p_1} d\theta = \frac{1}{2\pi} \int_0^{2\pi} |f(e^{i\theta})|^{p_1} \cdot 1 \, d\theta$$

$$\le \left(\frac{1}{2\pi} \int_0^{2\pi} (|f(e^{i\theta})|^{p_1})^p d\theta \right)^{1/p} \left(\frac{1}{2\pi} \int_0^{2\pi} 1 \, d\theta \right)^{1/q}$$

$$\left(\text{we use Hölder with } p = \frac{p_2}{p_1} > 1 \text{ and } \frac{1}{q} = \frac{p-1}{p} \right)$$

$$= \left(\frac{1}{2\pi} \int_0^{2\pi} (|f(e^{i\theta})|^{p_1})^{p_2/p_1} d\theta \right)^{p_1/p_2} \cdot (1) = \left(\frac{1}{2\pi} \int_0^{2\pi} |f(e^{i\theta})|^{p_2} \right)^{p_1/p_2}.$$

So,

$$\left(\frac{1}{2\pi} \int_0^{2\pi} |f(e^{i\theta})|^{p_1} \right)^{1/p_1} \le \left(\frac{1}{2\pi} \int_0^{2\pi} |f(e^{i\theta})|^{p_2} \right)^{1/p_2} < \infty.$$

There is equality if and only if $|f(e^{i\theta})|^{p_1}$ and 1 are effectively proportional, i.e., $|f(e^{i\theta})|$ is effectively constant. In particular, in relation to Proposition 53.0.114, $\|F\|_2 \le \|F\|_\infty$ and equality occurs if and only if $|F(e^{i\theta})|$ is effectively a constant.

Conclusion. Proposition 53.0.114 does not apply to an inner function F, for in that case $c_F = \|F\|_2 = \|F\|_\infty = 1$. That is unfortunate because it is the case of interest in relation to the Krzyż conjecture, as noted in Chapter 52.

Remark 53.0.115. We should note that there is nothing special about the 2-norm in relation to Proposition 53.0.114. For clearly we have, more generally, the following.

Proposition 53.0.116. *If $F(z) \in B(H^\infty(U))$ satisfies $\|F\|_p < 1$ for some $1 \le p \le \infty$, then F is a $B(H^\infty(U))$-multiplier. We can take $c_f = \|F\|_p$.*

In the proof, we use Hölder's inequality instead of the special case of the Cauchy–Schwarz inequality. We recall Goluzin's results.

Theorem 1 ([37], p. 515). *Let m denote a non-negative integer and let γ_n ($n = m, m+1, \ldots$) denote complex numbers such that $\gamma_m \neq 0$ and $\sum_{n=m}^{\infty} |\gamma_n| < \infty$. Then*
(1) If $m = 0$, then a necessary and sufficient condition for the inequality

$$\left| \sum_{n=0}^{\infty} \gamma_n c_n \right| \leq |\gamma_0|, \tag{53.0.1}$$

to hold for all functions $f(z) = \sum_{n=0}^{\infty} c_n z^n \in B(H^{\infty}(U))$ is that

$$\Re\left(\frac{1}{\gamma_0} \sum_{n=0}^{\infty} \gamma_n \xi^n \right) \geq \frac{1}{2} \ \text{ on } |\xi| = 1. \tag{53.0.2}$$

Equality holds in inequality (53.0.1) only for $f(z) = c_0$, where $|c_0| = 1$.
(2) If $m > 0$, then a necessary and sufficient condition for the inequality

$$\left| \frac{1}{\gamma_m} \sum_{n=m}^{\infty} \gamma_n c_n + \frac{1}{\overline{\gamma}_m} \sum_{n=0}^{m-1} \overline{\gamma}_{2m-n} c_n \right| \leq 1, \tag{53.0.3}$$

to hold for all the functions $f(z) = \sum_{n=0}^{\infty} c_n z^n \in B(H^{\infty}(U))$ is that

$$\Re\left(\frac{1}{\gamma_m} \sum_{n=m}^{\infty} \gamma_n \xi^{n-m} \right) \geq \frac{1}{2} \quad \text{on} \quad |\xi| = 1. \tag{53.0.4}$$

Equality holds in inequality (53.0.3) only for $f(z) = c_m z^m$, where $|c_m| = 1$.

We might try to use Theorem 1 of Goluzin in a straightforward manner as follows: Let $F(z) = \sum_{n=0}^{\infty} \gamma_n z^n \in H(U)$ satisfy $\gamma_n \neq 0$ for all $n \in \mathbb{Z}^+\{0\}$. We would like to find out what should $F(z)$ satisfy for condition (2) of Definition 53.0.109 to be satisfied in the sense that all the functions in $F(z) \cdot B(H^{\infty}(U))$ have coefficients smaller in absolute value than some fixed $c_F < 1$. Taking any $f(z) = \sum_{n=0}^{\infty} c_n z^n \in B(H^{\infty}(U))$, we have

$$F(z) \cdot f(z) = \sum_{n=0}^{\infty} \gamma_n z^n \cdot \sum_{n=0}^{\infty} c_n z^n = \sum_{n=0}^{\infty} \left(\sum_{j=0}^{n} \gamma_j c_{n-j} \right) z^n.$$

For the inequality (53.0.1) in Theorem 1 to be satisfied, we need

$$|\gamma_0 c_n + \gamma_1 c_{n-1} + \cdots + \gamma_n c_0| \leq |\gamma_n|. \tag{53.0.5}$$

Here, we take 0 for the coefficients $\gamma_{-1}, \gamma_{-2}, \gamma_{-3}, \ldots$. For that to be satisfied for each $f \in B(H^\infty(U))$, we need condition (53.0.2) of Goluzin's Theorem 1 to hold true for each n. Thus,

$$\Re\left(\frac{1}{\gamma_n}\sum_{m=0}^{n}\gamma_m\xi^{n-m}\right) \geq \frac{1}{2} \text{ on } |\xi| = 1. \tag{53.0.6}$$

Let us denote the partial sums of the power series expansion of F by $F_n(z) = \sum_{m=0}^{n}\gamma_m z^m$. Then we have the identity:

$$\sum_{m=0}^{n}\gamma_m\xi^{n-m} = \xi^n\sum_{m=0}^{n}\gamma_m\left(\frac{1}{\xi}\right)^m = \xi^n F_n\left(\frac{1}{\xi}\right) = \xi^n F_n(\bar{\xi}).$$

Hence, Goluzin's necessary and sufficient condition (53.0.6) for the inequality (53.0.5) to hold for every $n \in \mathbb{Z}^+ \cup \{0\}$ is

$$\Re\left(\frac{\xi^n}{\gamma_n}F_n\left(\frac{1}{\xi}\right)\right) \geq \frac{1}{2} \text{ on } |\xi| = 1 \text{ and } \forall\, n \in \mathbb{Z}^+ \cup \{0\}.$$

Equivalently,

$$\Re\left(\frac{\xi^n}{\gamma_n}F_n\left(\bar{\xi}\right)\right) \geq \frac{1}{2} \text{ on } |\xi| = 1 \text{ and } \forall\, n \in \mathbb{Z}^+ \cup \{0\}.$$

Also,

$$\Re\left(\frac{\xi^n}{\gamma_n}F_n\left(\frac{1}{\xi}\right) - \frac{1}{2}\right) \geq 0 \text{ on } |\xi| = 1 \text{ and } \forall\, n \in \mathbb{Z}^+ \cup \{0\}.$$

Equivalently,

$$\Re\left(1 + 2\left(\left(\frac{\gamma_{n-1}}{\gamma_n}\right)\xi + \cdots + \left(\frac{\gamma_0}{\gamma_n}\right)\xi^n\right)\right) \geq 0$$

$$\text{on } |\xi| = 1 \text{ and } \forall\, n \in \mathbb{Z}^+ \cup \{0\}.$$

We recall that $\gamma_n = \frac{1}{n!}F^{(n)}(0)$. Hence, the real trigonometric polynomials

$$\Re\left(n!\xi^n\frac{F_n(\bar{\xi})}{F^{(n)}(0)} - \frac{1}{2}\right)$$

are non-negative for all non-negative integers n. We recall few facts from the classical theory of real, non-negative trigonometric polynomials.

Theorem 2 (E.3, c on p. 85 of [7], also p. 5 of [100]). *If $g(\theta)$ is a trigonometric polynomial with real coefficients and of degree n, such that $g(\theta) \geq 0$ for all θ, then there exists an algebraic polynomial of degree n, $q(z) \in \mathbb{C}[z]$, $\deg q = n$, so that $g(\theta) = |q(e^{i\theta})|^2$. If $g(\theta) \not\equiv 0$ then, we can choose such a $q(z)$, so that $q(z) \neq 0$ for all $z \in U$ and $q(0) > 0$ in which* *case $q(z)$ is uniquely determined. Also, if $g(\theta)$ is an even function, then we can choose $q(z) \in \mathbb{R}[z]$.*

Remark 53.0.117. If $t(\theta) \in \tau_n - \tau_{n-1}$ (a real trigonometric polynomial of exact degree n), then there are complex numbers $z_1, \ldots, z_{2n}$, and $0 \neq c \in \mathbb{C}$ such that

$$t(z) = c \prod_{j=1}^{2n} \sin\left(\frac{z - z_j}{2}\right),$$

and the non-real zeros z_j of $t(z)$ form conjugate pairs.

Definition 53.0.118. Let $G(z) = \sum_{n=0}^{\infty} d_n z^n$, $H(z) = \sum_{n=0}^{\infty} e_n z^n$ in $H(U)$. We say that $H(z)$ dominates $G(z)$ coefficientwise if $|d_n| \leq |e_n|$ for all $n \in \mathbb{Z}^+ \cup \{0\}$. We will designate this relation by writing $G \ll H$.

Summing up what we proved, we have the following theorem:

Theorem 53.0.119. *The function $F(z) \in B(H^\infty(U))$ satisfies $G(z) \ll F(z)$ for all functions $G(z) \in F(z) \cdot B(H^\infty(U))$, if and only if, for all $n \in \mathbb{Z}^+ \cup \{0\}$, there exist a $q_n(z) \in \mathbb{C}[z]$, $\deg q_n = n$ such that*

$$\Re\left(n! \xi^n \frac{F_n\left(\frac{1}{\xi}\right)}{F^{(n)}(0)}\right) = \frac{1}{2} + |q_n(\xi)|^2 \quad \text{on} \quad |\xi| = 1,$$

where we assumed further the following:

(1) $F^{(n)}(0) \neq 0$ *for all* $n \in \mathbb{Z}^+ \cup \{0\}$.
(2) $F_n(z) = \sum_{m=0}^{n} \frac{F^{(m)}(0)}{m!} z^m$, *the $(n+1)$-st partial sum of the power series expansion of $F(z)$.*

Remark 53.0.120. We do not have interesting examples of functions $F(z) \in B(H^\infty(U))$ that satisfy

$$\Re\left(\frac{\xi^n F_n\left(\frac{1}{\xi}\right)}{F^{(n)}(0)}\right) = \frac{1}{2n!} + |q_n(\xi)|^2 \text{ on } |\xi| = 1$$

for all $n \in \mathbb{Z}^+ \cup \{0\}$. (We note that the $q_n(z)$ in the current remark equals $(\sqrt{n!})^{-1} \cdot q_n(z)$, where the last q_n is the one in Theorem 53.0.119.) For the reader to think about.

Chapter 54

Goluzin's Proofs and Goluzin-type Theorems

We will bring Goluzin's proofs for the claims in [37, Theorem 1, p. 515]. After that, we will make variations on his computations in order to obtain similar results (Goluzin-type theorems).

Proof (Goluzin). The functions $f(z) = \sum_{n=0}^{\infty} c_n z^n \in B(H^{\infty}(U))$ have limiting values along radial paths almost everywhere on the circle $|z| = 1$, if we denote these limiting values on $|\xi| = 1$ by $f(z)$ and take the limit of r approaches 1 from below in the formulas

$$c_n = \frac{1}{2\pi i} \int_{|\xi|=r} \frac{f(\xi)}{\xi^{n+1}} d\xi, \quad \int_{|\xi|=r} f(\xi)\xi^n d\xi = 0 \ (n = 0, 1, 2, \ldots),$$

which hold for $0 < r < 1$, we obtain the formulas

$$c_n = \frac{1}{2\pi i} \int_{\mathbb{T}} \frac{f(\xi)}{\xi^{n+1}} d\xi, \quad \int_{\mathbb{T}} f(\xi)\xi^n d\xi = 0 \ (n = 0, 1, 2, \ldots),$$

where $\mathbb{T} = \{|\xi| = 1\}$ is the unit circle and the integrals are Lebesgue. On the basis of this and the uniform convergence of the series $\sum_{n=m}^{\infty} \gamma_n \xi^n$ on the circle $|\xi| = 1$, we have

$$\frac{1}{\gamma_m} \sum_{n=m}^{\infty} \gamma_n c_n = \frac{1}{\gamma_m} \sum_{n=m}^{\infty} \gamma_n \left(\frac{1}{2\pi i} \int_{\mathbb{T}} \frac{f(\xi)}{\xi^{n+1}} d\xi \right)$$

$$= \frac{1}{\gamma_m} \sum_{n=m}^{\infty} \left(\frac{1}{2\pi i} \int_{\mathbb{T}} \frac{f(\xi)}{\xi^{m+1}} d\xi \right) \frac{\gamma_n}{\xi^{n-m}}$$

$$= \frac{1}{2\pi i} \int_{\mathbb{T}} \frac{f(\xi)}{\xi^{m+1}} \left(\sum_{n=m}^{\infty} \frac{\gamma_n}{\gamma_m \xi^{n-m}} \right) d\xi$$

$$= \frac{1}{2\pi i} \int_{\mathbb{T}} \frac{f(\xi)}{\xi^{m+1}} \left(\sum_{n=m}^{\infty} \frac{\gamma_n \xi^{m-n}}{\gamma_m} + \sum_{n=m}^{\infty} \frac{\overline{\gamma}_n \overline{\xi}^{m-n}}{\overline{\gamma}_m} \right) d\xi$$

$$- \frac{1}{2\pi i} \int_{\mathbb{T}} \frac{f(\xi)}{\xi^{m+1}} \left(\sum_{n=m}^{\infty} \frac{\overline{\gamma}_n \overline{\xi}^{m-n}}{\overline{\gamma}_m} \right) d\xi$$

$$= \frac{1}{2\pi i} \int_{\mathbb{T}} \frac{f(\xi)}{\xi^{m+1}} 2 \cdot \Re \left(\frac{1}{\gamma_m} \sum_{n=m}^{\infty} \gamma_n \xi^{m-n} \right) d\xi$$

$$- \frac{1}{2\pi i} \int_{\mathbb{T}} \frac{f(\xi)}{\xi^{m+1}} \left(\sum_{n=m}^{\infty} \frac{\overline{\gamma}_n \overline{\xi}^{m-n}}{\overline{\gamma}_m} \right) d\xi$$

$$= \frac{1}{2\pi i} \int_{\mathbb{T}} \frac{f(\xi)}{\xi^{m+1}} \left[2 \cdot \Re \left(\frac{1}{\gamma_m} \sum_{n=m}^{\infty} \gamma_n \xi^{m-n} \right) - 1 \right] d\xi$$

$$- \frac{1}{\overline{\gamma}_m} \sum_{n=m+1}^{2m} \overline{\gamma}_n c_{2m-n},$$

where, in the last sum, we note that for $n > 2m$, we have

$$\int_{\mathbb{T}} \frac{f(\xi)}{\xi^{m+1}} \left(\sum_{n=2m+1}^{\infty} \frac{\overline{\gamma}_n}{\overline{\gamma}_m} \xi^{n-m} \right) d\xi = \sum_{n=2m+1}^{\infty} \frac{\overline{\gamma}_n}{\overline{\gamma}_m} \left(\int_{\mathbb{T}} f(\xi) \xi^{n-2m-1} d\xi \right) = 0$$

by $n - 2m - 1 \geq 0$, so the sum is really $\sum_{n=m+1}^{2m}$, and

$$\frac{1}{2\pi i} \int_{\mathbb{T}} \frac{f(\xi)}{\xi^{2m-n+1}} d\xi = c_{2m-n} \quad (2m - n \geq 0).$$

That is,

$$\frac{1}{\gamma_m} \sum_{n=m}^{\infty} \gamma_n c_n + \frac{1}{\overline{\gamma}_m} \sum_{n=0}^{m-1} \overline{\gamma}_{2m-n} c_n$$

$$= \frac{1}{2\pi i} \int_{\mathbb{T}} \frac{f(\xi)}{\xi^{m+1}} \left[2 \cdot \Re \left(\frac{1}{\gamma_m} \sum_{n=m}^{\infty} \gamma_n \xi^{m-n} \right) - 1 \right] d\xi. \quad (54.0.1)$$

For $m = 0$, the second summation on the left (of equation (54.0.1)) disappears. If the condition in equation (53.0.4) is satisfied on the circle $|\xi| = 1$, then the expression $\Re \left(\frac{1}{\gamma_m} \sum_{n=m}^{\infty} \gamma_n \xi^{m-n} \right) - 1 \geq 0$ and using the triangle inequality on equation (54.0.1), we get

$$\left| \frac{1}{\gamma_m} \sum_{n=m}^{\infty} \gamma_n c_n + \frac{1}{\overline{\gamma}_m} \sum_{n=0}^{m-1} \overline{\gamma}_{2m-n} c_n \right|$$

$$\leq \frac{1}{2\pi} \int_{\mathbb{T}} \left[2 \cdot \Re \left(\frac{1}{\gamma_m} \sum_{n=m}^{\infty} \gamma_n \xi^{m-n} \right) - 1 \right] d\phi = 1,$$

(by $\left| \frac{f(\xi)}{\xi^{m+1}} \right| \leq 1$, $\xi = e^{i\phi}$, $d\xi = ie^{i\phi}d\phi$, and the fact that for $m > n$, $m - n > 0$ so $\int_{\mathbb{T}} \Re(\gamma_n \xi^{m-n})d\phi = 0$ while for $m - n = 0$,

$$\frac{1}{2\pi} \int_{\mathbb{T}} \left[2 \cdot \Re \left(\frac{1}{\gamma_m} \cdot \gamma_m \xi^{m-m} \right) - 1 \right] d\phi = 2 - 1 = 1).$$

By setting $m = 0$, we get the inequality (53.0.1). Since the function in the square brackets in equation (54.0.1) is not identically equal to zero on the circle $|\xi| = 1$, equality holds in the relation (53.0.3) or in (53.0.1) only when $|f(\xi)| = 1$ and $\arg \left(\frac{f(\xi)}{\xi^m} \right) = \text{Const.}$ on a set of points of positive measure on the circle $|\xi| = 1$. But then $\frac{f(\xi)}{\xi^m} = \text{Const.}$ on the set mentioned, and $(|f(\xi)| = 1)$ this can be the case only if $f(z) = e^{i\alpha} z^m$ in the disk $|z| < 1$. Then indeed equality holds in (53.0.3) or (53.0.1). Thus, the sufficiency of the conditions of Theorem 1 is proved.

Let us suppose now, conversely, that inequality (53.0.3), or inequality (53.0.1) if $m = 0$, holds for all $f(z) \in B(H^{\infty}(U))$. If we apply this inequality to a function

$$f(z) = z^m \frac{z - \alpha}{1 - \overline{\alpha}z} = z^m \left(-\alpha + \sum_{n=1}^{\infty} (1 - |\alpha|^2)\overline{\alpha}^{n-1} z^n \right) \in B(H^{\infty}(U)),$$

for which $c_0 = c_1 = \cdots = c_{m-1} = 0$, $c_m = -\alpha$, $c_n = (1 - |\alpha|^2)\overline{\alpha}^{n-m-1}$ $(n = m + 1, m + 2, \ldots)$, $|\alpha| < 1$, we obtain

$$\left| -\alpha + \frac{1 - |\alpha|^2}{\gamma_m} \sum_{n=m+1}^{\infty} \gamma_n \overline{\alpha}^{n-m-1} \right| \leq 1,$$

or, by squaring,

$$|\alpha|^2 - 2\Re \left(\frac{1 - |\alpha|^2}{\gamma_m} \sum_{n=m+1}^{\infty} \gamma_n \overline{\alpha}^{n-m-1} \right) + \left| \frac{1 - |\alpha|^2}{\gamma_m} \sum_{n=m+1}^{\infty} \gamma_n \overline{\alpha}^{n-m-1} \right|^2 \leq 1.$$

If we move the term $|\alpha|^2$ to the right and divide both sides by the common factor $1 - |\alpha|^2$, we obtain

$$-2 \cdot \Re \left(\frac{1}{\gamma_m} \sum_{n=m+1}^{\infty} \overline{\alpha}^{n-m-1} \right) + (1 - |\alpha|^2) \left| \frac{1}{\gamma_m} \sum_{n=m+1}^{\infty} \gamma_n \overline{\alpha}^{n-m-1} \right|^2 \leq 1.$$

If we now let $|\alpha|$ approach 1, we obtain inequality (53.0.4) with $\xi = \overline{\alpha}$. Since $\arg \alpha$ is arbitrary, this proves the necessity of condition (53.0.4). This completes the proof of Theorem 1. $\qquad\qquad\square$

We now demonstrate how to use the technique of Goluzin in order to prove results that are of the type of Theorem 1.

Theorem 54.0.121. *Let* $n \in \mathbb{Z}^+ \cup \{0\}$, $\gamma_0, \ldots, \gamma_n \in \mathbb{C}$ *and suppose that* $\Re\left(\sum_{j=0}^n \gamma_j \xi^j\right) \geq 0$ *on* $|\xi| = 1$. *Then for all functions* $f(z) = \sum_{m=0}^\infty c_m z^m \in B(H^\infty(U))$ *we have the following:*

(1)

$$\sum_{j=0}^n \gamma_j c_{n-j} + \sum_{j=0}^n \overline{\gamma}_j c_{n+j} = \frac{1}{2\pi i} \int_{\mathbb{T}} \frac{f(\xi)}{\xi^{n+1}} \cdot 2 \cdot \Re\left(\sum_{j=0}^n \gamma_j \xi^j\right) d\xi.$$

(2)

$$\left| \sum_{j=0}^n \gamma_j c_{n-j} + \sum_{j=0}^n \overline{\gamma}_j c_{n+j} \right| \leq 2 \cdot \Re(\gamma_0).$$

Proof.

$$\sum_{j=0}^n \gamma_j c_{n-j} = \sum_{j=0}^n \gamma_j \left(\frac{1}{2\pi i} \int_{\mathbb{T}} \frac{f(\xi)}{\xi^{n-j+1}} d\xi \right) = \frac{1}{2\pi i} \int_{\mathbb{T}} \frac{f(\xi)}{\xi^{n+1}} \left(\sum_{j=0}^n \frac{\gamma_j}{\xi^{-j}} \right) d\xi$$

$$= \frac{1}{2\pi i} \int_{\mathbb{T}} \frac{f(\xi)}{\xi^{n+1}} \left(\sum_{j=0}^n \gamma_j \xi^j \right) d\xi = \frac{1}{2\pi i} \int_{\mathbb{T}} \frac{f(\xi)}{\xi^{n+1}} \left(\sum_{j=0}^n \gamma_j \xi^j + \sum_{j=0}^n \overline{\gamma}_j \overline{\xi}^j \right) d\xi$$

$$- \frac{1}{2\pi i} \int_{\mathbb{T}} \frac{f(\xi)}{\xi^{n+1}} \left(\sum_{j=0}^n \overline{\gamma}_j \overline{\xi}^j \right) d\xi$$

$$= \frac{1}{2\pi i} \int_{\mathbb{T}} \frac{f(\xi)}{\xi^{n+1}} 2 \cdot \Re\left(\sum_{j=0}^n \gamma_j \xi^j \right) d\xi - \frac{1}{2\pi i} \int_{\mathbb{T}} \frac{f(\xi)}{\xi^{n+1}} \left(\sum_{j=0}^n \frac{\overline{\gamma}_j}{\xi^j} \right) d\xi$$

$$= \frac{1}{2\pi i} \int_{\mathbb{T}} \frac{f(\xi)}{\xi^{n+1}} 2 \cdot \Re\left(\sum_{j=0}^n \gamma_j \xi^j \right) d\xi - \sum_{j=0}^n \overline{\gamma}_j \left(\frac{1}{2\pi i} \int_{\mathbb{T}} \frac{f(\xi)}{\xi^{n+j+1}} d\xi \right)$$

$$= \frac{1}{2\pi i} \int_{\mathbb{T}} \frac{f(\xi)}{\xi^{n+1}} 2 \cdot \Re\left(\sum_{j=0}^n \gamma_j \xi^j \right) d\xi - \sum_{j=0}^n \overline{\gamma}_j c_{n+j}.$$

Hence,

$$\sum_{j=0}^{n} \gamma_j c_{n-j} + \sum_{j=0}^{n} \overline{\gamma}_j c_{n+j} = \frac{1}{2\pi i} \int_{\mathbb{T}} \frac{f(\xi)}{\xi^{n+1}} 2 \cdot \Re\left(\sum_{j=0}^{n} \gamma_j \xi^j\right) d\xi.$$

This concludes the proof of identity (1). We now apply the triangle inequality.

$$\left|\sum_{j=0}^{n} \gamma_j c_{n-j} + \sum_{j=0}^{n} \overline{\gamma}_j c_{n+j}\right| \leq \frac{1}{2\pi} \int_{\mathbb{T}} \left|\frac{f(\xi)}{\xi^{n+1}}\right| \cdot 2 \cdot \left|\Re\left(\sum_{j=0}^{n} \gamma_j \xi^j\right)\right| d\phi \leq$$

where $\xi = e^{i\phi}$, $d\xi = ie^{i\phi}d\phi$, and by $|\xi^{n+1}| = 1$, $|f(\xi)| \leq 1$ and $\Re\left(\sum_{j=0}^{n} \gamma_j \xi^j\right) \geq 0$ on $|\xi| = 1$, we have the inequality

$$\leq \frac{1}{2\pi} \int_{\mathbb{T}} 2 \cdot \Re\left(\sum_{j=0}^{n} \gamma_j \xi^j\right) d\phi = 2 \cdot \Re(\gamma_0).$$

This completes the proof of (2) and of our theorem. $\qquad\square$

Let us generalize a bit Theorem 54.0.121.

Theorem 54.0.122. *Let* $n \in \mathbb{Z}^+ \cup \{0\}$, $\gamma_0, \ldots, \gamma_n \in \mathbb{C}$, $k \in \mathbb{Z}^+ \cup \{0\}$ *and suppose that* $\Re\left(\sum_{j=0}^{n} \gamma_j \xi^{j+k-n}\right) \geq 0$ *on* $|\xi| = 1$. *Then for all functions* $f(z) = \sum_{m=0}^{\infty} c_m z^m \in B(H^\infty(U))$ *we have the following:*

(1)

$$\sum_{j=0}^{n} \gamma_j c_{n-j} + \sum_{j=\max\{n-2k,0\}}^{n} \overline{\gamma}_j c_{2k-n+j}$$
$$= \frac{1}{2\pi i} \int_{\mathbb{T}} \frac{f(\xi)}{\xi^{k+1}} \cdot 2 \cdot \Re\left(\sum_{j=0}^{n} \gamma_j \xi^{j+k-n}\right) d\xi.$$

(2)

$$\left|\sum_{j=0}^{n} \gamma_j c_{n-j} + \sum_{j=\max\{n-2k,0\}}^{n} \overline{\gamma}_j c_{2k-n+j}\right| \leq \frac{1}{2\pi} \int_{\mathbb{T}} 2 \cdot \Re\left(\sum_{j=0}^{n} \gamma_j \xi^{j+k-n}\right) d\xi.$$

In particular, if $k = 0$ *then we get*

$$\left|\sum_{j=0}^{n} \gamma_j c_{n-j} + \overline{\gamma}_n c_0\right| \leq 2 \cdot \Re(\gamma_n),$$

so

$$|\overline{\gamma}_n| - 2\Re(\gamma_n) \leq \left|\sum_{j=0}^{n} \gamma_j c_{n-j}\right| \leq |\overline{\gamma}_n| + 2\Re(\gamma_n).$$

Proof.

$$\sum_{j=0}^{n}\gamma_j c_{n-j} = \sum_{j=0}^{n}\gamma_j\left(\frac{1}{2\pi i}\int_{\mathbb{T}}\frac{f(\xi)}{\xi^{n-j+1}}\right) = \frac{1}{2\pi i}\int_{\mathbb{T}}\frac{f(\xi)}{\xi^{k+1}}\left(\sum_{j=0}^{n}\frac{\gamma_j}{\xi^{n-j-k}}\right)d\xi$$

$$= \frac{1}{2\pi i}\int_{\mathbb{T}}\frac{f(\xi)}{\xi^{k+1}}\left(\sum_{j=0}^{n}\gamma_j\xi^{j+k-n} + \sum_{j=0}^{n}\overline{\gamma}_j\overline{\xi}^{j+k-n}\right)d\xi$$

$$-\frac{1}{2\pi i}\int_{\mathbb{T}}\frac{f(\xi)}{\xi^{k+1}}\left(\sum_{j=0}^{n}\overline{\gamma}_j\overline{\xi}^{j+k-n}\right)d\xi$$

$$= \frac{1}{2\pi i}\int_{\mathbb{T}}\frac{f(\xi)}{\xi^{k+1}}\cdot 2\cdot\Re\left(\sum_{j=0}^{n}\gamma_j\xi^{j+k-n}\right)d\xi$$

$$-\sum_{j=\max\{n-2k,0\}}^{n}\overline{\gamma}_j c_{2k-n+j},$$

using the fact that $c_m = 0$ for $m < 0$. Hence,

$$\sum_{j=0}^{n}\gamma_j c_{n-j} + \sum_{j=\max\{n-2k,0\}}^{n}\overline{\gamma}_j c_{2k-n+j}$$

$$= \frac{1}{2\pi i}\int_{\mathbb{T}}\frac{f(\xi)}{\xi^{k+1}}\cdot 2\cdot\Re\left(\sum_{j=0}^{n}\gamma_j\xi^{j+k-n}\right)d\xi.$$

This proves the identity (1). To prove the inequality (2), we use the triangle inequality and the assumption that

$$\Re\left(\sum_{j=0}^{n}\gamma_j\xi^{j+k-n}\right) \geq 0 \ \text{ on } \ |\xi| = 1.$$

$$\left|\sum_{j=0}^{n}\gamma_j c_{n-j} + \sum_{j=\max\{n-2k,0\}}^{n}\overline{\gamma}_j c_{2k-n+j}\right|$$

$$\leq \frac{1}{2\pi}\int_{\mathbb{T}}\left|\frac{f(\xi)}{\xi^{k+1}}\right|\cdot 2\cdot\Re\left(\sum_{j=0}^{n}\gamma_j\xi^{j+k-n}\right)d\xi$$

$$\leq \frac{1}{2\pi}\int_{\mathbb{T}}2\cdot\Re\left(\sum_{j=0}^{n}\gamma_j\xi^{j+k-n}\right)d\xi,$$

because $|\xi^{k+1}| = 1$, $|f(\xi)| \leq 1$. Thus, inequality (2) is proved. Finally, if $k = 0$, then our assumption is

$$\Re\left(\sum_{j=0}^{n}\gamma_j\xi^{j-n}\right) = \Re\left(\frac{1}{\xi^n}\sum_{j=0}^{n}\gamma_j\xi^{j}\right) \geq 0 \ \text{ on } \ |\xi| = 1.$$

The lower bound of the index of the second sum is $\max\{n - 2\cdot 0, o\} = n$ and so inequality (2) becomes

$$\left|\sum_{j=0}^{n}\gamma_j c_{n-j} + \overline{\gamma}_n c_0\right| \leq 2 \cdot \Re(\gamma_n),$$

i.e.,

$$|\overline{\gamma}_n c_0| - 2\Re(\gamma_n) \leq \left|\sum_{j=0}^{n}\gamma_j c_{n-j}\right| \leq |\overline{\gamma}_n c_0| + 2\Re(\gamma_n).$$

$\square$

Remark 54.0.123. 1. The case $k = n$ in Theorem 54.0.122 gives us Theorem 54.0.121 as a special case.

2. The condition

$$\Re\left(\frac{1}{\xi^n}\sum_{j=0}^{n}\gamma_j\xi^{j}\right) \geq 0$$

is equivalent to the condition $\Re(z^{-n}F_n(z)) \geq 0$ on $0 < |z| \leq 1$.

Chapter 55

The Goluzin Functions

We will discuss simple properties of the family of functions that are implied by Goluzin's Theorem 1 ([37], p. 515), part (1).

Definition 55.0.124. A Goluzin function is an analytic function $F(z) = \sum_{n=0}^{\infty} \gamma_n z^n \in H(U)$ that satisfies the following three conditions.

(a) $F(0) = \gamma_0 \neq 0$,
(b) $\sum_{n=0}^{\infty} |\gamma_n| < \infty$,
(c) for all the functions $f(z) = \sum_{n=0}^{\infty} c_n z^n \in B(H^{\infty}(U))$ the following inequality holds

$$\left| \sum_{n=0}^{\infty} \gamma_n c_n \right| \leq |\gamma_0|. \tag{55.0.1}$$

Theorem 1(1) of Goluzin asserts that $F(z)$ is a Goluzin function if and only if

$$\Re\left(\frac{1}{\gamma_0} \sum_{n=0}^{\infty} \gamma_n \xi^n \right) \geq \frac{1}{2} \quad \text{on} \ |\xi| = 1.$$

This could be written as

$$\Re\left(\frac{F(\xi)}{F(0)} \right) \geq \frac{1}{2} \quad \text{on} \ |\xi| = 1. \tag{55.0.2}$$

Moreover, equality holds in inequality (55.0.1), only for $f(z) = c_0$, where $|c_0| = 1$.

Remark 55.0.125. Both inequality (55.0.1) and equation (55.0.2) are homogeneous in the γ_n. This means that if we replace the sequence $\{\gamma_n\}_{n=0}^{\infty}$

by $\{c\gamma_n\}_{n=0}^{\infty}$ for some fixed $c \neq 0$, then both inequalities (55.0.1) and (55.0.2) are not altered. This means that whenever $F(z)$ is a Goluzin function and $c \neq 0$, then also $cF(z)$ is a Goluzin function. This flexibility allows us to normalize our Goluzin functions. Here are four possible normalizations.

(n_1) Taking $c = (\sum_{n=0}^{\infty} |\gamma_n|)^{-1}$ enables us to assume that $\sum_{n=0}^{\infty} |\gamma_n| = 1$.

(n_2) By Goluzin's Theorem 1, equation (55.0.2), we note on taking $\xi = 1$, that $F(1) = \sum_{n=0}^{\infty} \gamma_n \neq 0$. So taking $c = (\sum_{n=0}^{\infty} \gamma_n)^{-1} = (F(1))^{-1}$ enables us to assume that $F(1) = \sum_{n=0}^{\infty} \gamma_n = 1$.

(n_3) Taking $c = \exp(-i \arg \gamma_0)$ $(\gamma_0 \neq 0)$ and noting that $c\gamma_0 = |\gamma_0|$ enables us to assume that γ_0 is a real and positive number.

(n_4) We can simultaneously have both normalizations (n_1) and (n_3) by taking $c = (\sum_{n=0}^{\infty} |\gamma_n|)^{-1} \exp(-i \arg \gamma_0)$. This enables us to assume both that $\sum_{n=0}^{\infty} |\gamma_n| = 1$ and that $\gamma_0 > 0$.

Using (n_4), we can rewrite the Goluzin inequalities (55.0.1) and (55.0.2) as follows: $|\sum_{n=0}^{\infty} \gamma_n c_n| \leq \gamma_0$, $\Re\left(\sum_{n=0}^{\infty} \gamma_n \xi^n\right) \geq \frac{1}{2}\gamma_0$ on $|\xi| = 1$. In this case,

$$\overline{F(\overline{z})} = \sum_{n=0}^{\infty} \overline{\gamma}_n z^n \in B(H^{\infty}(U)),$$

(by $\sum_{n=0}^{\infty} |\gamma_n| = 1$), so taking $c_n = \overline{\gamma}_n$ we arrive at $\sum_{n=0}^{\infty} |\gamma_n|^2 \leq \gamma_0$, i.e.,

$$\|F\|_2^2 \leq F(0). \tag{55.0.3}$$

Also $\sum_{n=1}^{\infty} |\gamma_n|^2 \leq \gamma_0 - \gamma_0^2 = \gamma_0(1 - \gamma_0) \leq \frac{1}{4}$ (by $0 < \gamma_0 \leq 1$), i.e.,

$$\|F(z) - F(0)\|_2 \leq \frac{1}{2}. \tag{55.0.4}$$

As for conclusions from inequality (55.0.2), we have $\|F\|_{\infty} \geq \Re\left(\sum_{n=0}^{\infty} \gamma_n \xi^n\right) \geq \frac{1}{2}\gamma_0$. Even more is true:

$$\forall |\xi| = 1, \ |F(\xi)|^2 = (\Re F(\xi))^2 + (\Im F(\xi))^2 \geq (\Re F(\xi))^2 \geq \left(\frac{\gamma_0}{2}\right)^2,$$

so in fact $\min_{|\xi|=1} |F(\xi)| \geq \frac{\gamma_0}{2}$, and also $\|F\|_2 \geq \frac{\gamma_0}{2}$. Combining that and equation (55.0.3) we obtain

$$\frac{1}{2} F(0) \leq \|F\|_2 \leq \sqrt{F(0)}.$$

So, we have the following:

Theorem 55.0.126. *Let $F(z) = \sum_{n=0}^{\infty} \gamma_n z^n$ be a Goluzin function that is normalized by $\sum_{n=0}^{\infty} |\gamma_n| = 1$ and $F(0) > 0$. Then*

$$\frac{1}{2} F(0) \leq \|F\|_2 \leq \sqrt{F(0)},$$

$$\min_{|\xi|=1} |F'(\xi)| \geq \frac{F(0)}{2}$$

$$\|F(z) - F(0)\|_2 \leq \frac{1}{2}.$$

Also, $F(z)$ does not vanish in U. In fact, it is bounded from below by $\frac{1}{2} F(0)$, i.e.,

$$\min_{|z|\leq 1} |F(z)| \geq \frac{1}{2} F(0).$$

Proof. Only the last assertion needs an explanation. Since clearly a Goluzin function is not constant, it is an open mapping in U. So, the pre-image of the boundary of the image, $F^{-1}(\partial F(U))$ is a subset of $\mathbb{T}$ and now the result follows by $\min_{|\xi|=1} |F(\xi)| \geq \frac{1}{2} F(0)$. $\square$

The assumptions in Goluzin's Theorem 1 ([37], p. 515), part (1): $m \in \mathbb{Z}^+ \cup \{0\}$, $\gamma_n \in \mathbb{C}$ $(n = m, m+1, \ldots)$, $\gamma_m \neq 0$ and $\sum_{n=m}^{\infty} |\gamma_n| < \infty$. (1) $m = 0$. The inequality

$$\left| \sum_{n=0}^{\infty} \gamma_n c_n \right| \leq |\gamma_0|, \tag{55.0.5}$$

holds for all the functions $f(z) = \sum_{n=0}^{\infty} c_n z^n \in B\left(H^{\infty}(U)\right)$ if and only if the following inequality holds:

$$\Re \left(\frac{1}{\gamma_0} \sum_{n=0}^{\infty} \gamma_n \xi^n \right) \geq \frac{1}{2} \quad \text{on} \quad |\xi| = 1. \tag{55.0.6}$$

Equality holds in inequality (55.0.5) only for $f(z) = c_0$, where $|c_0| = 1$.

Remark 55.0.127. Both inequalities (55.0.5) and (55.0.6) are homogeneous in the γ_n. This means that we may replace $\{\gamma_n\}_{n=0}^{\infty}$ by $\{c\gamma_n\}_{n=0}^{\infty}$, $(c \neq 0)$, and both (55.0.5) and (55.0.6) will remain as they are with the new sequence.

(a) For example, taking $c = \left(\sum_{n=m=0}^{\infty} |\gamma_n| \right)^{-1}$ enables us to assume that $\sum_{n=m=0}^{\infty} |\gamma_n| = 1$.

(b) Also, we note by (55.0.6) at $\xi = 1$ that $\sum_{n=0}^{\infty} \gamma_n \neq 0$, because

$$\Re\left(\frac{1}{\gamma_0}\sum_{n=0}^{\infty}\gamma_n\right) \geq \frac{1}{2},$$

so taking $c = \left(\sum_{n=0}^{\infty}\gamma_n\right)^{-1}$ we may assume that $\sum_{n=0}^{\infty}\gamma_n = 1$.

(c) Another, a third, possible normalization is by taking $c = \exp\left(-i\arg\gamma_0\right)$ in this case $c\gamma_0 = |\gamma_0|$ which allows us to assume that γ_0 is a real and positive number. We note that we can have simultaneously the normalizations (a) and (c) by taking $c = \left(\sum_{n=m=0}^{\infty}|\gamma_n|\right)^{-1}\exp\left(-i\arg\gamma_0\right)$. Having the normalizations (a) and (c), we may assume that $\sum_{n=0}^{\infty}|\gamma_n| = 1$, $\gamma_0 > 0$. Then we can rewrite (55.0.5) and (55.0.6) as follows:

$$\left|\sum_{n=0}^{\infty}\gamma_n c_n\right| \leq \gamma_0,$$

$$\Re\left(\sum_{n=0}^{\infty}\gamma_n\xi^n\right) \geq \frac{1}{2}\gamma_0 \quad \text{on} \quad |\xi| = 1.$$

In this case, $\sum_{n=0}^{\infty}\overline{\gamma}_n z^n \in B\left(H^{\infty}(U)\right)$, by $\sum_{n=0}^{\infty}|\gamma_n| = 1$, so we may take in (55.0.5) $c_n = \overline{\gamma}_n$ and obtain $\sum_{n=0}^{\infty}|\gamma_n|^2 \leq \gamma_0$. If $F(z) = \sum_{n=0}^{\infty}\gamma_n z^n$, then this is

$$\|F\|_2 \leq F(0). \tag{55.0.7}$$

Also, $\sum_{n=1}^{\infty}|\gamma_n|^2 \leq \gamma_0 - \gamma_0^2 = \gamma_0(1-\gamma_0) \leq \frac{1}{4}$, by $0 < \gamma_0 < 1$, so

$$\|F(z) - F(0)\|_2^2 \leq \left(\frac{1}{2}\right)^2. \tag{55.0.8}$$

As for (55.0.6), we have $\|F\|_{\infty} \geq \Re\left(\sum_{n=0}^{\infty}\gamma_n\xi^n\right) \geq \frac{1}{2}\gamma_0$. Also:

$$|F(\xi)|^2 = (\Re F(\xi))^2 + (\Im F(\xi))^2 \geq (\Re F(\xi))^2 \geq \left(\frac{\gamma_0}{2}\right)^2.$$

Hence, $\|F\|_2 \geq \gamma_0/2$. Combining that and (55.0.7), we obtain

$$\frac{1}{2}F(0) \leq \|F\|_2 \leq \sqrt{F(0)}.$$

Chapter 50

Standard Ways to Estimate Power Series Coefficients

Suppose we want to estimate the size of the absolute values of power series expansion coefficients. What representations of coefficients do we have? Here is a short list of the most elementary representations.

(1) By derivatives, $\frac{1}{n!}f^{(n)}(a)$, $(n = 0, 1, 2, \ldots)$. Usually $a = 0$.

(2) Using integrals,

$$\frac{1}{2\pi}\int_0^{2\pi} f(e^{i\theta})e^{-in\theta}d\theta, \quad (n = 0, 1, 2, \ldots).$$

(3) Viete type (using zeros), in polynomials $a_n(z - \alpha_1)\ldots(z - \alpha_n)$

$$a_{n-m} = a_n \sum_{1 \le i_1 < \cdots < i_m \le n} (-\alpha_{i_1})\ldots(-\alpha_{i_m})$$

$$= a_n \cdot \sum_{1 \le i_1 < \cdots < i_m \le n} (-1)^m \prod_{j=1}^{m} \alpha_{i_j}.$$

For entire functions of order 0

$$a_0 \prod_{n=1}^{\infty} \left(1 - \frac{z}{\alpha_n}\right) = a_0 + a_1 z + \cdots,$$

$$a_n = a_0 \cdot \sum_{1 \le i_1 < i_2 < \cdots} \left(\frac{-1}{\alpha_{i_1}}\right)\cdots\left(\frac{-1}{\alpha_{i_n}}\right)$$

$$= a_0 \cdot \sum_{1 \le i_1 < i_2 < \cdots} (-1)^n \left(\prod_{j=1}^{n}\left(\frac{1}{\alpha_{i_j}}\right)\right).$$

(4) What if $f(z)$ has no zero? Then, pick a random $c \in \mathbb{C}^{\times}$ and consider $g(z) = f(z) - c$. Generically, g will have zeros and the coefficients of $f(z)$ and of $g(z)$ are identical starting from $n = 1$. This is interesting by itself, for this shows that the Viete-type formulas are independent of c, i.e.,

$$(-1)^m a_n \cdot \sum_{1 \leq i_1 < \cdots < i_m \leq n} \prod_{j=1}^{m} \alpha_{i_j}(c), \quad (m \geq 1),$$

does not depend on c and

$$(a_0 - c) \cdot \sum_{1 \leq i_1 < i_2 < \cdots} (-1)^n \prod_{j=1}^{n} \left(\frac{1}{\alpha_{i_j}(c)} \right),$$

does not depend on c. Thus, on differentiation with respect to c, we get

$$\frac{d}{dc} \left\{ \sum_{1 \leq i_1 < \cdots < i_m \leq n} \prod_{j=1}^{m} \alpha_{i_j}(c) \right\} = 0, \quad (m \geq 1),$$

or in the entire function of order zero case,

$$\frac{d}{dc} \left\{ (a_0 - c) \cdot \sum_{1 \leq i_1 < i_2 < \cdots} (-1)^n \prod_{j=1}^{n} \left(\frac{1}{\alpha_{i_j}(c)} \right) \right\} = 0, \quad (n \geq 1).$$

So,

$$\sum_{1 \leq i_1 < \cdots < i_m \leq n} \left[\sum_{k=1}^{m} \left(\alpha'_{i_k}(c) \prod_{j=1, j \neq k}^{m} \alpha_{i_j}(c) \right) \right] = 0, \quad (m \geq 1),$$

or

$$\sum_{1 \leq i_1 < i_2 < \cdots} \left[\sum_{k=1}^{n} \left(\frac{\alpha'_{i_k}(c)}{(\alpha_{i_k}(c))^2} \prod_{j=1, j \neq k}^{n} \left(\frac{1}{\alpha_{i_j}(c)} \right) \right) \right]$$

$$= \sum_{1 \leq i_1 < i_2 < \cdots} \prod_{j=1}^{n} \left(\frac{1}{\alpha_{i_j}(c)} \right),$$

which can be written as

$$\sum_{1 \leq i_1 < i_2 < \cdots} \left(\left(\sum_{k=1}^{n} \frac{\alpha'_{i_k}(c)}{\alpha_{i_k}(c)} \right) - 1 \right) \prod_{j=1}^{n} \left(\frac{1}{\alpha_{i_j}(c)} \right) = 0.$$

(5) Residue-type representations. We have

$$\frac{1}{2\pi i} \oint_{|z|=R} g(z)\frac{f'(z)}{f(z)}\,dz = \sum_{f(\alpha_k)=0,\,|\alpha_k|<R} g(\alpha_k),$$

provided that $f(z)$ has no zeros on $|z| = R$. A particular case is when $g(z) = f^{(n)}(z)$ in which case we obtain

$$\frac{1}{2\pi i} \oint_{|z|=R} f^{(n)}(z)\frac{f'(z)}{f(z)}\,dz = \sum_{f(\alpha_k)=0,\,|\alpha_k|<R} f^{(n)}(\alpha_k).$$

Hence, if $R > 0$ is small enough so that $f(z) - f(0)$ has no zero in $0 < |z| < R$, and the zero $z = 0$ has multiplicity $m \geq 1$, then

$$\frac{1}{m} \cdot \frac{1}{n!} \cdot \frac{1}{2\pi i} \oint_{|z|=R} f^{(n)}(z)\frac{f'(z)}{f(z)-f(0)}\,dz = \frac{f^{(n)}(0)}{n!},$$

our n'th coefficient.

(6) Cauchy type, based on

$$f(w) = \frac{1}{2\pi i} \oint_{|z-w|=R} \frac{f(z)}{z-w}\,dz.$$

On differentiating n times with respect to w, we obtain the well known

$$\frac{f^{(n)}(0)}{n!} = \frac{1}{2\pi i} \oint_{|z|=R} \frac{f(z)}{z^{n+1}}\,dz.$$

(7) Jensen's. We consider a function $w = w(z)$ which is meromorphic in the finite or the infinite disk $|z| < R \leq \infty$. Let the poles of w be denoted by b_ν and the zeros by a_μ. For simplicity, we assume that $w(0) \neq 0$. Let ρ be a number such that $0 < \rho < R$ and such that the function w is nonzero and finite on the circle $|z| = \rho$. Then the Poisson–Jensen formula is

$$\log\left|w(re^{i\phi})\right| = \frac{1}{2\pi} \int_0^{2\pi} \log\left|w(\rho e^{i\theta})\right| \frac{\rho^2 - r^2}{\rho^2 + r^2 - 2\rho r \cos(\theta - \phi)}\,d\theta$$

$$+ \sum_{|b_\nu|<\rho} \log\left|\frac{\rho^2 - \bar{b}_\nu z}{\rho(z - b_\nu)}\right| - \sum_{|a_\mu|<\rho} \log\left|\frac{\rho^2 - \bar{a}_\mu z}{\rho(z - a_\mu)}\right|,$$

for $z = re^{i\phi}$, $0 \leq r < \rho$. In particular, if we set $z = 0$, the formula becomes Jensen's formula

$$\log |w(0)| = \frac{1}{2\pi} \int_0^{2\pi} \log \left| w(\rho e^{i\theta}) \right| d\theta + \sum_{|b_\nu| < \rho} \log \frac{\rho}{|b_\nu|} - \sum_{|a_\mu| < \rho} \log \frac{\rho}{|a_\mu|}.$$

If, in addition, the harmonic conjugates with respect to the variable z, multiplied by i, are added to both sides of the Poisson–Jensen formula, the result is

$$\log w(z) = \frac{1}{2\pi} \int_0^{2\pi} \log \left| w(\rho e^{i\theta}) \right| \frac{\rho e^{i\theta} + z}{\rho e^{i\theta} - z} d\theta$$

$$+ \sum_{|b_\nu| < \rho} \log \frac{\rho^2 - \overline{b}_\nu z}{\rho(z - b_\nu)} - \sum_{|a_\mu| < \rho} \log \frac{\rho^2 - \overline{a}_\mu z}{\rho(z - a_\mu)} + iC,$$

where

$$C = \arg w(0) - \sum_{|b_\nu| < \rho} \arg \left(-\frac{\rho}{b_\nu} \right) + \sum_{|a_\mu| < \rho} \arg \left(-\frac{\rho}{a_\mu} \right) + 2n\pi.$$

If the origin $z = 0$ is either a zero or a pole of $w(z)$, then both sides of Jensen's formula become infinite. To remove this inconvenience, we apply Jensen's formula to the expression $w \cdot z^{-\lambda}$ if w has an expansion of the form $w(z) = c_\lambda z^\lambda + c_{\lambda+1} z^{\lambda+1} + \cdots$, $(c_\lambda \neq 0)$. One finds that

$$\log |c_\lambda| = \frac{1}{2\pi} \int_0^{2\pi} \log \left| w(\rho e^{i\theta}) \right| d\theta + \sum_{0 < |b_\nu| < \rho} \log \frac{\rho}{|b_\nu|}$$

$$- \sum_{0 < |a_\mu| < \rho} \log \frac{\rho}{|a_\mu|} - \lambda \log \rho.$$

We took (7) from [75, pp. 162–165].

Chapter 57

More on Inner Functions [59]

We start by quoting some results from [59].

Lemma ([59], p. 74). *Let $\Phi(x)/x \to \infty$ for $x \to \infty$, where $\Phi(x) \geq 0$. Let $f_n(t) \geq 0$, $\int_{-\pi}^{\pi} \Phi(f_n(t))dt \leq c$, and $f_n(t)dt \to_n d\mu(t)$ w^*, where μ is a measure on $[-\pi, \pi]$. Then μ is absolutely continuous.*

Lemma ([59], p. 75). *If $F(z)$ is regular in $\{|z| < 1\}$ and $\int_{-\pi}^{\pi} \log^+ |F(re^{i\theta})|d\theta$ is bounded for $0 \leq r < 1$, then, if $F \not\equiv 0$, also $\int_{-\pi}^{\pi} |\log |F(re^{i\theta})||rd\theta$ is bounded for $0 < r < 1$.*

Expressing an analytic function in terms of its real part ([59], p. 76):

If $F(z)$ is regular in a region including $\{|z| < 1\}$ in its interior and if $0 \leq r < 1$, then

$$F(re^{i\theta}) = \frac{1}{2\pi} \int_{-\pi}^{\pi} \frac{e^{it} + re^{i\theta}}{e^{it} - re^{i\theta}} \Re F(e^{it})dt + i\Im F(0).$$

Theorem ([59], p. 76). *Let $F(z) \not\equiv 0$ belong to $H^p(U)$, $p > 0$. Let $B(z)$ be the Blaschke product consisting of the zeros of $F(z)$. Then there is a singular measure $\sigma \geq 0$ on $[-\pi, \pi]$ with*

$$F(z) = B(z) \exp\left(-\frac{1}{2\pi} \int_{-\pi}^{\pi} \frac{e^{it} + z}{e^{it} - z} d\sigma(t)\right) e^{ic}$$

$$\times \exp\left(\frac{1}{2\pi} \int_{-\pi}^{\pi} \frac{e^{it} + z}{e^{it} - z} \log|F(e^{it})| dt\right),$$

for $|z| < 1$, where c is a real constant.

317

Definition 57.0.128. The formula in the last theorem is called the canonical representation of the function $F \not\equiv 0$ in $H^p(U)$. It is essentially unique. We call (after Beurling)

$$I_F(z) = e^{ic} B(z) \exp\left(-\frac{1}{2\pi} \int_{-\pi}^{\pi} \frac{e^{it} + z}{e^{it} - z} d\sigma(t) \right)$$

the inner factor of $F(z)$, and

$$O_F(z) = \exp\left(\frac{1}{2\pi} \int_{-\pi}^{\pi} \frac{e^{it} + z}{e^{it} - z} \log |F(e^{it})| \, dt \right),$$

the outer factor of $F(z)$. Note that $|I_F(z)| \leq 1$ for $|z| < 1$ $(d\sigma \geq 0)$ and $|I_F(\xi)| = 1$ a.e. $|\xi| = 1$.

Beurling's Theorem.

Lemma 1 ([59], p. 78). *Let* $p > 0$. *Then, if* $f \in H^p(U)$, *there exist polynomials* $P(z)$ *with* $\|F - P\|_p$ *arbitrarily small.*

Lemma 2 (Generalization of Smirnov's theorem, [59], p. 79). *Let* F *and* G *(both* $\not\equiv 0$*) belong to (perhaps different)* $H^p(U)$ *spaces. If the ratio of their outer factors* $k(z) = O_F(z)/O_G(z)$ *satisfies* $k(e^{i\theta}) \in L^{p'}(\mathbb{T})$, *say, then* $O_F(z)/O_G(z) \in H^{p'}(U)$.

Theorem 3 (General case due to Srinivasan and Wang; for $p=2$ this is due to Beurling. [59], p. 79). *Let* $F = I_F \cdot O_F \in H^p(U)$, $p > 0$. *The closure of* $F(z) \cdot \{\text{Polynomials in } z\}$ *in* $H^p(U)$ *is precisely* $I_F \cdot H^p(U)$.

Invariant subspaces.

Let $\mathcal{H}$ be a Hilbert space with orthonormal basis $\{e_n | n = 0, \pm 1, \pm 2, \ldots\}$, and consider the unitary transformation V defined on $\mathcal{H}$ by putting $V e_n = e_{n+1}$. The problem is to study the invariant subspaces of V. We map $\mathcal{H}$ isometrically onto $L^2(-\pi, \pi)$ by making e_n correspond to $e^{in\theta}/\sqrt{2\pi}$; then V corresponds to multiplication by $e^{i\theta}$.

Theorem ([59], p. 81). *Let a closed subspace* E *of* $L^2(-\pi, \pi)$ *be such that* $e^{i\theta} \cdot E = E$. *Then* $E = \chi_A \cdot L^2$ *where* χ_A *is the characteristic function of some measurable set* $A \subset (-\pi, \pi)$.

Theorem ([59], p. 82). *Let* E *be a closed subspace of* L^2 *and suppose that* $e^{i\theta} \cdot E \subset E$ *properly. Then* $E = \omega \cdot H^2$, *where* $|\omega(\theta)| = 1$ *a.e.*

Corollary ([59], p. 82). *Let E be a subspace of H^2 and satisfy $e^{i\theta} \cdot E \subseteq E$, $E \neq \{0\}$. Then $E = \omega \cdot H^2$, where ω is an inner function (i.e., one whose outer factor is identically 1).*

Corollary ([59], pp. 82–83). *Let $f_\alpha = I_\alpha O_\alpha$, with each I_α inner and each O_α outer in H^2, and suppose that*

$$I_\alpha(z) = B_\alpha(z) \exp\left(\frac{-1}{2\pi} \int_{-\pi}^{\pi} \frac{e^{it} + z}{e^{it} - z} d\sigma_\alpha(t) \right),$$

with Blaschke products B_α and singular $d\sigma_\alpha \geq 0$. Let E be the invariant subspace of H^2 generated by the f_α. In other words, let E be the smallest closed subspace of H^2 containing all the f_α such that $e^{i\theta} \cdot E \subset E$. Then $E = \omega \cdot H^2$, where

$$\omega(z) = B(z) \exp\left(\frac{-1}{2\pi} \int_{-\pi}^{\pi} \frac{e^{it} + z}{e^{it} - z} d\sigma(t) \right).$$

Here, $B(z)$ is the greatest common divisor of the $B_\alpha(z)$, and $d\sigma \geq 0$ is the largest measure $\leq$ all the $d\sigma_\alpha$.

Approximation of inner functions by Blaschke products.

Most of the previous results hold for H^∞ if $\|\cdot\|_p$ is replaced by $\|\cdot\|_\infty$. A notable exception is the theorem saying that $\|F(re^{i\theta}) - F(e^{i\theta})\|_p \to 0$ as $r \to 1^-$ if $F \in H^p$. This is, in general, false for $F \in H^\infty$ if we use the norm $\|\cdot\|_\infty$. Indeed, if $F \in H^\infty$ and we have $\|F(re^{i\theta}) - F(e^{i\theta})\|_\infty \to 0$ as $r \to 1^-$, then $F(e^{i\theta})$ is continuous, but there are plenty of $F \in H^\infty$ with $F(e^{i\theta})$ not continuous. We do have for $F \in H^\infty$, $\|F\|_\infty = \operatorname{ess\,sup}_\theta |F(e^{i\theta})|$. A special class of functions in H^∞ are the inner functions, i.e., those whose outer factor equals to 1. An inner function $\omega(z)$ has the property that $|\omega(e^{i\theta})| = 1$ a.e. An inner function $\omega(z)$ has the representation

$$\omega(z) = e^{ic} \cdot B(z) \cdot \exp\left(\frac{-1}{2\pi} \int_{-\pi}^{\pi} \frac{e^{it} + z}{e^{it} - z} d\sigma(t) \right),$$

where $c \in \mathbb{R}$, $B(z)$ is a Blaschke product, and $d\sigma \geq 0$ is a singular measure on $[-\pi, \pi]$.

Lemma ([59], p. 84). *An inner function $\omega(z)$ is a constant multiple of a Blaschke product if and only if $\int_{-\pi}^{\pi} \log|\omega(re^{i\theta})| d\theta \to 0$ as $r \to 1$.*

Theorem (Frostman. Rediscovered years later by Newman, [59], p. 85). *Let $\omega(z)$ be any inner function. Then, given any $\epsilon > 0$ there is a Blaschke product $B(z)$ and a real number c with $\|\omega - e^{ic} \cdot B\|_\infty < \epsilon$.*

Duality for H^p spaces.

A. H^p spaces and their duals.

1. Sarason's theorem. By looking at the boundary values $f(e^{i\theta})$ of $f \in H^p$, we see that H^p can be considered as a $\|\cdot\|_p$-closed subspace of $L^p(-\pi, \pi)$. We define $\mathcal{C} = \{f \text{ continuous on } [-\pi, \pi] | f(-\pi) = f(\pi)\}$. $\mathcal{A} = \mathcal{C} \cap H^\infty$. This is the set of functions in $\mathcal{C}$ which have analytic extension to $\{|z| < 1\}$, yielding continuous functions on the closed unit disk. We equip $\mathcal{C}$ and $\mathcal{A}$ with the sup-norm $\|\cdot\|_\infty$. $\mathcal{M}$ = set of finite complex-valued Radon measures on $\{|\xi| = 1\}$, equipped with the measure norm $\|\mu\| = \int_{-\pi}^{\pi} |d\mu(e^{i\theta})|$. By a classical theorem of F. Riesz, $\mathcal{M}$ is the dual of $\mathcal{C}$.

Notation.

$$H^p(0) = \left\{ f \in H^p \, \bigg| \, \int_{-\pi}^{\pi} f(e^{i\theta})d\theta = 0 \right\} = zH^p.$$

Theorem 1.1 ([59], p. 141). *If $1 < p < \infty$ and $\frac{1}{p} + \frac{1}{q} = 1$, L^q/H^q has dual $H^p(0)$ and H^p has dual $L^q/H^q(0)$.*

Theorem 1.2 ([59], p. 141). *The dual of L^1/H^1 is $H^\infty(0)$. The dual of H^1 is $L^\infty/H^\infty(0)$.*

Theorem 1.3 ([59], p. 142). *The dual of $\mathcal{C}/\mathcal{A}$ is $H^1(0)$. If we denote (naturally) $\mathcal{A}(0) = \{e^{i\theta}f(e^{i\theta})|f \in \mathcal{A}\}$, then: $\mathcal{C}/\mathcal{A}(0)$ has dual H^1 which has dual $L^\infty/H^\infty(0)$.*

2. Duality method of Khavinson and of Rogosinski–Shapiro. The duality results in 1 give us some theorems on approximation by H^p functions.

Theorem 2.1 ([59], p. 143). *Let $F \in L^p(-\pi, \pi)$, $1 < p < \infty$, and denote $\|F - H^p\|_p = \inf\{\|F - h\|_p | h \in H^p\}$. Then with $\frac{1}{q} = 1 - \frac{1}{p}$,*

(i)

$$\|F - H^p\|_p = \sup\left\{ \left| \int_{-\pi}^{\pi} F(e^{i\theta})g(e^{i\theta})d\theta \right| \, \bigg| \, g \in H^q(0) \wedge \|g\|_q = 1 \right\}.$$

(ii) *There is an $h_0 \in H^p$ with $\|F - H^p\|_p = \|F - h_0\|_p$ (i.e., the infimum is attained).*

(iii) *There is a $g_0 \in H_q(0)$, $\|g_0\|_q = 1$ with $\|F - H^p\|_p = \int_{-\pi}^{\pi} F(e^{i\theta})g_0(e^{i\theta})d\theta$ (i.e., the supremum is actually a maximum).*

Theorem 2.2 ([59], p. 144). *Let $F \in L^1(-\pi, \pi)$. Then there is an $h_0 \in H^1$ with $\|F - H^1\|_1 = \|F - h_0\|_1$, and there is a $g_0 \in H^\infty(0)$ with $\|g_0\|_\infty = 1$ and $(F(e^{i\theta}) - h_0(e^{i\theta})g_0(e^{i\theta})) = |F(e^{i\theta}) - h_0(e^{i\theta})|$ a.e.*

Theorem 2.3 ([59], p. 145). *Let $F \in \mathcal{C}$. Then*

(1)
$$\|F - \mathcal{A}\|_\infty = \|F - H^\infty\|_\infty$$
$$= \sup\left\{ \left| \int_{-\pi}^{\pi} F(e^{i\theta}) g(e^{i\theta}) d\theta \right| \,\Big|\, g \in H^1(0) \wedge \|g\|_1 = 1 \right\}.$$

(ii) *There is a $g_0 \in H^1(0)$, $\|g_0\|_1 = 1$, with $\|F - \mathcal{A}\|_\infty = \int_{-\infty}^{\infty} F(e^{i\theta}) g_0(e^{i\theta}) d\theta$.*

(iii) *There is an $h_0 \in H^\infty$ with $|F(e^{i\theta}) - h_0(e^{i\theta})| = \|F - \mathcal{A}\|_\infty$ a.e.*

If $F \notin \mathcal{C}$, all we have is the following,

Theorem 2.4 ([59], p. 146). *Let $F \in L^\infty$. Then there is an $h_0 \in H^\infty$ with*

$$\|F - h_0\|_\infty = \|F - H^\infty\|_\infty$$
$$= \sup\left\{ \left| \int_{-\pi}^{\pi} F(e^{i\theta}) g(e^{i\theta}) d\theta \right| \,\Big|\, g \in H^1(0) \wedge \|g\|_1 = 1 \right\}.$$

In general, the supremum is not attained if $F \in H^\infty$ is not continuous.

3. **Sarason's theorem (finally).** Let's go back to the chain: $\mathcal{C}/\mathcal{A}(0)$ has dual H^1 which has dual $L^\infty/H^\infty(0)$. Thus, if B is the Banach space $\mathcal{C}/\mathcal{A}(0)$, then B^{**} is $L^\infty/H^\infty(0)$. B has a canonical isometric image in B^{**} obtained by identifying linear functionals over B^*. In the present case, the element of $L^\infty/H^\infty(0)$ corresponding to $F + \mathcal{A}(0)$, $F \in \mathcal{C}$ is the coset $\Phi + H^\infty(0)$, where $\Phi \in L^\infty$ determined by

$$\int_{-\pi}^{\pi} F(e^{i\theta}) g(e^{i\theta}) d\theta = \int_{-\pi}^{\pi} g(e^{i\theta}) \Phi(e^{i\theta}) d\theta$$

for all $g \in H^1$. We see that this holds if and only if $\Phi \in F + H^\infty(0)$, which is to say that in the canonical embedding of $\mathcal{C}/\mathcal{A}(0)$ in $L^\infty/H^\infty(0)$, $F + \mathcal{A}(0)$, $F \in \mathcal{C}$, corresponds to $F + H^\infty(0)$. The image of $\mathcal{C}/\mathcal{A}(0)$ in $L^\infty/H^\infty(0)$ under this embedding is thus $\mathcal{E} = \{F + H^\infty(0) \,|\, F \in \mathcal{C}\}$. In particular, $\mathcal{E}$ is $\|\cdot\|_\infty$-closed in the quotient space $L^\infty/H^\infty(0)$. Now, the canonical homomorphism $\Theta : L^\infty \to L^\infty/H^\infty(0)$ is continuous. Therefore, $\Theta^{-1}(\mathcal{E})$ is $\|\cdot\|_\infty$-closed in L^∞. But $\Theta^{-1}(\mathcal{E})$ is $\mathcal{C} + H^\infty(0)$ which also equals $\mathcal{C} + H^\infty$ since $1 \in \mathcal{C}$. Therefore, we prove the following,

Theorem 3.1 (Sarason, [59], p. 148). $\mathcal{C} + H^\infty$ *is* $\|\cdot\|_\infty$*-closed.*

From this, one can prove the following:

Theorem 3.2 (Sarason, [59], p. 148). *If F and $G \in \mathcal{C} + H^\infty$, $FG \in \mathcal{C} + H^\infty$ i.e., $\mathcal{C} + H^\infty$ is an algebra.*

B. Elements of constants modulus in cosets of L^∞ / H^∞.

Marshal's theorem.

1. A result of Adamian, Arov and Krein. Given $F \in L^\infty$ we have seen in Section A that under certain circumstances (e.g., F continuous), the coset $F + H^\infty$ contains an element of constant modulus equal to $\| - H^\infty\|_\infty$. We are interested in seeing the extent to which elements of constant modulus occur in $F + H^\infty$. Around 1920, Nevanlinna proved some deep results about this, using very "hard" methods. Complete and definitive results were arrived some 40 years later by Adamian, Arov and Krein with the help of operator theory. Garnett found functional analytic proofs of some of these results towards the end of the 1970's. Here is one of these results.

Theorem 1.1 (proof of Garnett, [59], p. 150). *If $F \in L^\infty$ and $\|F - H^\infty\|_\infty < 1$, then $F + H^\infty$ contains an element w with $|w(e^{i\theta})| = 1$ a.e.*

Corollary 1.2 ([59], p. 152). *Let $f \in H^\infty$, $\|f\|_\infty < 1$, and let Ω be any inner function. Then there is another inner function, $\omega \in f + \Omega \cdot H^\infty$.*

Remark 57.0.129. Adamian, Arov and Krein also proved that if $F \in L^\infty$ and $\|F - H^\infty\|_\infty = 1$ but $F - H^\infty$ contains more than one element of norm 1, then there is an $\omega \in F - H^\infty$ with $|\omega(e^{i\theta})| = 1$ a.e. Garnett has a functional-theoretic proof of this fact.

2. Marshall's theorem. Around 1975, D. Marshall verified a long-standing conjecture about uniform approximation of H^∞ functions by linear combinations of Blaschke products. Let us first mention a much easier case about L^∞.

Theorem 2.1 ([59], p. 152). *Let $f \in L^\infty(-\pi, \pi)$ and $\|f\|_\infty \le 1$. Then, given any $\epsilon > 0$ we can find $u_1, \ldots, u_n \in L^\infty$ with $|u_k(\theta)| = 1$ a.e. (so-called uni-modular functions) and numbers $\lambda_1, \ldots, \lambda_n \ge 0$, $\sum_k \lambda_k = 1$, with $\|f - \sum_k \lambda_k u_k\|_\infty < \epsilon$. In other words, the norm-closed convex hull of the set of uni-modular functions in L^∞ is precisely the unit ball of L^∞.*

The uni-modular functions in H^∞ are the inner functions. It is natural to conjecture that, in H^∞, the norm-closed convex hull of the set of inner functions is the unit ball of H^∞. Marshall proved this.

Lemma 2.2 (Douglas and Rudin, [59], p. 153). *Let $u \in L^\infty$, $|u(\xi)| = 1$ a.e. on $|\xi| = 1$. Then there are inner functions ω and Ω in H^∞ with*

$$\left| u(\xi) - \frac{\omega(\xi)}{\Omega(\xi)} \right| < \epsilon \quad \text{a.e. on } |\xi| = 1.$$

Corollary 2.3 ([59], p. 156). *Let $f \in L^\infty$, $\|f\|_\infty \leq 1$ and let $\epsilon > 0$. Then we can find inner functions $\omega_1, \ldots, \omega_n$, $\Omega_1, \ldots, \Omega_n$, and numbers $\lambda_k > 0$, $\sum_{k=1}^n \lambda_k = 1$ with*

$$\left| f(\xi) - \sum_{k=1}^n \lambda_k \frac{\omega_k(\xi)}{\Omega_k(\xi)} \right| < \epsilon \quad \text{a.e. on } |\xi| = 1.$$

Remark 57.0.130. All the Ω_k can be taken equal. Just take the common denominator.

Lemma 2.4 ([59], p. 156). *Let $f \in H^\infty$. Then we can find inner functions $\Omega, \omega, \omega_1, \ldots, \omega_n$ and real constants $a, a_1, \ldots, a_n$ such that*

(i) $g = (a\omega + a_1\omega_1 + \cdots + a_n\omega_n)/\Omega$ *is in H^∞.*
(ii) $\|f - g\|_\infty < 2\epsilon$.

Theorem 2.5 (Marshall, [59], p. 156). *Let $f \in H^\infty$ and $\|f\|_\infty \leq 1$. Given $\epsilon > 0$, we can find inner functions $u_1, \ldots, u_n$ and positive numbers $\lambda_1, \ldots, \lambda_n$, $\sum_{k=1}^n \lambda_k = 1$, with $\|f - \sum_{k=1}^n \lambda_k u_k\|_\infty < \epsilon$.*

We recall a theorem of Frostman that says that any inner function can be uniformly approximated by Blaschke products (these are understood to contain arbitrary constant factors of modulus 1).

Marshall's Theorem ([59], p. 157). *Let $f \in H^\infty$, $\|f\|_\infty \leq 1$. Then there are Blaschke products $B_1, \ldots, B_n$ and positive numbers λ_k, $\sum_{k=1}^n \lambda_k = 1$, with $\|f - \sum_{k=1}^n \lambda_k B_k\|_\infty < \epsilon$, $\epsilon > 0$ being arbitrary.*

C. Szegő's theorem.

Let μ be a finite positive measure on $[-\pi, \pi]$ and let $\mathcal{P}(0)$ be the class of polynomials $P(z)$ with $P(0) = 0$. If $1 \leq p < \infty$, we are interested in how small we can make $\int_{-\pi}^\pi |1 - P(e^{i\theta})|^p d\mu(\theta)$ for $P \in \mathcal{P}(0)$.

Theorem 1.1 ([59], p. 158). *If σ is a positive singular measure, then*

$$\inf \left\{ \int_{-\pi}^\pi \left| 1 - P(e^{i\theta}) \right|^p d\sigma(\theta) \,|\, P \in \mathcal{P}(0) \right\} = 0.$$

Theorem 1.2 (Kolmogorov, [59], p. 159). *Let $d\mu(\theta) = w(\theta)d\theta + d\sigma(\theta)$, with $w \in L^1$, $w \geq 0$, $d\sigma \geq 0$ and σ singular. Let $1 \leq p < \infty$. Then*

$$\inf_{P \in \mathcal{P}(0)} \int_{-\pi}^{\pi} \left| 1 - P(e^{i\theta}) \right|^p w(\theta)d\theta = \inf_{P \in \mathcal{P}(0)} \int_{-\pi}^{\pi} \left| 1 - P(e^{i\theta}) \right|^p d\mu(\theta).$$

(*Thus, only the absolutely continuous part of μ matters.*)

Theorem 1.3 (Szegö, [59], p. 159). *If $1 \leq p < \infty$, then*

$$\inf_{P \in \mathcal{P}(0)} \frac{1}{2\pi} \int_{-\pi}^{\pi} \left| 1 - P(e^{i\theta}) \right|^p w(\theta)d\theta = \exp\left(\frac{1}{2\pi} \int_{-\pi}^{\pi} \log w(\theta)d\theta \right).$$

D. The Helson–Szegö's theorem.

In a 1960 Bologna Annali paper, Helson and Szegö gave a characterization of the finite positive measures μ on $[-\pi, \pi]$ having the property that

$$\int_{-\pi}^{\pi} \left| \widetilde{T}(\theta) \right|^2 d\mu(\theta) \leq \text{Const.} \int_{-\pi}^{\pi} |T(\theta)|^2 d\mu(\theta),$$

for all trigonometric polynomials $T(\theta)$. A trigonometric polynomial is a finite sum of the form $T(\theta) = \sum_n a_n e^{in\theta}$ and the harmonic conjugate of $T(\theta)$ is $(\widetilde{T})(\theta) = -i \sum_n (\text{sgn} n) a_n e^{in\theta}$ where we put $\text{sgn} 0 = 0$.

Definition 57.0.131. A positive measure μ is called a Helson–Szegö measure if it satisfies

$$\int_{-\pi}^{\pi} \left| \widetilde{T}(\theta) \right|^2 d\mu(\theta) \leq \text{Const.} \int_{-\pi}^{\pi} |T(\theta)|^2 d\mu(\theta),$$

for all trigonometric polynomials T.

A simple direct computation shows that $d\mu(\theta) = d\theta$ is a Helson–Sezgö measure.

Theorem 1.1 ([59], p. 163). *A Helson–Szegö measure is absolutely continuous.*

Theorem 1.2 ([59], p. 164). *A non-zero Helson–Szegö measure is necessarily of the form $d\mu(\theta) = w(\theta)d\theta$ with $w \geq 0$ in $L^1(-\pi, \pi)$ and $\int_{-\pi}^{\pi} \log w(\theta)d\theta > -\infty$.*

This reduces the determination of all Helson–Szegö measures to those $w(\theta)d\theta$ described in Theorem 1.2. Helson and Szegö introduced an auxiliary operation related to $\widetilde{\cdot}$, harmonic conjugation.

Definition 57.0.132. If $T(\theta) = \sum_n a_n e^{in\theta}$ is a trigonometric polynomial, we denote the projection of T by $(\Pi T)(\theta) = \sum_{n>0} a_n e^{in\theta}$. Also, we will use

$$< f, g >_w = \int_{-\pi}^{\pi} f(\theta)\overline{g(\theta)}w(\theta)d\theta, \quad \|f\|_w = \sqrt{\int_{-\pi}^{\pi} |f(\theta)|^2 w(\theta)d\theta}.$$

Lemma 1.3 ([59], p. 165). $\|\widetilde{T}\|_w \leq k \cdot \|T\|_w$ *for all trigonometric polynomials T and some k if and only if $\|\Pi T\|_w \leq c \cdot \|T\|_w$ for all such T and some c.*

Lemma 1.4 ([59], p. 165). $w(\theta)d\theta$ *is a Helson–Szegö measure if and only if there is a $\rho < 1$ such that, whenever $P, Q \in \mathcal{P}(0)$,*

$$\left| \Re \int_{-\pi}^{\pi} P(e^{i\theta}) \cdot e^{-i\theta} Q(e^{i\theta}) w(\theta)d\theta \right| \leq \rho \|P\|_w \|Q\|_w.$$

Theorem 1.5 (Helson and Szegö, [59], p. 166). *A measure $d\mu$ such that*

$$\int_{-\pi}^{\pi} \left| \widetilde{T}(\theta) \right|^2 d\mu(\theta) \leq C \cdot \int_{-\pi}^{\pi} |T(\theta)|^2 d\mu(\theta),$$

for all trigonometric polynomials T is necessarily of the form $d\mu(\theta) = \exp(u(\theta) + \widetilde{v}(\theta))d\theta$ where u and v are real-valued, $|u(\theta)|$ is bounded, and $|v(\theta)| \leq \frac{\pi}{2} - \epsilon$, $\epsilon > 0$. Conversely, if $d\mu(\theta) = \exp(u(\theta) + \widetilde{v}(\theta))d\theta$ with u and v as stated, the above inequality holds with some C.

Remark 57.0.133. In 1970 or 1971, Hunt, Muckenhoupt and Weeden determined completely all weights $w(\theta)$ for which

$$\int_{-\pi}^{\pi} \left| \widetilde{T}(\theta) \right|^p d\mu(\theta) \leq C \cdot \int_{-\pi}^{\pi} |T(\theta)|^p d\mu(\theta),$$

where $1 < p < \infty$. Their solution looks entirely different from that of Helson and Szegö, and, for $p = 2$, is as follows: $w(\theta)d\theta$ is a Helson–Szegö measure if and only if, for all intervals I, we have

$$\left\{ \frac{1}{|I|} \int_I w(\theta)d\theta \right\} \cdot \left\{ \frac{1}{|I|} \int_I \frac{d\theta}{w(\theta)} \right\} \leq C,$$

a finite constant independent of I.

Problem 7 ([59], p. 169). Let $\omega(\theta) \in L^\infty(-\pi, \pi)$ and $|\omega(\theta)| = 1$ a.e. The problem is to show that there is a non-zero $h \in H^\infty$ such that **(1)** $|\omega(\theta) - h(\theta)| \leq 1$ a.e. if and only if there is a non-zero $f \in H^1$ with **(2)** $\omega(\theta) = f(\theta)/|f(\theta)|$ a.e.

(a) if there is an $f \in H^1$ satisfying **(2)** above, show that $h = f/(P + \widetilde{i(P)})$ is in H^∞ and satisfies **(1)**, where $P(\theta) = |f(\theta)|$.

(b) *If there is a non-zero $h \in H^\infty$ satisfying **(1)** above, show that $h \exp(\widetilde{\Psi} - i\Psi) \in H^1$, where Ψ, $-\frac{\pi}{2} \leq \Psi \leq \frac{\pi}{2}$, is such that $e^{-i\Psi(\theta)}\overline{\omega(\theta)}h(\theta) > 0$ a.e. Hence, we get an $f \in H^1$, satisfying **(2)**.

Chapter 58

An Identity

We found it in [9], in Chapter 4 "Convergence results for Fourier Series", (p. 159). To any $f : [0,1] \to \mathbb{C}$ that is integrable (in the sense of Lebesgue), we assign its Fourier coefficients $\{\widehat{f}(k)\}_{k \in \mathbb{Z}}$ by the prescription

$$\widehat{f}(k) = \int_0^1 f(x)e^{-2\pi i k x}dx, \quad k \in \mathbb{Z}. \tag{58.0.1}$$

We denote by $\widehat{f}$ the function $k \to \widehat{f}(k)$, defined on $\mathbb{Z}$ which generates the two-way infinite sequence of Fourier coefficients $\{\widehat{f}(k)\}_{k \in \mathbb{Z}}$ of $f \in L^1[0,1]$. Here are some basic properties of $\widehat{f}$.

(P_1) The mapping $f \to \widehat{f}$ is linear on $L^1[0,1]$.

(P_2) Given $f \in L^1[0,1]$, we have $|\widehat{f}(k)| \leq \|f\|_{L^1}$ for all $k \in \mathbb{Z}$.

(P_3) Given $a \in \mathbb{R}$, denote by T_a the translation operator, acting on the linear space of the functions $f : \mathbb{R} \to \mathbb{C}$ by $(T_a f)(x) = f(x - a)$ for $x \in \mathbb{R}$. For any $f \in L^1[0,1]$, we have $\widehat{(T_a f)}(k) = e^{-2\pi i k a}\widehat{f}(k)$ for all $k \in \mathbb{Z}$.

(P_4) For $f : \mathbb{R} \to \mathbb{C}$ locally absolutely continuous and periodic of period 1, let $f' \in L^1[0,1]$ be equal almost everywhere to the derivative of f. Then $\widehat{(f')}(k) = 2\pi i k \cdot \widehat{f}(k)$ for all $k \in \mathbb{Z}$.

(P_5) (Riemann–Lebesgue lemma) $\lim_{|k| \to \infty} \widehat{f}(k) = 0$ for any $f \in L^1[0,1]$.

(P_6) Let $f, g \in L^1[0,1]$ then $\|f \star g\|_{L^1} \leq \|f\|_{L^1} \cdot \|g\|_{L^1}$ and for any $k \in \mathbb{Z}$ $\widehat{f \star g}(k) = \widehat{f}(k) \cdot \widehat{g}(k)$. Here, $\star$ is the convolution operator.

Theorem 4.1 ([9], p. 166). *Let $f \in L^1[0,1]$ and suppose that f is differentiable at the point x_0. Then $\sum_{k=-m}^{n} \widehat{f}(k)e^{2\pi i x_0} \to f(x_0)$ as $m, n \to \infty$.*

Proof. By subtracting a constant from f and shifting the origin, if needed, we may suppose that $x_0 = 0$ and $f(x_0) = 0$. Set $g(0) = g(1) = f'(0)$ and

$$g(x) = \frac{f(x)}{e^{2\pi i x} - 1},$$

for $x \in (0, 1)$. The differentiability of the periodic function $f(x)$ at $x = 0$ ensures the existence of some $m, \epsilon > 0$ such that

$$\left| \frac{f(x)}{e^{2\pi i x} - 1} \right| \le m \text{ for } x \in (0, \epsilon) \cup (1 - \epsilon, 1).$$

Since

$$\frac{1}{|e^{2\pi i x} - 1|} = \frac{1}{2 \sin(\pi x)} \le \frac{1}{\sin(\pi \epsilon)} \text{ for } x \in (\epsilon, 1 - \epsilon),$$

setting $M = \max \left\{ m, (\sin(\pi \epsilon))^{-1} \right\}$, we get that the measurable function g satisfies $|g(x)| \le M \cdot (1 + |f(x)|)$ for $x \in (0, 1)$. Thus, $g \in L^1[0, 1]$. Since $f(x) = (e^{2\pi i x} - 1)g(x)$ for all $x \in [0, 1]$, from equation (58.0.1) that $\widehat{f}(k) = \widehat{g}(k - 1) - \widehat{g}(k)$ for all $k \in \mathbb{Z}$. This telescopic effect yields

$$\sum_{k=-m}^{n} \widehat{f}(k) = \widehat{g}(-m - 1) - \widehat{g}(n) \to 0 \text{ for } n, m \to \infty,$$

due to (P_5). $\qquad\square$

The identity we meant (in the title of this chapter) is explained here. Let us define for a function $h \in L^1[0, 1]$ the Fourier coefficients $\widehat{h}(k) = \int_0^1 h(x) e^{-2\pi i k x} dx$, $k \in \mathbb{Z}$. Let $f \in L^1[0, 1]$ satisfy $f(0) = 0$, and f is differentiable at $x = 0$. Let

$$g(x) = \frac{f(x)}{e^{2\pi i x} - 1} \text{ for } x \in (0, 1) \text{ and } g(0) = g(1) = f'(0).$$

Then $\widehat{f}(k) = \widehat{g}(k-1) - \widehat{g}(k)$ for all $k \in \mathbb{Z}$. So also $\widehat{f}(k+1) = \widehat{g}(k) - \widehat{g}(k+1)$ and on adding the last two identities we get $\widehat{f}(k+1) + \widehat{f}(k) = \widehat{g}(k-1) - \widehat{g}(k+1)$. Using induction on N a non-negative integer, we obtain the following first identity (not yet the identity referred to above):

$$\sum_{j=k}^{k+2^N-1} \widehat{f}(j) = \widehat{g}(k - 1) - \widehat{g}(k + 2^N - 1).$$

On letting $N \to \infty$ and using the Riemann–Lebesgue lemma, we obtain

$$\sum_{j=k}^{\infty} \widehat{f}(j) = \widehat{g}(k - 1).$$

However, we take a different route. We can deduce (under the corresponding assumption) that if

$$h(x) = \frac{f(x)}{(e^{2\pi ix} - 1)^2} = \frac{g(x)}{e^{2\pi ix} - 1}$$

then $\widehat{g}(k) = h(k-1) - h(k)$ and $\widehat{g}(k-1) = h(k-2) - h(k-1)$. On subtraction, we get $\widehat{f}(k) = \widehat{g}(k-1) - \widehat{g}(k) = (\widehat{h}(k-2) - \widehat{h}(k-1)) - (\widehat{h}(k-1) - \widehat{h}(k))$, so that

$$\widehat{f}(k) = \widehat{h}(k-2) - 2\widehat{h}(k-1) + \widehat{h}(k).$$

Next, we let

$$l(x) = \frac{f(x)}{(e^{2\pi ix} - 1)^3} = \frac{h(x)}{e^{2\pi ix} - 1}.$$

Then $\widehat{h}(k) = (\widehat{l})(k-1) - \widehat{l}(k)$, $\widehat{h}(k-1) = (\widehat{l})(k-2) - \widehat{l}(k-1)$, $\widehat{h}(k-2) = (\widehat{l})(k-3) - \widehat{l}(k-2)$. Substituting, we get

$$\widehat{f}(k) = \widehat{h}(k-2) - 2\widehat{h}(k-1) + \widehat{h}(k)$$
$$= \widehat{l}(k-3) - 3\widehat{l}(k-2) + 3\widehat{l}(k-1) - \widehat{l}(k).$$

Using induction and the recursion for the binomial coefficients used in the standard construction of the Pascal triangle, we can prove the following: Let $f \in L^1[0,1]$ satisfy that

(1) For some $n \in \mathbb{Z}^+$ f is n-times differentiable at 0.
(2) $f(0) = \cdots = f^{(n-1)}(0) = 0$.

Let

$$g(x) = \frac{f(x)}{(e^{2\pi ix} - 1)^n} \text{ for } x \in (0,1) \text{ and } g(0) = g(1) = f^{(n)}(0).$$

Then

$$\widehat{f}(k) = \sum_{j=0}^{n} (-1)^j \binom{n}{j} \widehat{g}(k - n + j).$$

Using the identity $e^{2\pi ix} - 1 = 2i\sin(\pi x)e^{\pi ix}$ we get

$$g(x) = \frac{f(x)}{(2i\sin(\pi x)e^{\pi ix})^n} = \left(\frac{1}{2i}\right)^n (\csc(\pi x))^n e^{-in\pi x} \cdot f(x).$$

It follows similarly that: Let $f \in L^1[0,1]$ satisfy

(1) For some $n \in \mathbb{Z}^+$ f is n-times differentiable at $\frac{1}{2}$.
(2) $f(1/2) = \cdots = f^{(n-1)}(1/2) = 0$.

Let

$$h(x) = \frac{f(x)}{(e^{2\pi i x} + 1)^n} \text{ for } x \in (0,1) - \{1/2\} \text{ and } h(1/2) = f^{(n)}(1/2).$$

Then

$$\widehat{f}(k) = \sum_{j=0}^{n} \binom{n}{j} \widehat{h}(k - n + j).$$

Using the identity $e^{2\pi i x} + 1 = 2\cos(\pi x)e^{\pi i x}$ we get

$$h(x) = \frac{f(x)}{(2\cos(\pi x)e^{\pi i x})^n} = \left(\frac{1}{2}\right)^n (\sec(\pi x))^n e^{-in\pi x} \cdot f(x).$$

Chapter 59

More on Inner Functions [22]

In this chapter, we quote the following from [22, p. 30]:

6. A function $f(z)$ analytic in $|z| < 1$ is a Blaschke product (aside from a constant factor of modulus one) if and only if $\lim_{r \to 1^-} \int_0^{2\pi} |\log |f(re^{i\theta})|| d\theta = 0$.
7. If $f(z)$ is an inner function and $\phi(z)$ is a conformal mapping of the unit disk onto itself, then $f(\phi(z))$ and $\phi(f(z))$ are inner functions.
8. Let $f(z)$ be an inner function, and suppose that $|\alpha| < 1$. Then for all such α outside a set of planar measure zero, the function

$$\frac{f(z) - \alpha}{1 - \overline{\alpha} f(z)},$$

is a Blaschke product (Frostman. He proved a stronger claim).

Quotient spaces and annihilators (Section 7.1 in [22], pp. 110–111).

Let X be a Banach space, and let S be a (closed) subspace. A coset of X modulo S is a subset $\xi = x + S$ consisting of all elements of the form $x + y$, where x is some fixed member of X and y ranges over S. Two cosets are either identical or disjoint. The quotient space X/S has its elements all distinct cosets of X modulo S. With the natural definitions of addition and scalar multiplication, X/S is a linear space. To be specific, $(x_1 + S) + (x_2 + S) = (x_1 + x_2) + S$ and $\alpha(x + S) = (\alpha x) + S$. The zero element of X/S is the coset S. Finally, the norm of a coset $\xi = x + S$ is defined by $\|\xi\| = \inf_{y \in S} \|x + y\|$. Since S is closed, $\|\xi\| = 0$ implies $\xi = S$. Under this norm, X/S is complete, and therefore is itself a Banach space. To see this, let $\{\xi_n\}$ be a Cauchy sequence of cosets. Choose a subsequence of integers $0 < n_1 < n_2 < \cdots$ such that $\|\xi_{n_k} - \xi_{n_{k+1}}\| \le 2^{-k}$, $k = 1, 2, \ldots$. Then choose $x_k \in \xi_{n_k}$ $(k = 1, 2, \ldots)$ such that $\|x_k - x_{k+1}\| < 2\|\xi_{n_k} - \xi_{n_{k+1}}\|$. With this construction, $\{x_k\}_k$ is a Cauchy

331

sequence, so x_k tends to a limit $x \in X$. Let $\xi = x + S$, then $\xi_{n_k} \to \xi$, because (by definition of the norm) $\|\xi_{n_k} - \xi\| \leq \|x_k - x\|$. Finally, since $\{\xi_n\}$ is a Cauchy sequence, this implies that $\xi_n \to \xi$. The annihilator of the subspace S is the set $S^\perp$ of all linear functionals $\Phi \in X^*$ such that $\Phi(x) = 0$ for all $x \in S$. $S^\perp$ is a subspace of X^*. The following results play an essential role in the theory of extremal problems.

Theorem 7.1 ([22], p. 110). *The quotient space $X^*/S^\perp$ is isometrically isomorphic to S^*. Furthermore, for each fixed $\Phi \in X^*$: $\sup_{x \in S, \|x\| \leq 1} |\Phi(x)| = \min_{\Psi \in S^\perp} \|\Phi + \Psi\|$, where "min" indicates that the infimum is attained.*

Theorem 7.2 ([22], p. 111). *The space $(X/S)^*$ is isometrically isomorphic to $S^\perp$. Furthermore, for each fixed $x \in X$: $\max_{\Psi \in S^\perp, \|\Psi\| \leq 1} |\psi(x)| = \inf_{y \in S} \|x + y\|$, where "max" indicates that the supremum is attained.*

Representation of linear functionals (Section 7.2 in [22], pp. 112–113).

H^p is a Banach space if $1 \leq p \leq \infty$, with norm $\|f\| = M_p(1, f)$. Here, $M_p(r, f) = \left\{ \frac{1}{2\pi} \int_0^{2\pi} |f(re^{i\theta})|^p d\theta \right\}^{1/p}$ if $0 < p < \infty$, and $M_\infty(r, f) = \max_{0 \leq \theta < 2\pi} |f(re^{i\theta})|$ for $0 \leq r < 1$. $M_p(1, f) = \lim_{r \to 1^-} M_p(r, f)$. The polynomials are dense in H^p if $0 < p < \infty$. If $1 \leq p \leq \infty$, the set of boundary functions of H^p is the subspace of L^p for which $\int_0^{2\pi} e^{in\theta} f(e^{i\theta}) d\theta = 0$, $n = 1, 2, \ldots$. In particular, if each $f \in H^p$ is identified with its boundary function, H^p can be regarded as a subspace of L^p, $0 < p \leq \infty$. According to Riesz representation theorem, every bounded linear functional Φ on L^p $(1 \leq p < \infty)$ has a unique representation

$$\Phi(f) = \frac{1}{2\pi} \int_0^{2\pi} f(e^{i\theta}) g(e^{i\theta}) d\theta, \quad g \in L^q, \text{ where } \frac{1}{p} + \frac{1}{q} = 1. \qquad (59.0.1)$$

In fact, $\|\Phi\| = \|g\|_q$, and $(L^p)^*$ is isometrically isomorphic to L^q. Since H^p is a subspace of L^p, Theorem 7.1 can be used to describe $(H^p)^*$ if the annihilator of H^p in $(L^p)^*$ can be determined. But if $g \in L^q$ annihilates every H^p function, then surely $\int_0^{2\pi} e^{in\theta} g(e^{i\theta}) d\theta = 0$, $n = 0, 1, 2, \ldots$. Therefore, $g(e^{i\theta})$ is the boundary function of some $g(z) \in H^q$ and $g(0) = 0$. Call this class H_0^q. Conversely, if $g \in H_0^q$, it is clear that $\int_0^{2\pi} f(e^{i\theta}) g(e^{i\theta}) d\theta = 0$ for every $f \in H^p$. Hence, H_0^q is the annihilator of H^p and it follows from Theorem 7.1 that $(H^p)^*$ is isometrically isomorphic to L^q/H_0^q. We may replace L^q/H_0^q by L^q/H^q since the correspondence $\xi \leftrightarrow e^{i\theta}\xi$ between cosets of the two spaces is an isometric isomorphism. It is even possible to give a canonical representation of the bounded linear functionals on H^p. Any $\Phi \in (H^p)^*$ can be extended (by the Hahn–Banach theorem) to a functional on L^p, and hence maybe represented in the form of equation (59.0.1) for some $g \in L^q$. This representation is not unique. Two functions g_1, g_2 belonging to the same coset of L^q/H_0^q, so that $g_1(e^{i\theta}) - g_2(e^{i\theta}) \sim \sum_{n=1}^{\infty} c_n e^{in\theta}$, obviously represent the same functional Φ on H^p. Functions in different cosets, however, generate different functionals. Thus, for $p > 1$, the representation becomes unique if we distinguish in each coset that function g for which $\int_0^{2\pi} e^{-in\theta} g(e^{i\theta}) d\theta = 0$, $n = 1, 2, \ldots$. Equivalently, there is a unique function $g \in H^q$ for which

$$\Phi(f) = \frac{1}{2\pi} \int_0^{2\pi} f(e^{i\theta}) \overline{g(e^{i\theta})} d\theta \text{ for all } f \in H^p,\ 1 < p < \infty. \tag{59.0.2}$$

Since $1 < q < \infty$, the Riesz theorem guarantees that the "analytic projection" g of the original L^q function is in H^q. In summary, we have the following:

Theorem 7.3 ([22], p. 113). *For $1 \le p < \infty$, the space $(H^p)^*$ is isometrically isomorphic to L^q/H^q, where $\frac{1}{p} + \frac{1}{q} = 1$. Furthermore, if $1 < p < \infty$, each $\Phi \in (H^p)^*$ is representable in the form of equation (59.0.2) by a unique function $g \in H^q$, while each $\Phi \in (H^1)^*$ can be represented in the form of equation (59.0.2) by some $g \in L^\infty$.*

We might have chosen to put $g(e^{-i\theta})$ instead of $\overline{g(e^{i\theta})}$ in equation (59.0.2). This would have the advantage of setting up an isomorphism between $(H^p)^*$ and H^q. But in either case the correspondence need not be an isometry. In fact, $\|\Phi\|$ is equal to the norm of the coset determined by $\overline{g(e^{i\theta})}$ (or by $g(e^{-i\theta})$). So, only the inequality $\|\Phi\| \le \|g\| \le A_p \|\Phi\|$, $1 < p < \infty$, is true generally. The right-hand inequality comes from the Riesz theorem. A_p is a constant independent of Φ. In the case $p = 2$, it is easy to see that $\|\Phi\| = \|g\|$.

Beurling's Approximation Theorem (Section 7.3 in [22], p. 113).

We recall that a function $f \in H^p$ has a canonical factorization $f(z) = B(z)S(z)F(z)$, where $B(z)$ is a Blaschke product, $S(z)$ is a singular inner function generated by a non-decreasing singular function $\mu(t)$, and $F(z)$ is an outer function. The inner function $f_0(z) = B(z)S(z)$ is called the inner factor of $f(z)$. Let $g_0(z)$ be another inner function with an associated singular function $\nu(t)$. f_0 is said to be a divisor of g_0 if $g_0(z)/f_0(z)$ is an inner function. This is clearly the case if and only if every zero of $f_0(z)$ in $|z| < 1$ is also a zero of $g_0(z)$ (with the same or higher multiplicity) and $[\nu(t) - \mu(t)]$ is non-decreasing. For fixed $f \in H^p$, let $\mathcal{P}[f]$ denote the (always closed) subspace generated by the functions $z^n f(z)$, $n = 0, 1, 2, \ldots$. Thus $\mathcal{P}[f]$ consists of all

H^p functions which can be approximated by polynomial multiples of f. The problem is to characterize $\mathcal{P}[f]$. It is already known that $\mathcal{P}[1] = H^p$, since the polynomials are dense in H^p.

Theorem 7.4 (Beurling, [22], p. 114). *Let f and g be H^p functions $(1 \leq p < \infty)$, not identically zero, with inner factors f_0 and g_0, respectively. Then $g \in \mathcal{P}[f]$ if and only if f_0 is a divisor of g_0.*

Corollary 1. $\mathcal{P}[f] = \mathcal{P}[f_0]$. *If* $\mathcal{P}[f_0] = \mathcal{P}[g_0]$, *then* $f_0 = g_0$.

Corollary 2. *Let $0 < p < \infty$. Then for any inner function f_0, $\mathcal{P}[f_0] = f_0 \cdot H^p$.*

Linear functionals on H^p, $0 < p < 1$ (Section 7.4 in [22], p. 115).

H^p is not normable in the case $p < 1$. However, its bounded linear functionals can still be defined in the usual manner. Thus, a linear functional Φ on H^p is said to be bounded if $\|\Phi\| = \sup_{\|f\|_p = 1} |\Phi(f)| < \infty$. A linear functional Φ on H^p is bounded if and only if it is continuous. Also, the bounded linear functionals on H^p form a Banach space under the given norm. H^p is an F-space even for $p < 1$. Hence, the principle of uniform boundedness (Banach–Steinhaus) still applies: every pointwise bounded sequence of bounded linear functionals on H^p is uniformly bounded. We want to derive a complete representation of the bounded linear functionals on H^p, $p < 1$. We need to introduce some notation. Let A denote the class of functions, analytic in $|z| < 1$ and continuous in $|z| \leq 1$. It is convenient to write $f \in \Lambda_\alpha$ to indicate that $f \in A$ admits boundary function $f(e^{i\theta})$ that belongs to the Lipschitz class Λ_α, $0 < \alpha \leq 1$. Similarly, $f \in \Lambda_*$ will mean $f \in A$ and $f(e^{i\theta}) \in \Lambda_*$. We recall that a continuous function $\phi(x)$ is said to be of class Λ_* if there is a constant B such that

$$|\phi(x + h) - 2\phi(x) + \phi(x - h)| \leq B \cdot h, \quad \forall x, \quad \forall h > 0.$$

Here, $\phi(x)$, $-\infty < x < \infty$ is complex valued. Also, it is known that for any $\alpha < 1$, $\Lambda_1 \subset \Lambda_* \subset \Lambda_\alpha$.

Theorem 7.5 ([22], p. 115). *To each bounded linear functional Φ on H^p, $0 < p < 1$, there corresponds a unique function $g \in A$ such that*

$$\Phi(f) = \lim_{r \to 1^-} \frac{1}{2\pi} \int_0^{2\pi} f(re^{i\theta})g(e^{-i\theta})d\theta, \quad f \in H^p. \tag{59.0.3}$$

If $(n + 1)^{-1} < p < n^{-1}$ $(n = 1, 2, \ldots)$, then $g^{(n-1)} \in \Lambda_\alpha$ where $\alpha = \frac{1}{p} - n$. Conversely, for any g with $g^{(n-1)} \in \Lambda_\alpha$, the limit in equation (59.0.3) exists *for all $f \in H^p$ and defines a bounded linear functional. If $p = (n + 1)^{-1}$, then $g^{(n-1)} \in \Lambda_*$; and conversely, any g with $g^{(n-1)} \in \Lambda_*$ defines through equation (59.0.3) a bounded linear functional on H^p.*

The function $g(z) = (1 - \xi z)^{-1}$, $|\xi| < 1$ provides an interesting example. Here the integral in (59.0.3) reduces to $f(r\xi)$, so $\Phi(f) = f(\xi)$. Point evaluation is therefore a bounded linear functional on H^p, $p < 1$. In particular, there are enough linear functionals on H^p to distinguish elements of the space. This contrasts sharply with the situation for L^p, $(p < 1)$, where there are no bounded linear functionals at all, except the zero functional.

Failure of the Hahn–Banach Theorem (Section 7.5 in [22], p. 119).

We saw in Section 7.4 that H^p has enough continuous functionals to distinguish elements of the space even if $p < 1$. On the other hand, we shall now see that there are not always enough functionals to separate points from subspaces. That is, given a proper subspace M of H^p $(p < 1)$ and an H^p function $f \notin M$, there may not be any functional $\Phi \in (H^p)^*$ such that $\Phi(M) = 0$ and $\Phi(f) \neq 0$. We shall see a construction of a proper subspace of H^p which is annihilated by no functional $\Phi \in (H^p)^*$ except the zero functional. We need some background material. Let

$$S(z) = \exp\left\{-\int_0^{2\pi} \frac{e^{it} + z}{e^{it} - z} d\mu(t)\right\}$$

be a singular inner function and let $\omega(t; \mu)$ denote the modulus of continuity of μ.

Lemma 1 ([22], p. 119). *If $\omega(t; \mu) = O(t \log \frac{1}{t})$, then there exist positive constants α and C such that $|S(z)| \geq C(1 - r)^\alpha$, $|z| = r < 1$.*

Reminder: Let $\phi(x)$ be a complex-valued function defined on $-\infty < x < \infty$ and periodic with period 2π. The modulus of continuity of ϕ is the function $\omega(t) = \omega(t; \phi) = \sup_{|x-y| \leq t} |\phi(x) - \phi(y)|$. For example, if ϕ is a function with only discontinuities of the first type (finite jumps) and which has jumps tightly bounded by some finite $j > 0$, then $\omega(t) \equiv j$ $(t > 0)$.

We introduce the space B_α $(\alpha > 0)$ of functions $f(z) = \sum_n a_n z^n$ analytic in $|z| < 1$, such that $\|f\|_\alpha^2 = \sum_{n=0}^\infty (n + 1)^{-\alpha} |a_n|^2 < \infty$. B_α is a Hilbert space with the inner product $< f, g > = \sum_{n=0}^\infty (n + 1)^{-\alpha} a_n \overline{b_n}$, where $g(z) = \sum_n b_n z^n$. The polynomials are dense in B_α, for each $\alpha > 0$. A straightforward calculation shows that

$$C_1 \|f\|_\alpha^2 \leq \int_0^{2\pi} \int_0^1 (1 - r)^{\alpha - 1} |f(re^{i\theta})|^2 dr d\theta \leq C_2 \|f\|_\alpha^2,$$

for some positive constants C_1 and C_2 depending only on α. So $B_\alpha \subseteq \mathcal{A}^2_{\alpha-1}$ a weighted Bergman space. A function $f \in B_\alpha$ will be called an α-outer function if the set of all polynomial multiples $P \cdot f$ is dense in B_α. Clearly, f is an α-outer function if and only if 1 belongs to the B_α closure of the polynomial multiples of f.

Lemma 2 ([22], p. 120). *If $\omega(t; \mu) = O(t \log \frac{1}{t})$, then S (above) is an α-outer function for every $\alpha > 0$.*

Theorem 7.6 ([22], p. 121). *Let S be a singular inner function such that $\omega(t; \mu) = O(t \log \frac{1}{t})$. Then $(S \cdot H^2)^{\perp}$ contains no non-null function $g(z) = \sum_n b_n z^n$ such that*

$$\sum_{n=1}^{\infty} n^{\gamma} |b_n|^2 < \infty, \quad \gamma > 0. \tag{59.0.4}$$

Proof. Let $S(z) = \sum_n a_n z^n$. Suppose $g \in H^2$ is orthogonal to $S \cdot H^2$. Then g is orthogonal to $z^k S(z)$, $k = 0, 1, 2, \ldots$. So,

$$\sum_{n-k}^{\infty} a_{n-k} \overline{b_n} = 0, \quad k = 0, 1, 2, \ldots. \tag{59.0.5}$$

Now define $c_n = (n+1)^{\gamma} b_n$. If (59.0.4) is satisfied, then $h(z) = \sum_{n=0}^{\infty} c_n z^n \in B_\gamma$ while the conditions (59.0.5) are equivalent to

$$\sum_{n=k}^{\infty} \frac{a_{n-k} \overline{c_n}}{(n+1)^{\gamma}} = 0, \quad k = 0, 1, 2, \ldots.$$

This says that h is orthogonal to $z^k S(z)$ ($k = 0, 1, 2, \ldots$) in the space B_γ. But by Lemma 2, the functions $z^k S(z)$ span B_γ. Hence, $h = 0$, which implies $g = 0$. $\qquad \square$

Lemma 3 ([22], p. 121). *If $g(z) = \sum_n b_n z^n \in \Lambda_\alpha$ for some $\alpha > 0$, then $\sum_{n=1}^{\infty} n^{\gamma} |b_n|^2 < \infty$ for all $\gamma < 2\alpha$.*

Theorem 7.7 ([22], p. 122). *Let S be a singular inner function with $\omega(t; \mu) = O(t \log \frac{1}{t})$. Then for each p $(0 < p < 1)$, the only bounded linear functional on H^p which annihilates the subspace $S \cdot H^p$ is the zero functional.*

Proof. Suppose $\Phi \in (H^p)^*$ annihilates $S \cdot H^p$. Then, in particular,

$$\Phi(z^n S(z)) = \frac{1}{2\pi} \int_0^{2\pi} e^{in\theta} S(e^{i\theta}) \overline{g(e^{i\theta})} d\theta = 0, \quad n = 0, 1, 2, \ldots,$$

where $g(z) = \sum_n b_n z^n \in \Lambda_\alpha$ for some $\alpha > 0$. This implies that g, regarded as an element of H^2, is orthogonal to the subspace $S \cdot H^2$. Therefore, by Theorem 7.6, $\sum_{n=1}^\infty n^\gamma |b_n|^2 = \infty$ for each $\gamma > 0$, unless $g \equiv 0$. But this contradicts Lemma 3, since $g \in \Lambda_\alpha$. $\qquad\square$

We remark that $S \cdot H^p$ is a proper subspace unless $S(z) \equiv 1$.

Corollary 1 ([22], p. 122). *If S is a singular inner function as described in Theorem 7.7, the quotient space $H^p/(S \cdot H^p)$ has no continuous linear functional except the zero functional, for each $p < 1$.*

Corollary 2 ([22], p. 123). *If $p < 1$, there is a subspace M of H^p and a bounded linear functional on M which cannot be extended to a bounded linear functional on H^p.*

Extreme points (Section 7.6 in [22], p. 123).

A set S in a linear space X is said to be convex if whenever x_1 and x_2 are in S, every proper convex combination, $ax_1 + (1 - a)x_2$, $0 < a < 1$, is also in S. An element $x \in S$ is called an extreme point of S if it is not a proper convex combination of any two distinct points in S. One reason for the interest in finding the extreme points of a convex set originates in the Krein–Milman theorem, which states that a compact convex set in a locally convex topological vector space is the closed convex hull of its extreme points. In particular, such a set has extreme points. In any Banach space X, the problem arises to describe the extreme points of the unit ball $B = \{x \in X \mid \|x\| \leq 1\}$. Every extreme point lies on the boundary of B, that is, it has a unit norm. It is useful to observe that if a point x with $\|x\| = 1$ is not an extreme point, it can be represented as the midpoint of the segment joining two distinct points $(x + y)$ and $(x - y)$ in B. Furthermore,

$$\|x + y\| = \|x - y\| = 1, \tag{59.0.6}$$

since $1 = \|x\| = \frac{1}{2}\{\|x + y\| + \|x - y\|\} \leq 1$. Hence, a point x with $\|x\| = 1$ is an extreme point if and only if (59.0.6) implies $y = 0$.

If X is a Hilbert space, it is not hard to show that every boundary point of B is an extreme point. The same is true if X is a uniformly convex space. That is, if for $\epsilon > 0$ there is a $\delta > 0$ such that $\|x\| = \|y\| = 1$ and $\|x - y\| > \epsilon$ imply $\|\frac{x+y}{2}\| < 1 - \delta$. If $1 < p < \infty$, the space L^p, and therefore H^p, is uniformly convex, so every boundary point of B is an extreme point of B. On the other hand, the unit ball in L^1 has no extreme points at all. What is the situation in H^1?

Theorem 7.8 (de Leeuw–Rudin, [22], p. 124). *A function $f \in H^1$ with $\|f\| = 1$ is an extreme point of the unit ball in H^1 if and only if f is an outer function.*

Corollary ([22], p. 125). (i) *If $f \in H^1$, $\|f\| = 1$, and f is not an extreme point of B, then there are two extreme points f_1, f_2 such that $f = \frac{1}{2}(f_1 + f_2)$.* (ii) *If $f \in H^1$, $\|f\| < 1$, then f is a convex combination of two extreme points.*

Theorem 7.9 ([22], p. 125). *A function $f \in H^\infty$ with $\|f\| = 1$ is an extreme point of the unit ball in H^∞, if and only if*

$$\int_0^{2\pi} \log\left(1 - \left|f(e^{i\theta})\right|\right) d\theta = -\infty. \tag{59.0.7}$$

We recall that an inner function $B(z)$ is a Blaschke product (multiplied by a uni-modular constant) if and only if

$$\lim_{r \to 1^-} \int_0^{2\pi} \log\left|B(re^{i\theta})\right| d\theta = 0.$$

Since $H^\infty \subset H^1$ this is compatible with the fact that B is extreme for the unit ball of H^2 and with the integral identity in Theorem 7.9.

$$\text{Chapter 60}$$

Extremal Problems ([22], p. 129)

Extremal problems come in pairs that are called duals.

The extremal problem and its dual (Section 8.1 in [22], p. 129).

The most general bounded linear functional on H^p $(1 \leq p < \infty)$ can be expressed in the form

$$\Phi(f) = \frac{1}{2\pi i} \int_{|z|=1} f(z)k(z)dz, \tag{60.0.1}$$

where $k(e^{i\theta}) \in L^q$, $\frac{1}{p} + \frac{1}{q} = 1$. The functionals of greatest interest are generated by kernels $k(e^{i\theta})$ which are boundary values of rational functions. Some examples are given in what follows.

(i)

$$k(z) = \sum_{j=1}^{n} \frac{c_j}{z - \beta_j}, \quad |\beta_j| < 1; \quad \Phi(f) = \sum_{j=1}^{n} c_j f(\beta_j).$$

(ii)

$$k(z) = \frac{n!}{(z - \beta)^{n+1}}, \quad |\beta| < 1; \quad \Phi(f) = f^{(n)}(\beta).$$

(iii)

$$k(z) = \sum_{j=0}^{n} \frac{c_j}{z^{j+1}}; \quad \Phi(f) = \sum_{j=0}^{n} c_j a_j, \quad \text{where } f(z) = \sum_{j=0}^{\infty} a_j z^j.$$

For a fixed $k \in L^q$, the typical extremal problem is to find

$$\|\Phi\| = \sup_{f \in H^p, \|f\|_p \leq 1} |\Phi(f)|. \tag{60.0.2}$$

In addition to finding the value of the supremum, several questions are to be considered. Is the supremum attained? If so, is the extremal function unique? What are the extremal functions?

A function $h \in L^q$ is said to be equivalent to the given kernel k (written $h \sim k$) if h and k belong to the same coset of L^q/H^q, that is, if $h - k \in H^q$. Thus, h and k determine the same functional on H^p if and only if $h \sim k$. An application of Theorem 7.1 gives at once the important relation

$$\sup_{f \in H^p, \|f\|_p \leq 1} \frac{1}{2\pi} \left| \int_{|z|=1} f(z)k(z)dz \right| = \min_{g \in H^q} \|k - g\|_q \tag{60.0.3}$$

for $1 \leq p < \infty$. This is called the duality relation. It connects the original extremal problem with what is called the dual extremal problem: to find the function $g \in H^q$ which is closest to the given kernel $k \in L^q$ (or, equivalently to find the function $h \sim k$ of minimal norm). According to Theorem 7.1, the minimum is actually attained, that is, the dual extremal problem always has a solution if $1 < q \leq \infty$.

The argument breaks down for $p = \infty$ ($q = 1$), since the conjugate space of L^∞ is larger than L^1. However, the dual problem still has a solution in this case, and the duality relation remains true.

Uniqueness of solutions (Section 8.2 in [22], p. 132).

It will be convenient to reserve the term extremal function to indicate a solution to the original problem (60.0.2). A function $K(e^{i\theta}) \sim k(e^{i\theta})$ for which $\|K\|_q = \inf_{h \sim k} \|h\|_q = \inf_{g \in H^q} \|k - g\|_q$, will be called an extremal kernel. It is natural to ask about the uniqueness of the extremal function and of the extremal kernel. Since $|\Phi(e^{i\alpha} f)| = |\Phi(f)|$ for every real α, some normalization clearly must be imposed before an extremal function can be unique. It is convenient to single out the extremal functions for which $\Phi(f) = \|\Phi\|$. These will be called normalized extremal functions. to avoid trivial complications, it is always assumed that Φ is not the zero functional, i.e., $k \notin H^q$.

Theorem 8.1 (Main existence and uniqueness theorem, [22], p. 132). *For each p $(1 \leq p \leq \infty)$ and for each function $k(e^{i\theta}) \in L^q(\frac{1}{p} + \frac{1}{q} = 1)$ with $k \notin H^q$, the duality relation $\sup_{f \in H^p, \|f\|_p \leq 1} |\Phi(f)| = \inf_{g \in H^q} \|k - g\|_q$ holds, where $\Phi(f)$ is defined by (60.0.1). If $p > 1$, there is a unique extremal* *function f for which $\Phi(f) > 0$. If $p = 1$ and $k(e^{i\theta})$ is continuous, at least one extremal function exists. If $p > 1$ $(q < \infty)$, the dual extremal problem has a unique solution. If $p = 1$ $(q = \infty)$, the dual extremal problem has at least one solution. It is unique if an extremal function exists.*

The uniqueness is based upon the following simple observation.

Lemma ([22], p. 133). *In order that a function $F \in H^p$, with $\|F\| = 1$ and $\Phi(F) > 0$, be an extremal function and that K $(K \sim k)$ be an extremal kernel, it is necessary and sufficient that*

$$e^{i\theta} F(e^{i\theta}) K(e^{i\theta}) \geq 0 \text{ a.e.,} \tag{60.0.4}$$

and that

$$\begin{cases} |K(e^{i\theta})| = \|K\| \text{ a.e.} & \text{if } p = 1; \\ |F(e^{i\theta})|^p = \|K\|^{-q}|K(e^{i\theta})|^q \text{ a.e.} & \text{if } 1 < p < \infty; \\ |F(e^{i\theta})| = 1 \text{ a.e. on } \{\theta \mid K(e^{i\theta}) \neq 0\} & \text{if } p = \infty. \end{cases} \tag{60.0.5}$$

Counterexamples in the case $p = 1$ (Section 8.3 in [22], p. 134).

The existence and uniqueness of extremal functions and extremal kernels have now been established for $1 < p \leq \infty$. The situation is entirely different in the case $p = 1$. An extremal function need not exist in H^1, and if it does, it need not be unique up to normalization. At least one extremal kernel must exist, but it is not in general unique.

Example (i) $\Phi(f) = \frac{1}{2\pi i} \int_{-1}^{1} f(x)dx$, $f \in H^1$. This has no extremal function.

Example (ii) The elementary example $\Phi(f) = \frac{1}{2\pi} \int_0^{2\pi} f(e^{i\theta})e^{-i\theta}d\theta$, $f \in H^1$, shows that a normalized extremal function need not be unique. $f(z) = z$ is extremal. Also, for every $\alpha \in \mathbb{C}$,

$$f(z) = \frac{(z + \alpha)(1 + \overline{\alpha}z)}{1 + |\alpha|^2}$$

is also extremal.

Rational kernels (Section 8.4 in [22], p. 136).

If the kernel $k(e^{i\theta})$ is the restriction to the unit circle of a rational function $k(z)$, it is possible to describe the extremal functions and the extremal kernels more or less explicitly. It looks like the following:

$$F(z) = A \prod_{i=s+1}^{\sigma} \frac{z - \alpha_i}{1 - \overline{\alpha_i}z} \prod_{i=1}^{n-1}(1 - \overline{\alpha_i}z)^{2/p} \prod_{i=1}^{n}(1 - \overline{\beta_i}z)^{-2/p} \tag{60.0.6}$$

and

$$K(z) = B \prod_{i=1}^{s} \frac{z - \alpha_i}{1 - \overline{\alpha_i} z} \prod_{i=1}^{n-1} (1 - \overline{\alpha_i} z)^{2/q} \prod_{i=1}^{n} \frac{(1 - \overline{\beta_i} z)^{1-2/q}}{z - \beta_i}. \tag{60.0.7}$$

These structural formulas remain valid in the cases $p = 1$ and $p = \infty$ with the understanding that $1/\infty = 0$.

Interesting exercises ([22], p. 144).

4. If $f \in H^p$ $(0 < p \leq \infty)$, then $|f(z)| \leq (1 - |z|^2)^{-1/p} \|f\|_p$, $|z| < 1$. This is sharp for each fixed z.

6. For $f \in H^2$ and $|z_1| < 1$, $|z_2| < 1$, and $z_1 \neq z_2$, we have

$$\left| \frac{f(z_1) - f(z_2)}{z_1 - z_2} \right| \leq \left\{ \frac{1 - |z_1 z_2|^2}{|1 - z_1 \overline{z_2}|^2 (1 - |z_1|^2)(1 - |z_2|^2)} \right\}^{1/2} \|f\|_2, \quad |z| < 1.$$

The constant that multiplies $\|f\|_2$ on the right-hand side is the best possible one.

7. For each fixed z, $|z| < 1$, and for each odd integer $n = 2m + 1$, we have the sharp inequality

$$|f^{(n)}(z)| \leq \frac{n!}{(1 - |z|^2)^n} \left\{ \sum_{k=0}^{m} \binom{m}{k}^2 |z|^{2k} \right\} \|f\|_\infty, \quad f \in H^\infty.$$

8. Let M_n denote the maximum of $|a_0 + a_1 + \cdots + a_n|$ among all H^∞ functions $f(z) = \sum_k a_k z^k$ with $\|f\|_\infty \leq 1$. Then

$$M_n = \sum_{k=0}^{n} (\lambda_k)^2, \quad \lambda_k = (-1)^k \binom{-\frac{1}{2}}{k} = \frac{(2k)!}{4^k (k!)^2}.$$

Hence, $M_n \sim \frac{1}{\pi} \log n$ as $n \to \infty$. Observe that

$$\left\{ \sum_{k=0}^{n} \lambda_k z^k \right\}^2 = 1 + z + \cdots + z^n + b_{n+1} z^{n+1} + \cdots.$$

10. For $f(z) = \sum_k a_k z^k \in H^1$,

$$\max_{\|f\|_1 \leq 1} |a_0 + a_1 + \cdots + a_n| = \frac{1}{2} \sec \frac{(n+1)\pi}{2n+3}.$$

Observe that the eigenvalue problem: $x_{n-k} + x_{n-k+1} + \cdots + x_n = \lambda x_k$, $k = 0, 1, \ldots, n$, has the solution

$$\lambda = \frac{1}{2} \sec \frac{(n+1)\pi}{2n+3}, \quad x_k = \sin \frac{(k+1)\pi}{2n+3}, \quad k = 0, 1, \ldots, n.$$

Chapter 61

Iterating the Corollary ([59], p. 152)

We recall this result.

Corollary ([59], p. 152). *Let $f \in H^\infty$, let $\|f\|_\infty < 1$, and let Ω be any inner function. Then there is another inner function, $\omega \in f + \Omega H^\infty$.*

We give few conclusions. These are the results of the author of this book.

(I) Interpolating functions of the open unit disk of H^∞ by inner functions:

Proposition 61.0.134. *If $f \in H^\infty$ and $\|f\|_\infty < 1$ and $\{z_n\}_{n=1}^\infty$ is a Blaschke sequence, i.e., $\sum_{n=1}^\infty (1 - |z_n|) < \infty$, then there exists an inner function $\omega(z)$ that interpolates f on the Blaschke sequence, i.e., $\omega(z_n) = f(z_n)$, $n = 1, 2, \ldots$.*

Proof. Let $B(z)$ be the Blaschke product with zeros $\{z_n\}_{n=1}^\infty$. Then by the corollary above, there is $h \in H^\infty$ and an inner function $\omega(z)$ such that $\omega(z) = f(z) + B(z)h(z)$, $|z| < 1$. Hence, for $n = 1, 2, \ldots$, $\omega(z_n) = f(z_n) + B(z_n)h(z_n) = f(z_n)$. $\qquad\square$

That result is about the interpolation problem in the open unit disk, $|z| < 1$, of functions in the open unit disk of H^∞, i.e., $B_{H^\infty}(U) = \{f \in H^\infty(U) | \|f\|_\infty < 1\}$ by inner functions. There is no uniqueness here for we can choose an arbitrary natural number $N \in \mathbb{Z}^+$, and obtain a corresponding interpolating inner function $\omega_N(z) = f(z) + (B(z))^N h_N(z)$, $|z| < 1$ and corresponding $h_N(z) \in H^\infty(U)$. This problem is different from the classical one on universal interpolating sequences ([21], pp. 147–149). We recall that a sequence $\{z_n\}$ is said to be uniformly separated if there is a number $\delta > 0$

343

such that

$$\prod_{j=1,j\neq n}^{\infty}\left|\frac{z_n - z_j}{1 - \overline{z_j}z_n}\right| \geq \delta, \quad n = 1, 2, \ldots.$$

The convergence of the Blaschke product implies, in particular, that $\sum_n(1 - |z_n|) < \infty$, i.e., a uniformly separated sequence must be a Blaschke sequence. The universal interpolation sequences are the sequences $\{z_n\}_n$ which admit an H^∞ interpolation for every bounded sequence $\{w_n\}_n$. This problem has a natural generalization to any H^p ($0 < p < \infty$). It turns out that $\{z_n\}_n$ is a universal interpolation sequence if and only if it is uniformly separated. The condition of uniform separation is difficult both to understand intuitively and to verify in practice. So, it is needed to find more accessible characterizations. In fact, the very existence of uniformly separated sequences is not obvious.

Theorem 9.2 ([21], p. 155). *If there is a constant $c < 1$ such that*

$$1 - |z_{n+1}| \leq c(1 - |z_n|), \quad n = 1, 2, \ldots, \tag{61.0.1}$$

then $\{z_n\}_n$ is uniformly separated. Condition (61.0.1) is also necessary if $0 \leq z_1 < z_2 < \cdots$.

(II) **The geometry of Blaschke sequences whose derived set is $\partial U = \{z \in \mathbb{C}\,|\,|z| = 1\}$:**

Consider a Blaschke sequence $\{z_n\}_n$, i.e., $\sum_n(1 - |z_n|) < \infty$, whose derived set $\{z_n\}'_n$, that is the set of its accumulation points is the unit circle. Thus, $\{z_n\}'_n = \partial U$. Can it have the following property? For each $\xi \in \partial U$, there is a Stolz angle S_ξ: $|\theta - \phi_0| \leq c(1 - r)$ with the vertex $\xi = e^{i\phi_0}$ where $z = re^{i\theta}$, $0 < r < 1$, such that $S_\xi \cap \{z_n\}_n$ is an infinite set. The answer is negative.

Proof. Suppose to the contrary, that such a Blaschke sequence exists. Consider a Jordan domain D with a smooth boundary ∂D that is properly contained in U. Let f be a Riemann mapping of U onto D. By a result of Caratheodory, f can be continuously continued to ∂U, so that $f(\partial U) = \partial D$ bijectively. Clearly $\|f\|_\infty < 1$. Let $\omega(z)$ be an inner function that interpolates f on the Blaschke sequence $\{z_n\}_n$. By the assumption on $\{z_n\}_n$ and by the theorem of Fatou on non-tangential limits of ω on ∂U, it follows that for each $\xi \in \partial U$, $\lim_{z\to\xi, \angle}\omega(z)$ exists and so must equal the limit of the sequence $\{f(z_{n_k})\}_k$, where $\forall k$, $z_{n_k} \in S_\xi$. But by Caratheodory $\lim_{k\to\infty} f(z_{n_k}) = f(\xi) \in \partial D \subset U$. Thus, all the non-tangential limits of $\omega(z)$ at every $\xi \in \partial U$ fill the Jordan curve ∂D which is contained in $|z| < 1$. But $\omega(z)$ is inner and so those limits, $\omega(\xi)$ satisfy $|\omega(\xi)| = 1$ a.e. on $\mathbb{T} = \partial U$. This contradiction proves the assertion. $\qquad\square$

The above proof shows that we get a contradiction to the theorem of Fatou also when we weaken the assumption on the Blaschke sequence that $S_\xi \cap \{z_n\}_n$ is an infinite set for every $\xi \in \partial U$, to $\xi \in A$ whose measure is positive. Thus, we obtain the following theorem.

Theorem 61.0.135. *Let $\{z_n\}_n$ be a Blaschke sequence $(\sum_n (1 - |z_n|) < \infty)$ whose derived set is $\{z_n\}'_n = A$. Then*

$$|\{\xi \in A \,|\, \text{the set } S_\xi \cap \{z_n\}_n \text{ is an infinite set}\}| = 0$$

Here, S_ξ is a Stolz angle with vertex at ξ, and for a Lebesgue measurable subset B of ∂U, $|B|$ denotes the Lebesgue measure of B.

Remark 61.0.136. Theorem 61.0.135 says that if a discrete sequence in U accumulates at each and every point of a subset A of $\mathbb{T}$, which has a positive measure and if the sequence approaches that set A fast enough (e.g., as fast as a Blaschke sequence), then it stays out of the Stolz angles at each boundary point $\xi \in A$ (it has only finitely many elements in that angle) except for a tiny (measure zero) exceptional subset of the boundary points A.

Remark 61.0.137. We do not know if the Blaschke rate of approaching the boundary, $\sum_n (1 - |z_n|)$ is sharp for Theorem 61.0.135.

(III) **Iterations of the Corollary in [59, p. 152].**

Proposition 61.0.138. *Let $f \in H^\infty$, $\|f\|_\infty < 1$, and let Ω be an inner function, and let $n \geq 2$ be an integer. Then there are H^∞ functions $h_1, \ldots, h_n$ and an inner function ω such that*

$$\omega = f \cdot \{1 + h_n + (h_{n-1} \cdot h_n) + \cdots + (h_2 \cdot \ldots \cdot h_n)\} + \Omega \cdot (h_1 \cdot h_2 \cdot \ldots \cdot h_n).$$

Proof. Let us denote $\Omega = \Omega_1$. By the Corollary in [59, p. 152], there is an $h_1 \in H^\infty$ and an inner function Ω_2 such that $\Omega_2 = f + \Omega_1 \cdot h_1$. Thus, we can generate recursively two sequences of functions: $\Omega_1, \Omega_2, \Omega_3, \ldots$ inner functions, and $h_1, h_2, h_3, \ldots$ functions in H^∞ using $\Omega_{k+1} = f + \Omega_k \cdot h_k$, $k \in \mathbb{Z}^+$. We can easily solve for the sequence of the inner functions by repeated substitutions. Namely,

$$\Omega_{n+1} = f + \Omega_n \cdot h_n = f + (f + \Omega_{n-1} \cdot h_{n-1}) \cdot h_n$$

$$= f + f h_n + \Omega_{n-1}(h_{n-1} h_n) = \cdots$$

$$= f + f \cdot h_n + f \cdot (h_{n-1} \cdot h_n) + \cdots + f \cdot (h_{n-k} h_{n-k+1} \ldots h_n)$$

$$+ \Omega_{n-k-1}(h_{n-k-1} h_{n-k} \ldots h_n).$$

Taking $k = n - 2$, we obtain

$$\Omega_{n+1} = f \cdot \{1 + h_n + (h_{n-1} \cdot h_n) + \cdots + (h_2 \cdot \ldots \cdot h_n)\} + \Omega \cdot (h_1 \cdot h_2 \cdot \ldots \cdot h_n).$$

Now, we denote $\omega = \Omega_{n+1}$ to conclude our proof. $\qquad\square$

We can generalize this proposition.

Proposition 61.0.139. *Let $n \geq 2$ be an integer, and $f_2, \ldots, f_{n+1} \in H^\infty$ satisfy $\|f_k\|_\infty < 1$ ($k = 2, \ldots, n+1$), and let Ω be an inner function. Then there are H^∞ functions $h_1, h_2, \ldots, h_n$, and an inner function ω such that,*

$$\omega = f_{n+1} + f_n \cdot h_n + f_{n-1} \cdot (h_{n-1} h_n) + \cdots + f_2 \cdot (h_2 \ldots h_n) + \Omega \cdot (h_1 h_2 \ldots h_n).$$

Proof. We use the recursion $\Omega_{k+1} = f_{k+1} + \Omega_k \cdot h_k$. $\qquad\square$

Chapter 62

The Condition $\omega(\phi; t) = O(t \log\frac{1}{t})$

This condition on the modulus of continuity appeared, for example, in [24]. It appeared in another context in a more recent paper [54], which we are going to describe in this chapter. Consider the toy example of a Blaschke product of degree 2,

$$F(z) = \left(\frac{z - a}{1 - \overline{a}z}\right)\left(\frac{z - b}{1 - \overline{b}z}\right) \quad \text{where } |a|, |b| < 1.$$

It is elementary to compute the critical points of F in $|z| < 1$ (there is exactly one), but it is a tedious and relatively long computation. That hints that the problem of studying the critical points of Blaschke products, or more generally, of inner functions might be difficult.

Introduction for paper (Section 1 in [54], p. 1).

We denote by $\mathcal{J}$ the set of inner functions whose derivatives lie in the Nevanlinna class. A finite Blaschke product can be defined as a proper holomorphic self-map of the unit disk. It is uniquely defined up to a rotation by its zero set. That gives the more standard definition of such a finite Blaschke product

$$F(z) = e^{i\psi} \prod_{j=1}^{d} \left(\frac{z - a_j}{1 - \overline{a_j}z}\right), \quad \text{where } a_1, \ldots, a_d \in U$$

the number $d \geq 1$ is called the degree of F. F has $d - 1$ critical points (i.e., zeros of F') lying in U. A classical result of Heinz asserts that a finite Blaschke product is uniquely determined by the set of its critical points, up to post-composition with a Möbius transformation $m \in \text{Aut}(U)$, and furthermore, any set of $d - 1$ points in U (multiplicity included) arises as the critical set of some Blaschke product of degree d. We can rephrase this

347

result of Heinz as follows: to a Blaschke product F of degree $d \geq 2$, we assign a Blaschke product of degree $d - 1$ whose zeros are located at the critical points of F in U. Heinz' theorem says that this correspondence is a bijection (between Blaschke products of degree d and Blaschke products of degree $d - 1$), provided that one considers F modulo post-composition with a Möbius transformation (as not to change its critical set) and B the image of F up to rotations (which preserve the zero set). Paper [54] studies an analogue of this assignment in infinite degree. Let Inn denote the space of all inner functions. Consider the subspace $\mathcal{J}$ of inner functions whose first derivatives lie in the Nevanlinna class $\mathcal{N}$, i.e., which satisfy

$$\lim_{r \to 1^-} \frac{1}{2\pi} \int_0^{2\pi} \log^+ |F'(re^{i\theta})| d\theta < \infty. \tag{62.0.1}$$

By results of Ahern and Clark, [2], F' has an "inner-outer" decomposition $F' = Inn\, F' \cdot Out\, F'$. Intuitively, $Inn\, F' = B \cdot S$ describes the "critical structure" of the map F. The Blaschke product, B, records the location of the critical points of F in U, while the singular inner function S describes the "boundary critical structure". The mapping $F \to Inn\, F'$ from $\mathcal{J}/\mathrm{Aut}(U)$ to Inn/S^1 generalizes the Heinz assignment for finite degree Blaschke products. The presence of the singular factor allows us to distinguish different inner functions with the same critical set, for instance, it separates the universal cover map

$$z \to \exp\left(\frac{z+1}{z-1}\right)$$

of the punctured disk $U - \{0\}$ from the identity mapping. The following is proved in [54].

Theorem 1.1. *Let $\mathcal{J}$ be the set of inner functions whose first derivatives lie in the Nevanlinna class. The natural map*

$$F \to Inn(F') : \quad \mathcal{J}/\mathrm{Aut}(U) \to Inn/S^1,$$

is injective. The image consists of all inner functions of the form $B \cdot S_\mu$ where B is a Blaschke product and S_μ is the inner factor associated to a measure μ whose support is contained in a countable union of Beurling–Carleson sets.

Definition 62.0.140. A Beurling–Carleson set is a closed subset of the unit circle of zero Lebesgue measure whose complement is a union of arcs $\bigcup_k I_k$ with $\sum_k |I_k| \log \frac{1}{|I_k|} < \infty$.

Lemma 1.2. *The map in Theorem 1.1 is well defined: If $F \in \mathcal{J}$ is a inner function, then for any Möbius transformation $T \in \mathrm{Aut}(U)$, the Frostman shift $T \circ F \in \mathcal{J}$ and $Inn(T \circ F)' = Inn\, F'$.*

Proof. From the chain rule, we have $(T \circ F)'(z) = T'(F(z)) \cdot F'(z)$. Since $\log |T'|$ is bounded, $T \circ F \in \mathcal{J}$. The above identity tells us that the inner part $Inn(T \circ F)'$ is divisible by $Inn\, F'$. Using T^{-1} in place of T, we see that $Inn\, F'$ is divisible by $Inn(T \circ F)'$. Hence, $Inn(T \circ F)' = Inn\, F'$ (up to a uni-modular constant). $\qquad\qquad\square$

Gauss curvature equation (Section 1.1 in [54], pp. 3–4).

There is a connection between complex analysis and nonlinear elliptic PDE. It is given by **Liouville's Theorem.**

Up to post-composition with Möbius transformation in $\mathrm{Aut}(U)$, holomorphic self-maps of U without critical points are in bijection with conformal metrics of constant curvature -4. Let

$$\lambda_U = \frac{1}{1 - |z|^2},$$

denote the Poincaré metric on the unit disk U. The λ_U is a solution of the Gauss curvature equation,

$$k_\lambda := -\frac{\triangle \log \lambda}{\lambda^2} = -4, \qquad\qquad (62.0.2)$$

which may be alternatively written as follows:

$$\triangle u = 4e^{2u}, \quad u : U \to \mathbb{R}, \qquad\qquad (62.0.3)$$

after the change of variables $u = \log \lambda$. Given a holomorphic self-map of the disk, $F : U \to U$, set

$$\lambda_F := F^* \lambda_U = \frac{|F'|}{1 - |F|^2}, \quad u_F := \log \lambda_F.$$

The Gaussian curvature is a conformal invariant, hence λ_F also has curvature -4. Liouville's theorem says that all solutions of (62.0.2) arise in this way. Also, by the Schwarz lemma, λ_U is the maximal solution of (62.0.2), in the sense that if λ is any other solution, then $\lambda < \lambda_U$ pointwise. We have the following,

Theorem 1.3 ([54], p. 4). *The mapping $F \to u_F$ is a bijection between locally univalent inner functions in $\mathcal{J}/\mathrm{Aut}(U)$ and nearly-maximal solutions of (62.0.3) satisfying*

$$\limsup_{r \to 1^-} \int_{|z|=r} (u_U - u)\, d\theta < \infty. \tag{62.0.4}$$

For each $0 < r < 1$, we may view $(u_U - u)d\theta$ as a positive measure on the circle of radius r. Subharmoniticity guarantees the existence of a weak limit as $r \to 1^-$, which we denote by $\mu[u]$ or $\mu(u)$.

Theorem 1.4 ([54], p. 4). *The mapping $u \to \mu[u]$ is injective. Its image consists of all finite measures whose support is contained in a countable union of Beurling–Carleson sets. If $u = u_F$, the singular measure $\sigma(F') = \mu(u_F)$.*

The proof of Theorem 1.1 in [54] (above), uses the following.

Lemma 1.5 ([54], p. 5) (Decomposition rule). *An inner function $B_c \cdot S_\mu$ lies in the image of $F \to \mathrm{Inn}\, F'$ if and only if its singular part S_μ does.*

Definition 62.0.141. A measure μ is constructable if there is a $F_\mu \in \mathcal{J}$ such that $S_\mu = \mathrm{Inn}\, F'_\mu$. μ is invisible if for any measure $0 < \nu \leq \mu$, there does not exist a function $F_\nu \in \mathcal{J}$ with $\mathrm{Inn}\, F'_\nu = S_\nu$.

Any singular measure on the unit circle, μ, can be decomposed into a constructable part and an invisible part: $\mu = \mu_{con} + \mu_{inv}$.

Lemma 1.6 ([54], p. 5) (product rule). *Suppose measures μ_j, $j = 1, 2, \ldots$ are constructable. If their sum $\mu = \sum_{j=1}^{\infty} \mu_j$ is finite, then μ is also constructable.*

Lemma 1.7 ([54], p. 5) (division rule). *If a measure μ is constructable, then any $\nu \leq \mu$ is also constructable.*

Lemma 1.8 (([17]) ([54], p. 5)). *Suppose that the support of μ is contained in a Beurling–Carlson set. Then $S'_\mu \in \mathcal{N}$.*

Since S_μ divides S'_μ, the division rule implies that any measure μ supported on a Beurling–Carleson set is constructable. By the product rule, any measure supported on a countable union of Beurling–Carleson sets is also constructable. According to Theorem 1.1, any constructable measure is of this form. As a consequence, we see that Cullen's theorem, [17], is essentially sharp.

Corollary 1.9 ([54], p. 5). *Suppose μ is a measure on the unit circle with $S'_\mu \in \mathcal{N}$. Then, the support of μ is contained in a countable union of Beurling–Carleson sets.*

To complete the proof of Theorem 1.1, [54] gives a criterion for a measure to be invisible.

Theorem 1.10 ([54], p. 6). *Suppose μ is a measure on the unit circle which does not charge Beurling–Carleson sets. Then, it is invisible.*

In the late 1970's, Korenblum [60] and Roberts [88] independently showed that the set of all the singular inner functions S_μ, such that μ does not charge Beurling–Carleson sets, is exactly the collection of inner functions which are cyclic in Bergman space. To prove Theorem 1.10, it is first shown in [54], that any measure μ with modulus of continuity

$$\omega_\mu(t) \leq C \cdot t \log \left(\frac{1}{t} \right),$$

is invisible. These are exactly the measures dealt with, or used by [21], [24]. To obtain the full result, the authors of [54] adopt the argument from Roberts, [88], to their setting. This involves an iterative procedure based on the decomposition of a measure that does not charge Beurling–Carleson sets into "$t \log(1/t)$"-pieces.

Background on conformal metrics (Section 2 in [54], pp. 6–9).

Given an at most countable set C in the unit disk (counted with multiplicities), the machinery of Kraus and Roth, [62], seeks to construct a Blaschke product with critical set C. If such a Blaschke product exists, then the machinery produces the optimal or maximal Blaschke product F_C. Instead of constructing F_C directly, Kraus and Roth construct the conformal pseudo-metric $\lambda_{F_C} = F *_C \lambda_U$, the pullback of the Poincaré metric on U. To explain their construction, we need several concepts:

(i) **SK-metrics.** We will say that a conformal pseudo-metric λ on a domain D has "constant curvature -4" if it vanishes on a discrete set of points $C \subset D$ (which might be empty) and satisfies

$$k_\lambda = \frac{\triangle \log \lambda}{\lambda^2} = -4, \quad \lambda \in C^2(D - C), \text{ on } D - C.$$

More generally, if $k_\lambda \leq -4$ on $D - C$, then following Heinz, we say that $\lambda(z)$ is a (regular) SK-metric. In reality, this is a slight abuse of notation since the distributional Laplacian may have point masses at points of C.

(ii) **Perron families.** A collection Φ of SK-metrics is a Perron family if it is closed under "modifications" and "taking maxima". The first condition says that given a round disk $E \subset D$ and a metric $\lambda \in \Phi$, the unique SK-metric $M_E\lambda$, which agrees with λ on $D - E$, is non-vanishing and has a constant curvature -4 in E, which lies in Φ; while the second condition says that for any $\lambda_1, \lambda_2 \in \Phi$, their maximum, $\max\{\lambda_1, \lambda_2\}$ is also in Φ. Heinz proved that if a Perron family is non-empty, then the supremum of all the metrics in Φ is a non-vanishing conformal metric of curvature -4.

(iii) **Liouville's theorem.** Suppose λ is a conformal pseudo-metric defined on a simply-connected domain D. We say that λ vanishes at $c_i \in C$ with multiplicity m_i if

$$\lim_{z \to c_i} \frac{\lambda(z)}{|z - c_i|^{m_i}} = L_i \quad \text{for some} \quad 0 < L_i < \infty.$$

Liouville's theorem says that if λ has constant curvature -4 and all its zeros have integral multiplicities, then $\lambda = \lambda_F = F^*\lambda_U$ for some holomorphic function $F : D \to U$. Furthermore, the function F is unique up to post-composition with a Möbius transformation in $\mathrm{Aut}(U)$.

For a set C in U, let Φ_C be the collection of all SK-metrics on U which vanish on C. It clearly satisfies the two axioms of being a Perron family on the domain $U - C$. If Φ_C is non-empty, we obtain (sup) a metric of a constant curvature -4 and a holomorphic function $F_C : U \to U$ which vanishes on C to the correct order. Using the maximality of the metric λ_{F_C}, Kraus proved that the outer and singular inner factors of F_C are trivial, so that F_C is a Blaschke product. In the case when the critical set C is a Blaschke sequence, Kraus observed that $|B_C|\lambda_U$ is a SK-metric which guarantees that the Perron family Φ_C is non-empty. More generally, given a holomorphic function H with $\|H\|_\infty \le 1$ and a metric λ of curvature -4, $|H| \cdot \lambda$ is an SK-metric. Exploiting the lower bound $\lambda_{F_C} \ge |B_C|\lambda_U$, Kraus obtained the following theorem.

Theorem 2.1 (Kraus [54] p. 8). *Suppose C is a Blaschke sequence in the disk and λ is a metric of constant curvature -4, which vanishes precisely on C with the correct multiplicities. Then $\lambda = \lambda_{F_C}$ if and only if*

$$\lim_{r \to 1^-} \int_{|z|=r} \log\left(\frac{\lambda}{\lambda_U}\right) d\theta = 0. \tag{62.0.5}$$

In [54, Section 3], the authors use ideas of Ahern and Clark to show that the above theorem can be alternatively phrased as follows:

Corollary 2.2 ([54], p. 8). *Suppose C is a Blaschke sequence in U. An infinite Blaschke product $F \in \mathcal{J}$ is the maximal Blaschke product* *associated with C if and only if the singular factor of $Inn\, F'$ is trivial, i.e., if $Inn\, F' = B_C$.*

In order to generalize the arguments of Kraus and Roth to allow for singular factors, there is a need in the following:

Lemma 2.3 (Fundamental Lemma) ([54], p. 8). *For any inner function $F \in \mathcal{J}$,*

$$\lambda_F \geq |Inn\, F'| \cdot \lambda_U. \tag{62.0.6}$$

In fact, λ_F is the smallest metric of constant curvature -4 with the property expressed in equation (62.0.6).

Note that the minimality of the metric λ_F implies that the map $F \to Inn\, F'$ of Theorem 1.1 is injective. using the factorization $F' = Inn\, F' \cdot Out\, F'$, one can rewrite equation (62.0.6) as follows:

$$\frac{1 - |F(z)|^2}{1 - |z|^2} \leq |Out\, F'(z)|, \tag{62.0.7}$$

which was proved using Julia's lemma in [26].

Hulls and Wedges ([54], p. 9).

Wedge of two metrics. Given two inner functions $F, G \in \mathcal{J}$, consider the family $\Phi_{F,G}$ of SK-metrics that are pointwise less than $\min\{\lambda_F, \lambda_G\}$. This family is not empty; the metric $|Inn\, F'| \cdot |Inn\, G'| \cdot \lambda_U$ is in it by Lemma 2.3. Taking the supremum of conformal metrics in $\Phi_{F,G}$, we get a regular conformal metric of constant curvature -4. By Liouville's theorem, it is the pullback of λ_U by a holomorphic function which we denote by $F \wedge G$. It is shown later that $F \wedge G \in \mathcal{J}$.

Hull of a conformal metric. For an SK-metric K, let Ψ_K be the collection of all metrics of constant curvature -4 which are greater than K and Φ_K be the collection of all SK-metrics that are less than all the metrics in Ψ_K. Since Φ_K is a Perron family, its supremum is a metric $\widehat{K}$ of curvature -4. We call $\widehat{K}$, the hull of K. By the definition, it is clear that $\widehat{K}$ is the smallest metric of curvature -4 which exceeds K. Using this terminology, Lemma 2.3 says that λ_F is the hull of $|Inn\, F'| \cdot \lambda_U$.

Gap of a Nevanlinna function (Section 3 in [54], p. 9).

By definition, the Nevanlinna class $\mathcal{N}$ consists of holomorphic functions on U for which

$$\sup_{0<r<1} \frac{1}{2\pi} \int_{|z|=r} \log^+ |f(z)|\, d\theta < \infty, \quad z = re^{i\theta}. \tag{62.0.8}$$

It is well known that if $f(z)$ is not identically zero, condition (62.0.8) is equivalent to the boundedness of

$$\sup_{0<r<1} \frac{1}{2\pi} \int_{|z|=r} \big|\log|f(z)|\big|\, d\theta < \infty, \quad z = re^{i\theta}.$$

Since $\log|f(z)|$ is a subharmonic function, $\lim_{r\to 1^-} \frac{1}{2\pi}\int_{|z|=r} \log|f(z)|d\theta$ exists and is finite. However, unlike the Hardy norms, it is not the case that

$$\lim_{r\to 1^-} \frac{1}{2\pi} \int_{|z|=r} \log|f(z)|d\theta = \frac{1}{2\pi} \int_{|z|=1} \log|f(z)|d\theta,$$

where in the integral on the right-hand side, we consider the radial boundary values of f which are known to exist a.e. To understand the cause of the discrepancy, we consider the canonical decomposition $f = B(S/S_1) \cdot O$ into a Blaschke product, a quotient of singular inner functions and an outer function.

$$B(z) = \prod_j \left(-\frac{\overline{a_j}}{|a_j|}\right) \cdot \left(\frac{z - a_j}{1 - \overline{a_j}z}\right),$$

$$\left(\frac{S}{S_1}\right)(z) = \exp\left(-\int_{S^1} \frac{\xi + z}{\xi - z}\right), \quad \sigma \perp m,$$

$$O(z) = \exp\left(\int_{S^1} \frac{\xi + z}{\xi - z} \log|f(\xi)|\, dm_\xi\right),$$

where m is the Lebesgue measure on the unit circle S^1. Given an interval (an arc) I on S^1, rI denote its radial projection on the circle $S_r = \{z\,|\,|z| = r\}$.

Lemma 3.1 ([54], p. 10). *We define the gap of f and have the following identity:*

$$\mathrm{gap}(f) := \frac{1}{2\pi}\int_{|z|=1} \log|f(z)|\, d\theta - \lim_{r\to 1^-}\left\{\frac{1}{2\pi}\int_{|z|=r} \log|f(z)|\, d\theta\right\} = \sigma(S^1).$$

More generally, if I is an interval on S^1, then we have

$$\mathrm{gap}_I(f) := \frac{1}{2\pi}\int_I \log|f(z)|\, d\theta - \lim_{r\to 1^-}\left\{\frac{1}{2\pi}\int_{rI} \log|f(z)|\, d\theta\right\} = \sigma(I),$$

provided that the endpoints of I do not charge σ.

Proof. Using $f = B \cdot (S/S_1) \cdot O$, we will analyze the three components separately. We begin with the Blaschke factor. We claim that

$$\lim_{r \to 1^-} \left\{ \frac{1}{2\pi} \int_{|z|=r} \log |B(z)| d\theta \right\} = 0,$$

factoring out z^m, we may assume that $B(0) \neq 0$. Let $a_1, a_2, \ldots$ be an enumeration of the zeros of B in U. By Jensen's formula and by $|B(0)| = \prod_j |a_j|$, we have

$$\frac{1}{2\pi} \int_{|z|=r} \log |B(re^{i\theta})| d\theta = \sum_{|a_j|<r} \log \frac{r}{|a_j|} - \sum_{|a_j|<1} \log \frac{1}{|a_j|},$$

which tends to zero as $r \to 1^-$. This proves the claim for S^1. Since $\log |B(z)| < 0$, we have for any interval $I \subset S^1$

$$\lim_{r \to 1^-} \left\{ \frac{1}{2\pi} \int_{rI} \log |B(z)| d\theta \right\} = 0.$$

Clearly,

$$\frac{1}{2\pi} \int_I \log |B(z)| \, d\theta = 0 \text{ as well.}$$

Since $\log |O(z)|$ is harmonic on U which is the Poisson extension of its radial boundary values,

$$\lim_{r \to 1^-} \frac{1}{2\pi} \int_{rI} \log |O(z)| \, d\theta = \frac{1}{2\pi} \int_I \log |O(z)| \, d\theta.$$

Thus, the outer part also behaves (with no gap) as expected. The singular part exhibits the most interesting behavior. It is the source of the gap. Since $\log |(S/S_1)(z)|$ is the Poisson extension of the singular measure σ, we have

$$\lim_{r \to 1^-} \frac{1}{2\pi} \int_{rI} \log \left| \left(\frac{S}{S_1} \right) (z) \right| d\theta = \sigma(I),$$

if the endpoints of I do not charge σ. On the other hand,

$$\frac{1}{2\pi} \int_I \log \left| \left(\frac{S}{S_1} \right) (z) \right| d\theta = 0,$$

as the radial boundary values of $\log |(S/S_1)(z)|$ are zero a.e. on the unit circle. $\qquad \square$

Applications to inner functions (Section 3.1 in [54], p. 11).

We first give a slightly different perspective on a classical theorem due to Ahern and Clark.

Lemma 3.2 (([2]), ([54], p. 11)). *For an inner function $F \in \mathcal{J}$, its derivative admits a $B \cdot S \cdot O$ decomposition. In other words, the singular measure $\sigma(F') \geq 0$.*

Proof. By Lemma 1.2 ([54]), we may assume that $F(0) = 0$. Then $|F'(x)| \geq 1$ on S^1 (see Theorem 4.15 in [71]). In view of the fundamental inequality $|F'(rx)| \leq 4|F'(x)|$, $x \in S^1$, $0 < r < 1$, of Ahern and Clark, the dominated convergence theorem shows

$$\int_I \log^+ |F'(z)| dm - \lim_{r \to 1^-} \int_{rI} \log^+ |F'(z)| dm = 0,$$

for any interval $I \subset S^1$. However, by Fatou's lemma, the negative part of the logarithm can only dissipate and therefore

$$\mathrm{gap}_I(F') = \int_I \log |F'(z)| dm - \lim_{r \to 1^-} \frac{1}{2\pi} \int_{rI} \log |F'(z)| dm \geq 0.$$

$\square$

The following lemma says that as $r \to 1^-$, the measures $\log \frac{\lambda_U}{\lambda_F}(re^{i\theta}) dm$ converge weakly to $\sigma(F')$.

Lemma 3.3 ([54], p. 12). *Let $I \subset S^1$ be an interval. If $F \in \mathcal{J}$, then*

$$\lim_{r \to 1^-} \frac{1}{2\pi} \int_{rI} \log \frac{1 - |F(z)|^2}{1 - |z|^2} d\theta = \frac{1}{2\pi} \int_I \log |F'(z)| d\theta. \qquad (62.0.9)$$

Proof. The hyperbolic distance is contracted when a holomorphic mapping is applied, $d_U(F(0), F(z)) \leq d_U(0, z)$. It follows that the quotient

$$\frac{1 - |F(z)|^2}{1 - |z|^2} \geq C_{F(0)}$$

is bounded below by a positive constant. By the Schwarz lemma,

$$\frac{1}{2\pi} \int_{rI} \max \left\{ \log |F'(z)|, \log C_{F(0)} \right\} d\theta \leq \frac{1}{2\pi} \int_{rI} \log \frac{1 - |F(z)|^2}{1 - |z|^2} d\theta.$$

We note that

$$|F'(z)| \leq \frac{1 - |F(z)|^2}{1 - |z|^2}.$$

Applying the dominated convergence theorem as in the proof of Lemma 3.2 (above) gives the $\leq$ inequality in (62.0.9). For the $\geq$ direction, we average Dyakonov's inequality (2.3) over $z \in rI$:

$$\frac{1}{2\pi} \int_{rI} \log \left| Out\, F'(z) \right| d\theta \geq \frac{1}{2\pi} \int_{rI} \log \frac{1 - |F(z)|^2}{1 - |z|^2} d\theta.$$

The lemma follows after taking $r \to 1^-$ since $\log |Out\, F'(z)|$ is the harmonic extension of $\log |F'(z)|$ considered as a function on S^1. $\square$

One may compare the above lemma with Theorem 3 in [1]: If F is a locally univalent inner function (no critical points), then

$$F' \in \mathcal{N} \Leftrightarrow \lim_{r \to 1^-} \int_{|z|=r} \log \frac{1 - |F(z)|^2}{1 - |z|^2} d\theta < \infty \Leftrightarrow \lim_{r \to 1^-} \int_{|z|=r} \log \frac{\lambda_U}{\lambda_F} d\theta < \infty,$$

which is the statement of Theorem 1.3 (here in [54]). For the second equivalence, one uses

$$\lim_{r \to 1^-} \frac{1}{2\pi} \int_{|z|=r} \log |F'(z)| d\theta = \log |F'(0)|,$$

which follows by the fact that $\log |F'(z)|$ is harmonic on U (by the assumption $F'(z) \neq 0$). This is finite.

Applications to conformal metrics (Section 3.2 in [54], p. 13).

Lemma 3.4 ([54], p. 13). *Suppose that $F \in \mathcal{J}$ is an inner function for which*

$$\lambda_F \geq |B_C \cdot S_\mu| \cdot \lambda_U. \tag{62.0.10}$$

Then, the singular measure $\sigma(F') \leq \mu$.

Proof. Let $I \subset S^1$ be an interval. From the definition of λ_F,

$$\int_{rI} \log \frac{\lambda_F}{|B_C \cdot S_\mu| \cdot \lambda_U} dm = \int_{rI} \log \left(\frac{|F'|(1 - |z|^2)}{1 - |F|^2} \right) dm - \int_{rI} \log |B_C \cdot S_\mu| dm.$$

By Lemma 3.1 (here in [54]) and the easy part of Lemma 3.3, as $r \to 1^-$, this leads to

$$0 \leq -\sigma(F')(I) + \sigma(S_\mu)(I),$$

at least if I is generic. $\square$

Remark 62.0.142. The same conclusion holds under the seemingly weaker assumption $\lambda_F \geq |B_C \cdot S_\mu \cdot O_h|$, where

$$O_h(z) = \exp\left(\int_{S^1} \frac{\xi + z}{\xi - z} h(\xi) dm_\xi\right), \quad h : S^1 \to \mathbb{R},$$

is an arbitrary outer function: the above computation results in $\sigma(F') \leq \mu - h \cdot dm$. Since $\sigma(F') \perp h \cdot dm$ are mutually singular, we have $\sigma(F') \leq \mu$ and $h \leq 0$.

Similar considerations show the following:

Lemma 3.5 ([54], p. 14). *For any $F, G \in \mathcal{J}$ and interval $I \subset S^1$,*

$$\lim_{r \to 1^-} \int_{|z|=r} \log\left(\frac{\lambda_F}{\lambda_G}\right) dm = -\sigma(F')(I) + \sigma(G')(I). \tag{62.0.11}$$

In particular, if $\lambda_F \geq \lambda_G$ then $\sigma(F') \leq \sigma(G')$.

Combining the last lemma with Theorem 2.1 gives Corollary 2.2.

Lemma 3.6 ([54], p. 14). *If λ_G is a metric of curvature -4 such that $\lambda_G \geq |H| \cdot \lambda_U$ for some bounded holomorphic function $H \not\equiv 0$, then $G \in \mathcal{J}$.*

Proof. Since H is a bounded holomorphic function, $\gamma_1 = \lim_{r \to 1^-} \int_{rS^1} \log |H| dm$ is finite. The condition $\lambda_G \geq |H| \cdot \lambda_U$ implies that the zeros of G' form a Blaschke sequence, which in turn implies that the integral $\gamma_2 = \lim_{r \to 1^-} \int_{rS^1} \log |G'| dm$ is also finite. An inspection of the inequality

$$0 \leq \liminf_{r \to 1^-} \int_{rS^1} \log \frac{\lambda_G}{|H|\lambda_U} dm \int -\gamma_1 + \gamma_2 - \limsup_{r \to 1^-} \int_{rS^1} \log^+ |G'| dm,$$

then shows that G' satisfies the Nevanlinna condition (62.0.8). It remains to prove that the outer part of G is trivial, so that G is an inner function. If this were not the case, then for a positive measure set of directions $\theta \in [0, 2\pi)$, $\limsup_{r \to 1^-} \lambda_G(re^{i\theta})$ would be finite. However, this contradicts the assumption $\lambda_G \geq |H|\lambda_U$, since by the Lusin–Privalov theorem, the radial limit of $H(re^{i\theta})$ is non-zero almost everywhere. $\square$

Injectivity and minimality (Section 3.3 in [54], p. 14).

Now, it is possible to show the injectivity statement of Theorem 1.1. If there were two functions $F, G \in \mathcal{J}$ with $Inn\, F' = Inn\, G' = B_C \cdot S_\mu$, then

$$\lambda_F \geq \lambda_{F \wedge G} \geq |B_C \cdot S_\mu| \cdot \lambda_U. \tag{62.0.12}$$

Lemmas 3.4 and 3.5 imply that $(F \wedge G)'$ has the same inner part as F'. From the definition of curvature,

$$\triangle \log(\lambda_F / \lambda_{F \wedge G}) = 4(\lambda_F^2 - \lambda_{F \wedge G}^2).$$

Hence, $\log(\lambda_F / \lambda_{F \wedge G})$ is subharmonic and non-negative. Yet,

$$\lim_{r \to 1^-} \int_{|z|=r} \log(\lambda_F / \lambda_{F \wedge G}) d\theta = 0, \tag{62.0.13}$$

which forces $\log(\lambda_F / \lambda_{F \wedge G}) = 0$. We deduce $\lambda_F = \lambda_{F \wedge G} = \lambda_G$ and therefore $F = G$ up to post composition with a Möbius transformation, by Liouville's theorem. Given an inner function $F \in \mathcal{J}$, we now show that λ_F is the smallest metric of constant curvature -4 that exceeds $|Inn\, F'| \cdot \lambda_U$. Following Section 2 in this chapter, we consider the hull λ of the metric $|Inn\, F'| \cdot \lambda_U$. The inequalities

$$\lambda_F \geq \lambda \geq |Inn\, F'| \cdot \lambda_U \tag{62.0.14}$$

reveal that λ has exactly the same zero set as λ_F (counted with multiplicities). In particular, all the zeros of λ have integral multiplicities. Proceeding as in the proof of injectivity, we obtain $\lim_{r \to 1^-} \int_{|z|=r} \log(\lambda_F / \lambda) d\theta = 0$ and $\lambda = \lambda_F$. $\qquad \square$

Stable approximations (Section 4 in [54], p. 15).

We study convergent sequences of inner functions.

Definition 62.0.143. Suppose that $\{F_n\} \subset \mathcal{J}$ is a sequence of inner functions which converges uniformly on compact subsets of U to an inner function F. We say that F_n is a (Nevanlinna) stable approximation of F if

$$Inn\, F' = \lim Inn\, F'_n, \quad Out\, F' = \lim Out\, F'_n. \tag{62.0.15}$$

In general, we have inequalities in one direction.

Theorem 4.1 ([54], p. 15). *Suppose $\{F_n\} \subset \mathcal{J}$ is a sequence of inner functions which converges uniformly on compact subsets of U to a holomorphic function $F : U \to U$. Also assume that the $I_n = Inm\, F'_n$ converges to an inner function I. Then $F \in \mathcal{J}$ and the following inequalities hold:*

$$\sigma(F') \leq \sigma(I), \tag{62.0.16}$$

$$|Inn\, F'| \geq |I|, \tag{62.0.17}$$

$$\int_{S^1} \log|F'| dm \leq \lim_{n \to \infty} \int_{S^1} \log|F'_n| dm. \tag{62.0.18}$$

Furthermore, either all of the above inequalities are equalities or none of them are.

Proof. Taking $n \to \infty$ in $\lambda_{F_n} \geq |I_n|\lambda_U$ gives $\lambda_F \geq |I|\lambda_U$. Inequality (62.0.16) follows by Lemma 3.4. Clearly, (Hurwitz) $Inn\, F'$ and $I = \lim(Inn\, F_n')$ have the same zeros in U but may have different singular factors. It is not hard to check that for singular inner functions, one has the inequality $|S_1| \leq |S_2|$ if and only if $\sigma(S_1) \geq \sigma(S_2)$. The "if" direction is obvious, while the "only if" direction follows from the identity

$$0 \leq \lim_{r \to 1^-} \int_{rE} \log \left| \frac{S_2}{S_1} \right| dm = -\sigma(S_2)(E) + \sigma(S_1)(E),$$

valid for any generic interval $E \subset S^1$ whose endpoints do not charge the measures $\sigma(S_1)$ and $\sigma(S_2)$. This proves (62.0.17), and shows that the equality case in (62.0.16) and (62.0.17) coincide. Since $F_n' \to F'$ uniformly on compact subsets of U, (62.0.17) is equivalent to the inequality $|Out\, F'(z)| \leq |\lim_{n \to \infty} Out\, F_n'(z)|$. Setting $z = 0$ and taking logarithms gives (62.0.18). However, if (62.0.18) is equality, then by the maximum modulus principle applied to $Out\, F'(z)/(\lim Out\, F_n'(z))$, we must have $|Out\, F'(z)| = |\lim Out\, F_n'(z)|, \; \forall\, z \in U$. This completes the proof. $\qquad\square$

For some applications, we need a slightly more general version of Theorem 4.1.

Theorem 4.2 ([54], p. 16). *In the context of Theorem 4.1, suppose instead that I_n converges to a non-zero holomorphic function $H : U \to U$ with the inner-outer decomposition $H = I \cdot O$. Then, the inequalities (62.0.16)–(62.0.18) still hold. The equality case in (62.0.18) implies that $\{F_n\}$ is a stable sequence, in particular, the outer factor $O = 1$ is trivial and the B_n converge to an inner function.*

We only sketch the details in the proof. Since $\|H\|_\infty \leq 1$ we have $|O(z)| \leq 1$ and $|I(z)| \geq |H(z)|$ for $z \in U$. Following the proof of Theorem 4.1, we obtain the inequality $\lambda_F \geq |I \cdot O|\lambda_U$. We may still use Lemma 3.6 to conclude that $F \in \mathcal{J}$. Remark 62.0.142 allows us to conclude (62.0.16) and (62.0.17) in this more general case as well. We may weaken (62.0.17) to $|Inn\, F'| \geq |H|$, which is equivalent to (62.0.18). This time, equality case in (62.0.18) forces $I = H$ and $O = 1$.

Remark 62.0.144. In view of Lemma 3.6, if a sequence of inner functions F_n converges to a function F with $F' \notin \mathcal{N}$, then $H = \lim(Inn\, F_n')$ must be 0.

An argument from [15] (Lemma 5.4), shows:

Lemma 4.3 ([54], p. 17). *An inner function $F \in \mathcal{J}$ admits a stable approximation by finite Blaschke products.*

Proof. By a theorem of Frostman (Theorem 2.5 in [71]), if $\xi \in U$ avoids a set of zero logarithmic capacity, then the Frostman shift $T_\xi \circ F$ is a Blaschke product, where

$$T_\xi(z) = \frac{z - \xi}{1 - \bar{\xi}z}.$$

If ξ is not an exceptional point, we may choose a sequence $F_{n,\xi}$ of finite Blaschke products converging to F so that $T_\xi \circ F_{n,\xi}$ is a sequence of partial products of $T_\xi \circ F$. By Corollary 4.13 in [71], we have

$$|(T_\xi \circ F_{n,\xi})'(x)| \le |(T_\xi \circ F)'(x)|, \quad x \in S^1.$$

It follows that

$$|(F_{n,\xi})'(x)| \le \left(\frac{1 + |\xi|}{1 - |\xi|}\right)^2 |F'(x)|,$$

which leads to the estimate

$$\int_{S^1} \log |F'_{n,\xi}(x)|dm \le 2\log \frac{1 + |\xi|}{1 - |\xi|} + \int_{S^1} \log |F'(x)|dm.$$

Since we can choose ξ arbitrarily close to 0, we can diagonalize to find a sequence F_n of finite Blaschke products converging to F for which

$$\limsup_{n \to \infty} \int_{S^1} \log |F'_n(x)|dm \le \int_{S^1} \log |F'(x)|dm.$$

However, by Theorem 4.2, the lower bound is automatic and the sequence $\{F_n\}$ is stable. $\square$

Suppose $F \in \mathcal{J}$ is an inner function. Lemma 4.3 provides a Nevanlinna stable approximation $F_n \to F$ by finite Blaschke products. Since finite Blaschke products are maximal, we have $\lambda_{F_n} \ge |Inn\, F'_n| \cdot \lambda_U$ for any $n \ge 1$. Taking $n \to \infty$ gives $\lambda_F \ge |Inn\, F'| \cdot \lambda_U$. Note that there is no circular reasoning since the proof of Theorem 4.2 only relied on the easy part of Lemma 3.3.

Remark 62.0.145. We can endow the space of analytic functions $\mathcal{E} = \{f | f' \in \mathcal{J}\}$ with the stable topology by specifying that $f_n \to f$ if the f_n converge uniformly on compact sets to f and $\log |f'_n|dm \to \log |f'|dm$ weakly. Lemma 4.3 shows that finite Blaschke products are dense in $\mathcal{J}$ while Theorem 4.2 implies that the subset $\mathcal{J} \subset \mathcal{E}$ is closed.

Example of an unstable approximation (Section 4.1 in [54], p. 18).

Let F_n be the Blaschke product of degree $n+1$ with zeros at the origin and at

$$z_j = e^{j(2\pi i/n)} \cdot \left(1 - \frac{1}{n^2}\right), \quad j = 1, 2, \ldots, n.$$

With the normalization $F_n'(0) > 0$, the maps F_n converge to the identity since $\sum_{j=1}^{n}(1 - |z_j|^2) \to_{n\to\infty} 0$. Recall that for $x \in S^1$, one has the formula $|F_n'(x)| = 1 + \sum_{j=1}^{n} P_{z_j}(x)$, where P_z is the Poisson kernel as viewed from $z \in U$ (see Theorem 4.15 in [71]). Computations show $\int_{I_j} \log|F_n'|\,dm \geq \int_{I_j} \log|1 + P_{z_j}|\,dm \gtrsim \frac{1}{n}$, where I_j consists of the points on S^1 for which the closest zero is z_j. Hence, $|\mathrm{Out}\,F_n'(0)| = \exp \int_{S^1} \log|F_n'|\,dm > c > 1$ for some constant c independent of $n \geq 1$. Since the outer parts $\mathrm{Out}\,F_n'$ do not converge to the constant 1, neither can the inner parts $\mathrm{Inn}\,F_n'$.

Understanding the image (Section 5 in [54], p. 18).

In this section, we consider the image of the map $F \to \mathrm{Inn}\,F'$ $(F \in \mathcal{J})$. A complete description of the image will be given in the following section.

Wedging F_μ with F_C (Section 5.1 in [54], p. 19).

Theorem 5.1 ([54], p. 19). (i) *Suppose $F_\mu \in \mathcal{J}$ is an inner function with $\mathrm{Inn}\,F_\mu' = S_\mu$. Let $F_{\mu,C} = F_\mu \wedge F_C$ where C is a Blaschke sequence. Then $\mathrm{Inn}\,F_{\mu,C}' = B_C S_\mu$.*
(ii) *Conversely, if $F_{\mu,C} \in \mathcal{J}$ is an inner function with $\mathrm{Inn}\,F_{\mu,C}' = B_C S_\mu$, then there exists an inner function F_μ with $\mathrm{Inn}\,F_\mu' = S_\mu$.*

Reminder. F_C is the maximal Blaschke product with critical set C. As for wedging: given two inner functions $F, G \in \mathcal{J}$, we consider the family $\Phi_{F,G}$ of SK-metrics that are pointwise less than $\min\{\lambda_F, \lambda_G\}$. This is a nonempty Perron family (the metric $|\mathrm{Inn}\,F'| \cdot |\mathrm{Inn}\,G'| \cdot \lambda_U$ is in $\Phi_{F,G}$). Taking the supremum of the conformal metrics in $\Phi_{F,G}$ (Heinz), we get a regular conformal metric of constant curvature -4. By Liouville's theorem, it is the pullback of λ_U by a holomorphic function $U \to U$ which is by the definition, the wedge $F \wedge G$. Lemma 3.6 shows that $F \wedge G \in \mathcal{J}$.

A proof of Theorem 5.1.

(i) Since $\lambda_{F_C} \geq \lambda_{F_{\mu,C}} \geq |B_C|\lambda_{F_\mu}$, the critical set of $F_{\mu,C}$ is precisely C with the correct multiplicities; while the inequalities $\lambda_{F_\mu} \geq \lambda_{F_{\mu,C}} \geq |B_C|\lambda_{F_\mu}$ show that $\sigma(F_{\mu,C}') = \mu$ (one divides by F_μ, integrates over rS^1, tends $r \to 1^-$ and applies Lemma 3.5). Hence, $\mathrm{Inn}\,F_{\mu,C}' = B_C \cdot S_\mu$.

(ii) Suppose $F_{\mu,C} \in \mathcal{J}$ is an inner function with $\operatorname{Inn} F'_{\mu,C} = B_C \cdot S_\mu$. Let F_n be a sequence of finite Blaschke products which converges $F_{\mu,C}$ (stability is not required in this proof). For any $0 < r < 1$, we can form the sequence of finite Blaschke products $F_{n,r}$ by removing the critical points from F_n that lie in the ball $r \cdot U$, and considering the maximal Blaschke product with the remaining critical points (normalized by $F_{n,r}(0) = 0$ and $F'_{n,r}(0) > 0$). For each r, we pick a subsequential limit F_r of $F_{n,r}$. We may then extract a further subsequential limit F by taking $r \to 1^-$. By construction, we have $|B_C|\lambda_F \leq \lambda_{F_{\mu,C}} \leq \lambda_F$. Since the limit F cannot be constant, by Hurwitz' theorem, F has no critical points. The above inequalities imply $\sigma(F') = \mu$ and therefore $\operatorname{Inn} F' = S_\mu$. $\qquad\square$

Division and product rules (Section 5.2 in [54], p. 19).

Lemma 5.2 ([54], p. 19). *Suppose that $F_{C_n} \to F_{\mu_1+\mu_2}$ is a stable sequence, $C_{1,n} \subset C_n$ is a subset such that $B_{C_{1,n}}$ converges to S_{μ_1}. Then, $F_{C_{1,n}} \to F_{\mu_1}$.*

Proof. Write $C_n = C_{1,n} \cup C_{2,n}$. From the assumptions, $B_{C_{1,n}} \to S_{\mu_1}$ and $B_{C_{2,n}} \to S_{\mu_2}$. After passing to a subsequence, we can ensure convergence: $F_{C_{1,n}} \to F_{\nu_1}$, $\nu_1 \leq \mu_1$, $F_{C_{2,n}} \to F_{\nu_2}$, $\nu_2 \leq \mu_2$. The monotonicity of measures follows from Theorem 4.1. For each n, we have $\lambda_{F_{C_n}} \geq |B_{1,n}| \cdot \lambda_{F_{C_{2,n}}}$ and therefore, after taking $n \to \infty$, we see that $\lambda_{F_{\mu_1+\mu_2}} \geq |S_{\mu_1}| \cdot \lambda_{F_{\nu_2}}$. As is now standard, we may deduce $\mu_1 + \mu_2 \leq \nu_1 + \nu_2$. By examining the inequality

$$0 \leq \lim_{r \to 1^-} \int_{rI} \log \frac{\lambda_{F_{\mu_1+\mu_2}}}{|S_{\mu_1}|\lambda_{F_{\nu_2}}} dm, \quad I \subset S^1.$$

Hence, $\nu_2 = \mu_2$ (and similarly $\nu_1 = \mu_1$). $\qquad\square$

Now a number of consequences.

Corollary 5.3 ([54], p. 20). *If a measure μ is constructable, i.e., if F_μ exists, then all $\nu \leq \mu$ are constructable. In particular, the image of the mapping $F \to \operatorname{Inn} F'$ is closed under taking divisors.*

(Indeed, given a stable approximation F_{C_n} to F_μ, we can select a $C_{1,n} \subset C_n$ so that $B_{C_{1,n}} \to S_\nu$.)

Corollary 5.4 ([54], p. 20). *If F_{μ_1} and F_{μ_2} are constructable, then $F_{\mu_1+\mu_2}$ is constructable.*

The proof relies on the Solynin-type estimate

$$\lambda_{F_{C_1}} \cdot \lambda_{F_{C_2}} \geq \lambda_{F_{C_1 \cup C_2}} \cdot \lambda_{F_{C_1 \cap C_2}}, \tag{62.0.19}$$

valid when C_1 and C_2 are finite subsets of the unit disk counted with multiplicities. The proof of the inequality (62.0.19) is essentially the proof of Lemma 2.8 in [63]. We sketch the details. Consider the function

$$u = \log^+ \left(\frac{\lambda_{F_{C_1 \cup C_2}} \cdot \lambda_{F_{C_1 \cap C_2}}}{\lambda_{F_{C_1}} \cdot \lambda_{F_{C_2}}} \right).$$

We claim that it is subharmonic and non-negative in U, yet tends to 0 as $|z| \to 1^-$. This will show it is equal to 0 identically. It is clearly non-negative by definition $(\log^+)$. To show that $u(z)$ is subharmonic, one can check that $\triangle u \geq 0$ (The reference above). For the last statement, note that by Lemma 2.3, for a Blaschke product F, the quotient $\lambda_F / \lambda_U \to 1$ uniformly as $|z| \to 1^-$.

A proof of Corollary 5.4. Choose approximations $F_{C_{1,n}} \to F_{\mu_1}$ and $F_{C_{2,n}} \to F_{\mu_2}$ by finite Blaschke products. Making a small perturbation if necessary, we can assume that the sets $C_{1,n}$ and $C_{2,n}$ are disjoint. Let $C_n = C_{1,n} \cup C_{2,n}$ be their union. Passing to a subsequence, we may assume that $F_{C_n} \to F_\mu$ for some measure on S^1. By Solynin's estimate (62.0.19), we have

$$\log \frac{\lambda_U}{\lambda_{F_{C_{1,n}}}} + \log \frac{\lambda_U}{\lambda_{F_{C_{2,n}}}} \leq \log \frac{\lambda_U}{\lambda_{F_{C_n}}}. \tag{62.0.20}$$

Taking $n \to \infty$ gives

$$\log \frac{\lambda_U}{\lambda_{F_{\mu_1}}} + \log \frac{\lambda_U}{\lambda_{F_{\mu_2}}} \leq \log \frac{\lambda_U}{\lambda_{F_\mu}}. \tag{62.0.21}$$

By examining averages over rI and taking $r \to 1^-$, we discover that $\mu \geq \mu_1 + \mu_2$. Applying Corollary 5.3 shows that the measure $\mu_1 + \mu_2$ is constructable. $\square$

Remark 62.0.146. Solynin's original estimate compares hyperbolic metrics on two domains in the plane with the hyperbolic metrics on their union and intersection:

$$\lambda_{\Omega_1}(z) \cdot \lambda_{\Omega_2}(z) \geq \lambda_{\Omega_1 \cup \Omega_2}(z) \cdot \lambda_{\Omega_1 \cap \Omega_2}(z), \quad z \in \Omega_1 \cap \Omega_2.$$

See [95, 96].

Corollary 5.5 ([54], p. 21). *If $S'_\mu \in \mathcal{N}$, then μ is constructable.*

This follows from the division rule (Corollary 5.3) and the fact that S_μ divides S'_μ.

Invisible measures (Section 5.3 in [54], p. 21).

Let μ be a finite positive measure on S^1, which is singular with respect to the Lebesgue measure. We say that μ is invisible if for any measure $0 < \nu \le \mu$, there does not exist a function $F_\nu \in \mathcal{J}$ with $Inn\, F'_\nu = S_\nu$.

Lemma 5.6 ([54], p. 22). *Either the map $F \to Inn\, F'$, $\mathcal{J}/\mathrm{Aut}(U) \to Inn/S^1$, is surjective or there exists an invisible measure.*

Proof. Suppose that F_μ is not constructable. Since the hull of the metric $|S_\mu| \cdot \lambda_U$ cannot vanish anywhere, it must be of the form λ_{F_ν} for some measure ν (Lemma 3.6 explains why F_ν must be inner). Applying Lemma 3.4, we see that $\nu < \mu$ since equality cannot hold. From the product rule (Corollary 5.4), it follows that the measure $\mu - \nu$ is invisible. More precisely, if $0 < \sigma \le \mu - \nu$ was constructable, then $\lambda_{F_\nu} > \lambda_{F_{\nu+\sigma/2}} > |S_\mu| \cdot \lambda_U$ would contradict the definition of ν. $\qquad\square$

The above argument shows a little more in the following:

Theorem 5.7 ([54], p. 22). *A measure μ is invisible if and only if the hull of $|S_\mu| \cdot \lambda_U$ is the Poincaré metric. More generally, any measure μ can be uniquely decomposed into a constructable part and an invisible part: $\mu = \mu_{con} + \mu_{inv}$, in which case, the hull of $|S_\mu| \cdot \lambda_U$ is $\lambda_{F_{\mu_{con}}}$.*

We can now prove the countable version of the product rule (Lemma 1.6). Suppose we are given countably many constructable measures μ_j, $j = 1, 2, \ldots$ such that their sum $\mu = \sum_{j=1}^{\infty} \mu_j$ is a finite measure. We claim that μ is constructable. According to Theorem 5.7, the hull of $|S_\mu| \cdot \lambda_U$ is of the form λ_{F_ν} for some measure $\nu \le \mu$. However, from Corollary 5.4, we know that $\widetilde{\mu_j} = \mu_1 + \cdots + \mu_j$ is constructable. This shows that $\nu \ge \widetilde{\mu_j}$ for any j, which forces $\nu = \mu$ (ν is μ_{con} by Theorem 5.7).

A criterion for invisibility (Section 5.4 in [54], p. 22).

In this section, we consider only conformal metrics with strictly positive densities, that is, genuine metrics instead of pseudo-metrics. Given a positive continuous function u on $S_r = rS^1$, $0 < r < 1$, let $\Lambda_r[u]$ denote the unique conformal metric of curvature -4 on $U_r = r \cdot U$ which agrees with u on S_r. $\Lambda_r[u]$ exists and is unique ([42], Section 12). For a non-vanishing SK-metric λ, we will write $\Lambda[\lambda] = \widehat{\lambda}$ for the minimal conformal metric of curvature -4 that exceeds λ.

Lemma 5.8 ([54], p. 23). *The operation $u \to \Lambda_r[u]$ is monotone in u, that is, if $u \ge v$ then $\Lambda_r[u] \ge \Lambda_r[v]$.*

To see this, note that the function $h = \log^+(\Lambda_r[v]/\Lambda_r[u])$ is non-negative ($\log^+$), subharmonic and identically 0 on S_r. To check that h is subharmonic, we use the definition of curvature and Kato's inequality (proposition 6.6 in [87]):

$$\triangle h \geq \left(4\Lambda_r[v]^2 - 4\Lambda_r[u]^2\right) \cdot \chi_{v>u} \geq 0.$$

A similar argument shows the following:

Lemma 5.9 ([54], p. 23). *Let λ be a non-vanishing conformal metric on the unit disk of curvature at most -4. For $0 < r < 1$, the metric $\Lambda_r[\lambda(re^{i\theta})]$ is the minimal conformal metric of curvature -4 that exceeds λ on U_r. The family of conformal metrics $\Lambda_r[\lambda(re^{i\theta})]$ is non-decreasing in r, and the limit*

$$\widehat{\lambda} = \Lambda[\lambda] = \lim_{r \to 1^-} \Lambda_r[\lambda(re^{i\theta})] \tag{62.0.22}$$

is the minimal conformal metric of curvature -4 that exceeds λ on U.

In general, it is difficult to evaluate $\Lambda_r[u]$ explicitly. In the following lemma, we do so when u is a constant function.

Lemma 5.10 ([54], p. 23). *Given any $0 < c \leq 1$, there exists a unique $0 < r' \leq r$ so that $\Lambda_r[c \cdot \lambda_U] = L * \lambda_U$ where $L(z) = \frac{r'}{r} \cdot z$ is the linear map $U_r \to U_{r'}$.*

This follows by observing that since the metrics $(L_{r'}) * \lambda_U$ are increasing in r', there is a unique value of r' for which $c \cdot \lambda_U = (L_{r'}) * \lambda_U$ on S_r.

Corollary 5.11 ([54], p. 23). *We have*

$$\lim_{c \to \infty} \left[\lim_{r \to 1^-} \frac{\Lambda_r[c]}{\lambda_U} \right] = 1$$

uniformly on compact subsets of U.

Finally, we have the following theorem.

Theorem 5.12 ([54], p. 24). *Suppose μ is a singular measure on S^1 which satisfies $\mu(I) \leq c|I| \log(1/|I|)$, for any interval $I \subset S^1$ and some constant $c > 0$. Then, μ is invisible.*

Proof. From the product rule (Corollary 5.4), it is possible to show that a measure μ is invisible if and only if $\epsilon \cdot \mu$ is for any $\epsilon > 0$. This allows us to assume that $\mu(I) \leq \epsilon|I| \log(1/|I|)$ which implies that the Poisson extension

$$P_\mu(z) \leq \epsilon \left(A \cdot \log \frac{1}{1-|z|} + B \right)$$

for some constants A and B. Hence, $|S_\mu| \cdot \lambda_U \to \infty$ as $|z| \to 1^-$. The theorem follows from the monotonicity principle (Lemma 5.8) and Corollary 5.11. $\qquad\square$

Roberts decompositions (Section 6 in [54], p. 24).

In this section, we show that if μ does not charge Beurling–Carleson sets, then it is invisible, that is, any measure $0 < \nu \leq \mu$ cannot be in the image of the map $(F \to Inn\, F')\, F \to \sigma(F')$. To upgrade the argument of Section 5.4, we will use the following theorem which is implicit in [88].

Theorem 6.1 ([54], p. 24). *Suppose μ is a measure on the unit circle which does not charge Beurling–Carleson sets. Given a real number $c > 0$ and an integer $j_0 \geq 1$, μ can be expressed as a countable sum*

$$\mu = \sum_{j=1}^{\infty} \mu_j, \tag{62.0.23}$$

where each μ_j enjoys the following estimate on the modulus of continuity:

$$\omega_{\mu_j}\left(\frac{1}{n_j}\right) \leq \frac{c}{n_j} \cdot \log n_j, \quad n_j = 2^{j+j_0}. \tag{62.0.24}$$

Here, $\omega_\mu(t) = \sup_{I \subset S^1} \mu(I)$, with the supremum taken over all intervals I of length t.

It will be important for us that the measure μ admits infinitely many decompositions with different parameters c and j_0, where c can be made arbitrarily small and j_0 arbitrarily large. We recall the proof.

Proof. For each $j = 1, 2, \ldots$, we can define a partition P_j of the unit circle into n_j arcs of equal length (we consider half-open intervals which contain only one of the endpoints, for example, the left endpoint). Since n_j divides n_{j+1}, each next partition can be chosen to be a refinement of the previous one. Given a measure μ on S^1, Roberts defines the grating of μ with respect to the sequence of partitions $\{P_j\}$. This procedure decomposes $\mu = \sum_{j=1}^{\infty} \mu_j + \nu$ so that equation (62.0.24) holds for each j, with the residual measure ν supported on a Beurling–Carleson set. To define μ_1, consider the intervals in the partition P_1. Call an interval $I \in P_1$ light if $\mu(I) \leq \frac{c}{n_1} \log n_1$ and heavy otherwise. On a light interval, take $\mu_1 = \mu$, while on a heavy interval, let μ_1 be a multiple of μ so that the mass $\mu_1(I) = \frac{c}{n_1} \log n_1$. Clearly, $\mu_1 \leq \mu$. Consider the difference $\mu - \mu_1$ and grant it with respect to the partition P_2 to form the measure μ_2. Then consider $\mu - \mu_1 - \mu_2$ and grant it with respect to P_3 to form μ_3, and so on. Continuing in this way, we obtain sequence of measures $\mu_1, \mu_2, \ldots$ where each next measure is supported on the heavy intervals of the previous generation. By construction, the bound (62.0.24) holds for all j. Since the residual measure ν is supported on the set of points which lie in heavy intervals at every stage, $\operatorname{supp}\nu \subset S^1 - \mathcal{L}$, where $\mathcal{L}$ is the union of interiors of light intervals of any generation. The relation $\log n_{j+1} = 2\log n_j$ implies that $S^1 - \mathcal{L}$ is a Beurling–Carleson set:

$$\sum_{\text{light}} |I| \log \frac{1}{|I|} \lesssim 2^{j_0} + \sum_{\text{heavy}} |J| \log \frac{1}{|J|}$$

$$\lesssim 2^{j_0} + \sum_{j=0}^{\infty} \sum_{J \in P_j \text{ heavy}} \mu_j(J) \le 2^{j_0} + \mu(S^1),$$

since any maximal light interval of generation $j \ge 2$ is contained in a heavy interval of the previous generation. $\qquad\square$

The estimate (62.0.24) on the modulus of continuity is easily seen to be equivalent to an estimate on the Poisson extension:

$$|P_{\mu_j}| \le c' \cdot \log \frac{1}{1 - |z|^2}, \quad z \in B\left(0, 1 - \frac{1}{n_j}\right). \tag{62.0.25}$$

Here, the constant c' can be taken to be $c \cdot c_1$ for some $c_1 > 0$. This is stated in Lemma 2.2 of Roberts paper [88]. We will also need a simple lemma on conformal metrics.

Lemma 6.2 ([54], p. 25). (i) *For any two singular measures μ_1 and μ_2 on the unit circle,*

$$\Lambda[|S_{\mu_1}| \cdot \Lambda[|S_{\mu_2}| \cdot \lambda_U]] = \Lambda[|S_{\mu_1}||S_{\mu_2}| \cdot \lambda_U].$$

(ii) *More generally,*

$$\Lambda[|S_{\mu_1}| \cdot \ldots \cdot \Lambda[|S_{\mu_{j-1}}| \cdot \Lambda[|S_{\mu_j}| \cdot \lambda_U]]\ldots] = \Lambda[|S_{\mu_1}| \cdot |S_{\mu_2}| \cdot \ldots \cdot |S_{\mu_j}| \cdot \lambda_U].$$

(iii) *For $\mu = \sum_{j=1}^{\infty} \mu_j$, we have*

$$\lim_{j \to \infty} \Lambda[|S_{\mu_1}| \cdot \ldots \cdot \Lambda[|S_{\mu_{j-1}}| \cdot \Lambda[|S_{\mu_j}| \cdot \lambda_U]]\ldots] = \Lambda[|S_\mu| \cdot \lambda_U].$$

Proof. (i) The "$\ge$" direction follows from the monotonicity of Λ. For the "$\le$" direction, it suffices to show that $|S_{\mu_1}| \cdot \Lambda[|S_{\mu_2}|\lambda_U] \le \Lambda[|S_{\mu_1}||S_{\mu_2}| \cdot \lambda_U]$, or $|S_{\mu_1}| \cdot \Lambda_r[|S_{\mu_2}|\lambda_U] \le \Lambda_r[|S_{\mu_1}||S_{\mu_2}| \cdot \lambda_U]$ for any $0 < r < 1$ (see Lemma 5.9). To this end, we form the function

$$u_r = \log^+\left(\frac{|S_{\mu_1}| \cdot \Lambda_r[|S_{\mu_2}|\lambda_U]}{\Lambda_r[|S_{\mu_1}||S_{\mu_2}| \cdot \lambda_U]}\right)$$

defined on $U_r = r \cdot U$. Since it is subharmonic and vanishes on $S_r = \partial U_r$, it must be identically 0. This proves the "$\leq$" direction.

(ii) Follows after applying (i) $j - 1$ times.

(iii) Let $\widetilde{\mu_j} = \mu_1 + \cdots + \mu_j$. By part (i), we have

$$|S_{\mu - \widetilde{\mu_j}}| \cdot \Lambda[|\widetilde{\mu_j}| \cdot \lambda_U] \leq \Lambda[|S_\mu| \cdot \lambda_U] \leq \Lambda[|\widetilde{\mu_j}| \cdot \lambda_U].$$

Since $|S_{\mu - \widetilde{\mu_j}}| \to 1$, $\Lambda[|S_{\widetilde{\mu_j}}| \cdot \lambda_U]$ are decreasing and converge to $\Lambda[|S_\mu| \cdot \lambda_U]$. $\qquad\qquad\square$

Now, we have the following:

A proof of Theorem 1.10.

Step 1. Let $\mu = \mu_j$ be the Roberts decomposition (62.0.23) with parameters c and j_0 to be chosen later. In view of the invisibility criterion (Theorem 5.7), it suffices to show that

$$\lambda_j := \Lambda_{1-(1/n_1)}[|S_{\mu_1}| \cdot \ldots \cdot \Lambda_{1-(1/n_{j-1})}[|S_{\mu_{j-1}}| \cdot \Lambda_{1-(1/n_j)}[S_{\mu_j}| \cdot \lambda_U]] \ldots]$$
$$(62.0.26)$$

is close to the hyperbolic metric at the origin, uniformly in $j \geq 1$. Indeed, by the monotonicity properties of Λ, we have

$$\lambda_j \leq \Lambda[|S_{\mu_1}| \cdot \ldots \cdot \Lambda[|S_{\mu_{j-1}}| \cdot \Lambda[|S_{\mu_j}| \cdot \lambda_U]] \ldots], \qquad (62.0.27)$$

so that if λ_j is close to λ_U, then so must $\Lambda[|S_{\mu_1}||S_{\mu_2}| \ldots |S_{\mu_j}|\lambda_U]$ be.

Step 2. The estimate on the modulus of continuity of μ_j implies that $|S_{\mu_j}|\lambda_U \geq \lambda_U^{4/5}$ on $S_{1-(1/n_j)}$. Here, we use the fact that we can choose $c' < \frac{1}{10}$ in (62.0.25). We claim that

$$\Lambda_{1-(1/n_j)}[|S_{\mu_j}| \cdot \lambda_U] \geq \left(\frac{1}{2}\right)\lambda_U \ \text{ on } S_{1-(1/n_{j-1})}. \qquad (62.0.28)$$

Assuming (62.0.28), we have $|S_{\mu_{j-1}}| \cdot \Lambda_{1-(1/n_j)}[|S_{\mu_j}|\lambda_U] \geq \lambda_U^{4/5}$ on $S_{1-(1/n_{j-1})}$. We could then inductively show that $\lambda_j \geq (\frac{1}{2})\lambda_U$ on $S_{1-(1/n_1)}$. By Corollary 5.11, this would mean that λ_j is very close to λ_U at the origin, provided n_1 is large (this is where we use that j_0 can be made arbitrarily large).

Step 3. Thus, we need to show (62.0.28). Define $\epsilon > 0$ by $1 - (1/n_j) = 1 - \epsilon$ so that $1 - (1/n_{j-1}) = 1 - \epsilon^{1/2}$. There exists a unique $0 < l < 1$ so that

$\Lambda_{1-(1/n_j)}[\lambda_U^{4/5}] = L*\lambda_U$ where $L(z) = l \cdot z$. Inspection shows that $1-l \asymp \epsilon^{4/5}$. Therefore,

$$\Lambda_{1-(1/n_j)}[\|S_{\mu_j}|\lambda_U] \geq \Lambda_{1-(1/n_j)}[\lambda_U^{4/5}] = \frac{l}{1 - |l \cdot z|^2} \geq \left(\frac{1}{2}\right) \cdot \lambda_U \text{ on } S_{1-(1/n_{j-1})},$$

as desired. $\qquad\qquad\qquad\qquad\qquad\qquad\qquad\qquad\qquad\qquad\qquad\qquad\qquad\square$

Chapter 03

Computations of the Modulus of Continuity of Inner Functions

The notion is used for functions and is used for measures (which are functions as well but the definition here is a bit different). We recall both. For functions we take the definition from [22, p. 71]: Let $\phi(x)$ be a complex-valued function defined on $-\infty < x < \infty$ and periodic with period 2π. The modulus of continuity of ϕ is the function

$$\omega(t) = \omega(t; \phi) = \sup_{|x-y| \leq t} |\phi(x) - \phi(y)|.$$

Thus, ϕ is continuous if and only if $\omega(t) \to 0$ as $t \to 0$. We say that ϕ is of class Λ_α $(0 < \alpha \leq 1)$ if $\omega(t) = O(t^\alpha)$ as $t \to 0$. Alternatively, Λ_α is the class of functions which satisfy a Lipschitz condition of order α: $|\phi(x) - \phi(y)| \leq A \cdot |x - y|^\alpha$. Clearly, $\Lambda_\beta \subset \Lambda_\alpha$ if $\alpha < \beta$.

For measures, we take the definition from [54, p. 24]: Suppose μ is a measure on S^1. The modulus of continuity of μ is the function

$$\omega_\mu(t) = \sup_{I \subset S^1} \mu(I),$$

with the supremum taken over all intervals I of length t. We would like to find the relation between the modulus of continuity of a singular inner function and that of its defining singular measure. Let $S(z)$ be analytic in $|z| < 1$ and with the properties $0 < |S(z)| \leq 1$, $|S(e^{i\theta})| = 1$ a.e., $S(0) > 0$. This implies that $-\log |S(z)|$ is a positive harmonic function which vanishes almost everywhere on the boundary of the unit disk. Thus, by the Herglotz representation and by the following well-known theorem, we get the following:

Theorem. *Let $u(z)$ be a Poisson–Stieltjes integral of the form*

$$u(z) = u(re^{i\theta}) = \frac{1}{2\pi} \int_0^{2\pi} P(r, \theta - t)d\mu(t), \quad P(r, \theta) = \frac{1 - r^2}{1 - 2r\cos\theta + r^2},$$

where $\mu(t)$ is of bounded variation on $[0, 2\pi]$. If the symmetric derivative $D\mu(\theta_0)$ exists at a point θ_0, then the radial limit $\lim_{r \to 1-} u(re^{i\theta_0})$ exists and has the value $D\mu(\theta_0)$.

The function $-\log|S(z)|$ can be represented as a Poisson–Stieltjes integral with respect to a bounded non-decreasing function $\mu(t)$ and $\mu'(t) = 0$ a.e. Since $S(0) > 0$, analytic completion gives

$$S(z) = \exp\left\{-\int_0^{2\pi} \frac{e^{it} + z}{e^{it} - z} d\mu(t)\right\}$$

$\mu(t)$ being bounded non-decreasing singular function ($\mu'(t) = 0$ a.e.), $S(z)$ is called a singular inner function. To make sense of $|S(x) - S(y)|$ for $|x - y| < t$, where x and y are real numbers, we naturally want to restrict our estimate to a circle concentric with S^1, say $|z| = r$ for a fixed radius $0 \le r \le 1$. Thus, we are thinking of $|S(re^{ix}) - S(re^{iy})|$ for $|x - y| < t \le 2\pi$.

The case $r = 1$: This radius is especially convenient because $|S(e^{it})| = 1$ a.e. That allows for the following computation.

$$|S(e^{ix}) - S(e^{iy})| = \left|S(e^{iy})\left(\frac{S(e^{ix})}{S(e^{iy})} - 1\right)\right| = \left|\frac{S(e^{ix})}{S(e^{iy})} - 1\right|$$

$$= \left|\exp\left\{-\int_0^{2\pi}\left(\frac{e^{it} + e^{ix}}{e^{it} - e^{ix}} - \frac{e^{it} + e^{iy}}{e^{it} - e^{iy}}\right)d\mu(t)\right\} - 1\right|$$

$$= \left|\exp\left\{-\int_0^{2\pi} \frac{2e^{it}(e^{ix} - e^{iy})}{(e^{it} - e^{ix})(e^{it} - e^{iy})} d\mu(t)\right\} - 1\right|.$$

A straightforward calculation gives the following identities:

$$\frac{2e^{it}(e^{ix} - e^{iy})}{(e^{it} - e^{ix})(e^{it} - e^{iy})} = i\left(\cot\left(\frac{x - t}{2}\right) - \cot\left(\frac{y - t}{2}\right)\right)$$

$$= i\frac{\sin\left(\frac{y - x}{2}\right)}{\sin\left(\frac{x - t}{2}\right)\sin\left(\frac{y - t}{2}\right)}.$$

So,

$$|S(e^{ix}) - S(e^{iy})| = \left| \exp\left\{ - \int_0^{2\pi} i \frac{\sin\left(\frac{y-x}{2}\right) d\mu(t)}{\sin\left(\frac{x-t}{2}\right) \sin\left(\frac{y-t}{2}\right)} \right\} - 1 \right|$$

$$= 2 \left| \sin\left\{ \frac{1}{2} \int_0^{2\pi} \frac{d\mu(t)}{\sin\left(\frac{x-t}{2}\right) \sin\left(\frac{y-t}{2}\right)} \cdot \sin\left(\frac{y-x}{2}\right) \right\} \right|$$

We proved the following:

Lemma 63.0.147. *Let the singular inner function $S(z)$ be given by*

$$S(z) = \exp\left\{ - \int_0^{2\pi} \frac{e^{it} + z}{e^{it} - z} d\mu(t) \right\},$$

where $\mu(t)$ is non-decreasing singular function ($\mu'(t) = 0$ a.e.). Then for $x, y \in [0, 2\pi]$ we have the identity

$$|S(e^{ix}) - S(e^{iy})| = 2 \left| \sin\left\{ \frac{1}{2} \int_0^{2\pi} \frac{d\mu(t)}{\sin\left(\frac{x-t}{2}\right) \sin\left(\frac{y-t}{2}\right)} \cdot \sin\left(\frac{y-x}{2}\right) \right\} \right|.$$

The computation above is an Eulerian computation in the sense that we completely ignored convergence of integrals, etc. so that the formula in Lemma 63.0.147 might be faulty as is. More concretely, for a general non-decreasing, singular $\mu(t)$ of a bounded variation, the expression within the outer sine

$$\frac{1}{2} \int_0^{2\pi} \frac{d\mu(t)}{\sin\left(\frac{x-t}{2}\right) \sin\left(\frac{y-t}{2}\right)} \cdot \sin\left(\frac{y-x}{2}\right)$$

will not tend to zero when $x - y \to 0$, for the singular inner function $S(z)$ can have a boundary function $S(e^{ix})$ which is not continuous. Thus, the integral

$$\int_0^{2\pi} \frac{d\mu(t)}{\sin\left(\frac{x-t}{2}\right) \sin\left(\frac{y-t}{2}\right)}$$

might diverge to infinity or otherwise, at certain points where $x - y \to 0$ in a rate greater than or equal to $\frac{1}{x-y}$.

The general case $r < 1$: In this case, we have by the integral representation of $S(z)$:

$$|S(re^{ix}) - S(re^{iy})| = \left| \exp\left\{ - \int_0^{2\pi} \frac{e^{it} + re^{ix}}{e^{it} - re^{ix}} d\mu(t) \right\} \right.$$

$$\left. - \exp\left\{ - \int_0^{2\pi} \frac{e^{it} + re^{iy}}{e^{it} - re^{iy}} d\mu(t) \right\} \right|.$$

Chapter 64

Inversion of Power Series

We recall a result of Frostman that any inner function on U can be approximated (uniformly on compact subsets of U) by Blaschke products. In the event that the approximated inner function is singular (i.e., does not have a zero in U) this looks, at first, surprising (given that Blaschke products have usually lots of zeros in U). In Chapters 64–70, we point out to a similar phenomenon, that originates in arithmetical reasons. In particular we indicate in Chapter 70 on an expansion of the function e^{-z} in U into an infinite product. The partial products have zeros and poles. The results in this topic are probably well known and old. Nevertheless, it is fun to go through the details from scratch. Let us suppose that in a neighborhood of the point $z = 0$ we have the following power series expansion:

$$f(z) = z + a_2 z^2 + a_3 z^3 + \cdots.$$

Then we can invert the power series in the sense that we will get the following expansion:

$$z = \alpha_1 f(z) + \alpha_2 f(z^2) + \alpha_3 f(z^3) + \cdots.$$

The coefficients α_n, $n = 1, 2, 3, \ldots$, satisfy the following recursion:

$$\begin{cases} \alpha_1 = 1, \\ \alpha_n = -\sum_{k=1}^{n-1} \alpha_k a_{n/k}^*, & n = 2, 3, \ldots. \end{cases}$$

Here,

$$a_{n/k}^* = \begin{cases} a_{n/k}, & k \mid n, \\ 0, & k \nmid n. \end{cases}$$

Equivalently, we have the recursion

$$\begin{cases} \alpha_1 = 1, \\ \alpha_n = -\sum_{\{d:\, 1 \leq d < n,\ d|n\}} \alpha_d a_{n/d}, \quad n = 2, 3, \ldots. \end{cases}$$

Let us write down explicitly the first few α_n coefficients, in terms of the original coefficients a_k:

$$\alpha_1 = 1 \qquad \alpha_{11} = -a_{11}$$

$$\alpha_2 = -a_2 \qquad \alpha_{12} = -a_{12} + 2a_2 a_6 + 2a_3 a_4 - 3a_2^2 a_3$$

$$\alpha_3 = -a_3 \qquad \alpha_{13} = -a_{13}$$

$$\alpha_4 = -a_4 + a_2^2 \qquad \alpha_{14} = -a_{14} + 2a_2 a_7$$

$$\alpha_5 = -a_4 \qquad \alpha_{15} = -a_{15} + 2a_3 a_5$$

$$\alpha_6 = -a_6 + 2a_2 a_3 \qquad \alpha_{16} = -a_{16} + 2a_2 a_8 + a_4^2 - 3a_2^2 a_4 + a_2^4$$

$$\alpha_7 = -a_7 \qquad \alpha_{17} = -a_{17}$$

$$\alpha_8 = -a_8 + 2a_2 a_4 - a_2^3 \qquad \alpha_{18} = -a_{18} + 2a_2 a_9 + 2a_3 a_6 - 3a_2 a_3^2$$

$$\alpha_9 = -a_9 + a_3^2 \qquad \alpha_{19} = -a_{19}$$

$$\alpha_{10} = -a_{10} + 2a_2 a_5 \qquad \alpha_{20} = -a_{20} + 2a_2 a_{10} + 2a_4 a_5 - 3a_2^2 a_5 \ldots.$$

We will now give few examples of inversions of power series:

Example 64.0.148.

$$f(z) = \frac{z}{1-z}.$$

In this case, we are forced to think of the domain $|z| < 1$. Otherwise, we do not deal with a power series that begins with z. Thus $a_n = 1$, $n = 1, 2, 3, \ldots$ and hence

$$\begin{cases} \alpha_1 = 1, \\ \alpha_n = -\sum_{\{d:\, 1 \leq d < n,\ d|n\}} \alpha_d, \quad n = 2, 3, \ldots. \end{cases}$$

Hence, $\alpha_n = \mu(n)$ the arithmetical function, known as the Möbius function. We proved the formal identity:

$$z = \sum_{n=1}^{\infty} \mu(n) \frac{z^n}{1 - z^n}.$$

Example 64.0.149.

$$f(z) = \frac{z}{1+z}.$$

We think of the domain $|z| < 1$. In this example, $a_n = (-1)^{n+1}$, $n = 1, 2, 3, \ldots$ and so

$$\begin{cases} \alpha_1 = 1, \\ \alpha_n = \sum_{\{d:\, 1 \le d < n,\, d|n\}} (-1)^{n/d} \alpha_d, \quad n = 2, 3, \ldots. \end{cases}$$

The following sequence of α coefficients is obtained:

$$1, 1, -1, 2, -1, -1, -1, 4, 0, -1, -1, -2, -1, -1, 1, \ldots.$$

If we denote by f_1 the function in example 64.0.148 and by f_2 the function in the current example, then $f_1(z) = -f_2(-z)$. Thus, we get the relation $\mu(n) = (-1)^{n+1} \alpha_n$, where α_n satisfies the recursion of the current example.

Example 64.0.150.

$$f(z) = -\log(1-z), \ |z| < 1.$$

Here, $a_n = \frac{1}{n}$, $n = 1, 2, 3, \ldots$ and we obtain

$$\begin{cases} \alpha_1 = 1, \\ \alpha_n = -\sum_{\{d:\, 1 \le d < n,\, d|n\}} \frac{d}{n} \alpha_d, \quad n = 2, 3, \ldots. \end{cases}$$

Using the recursion of example 64.0.148, we get $\alpha_n = \frac{\mu(n)}{n}$ and we proved the formal identity:

$$-z = \sum_{n=1}^{\infty} \frac{\mu(n)}{n} \log(1 - z^n).$$

Example 64.0.151.

$$f(z) = \frac{z - 2z^2}{1 - z}, \ |z| < 1.$$

In this example, $a_1 = 1$ and $a_n = -1$ for $n = 2, 3, \ldots$ so

$$\begin{cases} \alpha_1 = 1, \\ \alpha_n = \sum_{\{d:\, 1 \le d < n,\, d|n\}} \alpha_d, \quad n = 2, 3, \ldots. \end{cases}$$

This generates the following sequence of α_n:

$$1, 1, 1, 2, 1, 3, 1, 4, 2, 3, 1, \ldots.$$

We note that every α_n is a natural number and that for $n > 1$ we have $\alpha_n = 1$ if and only if $n = p$ a prime number. Also, $\alpha_n = 2$ if and only if $n = p^2$ the square of a prime number.

Remark 64.0.152. By example 64.0.150, it follows that

$$-\Re(z) = \sum_{n=1}^{\infty} \frac{\mu(n)}{n} \log|1 - z^n|.$$

That is an interesting equation indeed, for if we ignore convergence issues of the series involved and we take $z = i$ the imaginary unit, then $1 - z^n = 0$ if and only if $4|n$. But for values of n which are divisible by 4, we have $\mu(n) = 0$. So, the singular terms do not appear in the formal expansion. We obtain the following equation,

$$0 = \frac{1}{2} \log 2 \sum_{k=0}^{\infty} \frac{\mu(1 + 4k)}{1 + 4k} + \log 2 \sum_{k=0}^{\infty} \frac{\mu(2 + 4k)}{2 + 4k} + \frac{1}{2} \log 2 \sum_{k=0}^{\infty} \frac{\mu(3 + 4k)}{3 + 4k}.$$

Thus, we obtain

$$0 = \sum_{k=0}^{\infty} \frac{\mu(1 + 4k)}{1 + 4k} + 2 \sum_{k=0}^{\infty} \frac{\mu(2 + 4k)}{2 + 4k} + \sum_{k=0}^{\infty} \frac{\mu(3 + 4k)}{3 + 4k}.$$

Now, it is well known that

$$\sum_{n=1}^{\infty} \frac{\mu(n)}{n} = 0.$$

This is a non-trivial arithmetic theorem. Combined with the formal identities we displayed, it implies that

$$\sum_{k=1}^{\infty} \frac{\mu(4k)}{4k} = \sum_{k=1}^{\infty} \frac{\mu(4k - 2)}{4k - 2}. \tag{64.0.1}$$

We note that

$$\sum_{n=1}^{\infty} \frac{\mu(n)}{n} = \sum_{k=1}^{\infty} \frac{\mu(2k - 1)}{2k - 1} + \sum_{k=1}^{\infty} \frac{\mu(2k)}{2k}. \tag{64.0.2}$$

By the definition of the Möbius function, we have

$$\mu(2k) = \begin{cases} 0, & k = 2m, \\ -\mu(k), & k = 2m - 1. \end{cases}$$

So, we deduce that

$$\sum_{k=1}^{\infty} \frac{\mu(2k)}{2k} = -\frac{1}{2} \sum_{k=1}^{\infty} \frac{\mu(2k-1)}{2k-1}$$

and hence by that and by equation (64.0.2) we deduce that

$$0 = \sum_{n=1}^{\infty} \frac{\mu(n)}{n} = \frac{1}{2} \sum_{k=1}^{\infty} \frac{\mu(2k-1)}{2k-1},$$

so that

$$0 = \sum_{n=1}^{\infty} \frac{\mu(n)}{n} = \sum_{n=1}^{\infty} \frac{\mu(2n)}{2n} = \sum_{n=1}^{\infty} \frac{\mu(2n-1)}{2n-1}.$$

But

$$\sum_{n=1}^{\infty} \frac{\mu(2n)}{2n} = \sum_{n=1}^{\infty} \frac{\mu(4n)}{4n} + \sum_{n=1}^{\infty} \frac{\mu(4n-2)}{4n-2}$$

using equation (64.0.1) this implies that

$$0 = \sum_{n=1}^{\infty} \frac{\mu(4n)}{4n} = \sum_{n=1}^{\infty} \frac{\mu(4n-2)}{4n-2}$$

(of course $\mu(4n) = 0$ for every $n = 1, 2, 3, \ldots$). Finally, again, by example 64.0.150, we obtain

$$-Im(z) = \sum_{n=1}^{\infty} \frac{\mu(n)}{n} \arg|1 - z^n|.$$

Plugging into that once more $z = i$, we get formally

$$-1 = \sum_{k=1}^{\infty} \frac{\mu(4k-3)}{4k-3}(2m - \frac{1}{4})\pi + \sum_{k=1}^{\infty} \frac{\mu(4k-2)}{4k-2}(2m - 0)\pi$$

$$+ \sum_{k=1}^{\infty} \frac{\mu(4k-1)}{4k-1}(2m + \frac{1}{4})\pi,$$

where the integer m is determined by the branch of the *log* function. But

$$\sum_{k=1}^{\infty} \frac{\mu(4k-2)}{4k-2} = \sum_{k=1}^{\infty} \frac{\mu(4k-3)}{4k-3} + \sum_{k=1}^{\infty} \frac{\mu(4k-1)}{4k-1} = 0 \qquad (64.0.3)$$

and we have

$$-1 = \frac{\pi}{4} \sum_{k=1}^{\infty} \frac{\mu(4k-1)}{4k-1} - \frac{\pi}{4} \sum_{k=1}^{\infty} \frac{\mu(4k-3)}{4k-3}. \qquad (64.0.4)$$

These two equations (64.0.3) and (64.0.4) imply that

$$\sum_{k=1}^{\infty} \frac{\mu(4k-3)}{4k-3} = -\sum_{k=1}^{\infty} \frac{\mu(4k-1)}{4k-1} = \frac{2}{\pi}.$$

We sum up the above discussion in the following corollary.

Corollary 64.0.153. *Assuming the next series are convergent, we have*

$$\left\{ \sum_{n=1}^{\infty} \frac{\mu(n)}{n} = 0, \right.$$

(that is a well-known non-trivial theorem in number theory)

$$\begin{cases} \sum_{n=1}^{\infty} \frac{\mu(2n-1)}{2n-1} = 0, \\[2ex] \sum_{n=1}^{\infty} \frac{\mu(2n)}{2n} = 0, \end{cases}$$

$$\begin{cases} \sum_{n=1}^{\infty} \frac{\mu(4n-3)}{4n-3} = \frac{2}{\pi}, \\[2ex] \sum_{n=1}^{\infty} \frac{\mu(4n-2)}{4n-2} = 0, \\[2ex] \sum_{n=1}^{\infty} \frac{\mu(4n-1)}{4n-1} = -\frac{2}{\pi}, \\[2ex] \sum_{n=1}^{\infty} \frac{\mu(4n)}{4n} = 0. \end{cases}$$

This example indicates how to formally compute series of the form

$$\sum_{n=1}^{\infty} \frac{\mu(a \cdot n + b)}{a \cdot n + b},$$

for certain integers $a, b \in \mathbb{Z}$ by using example 64.0.150 with z equals the a'th-root of unity. Before we proceed, let us remark that when $f(z)$ is a polynomial, the recursion formula for the sequence $\{\alpha_n\}_n$ is simpler.

Example 64.0.154.

$$f(z) = z - xz^p.$$

In this case, $a_1 = 1$, $a_p = -x$ and for all $n \notin \{1, p\}$ we have $a_n = 0$. Then $\alpha_1 = 1$ and if $n \notin \{1, p, p^2, p^3, \ldots\}$ then $\alpha_n = 0$, while $\alpha_{p^n} = x^n$. This generates the telescopic series given by

$$z = \sum_{n=0}^{\infty} x^n (z^{p^n} - xz^{p^{n+1}}).$$

We note that this series does not (usually) converge in the full complex plane, $\mathbb{C}$. It always converges in the unit disk $|z| < 1$. That is in spite of the fact that we started from a simple entire function $f(z)$ (a polynomial).

Chapter 65

The Symmetry Between the a_n Coefficients and the α_n Coefficients

The recursion formula $\alpha_n = -\sum_{\{d:\, 1\leq d<n,\, d\mid n\}} \alpha_d a_{n/d}$, $n = 2, 3, \ldots$ can be written as follows:

$$\sum_{\{d:\, 1\leq d\leq n,\, d\mid n\}} \alpha_d a_{n/d} = 0, \quad n = 2, 3, \ldots.$$

This proves that the recursive relations between the a_n coefficients and the α_n coefficients are symmetric. Thus, we have two recursion formulas:

$$\begin{cases} \alpha_1 = 1, \\ \alpha_n = -\sum_{\{d:\, 1\leq d<n,\, d\mid n\}} \alpha_d a_{n/d}, \quad n = 2, 3, \ldots, \end{cases}$$

and

$$\begin{cases} a_1 = 1, \\ a_n = -\sum_{\{d:\, 1\leq d<n,\, d\mid n\}} a_d \alpha_{n/d}, \quad n = 2, 3, \ldots. \end{cases}$$

We deduce that if $f(z) = \sum_{n=1}^{\infty} a_n z^n$ and $g(z) = \sum_{n=1}^{\infty} \alpha_n z^n$, then $z = \sum_{n=1}^{\infty} \alpha_n f(z^n)$ and $z = \sum_{n=1}^{\infty} a_n g(z^n)$. If the last two identities make sense for $z = 1$, then $1 = \sum_{n=1}^{\infty} a_n g(1) = \sum_{n=1}^{\infty} \alpha_n f(1)$, i.e., $\left(\sum_{n=1}^{\infty} a_n\right)\left(\sum_{n=1}^{\infty} \alpha_n\right) = 1$. Also, if $F(z) = \sum_{k=1}^{\infty} b_k z^k$ and if the following series converge (which is true for small $|z|$), then

$$F(z) = \sum_{k=1}^{\infty} b_k \sum_{n=1}^{\infty} \alpha_n f(z^{kn}) = \sum_{m=1}^{\infty} f(z^m)\left(\sum_{kn=m} b_k \alpha_n\right).$$

By the symmetric relations between the two types of the coefficients, this proves that

$$\sum_{k=1}^{\infty} b_k z^k = \sum_{m=1}^{\infty} f(z^m) \left(\sum_{kn=m} b_k \alpha_n \right) = \sum_{m=1}^{\infty} g(z^m) \left(\sum_{kn=m} b_k a_n \right).$$

Chapter 66

Simple Estimates for the Remainder of Inverse-Series

We use the standard terminology of the remainder of an infinite series and agree that the Nth remainder of our inverse-series that is generated by the function $f(z)$ is the following series: $R_N(z) = \sum_{n=N+1}^{\infty} \alpha_n f(z^n)$. Then we have

$$R_N(z) = \sum_{N+1}^{\infty} \alpha_n \left(\sum_{k=1}^{\infty} a_k z^{nk} \right) = \sum_{n=N+1}^{\infty} z^m \left(\sum_{\{n \mid N+1 \leq n \leq m,\, n \mid m\}} \alpha_n a_{m/n} \right).$$

By the identity $\sum_{\{n \mid 1 \leq n \leq m,\, n \mid m\}} \alpha_n a_{m/n} = 0$, this implies that

$$R_N(z) = - \sum_{m=N+1}^{\infty} z^m \left(\sum_{\{n \mid 1 \leq n \leq N,\, n \mid m\}} \alpha_n a_{m/n} \right).$$

This formula enables us to estimate the Nth remainder. For example,

$$|R_N(z)| \leq \max_{1 \leq n \leq N} |\alpha_n| \cdot \sum_{n=1}^{N} |a_n| \cdot \sum_{m=N+1}^{\infty} |z|^m = \max_{1 \leq n \leq N} |\alpha_n| \cdot \sum_{n=1}^{N} |a_n| \cdot \frac{|z|^{N+1}}{1 - |z|},$$

for $|z| < 1$. This proves the following:

Proposition 66.0.155. *If there is r such that $0 \leq r < 1$ and such that* $\max_{1 \leq n \leq N} |\alpha_n| \cdot \sum_{n=1}^{N} |a_n| = O(r^{-N})$ *then* $\lim_{N \to \infty} R_N(z) = 0$, *uniformly on compact subsets of* $|z| < r$.

Example 66.0.156.

$$f(z) = \frac{z}{1-z}.$$

385

In this case, we have $\max_{1\leq n\leq N}|\alpha_n|\cdot\sum_{n=1}^{N}|a_n| = \max_{1\leq n\leq N}\cdot N \leq N$ and so by Proposition 66.0.155 the inverse-series of $f(z)$ converges uniformly on compact subsets of $|z| < 1$.

Example 66.0.157.

$$f(z) = -\log(1-z).$$

In this case, we have

$$\max_{1\leq n\leq N}|\alpha_n|\cdot\sum_{n=1}^{N}|a_n| = \max_{1\leq n\leq N}\left|\frac{\mu(n)}{n}\right|\sum_{n=1}^{N}\frac{1}{n} \leq \log(N+1)$$

and so once more by Proposition 66.0.155, the inverse-series of $f(z)$ converges uniformly on compact subsets of $|z| < 1$.

Chapter 67

Inverse-Series that Converge
for Roots of Unity

We recall that the basic formula for the inverse-series of $f(z)$ is given by $z = \sum_{n=1}^{\infty} \alpha_n f(z^n)$. We give examples for applications of this formula when the series converge for certain values of z on $|z| = 1$. For roots of unity, the formulas we get are as follows:

$$e^{2\pi i/k} = \sum_{n=1}^{k} f(e^{2n\pi i/k}) \left(\sum_{m=0}^{\infty} \alpha_{mk+n} \right), \quad k = 1, 2, 3, \ldots.$$

Let us write explicitly few of these formulas:

$$1 = f(1) \sum_{n=1}^{\infty} \alpha_n,$$

$$-1 = f(-1) \sum_{m=0}^{\infty} \alpha_{2m+1} + f(1) \sum_{m=0}^{\infty} \alpha_{2m+2},$$

$$i = f(i) \sum_{m=0}^{\infty} \alpha_{4m+1} + f(-1) \sum_{m=0}^{\infty} \alpha_{4m+2}$$

$$+ f(-i) \sum_{m=0}^{\infty} \alpha_{4m+3} + f(1) \sum_{m=0}^{\infty} \alpha_{4m+4}.$$

From this, it follows, for example, that it is impossible that the series $\sum \alpha$ converge if $f(z)$ satisfies $f(1) = f(e^{2n\pi i/k})$, $n = 1, 2, \ldots, k$ for some $k > 1$. For otherwise the k'th formula contradicts the first formula.

Chapter 68

Reconstruction of the Inverse Image
from $f(z^n)$

If $F(z) = \sum_{k=1}^{\infty} b_k z^k$, then $F(z) = \sum_{m=1}^{\infty} f(z^m) \left(\sum_{kn=m} b_k \alpha_n \right)$. This identity proves the following.

Proposition 68.0.158. *If z_1, z_2 are two points for which the series above converge, and if for each m, $m = 1, 2, 3, \ldots$ such that $\sum_{kn=m} b_k \alpha_n \neq 0$, we have $f(z_1^m) = f(z_2^m)$. Then $F(z_1) = F(z_2)$.*

Corollary 68.0.159. *If for each m, $m = 1, 2, 3, \ldots$, such that $\alpha_m \neq 0$ we have $f(z_1^m) = f(z_2^m)$ then $z_1 = z_2$.*

Remark 68.0.160. The assumption that the series converge for z_1, z_2 is necessary for Proposition 68.0.158 to be true. For example, let

$$ f(z) = \frac{1}{i\pi} \left(e^{i\pi z} - 1 \right) = z + \cdots . $$

If $z = 2n$ is an even integer then for any m, $m = 1, 2, 3, \ldots$, we have

$$ f((2n)^m) = \frac{1}{i\pi} \left(e^{i\pi (2n)^m} - 1 \right) = 0 $$

but, of course we cannot deduce that all the even integers are identical. It is not hard to see that the inverse-series of this particular $f(z)$ does not converge on every z, $|z| = 1$ because $f(1) = f(-1) = \frac{2}{\pi} i$.

The following proposition extends Proposition 68.0.158.

Proposition 68.0.161. *Let z_1, z_2 be two points, and let l be some fixed natural number ($l \in \{1, 2, 3, \ldots\}$). We assume the convergence of the series below and we also assume that for every $m \in \{1, 2, 3, \ldots\}$ such that*

389

$\sum_{kn=m} b_k \alpha_n \neq 0$ *we have* $f(z_1^{lm}) = f(z_2^{lm})$. *Then* $F(z_1^l) = F(z_2^l)$. *In partic- ular: If for every natural* m *we have* $f(z_1^m) = f(z_2^m)$, *then for each* m *we have* $F(z_1^m) = F(z_2^m)$. *(Here, as before, the* α_n *are the coefficients of the inverse-series of* $f(z)$ *and* $F(z) = \sum_{k=1}^{\infty} b_k z^k$.)*

Also, the last proposition, 68.0.161, can be generalized. Suppose that there is a neighborhood of the point $z = 0$ in which all the functions $\{f_m(z)\}_{m=1}^{\infty}$ are analytic and have the following power series expansions:

$$f_m(z) = z + a_{m2}z^2 + a_{m3}z^3 + \cdots, \quad m = 1, 2, 3, \ldots.$$

Then we can make the following inversion:

$$z = \alpha_1 f_1(z) + \alpha_2 f_2(z^2) + \alpha_3 f_3(z^3) + \cdots,$$

where the inversion coefficients α_n, $n = 1, 2, 3, \ldots$ satisfy the following recursion:

$$\begin{cases} \alpha_1 = 1, \\ \alpha_n = -\sum_{\{d \mid 1 \leq d < n,\, d \mid n\}} \alpha_d a_{d,n/d}, \quad n = 2, 3, \ldots. \end{cases}$$

In this case, if $F(z) = \sum_{k=1}^{\infty} b_k z^k$, then for a small enough $|z|$ we might have:

$$F(z) = \sum_{k=1}^{\infty} b_k \sum_{n=1}^{\infty} \alpha_n f_n(z^{nk}) = \sum_{m=1}^{\infty} \sum_{kn=m} b_k \alpha_n f_n(z^m).$$

This implies the following.

Proposition 68.0.162. *Let* z_1, z_2 *be two points for which the series below converge. If for each* m, $m = 1, 2, 3, \ldots$ *we have* $f_n(z_1^m) = f_n(z_2^m)$ *for every* n *such that* $n \mid m$, *then* $F(z_1) = F(z_2)$. *In particular, if for every* m, $m = 1, 2, 3, \ldots$ *we have* $f_m(z_1^m) = f_m(z_2^m)$, *then* $z_1 = z_2$.

Remark 68.0.163. We can think of two more definitions of inversion (of power series). However, we shall see that these two procedures do not lead us to functions that have interesting arithmetical properties. For this reason, we will only bring the definition but will not discuss these procedures further.

Definition 68.0.164. Let the function $f(z)$ be analytic in a neighborhood of $z = 0$. Suppose that we have the following power series expansion $f(z) = z + a_2 z^2 + \cdots$. Then we can perform inversion using the scheme

$$z = \alpha_1 f(z) + \alpha_2 f(z)z + \alpha_3 f(z)z^2 + \cdots.$$

The coefficients α_n, $n = 1, 2, 3, \ldots$ satisfy the following recursion:

$$\begin{cases} \alpha_1 = 1, \\ \alpha_n = -\sum_{k=1}^{n-1} \alpha_k a_{n-k+1}, \quad n = 2, 3, \ldots. \end{cases}$$

We note that by definition $z = (\alpha_1 + \alpha_2 z + \alpha_3 z^2 + \cdots) f(z)$. Hence, the coefficients α_n generate the function $\frac{z}{f(z)}$.

Definition 68.0.165. Let the function $f(z)$ be analytic in a neighborhood of $z = 0$. Suppose that we have the following power series expansion $f(z) = z + a_2 z^2 + \cdots$. Let us use Rogosinski's notation, and denote by $a_n^{(k)}$, $k \le n$, the coefficients of $f(z)^k$. Thus $f(z)^k = \sum_{n=k}^{\infty} a_n^{(k)} z^n$. Then we can perform inversion using the scheme

$$z = \alpha_1 f(z) + \alpha_2 f(z)^2 + \alpha_3 f(z)^3 + \cdots.$$

The coefficients α_n, $n = 1, 2, 3, \ldots$ satisfy the following recursion:

$$\begin{cases} \alpha_1 = 1, \\ \alpha_n = -\sum_{k=1}^{n-1} \alpha_k a_n^{(k)}, \quad n = 2, 3, 4, \ldots. \end{cases}$$

However, by definition, we have $f^{-1}(w) = \alpha_1 w + \alpha_2 w^2 + \alpha_3 w^3 + \cdots$. Hence, the coefficients α_n generate the inverse function f^{-1} near $w = 0$ such that $f^{-1}(0) = 0$.

As we said before, we will not discuss inversions of power series of the types specified in definition 68.0.164 and in definition 68.0.165.

Chapter 60

Identities that Involve Inversion-Series

Let $x \in \mathbb{C}$ be such that $|x| < 1$ is small. We consider the function $f(z) = z + a_2 z^2 + \cdots$ analytic in the vicinity of $z = 0$, and form the following elementary transformation:

$$\frac{1}{x} f(xz) = z + a_2 x z^2 + \cdots.$$

This makes sense also if $x = 0$. Let us denote the inversion-coefficients α of the transformed function by $\alpha_n(x)$, $n = 1, 2, 3, \ldots$. Then, as usual, we have

$$\begin{cases} \alpha_1(x) = 1, \\ \alpha_n(x) = -\sum_{\{d \,|\, 1 \le d < n,\, d|n\}} \alpha_d(x) a_{n/d} x^{n/d - 1}, \quad n = 2, 3, 4, \ldots. \end{cases}$$

Clearly, $\alpha_n(x)$ is a polynomial of degree $n - 1$ in x. Let us write explicitly the first few α coefficients:

$$\alpha_1(x) = 1, \qquad \alpha_2(x) = -a_2 x,$$
$$\alpha_3(x) = -a_3 x^2, \qquad \alpha_4(x) = -a_4 x^3 + a_2^2 x^2,$$
$$\alpha_5(x) = -a_5 x^4, \qquad \alpha_6(x) = -a_6 x^5 + 2a_2 a_3 x^3,$$
$$\alpha_7(x) = -a_7 x^6, \qquad \alpha_8(x) = -a_8 x^7 + 2a_2 a_4 x^4 - a_2^3 x^3,$$
$$\alpha_9(x) = -a_9 x^8 + a_3^2 x^4, \quad \alpha_{10}(x) = -a_{10} x^9 + 2a_2 a_5 x^5 \ldots.$$

By definition, we have $xz = \sum_{n=1}^{\infty} \alpha_n(x) f(xz^n)$. Let us substitute $z = 1$ and after that change the symbol x to the symbol z and obtain the following identity:

$$\frac{z}{f(z)} = \sum_{n=1}^{\infty} \alpha_n(z).$$

393

We note that it is not hard to find a domain of convergence for the series on the right-hand side. It is identical to the domain of convergence of the power series of $\frac{z}{f(z)}$ about $z = 0$. Also, for each n, $n = 1, 2, 3, \ldots$ we have $\lim_{z \to 1} \alpha_n(z) = \alpha_n$.

Remark 69.0.166. We can write the recursion for $\alpha_n(x)$ in a symmetric manner:

$$\begin{cases} \alpha_1(x) = 1, \\ \sum_{kn=m} a_k x^k \alpha_n(x) = 0, \quad m = 2, 3, 4, \ldots. \end{cases}$$

Additional identities are obtained by substituting roots of unity in the fundamental formula (that defines the scheme). For example, if we plug in $z = -1$, we obtain: $-x = \sum_{n=1}^{\infty} (\alpha_{2n-1}(x) f(-x) + \alpha_{2n}(x) f(x))$. From this, we easily derive the following:

$$\frac{2z}{f(z) - f(-z)} = \sum_{n=1}^{\infty} \alpha_{2n-1}(z),$$

$$\left(\frac{z}{f(z)} \right) \cdot \left(\frac{f(-z) + f(z)}{f(-z) - f(z)} \right) = \sum_{n=1}^{\infty} \alpha_{2n}(z).$$

Chapter 70

Expansion of e^{-z} to an Infinite Product

The function e^{-z} is that non-constant entire function which perhaps is the most typical zero free entire function. In our expansion of this function into an infinite product, there will be infinitely many of the factors that have zeros of fractional order on the unit circle $\mathbb{T} = \{z \in \mathbb{C} | |z| = 1\}$. That sounds a complete nonsense. But there will also be infinitely many of the factors that have poles of fractional order on $\mathbb{T}$. The secret in resolving that strange representation lies in the fact that there are cancellations of zeros and poles in a very exact arithmetical manner. We are going to expand e^{-z} into an infinite product of polynomials raised to fractional powers (positive and negative). Those polynomials have their zeros on the unit circle $\mathbb{T}$. It is clear right from the beginning that this interesting expansion can not converge all over the complex plane. It will converge within the unit disk, $U = \{z \in \mathbb{C} | |z| < 1\}$. Our starting point will be the inversion of the power series expansion of the function $-\log(1 - z)$. We described the inversion in example 64.0.150 and in example 66.0.157. The power series of $-\log(1 - z)$ converges uniformly on compact subsets of U. We saw that also the inverse-series of $-\log(1 - z)$ converges uniformly on compact subsets of U. The two series expansions are given by

$$-\log(1 - z) = \sum_{n=1}^{\infty} \frac{z^n}{n}, \quad |z| < 1,$$

and by

$$-z = \sum_{n=1}^{\infty} \frac{\mu(n)}{n} \log(1 - z^n), \quad |z| < 1.$$

We may exponentiate the last identity to obtain an infinite product which converges uniformly on compact subsets of U. Thus, we proved the following:

Proposition 70.0.167.

$$e^{-z} = \prod_{n=1}^{\infty} (1 - z^n)^{\mu(n)/n}, \quad |z| < 1,$$

and the convergence of the infinite product is uniform on compact subsets of $U = \{z \in \mathbb{C} \,||z| < 1\}.$

The convergence in the identity in Proposition 70.0.167 is uniform and absolute on compact subsets of U. Hence, we can rearrange the factors in the infinite product without altering the convergence and always to the same limiting function e^{-z}. That fact will assist us in obtaining more identities. Let $x \in \mathbb{R}$, $-1 < x < 1$. By Proposition 70.0.167, we have

$$e^{-ix} = \prod_{n=1}^{\infty} (1 - (ix)^n)^{\mu(n)/n}$$

$$= \prod_{k=1}^{\infty} (1 - ix^{4k-3})^{\mu(4k-3)/(4k-3)} \times (1 + x^{4k-2})^{\mu(4k-2)/(4k-2)}$$

$$\times (1 + ix^{4k-1})^{\mu(4k-1)/(4k-1)} \times (1 - x^{4k})^{\mu(4k)/(4k)}$$

$$= \prod_{k=1}^{\infty} (1 - ix^{4k-3})^{\mu(4k-3)/(4k-3)} \times (1 + ix^{4k-1})^{\mu(4k-1)/(4k-1)}$$

$$\times (1 + x^{4k-2})^{\mu(4k-2)/(4k-2)}.$$

Taking absolute values of both sides of our equation gives

$$1 = \prod_{k=1}^{\infty} (1 + x^{2(4k-3)})^{\mu(4k-3)/(2(4k-3))} \times (1 + x^{2(4k-1)})^{\mu(4k-1)/(2(4k-1))}$$

$$\times (1 + x^{4k-2})^{\mu(4k-2)/(4k-2)},$$

hence,

$$\prod_{k=1}^{\infty} (1 + x^{4k-2})^{\mu(4k-2)/(4k-2)} = \prod_{k=1}^{\infty} (1 + x^{2(4k-3)})^{-\mu(4k-3)/(2(4k-3))}$$

$$\times (1 + x^{2(4k-1)})^{-\mu(4k-1)/(2(4k-1))}.$$

We use the permanence principle in order to replace the real number x by the complex number z. This concludes the proof of the following:

Proposition 70.0.168.

$$e^{-iz} = \prod_{k=1}^{\infty} (1 - iz^{4k-3})^{\mu(4k-3)/(4k-3)} \times (1 + iz^{4k-1})^{\mu(4k-1)/(4k-1)}$$
$$\times (1 + z^{4k-2})^{\mu(4k-2)/(4k-2)}.$$

$$1 = \prod_{k=1}^{\infty} (1 + z^{2(4k-3)})^{\mu(4k-3)/(2(4k-3))} \times (1 + z^{2(4k-1)})^{\mu(4k-1)/(2(4k-1))}$$
$$\times (1 + z^{4k-2})^{\mu(4k-2)/(4k-2)},$$

$$e^{z} = \prod_{k=1}^{\infty} \left(\frac{1 + z^{4k-3}}{(1 - z^{2(4k-3)})^{1/2}} \right)^{\mu(4k-3)/(4k-3)}$$
$$\times \left(\frac{1 + z^{4k-1}}{(1 - z^{2(4k-1)})^{1/2}} \right)^{\mu(4k-1)/(4k-1)}.$$

$$e^{z} = \prod_{k=1}^{\infty} \left(\frac{1 + z^{4k-3}}{1 - z^{4k-3}} \right)^{\mu(4k-3)/(2(4k-3))} \left(\frac{1 + z^{4k-1}}{1 - z^{4k-1}} \right)^{\mu(4k-1)/(2(4k-1))}.$$

The convergence in all the identities above is absolute and uniform on compact subsets of $U = \{z \in \mathbb{C} \mid |z| < 1\}$.

Proposition 70.0.169.

$$1 = \prod_{n=1}^{\infty} (1 + z^{2(2n-1)})^{\mu(2n-1)/(2(2n-1))} \times (1 + z^{2n})^{\mu(2n)/(2n)}.$$

$$1 = \prod_{n=1}^{\infty} (1 - z^{2(2n-1)})^{\mu(2n-1)/(2(2n-1))} \times (1 + (-1)^n z^{2n})^{\mu(2n)/(2n)}.$$

$$e^{-z} = \prod_{n=1}^{\infty} \left(\frac{1 - z^{2n-1}}{1 + z^{2n-1}} \right)^{\mu(2n-1)/(2(2n-1))}.$$

The convergence in all the identities above is absolute and uniform on compact subsets of $U = \{z \in \mathbb{C} \mid |z| < 1\}$.

Proof. The second identity in Proposition 70.0.168 implies the first identity in Proposition 70.0.169. If we replace in this first identity z by iz, we obtain the second identity. This implies

$$e^{-z} = \prod_{n=1}^{\infty} \left(\frac{1 - z^{2n-1}}{(1 - z^{2(2n-1)})^{1/2}} \right)^{\mu(2n-1)/(2n-1)}.$$

That is just our third identity in this proposition. $\square$

Another type of identities emerges if we write

$$(1 - z^n)^{\mu(n)/n} = (1 - z)^{\mu(n)/n} \prod_{k=1}^{n-1} \left(z - e^{2\pi i \frac{k}{n}}\right)^{\mu(n)/n}, \quad n > 1.$$

Let us substitute that into the identity of e^{-z} given in Proposition 70.0.167. After changing the order of the terms, we get

$$e^{-z} = (1 - z)^{\sum_{n=1}^{\infty} \frac{\mu(n)}{n}} \prod_{m=2}^{\infty} \left(\prod_{k=1,\,(k,m)=1}^{m-1} \left(z - e^{2\pi i \frac{k}{m}}\right)\right)^{\sum_{n=1}^{\infty} \frac{\mu(nm)}{nm}}.$$

Using the theorem $\sum_{n=1}^{\infty} \frac{\mu(n)}{n} = 0$, we complete a proof of the following:

Proposition 70.0.170.

$$e^{-z} = \prod_{m=2}^{\infty} \left(\prod_{k=1,\,(k,m)=1}^{m-1} \left(z - e^{2\pi i \frac{k}{m}}\right)\right)^{\sum_{n=1}^{\infty} \frac{\mu(nm)}{nm}}.$$

This identity is merely a formal one because changing the order of the different factors of the $1 - z^n$ is not supported by absolute convergence, or otherwise.

The polynomials inside the infinite product are well known in number theory.

Definition 70.0.171. Let $m \geq 2$ be an integer. The mth cyclotomic polynomial is defined by

$$\Phi_m(z) = \prod_{k=1,\,(k,m)=1}^{m-1} \left(z - e^{2\pi i \frac{k}{m}}\right).$$

Remark 70.0.172. If p is a prime number, then

$$\Phi_p(z) = z^{p-1} + z^{p-2} + \cdots + z + 1.$$

Let us compute the first few cyclotomic polynomials.

$$\Phi_2(z) = z + 1, \qquad \Phi_3(z) = z^2 + z + 1,$$
$$\Phi_4(z) = z^2 + 1, \qquad \Phi_5(z) = z^4 + z^3 + z^2 + z + 1,$$
$$\Phi_6(z) = z^2 - z + 1, \ldots.$$

Using Definition 70.0.171 and Remark 70.0.172, we can write the identity in Proposition 70.0.170 as follows:

$$e^{-z} = \prod_{m=2}^{\infty} \Phi_m(z)^{\sum_{n=1}^{\infty} \frac{\mu(nm)}{nm}}$$

$$= (z+1)^{\sum_{n=1}^{\infty} \frac{\mu(2n)}{2n}} \times (z^2+z+1)^{\sum_{n=1}^{\infty} \frac{\mu(3n)}{3n}} \times (z^2+1)^{\sum_{n=1}^{\infty} \frac{\mu(4n)}{4n}}$$

$$\times (z^4+z^3+z^2+z+1)^{\sum_{n=1}^{\infty} \frac{\mu(5n)}{5n}} \times (z^2-z+1)^{\sum_{n=1}^{\infty} \frac{\mu(6n)}{6n}} \ldots.$$

Since e^{-z} is zero free and entire, the product above cannot effectively include $z+1$ and we deduce by that argument that

$$\sum_{n=1}^{\infty} \frac{\mu(2n)}{2n} = 0.$$

This is consistent with our Corollary 64.0.153. Also, by the definition of the Möbius function $\mu(n)$, it follows that if $\mu(m) = 0$, i.e., the integer m is divisible by a square of a prime number, then for any natural number n we have $\mu(nm) = 0$. Hence, the only cyclotomic polynomials $\Phi_m(z)$ that might effectively appear in the product must be of order m which is not divisible by a square of a prime, i.e., $\mu(m) = \pm 1$. For these values of m, the only nonzero terms in the sum

$$\sum_{n=1}^{\infty} \frac{\mu(nm)}{nm},$$

are those for which n satisfies two conditions. The first is that n and m are coprime, i.e., $(n, m) = 1$. The second condition is that also n is square free, i.e., $\mu(n) = \pm 1$. If n satisfies these two conditions, then $\mu(nm) = \mu(n)\mu(m)$. Hence, we have the following exponent for $\Phi_m(z)$:

$$\sum_{n=1}^{\infty} \frac{\mu(nm)}{nm} = \frac{\mu(m)}{m} \sum_{\{n\,|\,(n,m)=1,\mu(n)\neq 0\}} \frac{\mu(n)}{n}.$$

This proves the following refinement of Proposition 70.0.170.

Proposition 70.0.173.

$$e^{-z} = \prod_{\{m\,|\,m\geq 2,\mu(m)\neq 0\}} \left(\Phi_m(z)^{\frac{\mu(m)}{m}}\right)^{\sum_{\{n\,|\,(n,m)=1,\mu(n)\neq 0\}} \frac{\mu(n)}{n}}.$$

This is a formal manipulation!

Chapter 71

Composition of Inner Functions and Their Boundary Behavior

Let I_1, I_2 be inner functions in the unit disk $U = \{z \in \mathbb{C} \mid |z| < 1\}$. We agree that the constant function $f(z) \equiv \epsilon$, $z \in U$ where the constant ϵ is unimodular, $|\epsilon| = 1$, is not inner for our purpose in this chapter. Also, we assume that $I_1 \circ I_2$ is not a constant function. We inquire if the composition $I_1 \circ I_2$ is an inner function. Suppose we try to construct a counterexample. The strategy is to have the function $I_1 \circ I_2$ having the wrong boundary behavior, i.e., such that will contradict the possibility of the composition being an inner function. We note that $I_1 \circ I_2$ is analytic on U and is bounded by one, $|(I_1 \circ I_2)(z)| < 1$, $\forall\, z \in U$. In order for $I_1 \circ I_2$ not to be an inner function, we need it not to satisfy the condition $|(I_1 \circ I_2)(e^{i\theta})| = 1$ a.e. in $\theta \in [0, 2\pi]$.

Since I_2 is an inner function, it satisfies $|I_2(e^{i\theta})| = 1$ a.e. in $\theta \in [0, 2\pi]$ and the same goes for I_1, i.e., $|I_1(e^{i\theta})| = 1$ a.e. in $\theta \in [0, 2\pi]$. Here, $I_1(e^{i\theta}) = \lim_{r \to 1^-} I_1(e^{i\theta})$ and we know by a theorem of Fatou that any analytic function in U which is bounded (i.e., belongs to $H^\infty(U)$) has non-tangential limit at almost any point of the unit circle $\mathbb{T}$. Let us define the exceptional set of I_1 by

$$E(I_1) = \left\{ e^{i\theta} \in \mathbb{T} \,\middle|\, |I_1(e^{i\theta})| \neq 1 \right\}.$$

Then the measure (Lebesgue) $|E(I_1)| = 0$. If we could construct the inner function I_2 in such a manner that there was a set of positive measure $\mathcal{P}(I_2)$ on the unit circle $\mathbb{T}$ so that $I_2(\mathcal{P}(I_2)) \subseteq E(I_1)$, then the composition $I_1 \circ I_2$ would satisfy $E(I_1 \circ I_2) \supseteq \mathcal{P}(I_2)$ and so $|E(I_1 \circ I_2)| \geq |\mathcal{P}(I_2)| > 0$. Thus, in that case, $I_1 \circ I_2$ could not have been an inner function (for that implies $|E(I_1 \circ I_2)| = 0$).

Remark 71.0.174. Here is an example of a standard construction that looks like a first step in what we want. Let $\Phi : [0, 2\pi] \to [0, 2\pi]$ be the Cantor (ternary) function (the Devil's staircase). Let us define $g : [0, 2\pi] \to [0, 2\pi]$ by $g(x) = \frac{1}{2}(x + \Phi(x))$. Then g is continuous and strictly increasing and so it is a bijection, and hence a homeomorphism, because a closed interval is a compact Hausdorff space. Thus, we can consider the inverse function $f = g^{-1} : [0, 2\pi] \to [0, 2\pi]$ which is a homeomorphism. Let $\mathcal{P}$ be the image of the Cantor ternary set (in $[0, 2\pi]$) under g is $[0, 2\pi]$. Then the image of $\mathcal{P}$ under the homeomorphism f is the Cantor ternary set in $[0, 2\pi]$, which has measure zero.

To try and take advantage in the construction which is described in Remark 71.0.174, we try to construct a bounded analytic function $I_2 : U \to U$, such that $I_2(e^{i\theta}) = e^{if(\theta)}$, $0 \le \theta \le 2\pi$. Then the image of $e^{i\mathcal{P}}$ under I_2 is $e^{i(\text{Cantor})}$.

Next, we want to construct an inner function $I_1 : U \to U$ such that $E(I_1) \supseteq I_2(\mathcal{P}) = e^{i(\text{Cantor})}$ a set of measure zero on $\mathbb{T}$. If all of that were possible, we would have arrived at $I_1 \circ I_2$ which is not an inner function.

A fact: $I_1 \circ I_2$ is trivially an inner function (by elementary geometry and the fact that non-constant analytic functions are open mappings and so $\forall\, z \in U$, $(I_1 \circ I_2)(z) \notin \partial((I_1 \circ I_2)(U)))$.

So, a construction such as the one outlined above is an impossibility. This completes the proof of the following:

Theorem 71.0.175. *Let I_1, I_2 be inner functions in $H^\infty(U)$ such that $I_1 \circ I_2$ is not a constant function. Then there does not exist a set $\mathcal{P}(I_2) \subseteq \mathbb{T}$ such that (1) $|\mathcal{P}(I_2)| > 0$, and (2) $I_2(\mathcal{P}(I_2)) \subseteq E(I_1)$.*

The following celebrated result of Kolesnikov [58] seems to have some relations to Theorem 71.0.175.

Theorem (Kolesnikov, [58]). *Let $E \subset \mathbb{T}$. There exists an $f \in H^\infty(U)$ such that the radial limits of f exist exactly on the set $\mathbb{T} - E$ if and only if E is a $G_{\delta\sigma}$ of (Lebesgue) measure zero.*

Thus, if we restrict $E(I_1)$ to be the Kolesnikov set of I_1 (and not the larger set of all those $e^{i\theta} \in \mathbb{T}$ such that $|I_1(e^{i\theta})| \ne 1$), then $E(I_1)$ is a $G_{\delta\sigma}$ of measure zero. We remark that the set $E(I_1)^c = \mathbb{T} - E(I_1)$ in the sense of Kolesnikov was called by Arthur Danielyan [18], and in [14, pp. 21, 147] the Fatou set of I_1. We recall that if μ is a complex Borel measure on the unit circle, $\mathbb{T}$, then the Poisson integral of μ on the open unit disk, U, is defined by the formula

$$P_\mu(z) = \int_\mathbb{T} \frac{1 - |z|^2}{|z - \xi|^2} d\mu(\xi), \quad (z \in U).$$

If $d\mu(e^{i\theta}) = u(e^{i\theta})\frac{d\theta}{2\pi}$, where $u \in L^1(\mathbb{T})$, instead of P_μ we write P_u. It can be verified that $h = P_\mu$ is an harmonic function on U. Using Fubini's theorem and the identity

$$\frac{1}{2\pi} \int_0^{2\pi} \frac{1 - |z|^2}{|z - e^{i\theta}|^2} d\theta = 1, \quad (z \in U),$$

we see that

$$\frac{1}{2\pi} \int_0^{2\pi} \left| h(re^{i\theta}) \right| d\theta \le \int_\mathbb{T} \left(\frac{(1 - r^2)d\theta}{|re^{i\theta} - \xi|^2} \right) d|\mu|(\xi) = \|\mu\|,$$

where $\|\mu\|$ is the total variation of the measure μ on $\mathbb{T}$. Hence, h fulfills the following growth condition:

$$\sup_{0 \le r < 1} \int_0^{2\pi} \left| h(re^{i\theta}) \right| d\theta < \infty. \tag{71.0.1}$$

Hence, the Poisson integral of a Borel measure on $\mathbb{T}$ is a harmonic function on U, which satisfies condition (71.0.1). The converse of this assertion is also true and we have the following complete characterization.

Theorem (Plessner). *Let h be a function defined on U. Then the following assertions are equivalent.*

(i) *h is an harmonic function on U, which satisfies the growth condition (71.0.1).*

(ii) *There exists a (unique) Borel measure μ on $\mathbb{T}$ such that $h = P_\mu$.*

In the special case in which μ is positive, $h = P_\mu$ is a positive harmonic function on U which satisfies (71.0.1). If h is a positive harmonic function, it follows by the mean value property that

$$\int_0^{2\pi} \left| h(re^{i\theta}) \right| d\theta = \int_0^{2\pi} h(re^{i\theta})d\theta = 2\pi h(0), \quad (0 \le r < 1).$$

Therefore, in this case, the Plessner theorem can be rewritten as follows.

Corollary (Herglotz). *Let h be a function defined on U. Then the following assertions are equivalent.*

(i) *h is a positive harmonic function on U.*

(ii) *There exists a (unique) positive Borel measure μ on $\mathbb{T}$ such that $h = P_\mu$.*

Here is a celebrated result of Fatou that provides a sufficient condition for the existence of radial limits of P_μ.

Theorem (Fatou). *Let μ be a complex Borel measure on $\mathbb{T}$. Suppose that at $e^{i\theta} \in \mathbb{T}$ the symmetric derivative*

$$\mu'(e^{i\theta}) = \lim_{t \to 0^+} \frac{\mu(\{e^{is} \mid \theta - t < s < \theta + t\})}{2t},$$

exists. Then

$$\lim_{r \to 1^-} P_\mu(re^{i\theta}) = 2\pi\mu'(e^{i\theta}). \tag{71.0.2}$$

By Lebesgue's decomposition theorem, for each complex Borel measure μ, there is a function $u \in L^1(\mathbb{T})$ and a complex singular Borel measure σ, such that $d\mu(e^{i\theta}) = u(e^{i\theta}) \cdot \frac{d\theta}{2\pi} + d\sigma(e^{i\theta})$. Moreover,

$$\mu'(e^{i\theta}) = \lim_{t \to 0^+} \frac{\mu(\{e^{is} \mid \theta - t < s < \theta + t\})}{2t} = \frac{u(e^{i\theta})}{2\pi}.$$

Hence, we obtain the following two results. First, if $\mu = \sigma$ a complex singular Borel measure on $\mathbb{T}$, then

$$\lim_{r \to 1^-} P_\sigma(re^{i\theta}) = 0, \tag{71.0.3}$$

for almost all $e^{i\theta} \in \mathbb{T}$. Second, if $d\mu = u(e^{i\theta})\frac{d\theta}{2\pi}$ is absolutely continuous, then

$$\lim_{r \to 1^-} P_\mu(re^{i\theta}) = u(e^{i\theta}), \tag{71.0.4}$$

for almost all $e^{i\theta} \in \mathbb{T}$. Coming back to our construction, we would like to construct a bounded analytic function $I_2 : U \to U$, such that $I_2(e^{i\theta}) = e^{if(\theta)}$, $0 \leq \theta \leq 2\pi$, thus, in fact, a non-constant inner function on U. Here, $f = g^{-1} : [0, 2\pi] \to [0, 2\pi]$ is the inverse of $g(x) = \frac{1}{2}(x + \Phi(x)) : [0, 2\pi] \to [0, 2\pi]$, where Φ is Cantor's (ternary) function. Assuming we can use the Plessner theorem with $d\mu(e^{i\theta}) = \Re e^{if(\theta)}\frac{d\theta}{2\pi} = \frac{1}{2\pi}\cos f(\theta)d\theta$ to obtain $\Re I_2(z) = P_{\cos f(\theta)}(z)$ and obtain the analytic function itself by

$$I_2(re^{i\theta}) = \frac{1}{2\pi} \int_0^{2\pi} \frac{(1 - r^2)e^{if(\psi)}}{1 - 2r\cos(\theta - \psi) + r^2}d\psi,$$

so if all is correct, we obtain by the Fatou theorem $\lim_{r\to 1^-} I_2(re^{i\theta}) = e^{if(\theta)}$ for almost all $e^{i\theta} \in \mathbb{T}$. So, if $\mathcal{P}$ is the image of that Cantor set under g, it has a positive Lebesgue measure ($|\mathcal{P}| = 2\pi$). The image of $\mathcal{P}$ under f (recall that $f = g^{-1}$) is the Cantor (ternary) set in $[0, 2\pi]$, which has measure zero. We write $I_2(e^{i\mathcal{P}}) = e^{i(\text{Cantor})}$ a subset of $\mathbb{T}$ of measure zero. If the above construction is valid, we conclude the following:

Conclusion. There does not exist an inner function I on U whose exceptional set $E(I)$ contains the Cantor (ternary) set on $\mathbb{T}$, $e^{i(\text{Cantor})}$. Here, the exceptional set of I is the set:

$$E(I) = \left\{ e^{i\theta} \in \mathbb{T} \,\middle|\, |I(e^{i\theta})| \neq 1 \right\}.$$

It is not complement with respect to $\mathbb{T}$ of the Fatou set, namely the complement of the set

$$\left\{ e^{i\theta} \in \mathbb{T} \,\middle|\, \lim_{r\to 1^-} I(re^{i\theta}) = I(e^{i\theta}) \text{ exists} \right\}.$$

We note that the last set is a subset of the exceptional set if I, i.e.,

$$\left\{ e^{i\theta} \in \mathbb{T} \,\middle|\, \lim_{r\to 1^-} I(re^{i\theta}) \text{ does not exists} \right\} \subseteq E(I). \tag{71.0.5}$$

The exceptional set $E(I)$ can contain points $e^{i\theta} \in \mathbb{T}$ such that $\lim_{r\to 1^-} I(re^{i\theta}) = I(e^{i\theta})$ exists, but $|I(e^{i\theta}| \neq 1$. We quote a lemma that appears in [14, p. 24].

Lemma 1 ([14], p. 24). *Let E be a set on $|z| = 1$ of linear measure zero. Then there exists a function $f(z)$, analytic and bounded in $|z| < 1$, say $|f(z)| < 1$, whose radial limits $f(e^{i\theta})$ have modulus 1 for almost all points of $|z| = 1$, and such that*

$$\liminf_{r\to 1^-} |f(re^{i\theta})| = 0, \quad \limsup_{r\to 1^-} |f(re^{i\theta})| = 1, \tag{71.0.6}$$

for each $e^{i\theta}$ in E.

But then, this lemma implies that for any subset E of $\mathbb{T}$ of measure zero, there exists an inner function $I = f$ such that

$$E \subseteq \left\{ e^{i\theta} \in \mathbb{T} \,\middle|\, \lim_{r\to 1^-} I(re^{i\theta}) \text{ does not exists} \right\}.$$

Using equation (71.0.5), we deduce that $E \subseteq E(I)$. In particular, we can take the null measure set $E = e^{i(\text{Cantor})}$ and conclude that there exists an inner function I such that $e^{i(\text{Cantor})} \subseteq E(I)$. This contradicts the conclusion above. This contradiction implies that either there exists no inner function I_2 on U such that $\lim_{r \to 1^-} I_2(re^{i\theta}) = e^{if(\theta)}$ for almost all $e^{i\theta} \in \mathbb{T}$, or that Lemma 1 in [14, p. 24] is not correct. The first possibility implies that the integral (Poisson)

$$\frac{1}{2\pi} \int_0^{2\pi} \frac{(1 - r^2)e^{if(\theta)}}{1 - 2r\cos(\theta - \psi) + r^2} d\theta,$$

does not converge. In other words, the differential

$$\frac{1}{2\pi} e^{if(\theta)} d\theta$$

cannot be the differential of a complex Borel measure on $\mathbb{T}$ and in particular $e^{if(\theta)} \notin L^1(\mathbb{T})$. Otherwise, as mentioned above, Lemma 1 in [14, p. 24] is faulty. Leaving our particular case of the construction related to the Cantor set (of Lebesgue measure zero) on $\mathbb{T}$, exactly the same arguments (assuming the validity of Lemma 1 in [14, p. 24]) prove the following:

Theorem 71.0.176. *Let I be a non-invertible inner function on U. Consider the function $a : [0, 2\pi] \to [0, 2\pi]$ which is defined a.e. on $[0, 2\pi]$ by the boundary values of I:*

$$I(e^{i\theta}) = e^{ia(\theta)}, \quad a.e.\ \theta \in [0, 2\pi].$$

Then for any Lebesgue measurable set $\mathcal{P} \subseteq [0, 2\pi]$ of a positive measure $|\mathcal{P}| > 0$, also the measurable image of $-\mathcal{P}$ under a, is of a positive measure, i.e., $|a(\mathcal{P})| > 0$. Thus, the boundary function of any non-invertible inner function on U maps sets of positive measure on $\mathbb{T}$ onto sets of positive measure on $\mathbb{T}$.

Remark 71.0.177. For an invertible inner function, i.e., an automorphism of U,

$$I(z) = e^{i\gamma} \frac{z - \alpha}{1 - \overline{\alpha}z}, \quad (\alpha \in U, \gamma \in \mathbb{R} \text{ are fixed})$$

the claim in theorem 71.0.176 is true and not hard to prove. Thus, in Theorem 71.0.176, our assumption (the optimal assumption) should be: "Let I be a non-constant inner function".

We would like to quote a theorem of a similar flavor from the classical book [75]. The theorem is from [75, Chapter VII, Section 4, p. 205]. Rolf Nevanlinna introduces the theorem as follows: "From the theorem of Riesz one concludes that the set of boundary values of a non-constant function of bounded type (type N) cannot be countable; for such a function can have a countable set of boundary values only on a point set on $|z| = 1$ which as the union of countably many null sets is itself a null set. We now concentrate on the proof of a general theorem which completely answers the question about the measure of the of the set E". Here, E denotes the boundary values of a function of bounded type for $|z| < 1$. We know that such a function has well-determined non-tangential limits almost everywhere on $|z| = 1$. The theorem of Riesz referred to is the following:

Theorem of Riesz ([75], p. 205). *If a function of bounded type for $|z| < 1$ has a constant radial limit on a point set of positive measure on $|z| = 1$, then the function is identically equal to this constant.*

We finally arrive at the theorem we aimed at.

Theorem ([75], p. 205). *Let E_w be the set of radial boundary values that a non-constant function $w(z)$ of bounded type for $|z| < 1$ assumes on a given set E_z on the circle $|z| = 1$. If the (linear) measure of E_z is positive, then the inner harmonic measure of E_w is likewise positive; i.e., there is a closed subset of E_w that has positive harmonic measure.*

This theorem was proved independently by Frostman, [32], and by Rolf Nevanlinna, [77], as an extension of an older theorem, [76]. As for the harmonic measure $\omega(z; \theta_1, \theta_2)$, it is $\frac{1}{2\pi}$ times the length of the arc cut from the unit circle by the chords determined by z and the $e^{i\theta_1}$ and $e^{i\theta_2}$. If α is the angle formed by these chords, then $2\pi\omega = 2\alpha - (\theta_2 - \theta_1)$. The harmonic measure $\omega(z, \alpha)$ of an arc α of the circle $|z| = 1$ with respect to the unit disk has the following properties:

1. For $|z| < 1$, $\omega(z, \alpha)$ is harmonic and bounded.
2. At every interior point of α, ω takes the boundary value 1; it vanishes on the arc β complementary to α.

By the maximum principle and the minimum principle, the harmonic measure is uniquely determined by these two conditions. The harmonic measure is given by

$$\omega(z; \theta_1, \theta_2) = \frac{1}{2\pi} \int_{\theta_1}^{\theta_2} \frac{1 - r^2}{1 + r^2 - 2r\cos(\theta - \phi)} d\theta \quad (z = re^{i\phi}).$$

The harmonic measure $\omega(z, \alpha, G)$ of an arc α with respect to a region G bounded by finitely many Jordan arcs $(\alpha \cup \beta)$ at the point $z \in G$ is the measure which is uniquely determined by the following conditions:

1. $\omega(z, \alpha, G)$ is harmonic and bounded in G.
2. On α, ω assumes the value 1 and on the complementary arc β, the value 0.

The harmonic measure is invariant with respect to one-to-one, conformal mappings of the reference region.

Principle of monotoneity. *The harmonic measure $\omega(z, \alpha, G)$ increases when the region G is extended across the portion β of the boundary $\Gamma = \alpha \cup \beta$ complementary to α.*

The point set α is of harmonic measure zero with respect to G if the image point set α_x (the universal coverings surface G^∞ of G is mapped conformally onto the unit disk K, then the set α is in one-to-one correspondence with a certain set α_x of points on the circumference $|x| = 1$. Provided this set is measurable, we call α harmonically measurable with respect to G.) has (linear) measure zero, i.e., provided α_x can be covered by a sequence of arcs of arbitrarily small total length. Here is an extension of a well-known classical theorem of Liouville:

Theorem 3 ([75], p. 140). *If the function $w(z)$, which is single-valued and analytic outside a harmonic null set α_z, admits a set of values α_w of positive harmonic measure in the region $z \neq \alpha_z$, then the function reduces to a constant.*

Theorem of Lindberg ([75], p. 161). *A closed point set of finite logarithmic measure is of harmonic measure zero.*

Finally, we point at a theorem in [14, p. 38].

Theorem 2.16 ([14], p. 38). *If, on a set of positive measure on $|z| = 1$, the radial limit values of a bounded analytic function $w = f(z)$ lie in a set of capacity zero, then $f(z)$ is identically constant.*

The proof of Theorem 2.16. ([14], p. 38), follows the proof of the theorem that a Blaschke product can omit a non-countable set of values in U, of capacity zero. Indeed, it is shown that to an arbitrary closed set E_w of capacity zero in $|w| < 1$, there exists a function $w = f(z)$ which is bounded and analytic in $|z| < 1$, having radial limits of modulus 1 almost everywhere on $|z| = 1$ and not reducing to a finite Blaschke product or a constant, that assumes infinitely often every value of $|w| < 1$ except for a set of capacity zero which includes E_w. To prove Theorem 2.16, a use is made in a theorem of Egorov which implies that on a closed set of positive measure, the radial limits $f(e^{i\theta})$ lie in a closed set E_w of capacity zero. The potential

$$u_1(w) = \int_{E_w} \log \frac{2M}{|w - \xi|} d\mu(\xi),$$

where M is some bound on $|f(z)|$ in $|z| < 1$ and where $\mu(\xi)$ is a distribution of unit mass on E_w exists and tends to $+\infty$ as w approaches any point of E_w. Once we have that $u_1(w)$, we follow the arguments used before in [14].

Chapter 72

The Parametrization of the Extremals for the Krzyż Problem

Let $B = \{f \in H(U) \mid \forall\, |z| < 1,\ 0 < |f(z)| < 1\}$. The Krzyż problem is the coefficient problem for B. So, we can state it as follows:

$$\text{for } n \in \{2, 3, \ldots\}, \quad A_n = \sup\{|a_n| \,|\, f(z) = \sum_{k=0}^{\infty} a_k z^k \in B\}.$$

In fact, the supremum is attained and since the problem is invariant for rotations (both inner and outer rotations, $e^{i\alpha} f(e^{i\beta} z)$), we can write

$$\text{for } n \in \{2, 3, \ldots\}, \quad A_n = \sup\{\Re\, a_n \mid f(z) = \sum_{k=0}^{\infty} a_k z^k \in B\}.$$

The Krzyż conjecture states that: $A_n = \frac{2}{e}$, $n \in \{2, 3, \ldots\}$ and the extremal functions for A_n are the rotations of the function

$$\exp\left(-\frac{1 + z^n}{1 - z^n}\right).$$

It was proved by Hummel, Scheinberg, and Zalcman in [49], among other things, that the extremals for A_n belong to the following family of functions in B:

$$\exp\left(-\sum_{j=1}^{N} \lambda_j \frac{1 + k_j z}{1 - k_j z}\right), \text{ where } 1 \leq N \leq n, \text{ and for } j = 1, \ldots, N,$$

$\lambda_j > 0$ and $|k_j| = 1$, and the N parameters on the unit circle, $k_1, \ldots, k_N$ are different from one another. So, this subfamily of B is parametrized by $2N$

411

parameters. N parameters are positive numbers and the other N parameters are N distinct points on $\mathbb{T}$. Let us denote this subfamily of B by Extremals$_n$. We note that we restricted attention to the functions $f \in B$ for which $f(0) > 0$. We can assume that thanks to the rotations freedom we have in solving A_n. Is the parametrization of Extremals$_n$ a faithful parametrization? We recall that $e^\alpha = e^\beta$ if and only if $\alpha - \beta = 2\pi i k$ for some $k \in \mathbb{Z}$. Thus, the faithfulness question leads to inquire for necessary and sufficient condition for the following:

$$\sum_{j=1}^{N} \lambda_j \frac{1 + k_j z}{1 - k_j z} = \sum_{l=1}^{M} \delta_l \frac{1 + \epsilon_l z}{1 - \epsilon_l z} + 2\pi i k,$$

for $1 \leq M, N \leq n$, $\lambda_j, \delta_l > 0$ $(j = 1, \ldots, N,\ l = 1, \ldots, M)$, $|k_j| = |\epsilon_l| = 1$ where the $k_1, \ldots, k_N$ are distinct points on $\mathbb{T}$ and $\epsilon_1, \ldots, \epsilon_M$ are distinct too. Both sides of the above equation are the partial fraction representation of the same rational function over $\mathbb{C}$ from which we easily deduce that they are identical: $N = M$, and $k = 0$, and after a suitable re-ordering $\lambda_j = \delta_j$, $k_j = \epsilon_j$ $(j = 1, \ldots, N)$. Thus, the parametrization is faithful (injective and also surjective in this case). There is a natural splitting of Extremals$_n$ into the disjoint union of n subsets according to N. Say,

$$\text{Extremals}_n = \bigcup_{N=1}^{n} E_N,$$

where the functions in E_N depend on the vector of parameters $(\lambda_1, \ldots, \lambda_N; k_1, \ldots, k_N)$. There are some obvious relations among the families of the extremals.

1. $2 \leq n < m \Rightarrow$ Extremals$_n \subset$ Extremals$_m$.
2. $\forall 2 \leq n \forall m \in \mathbb{Z}^+$, $A_n \leq A_{nm}$ (for if $f(z) = a_0 + \cdots + a_n z^n + \cdots$ solves A_n, then by $f(z^m) = a_0 + \cdots + a_n z^{nm} + \cdots$ it follows that $|a_n| \leq A_{nm}$).
3. $\dim_{\mathbb{R}} E_N = 2N$ (since $k_j \in \mathbb{T}$ depends on a single real number arg k_j).
4. $\dim_{\mathbb{R}}$ Extremals$_n = 2n$. However, E_N is of pure dimension $2N$ while Extremals$_n$ is not of pure dimension $2n$ for $2 \leq n$.
5. E_N is parametrized by $(\mathbb{R}^+)^N \times \left(\mathbb{T}^N - \{(k_1, \ldots, k_N) | \exists\, i \neq j\ \text{s.t}\ k_i = k_j\}\right)$.

For example, E_1 is faithfully parametrized by $\mathbb{R}^+ \times \mathbb{T}$, E_2 is parametrized by $(\mathbb{R}^+)^2 \times (\mathbb{T}^2 - \text{diag}\,\mathbb{T}^2)$, and E_3 is faithfully parametrized by $(\mathbb{R}^+)^3 \times (\mathbb{T}^3 - \{(k_1, k_1, k_2), (k_1, k_2, k_1), (k_2, k_1, k_1), (k_1, k_1, k_1) | k_1, k_2 \in \mathbb{T},\ k_1 \neq k_2\})$.

Geometrically, if we identify E_1 with its faithful parametrization then it is half a cylinder and in particular it is connected and of a pure dimension 2. In E_2, the factor $\mathbb{T}^2 - \operatorname{diag} \mathbb{T}^2$ is not connected.

Can we give a geometric meaning to the complex valued function a_n on E_N?

Chapter 73

General Remarks

In [107, p. 275], we find the following facts. Let $h(z) \in H^1(U)$ satisfy $\|h\|_1 = 1$. Then we may write $h(z) = \sum_{n=0}^{\infty} b_n z^n$, where $b_n = \sum_{k=0}^{n} p_k q_{n-k}$, $\sum_{n=0}^{\infty} |p_n|^2 = \sum_{n=0}^{\infty} |q_n|^2 = 1$. In trying to generalize this, let $g(z) \in H^p(U)$ satisfy $\|g\|_p^p = 1$. Then $\frac{1}{2\pi} \int_0^{2\pi} |g(e^{i\theta})|^p d\theta = 1$ and hence $g(z)^p \in H^1(U)$ satisfy $\|g^p\|_1 = 1$. So, we may write $g(z)^p = \sum_{n=0}^{\infty} a_n z^n$ where $a_n = \sum_{k=0}^{n} p_k q_{n-k}$, $\sum_{n=0}^{\infty} |p_n|^2 = \sum_{n=0}^{\infty} |q_n|^2 = 1$. Here, we assume that $p \in \mathbb{Z}^+$, otherwise g^p is, in general, not analytic in U (it branches in zeros of g of multiplicity > 1).

Example 73.0.178. $p = 2$. Let $g(z) = \sum_{n=0}^{\infty} b_n z^n$, $g(z)^2 = \sum_{n=0}^{\infty} a_n z^n$. Then $a_n = \sum_{k=0}^{n} b_k b_{n-k}$ $\left(\sum_{n=0}^{\infty} a_n z^n = \left(\sum_{n=0}^{\infty} b_n z^n \right)^2 \right)$ where $\sum_{n=0}^{\infty} |b_n|^2 = \frac{1}{2\pi} \int_0^{2\pi} |g(e^{i\theta})|^2 d\theta = 1$.

Example 73.0.179. $p = 2N$ $(N \in \mathbb{Z}^+)$. Let $g(z) = \sum_{n=0}^{\infty} b_n z^n$, $g(z)^{2N} = \sum_{n=0}^{\infty} a_n z^n$. Then $a_n = \sum_{i_1 + \cdots + i_{2N} = n} b_{i_1} \ldots b_{i_{2N}} = \sum_{k=0}^{n} p_k q_{n-k}$, $\sum_{n=0}^{\infty} |p_n|^2 = \sum_{n=0}^{\infty} |q_n|^2 = 1$. Also, this case is easy. We take $g(z)^N = \sum_{n=0}^{\infty} p_n z^n$ and $p_n = q_n$.

Example 73.0.180. $p = 1$. Let $g(z) = \sum_{n=0}^{\infty} b_n z^n$ satisfy $g(z) \neq 0$, $|z| < 1$, and $\|g\|_1 = 1$. Then there are two analytic functions in U, $\pm\sqrt{g(z)}$. If $\sqrt{g(z)} = \sum_{n=0}^{\infty} a_n z^n$, then $\left(\sum_{n=0}^{\infty} a_n z^n \right)^2 = \sum_{n=0}^{\infty} b_n z^n$. We obtain an infinite system of equations in the infinitely many unknowns $a_0, a_1, \ldots$:

$$a_0^2 = b_0, \quad 2a_0 a_1 = b_1, \quad 2a_0 a_2 + a_1^2 = b_2, \quad 2a_0 a_3 + 2a_1 a_2 = b_3, \ldots.$$

So, we get the two solutions:

$$a_0 = \pm\sqrt{b_0}, \quad a_1 = \pm\frac{b_1}{2\sqrt{b_0}}, \quad a_2 = \pm\frac{2b_0 b_2 - b_1^2}{4b_0\sqrt{b_0}}, \ldots.$$

Thus, $\sqrt{g(z)} \in H^2(U)$ satisfies $\|(\sqrt{g(z)})^2\|_1 = 1$ and once more $p_n = q_n$ are the coefficients of $\pm\sqrt{g(z)}$. Going to the general case where $g \in H^1(U)$, $\|g\|_1 = 1$. Let $B(z)$ be the Blaschke product formed with all the zeros of g. Then $\frac{g}{B} \in H^1(U)$, $\|g/B\|_1 = 1$ and $(g/B)(z) \neq 0 \ \forall\, |z| < 1$. So, g/B reduces to the first part of the current example. By that first part, we obtain

$$\frac{g(z)}{B(z)} = \sum_{n=0}^{\infty} \left(\sum_{k=0}^{n} p_k p_{n-k} \right) z^n \text{ where } \sqrt{\frac{g(z)}{B(z)}} = \sum_{n=0}^{\infty} p_n z^n.$$

$$\Rightarrow g(z) = \sum_{n=0}^{\infty} b_n z^n = B(z) \sum_{n=0}^{\infty} \left(\sum_{k=0}^{n} p_k p_{n-k} \right) z^n.$$

Let $B(z) = \sum_{n=0}^{\infty} c_n z^n$ then

$$\sum_{n=0}^{\infty} b_n z^n = \left(\sum_{n=0}^{\infty} c_n z^n \right) \left(\sum_{n=0}^{\infty} \left(\sum_{k=0}^{n} p_k p_{n-k} \right) z^n \right).$$

This gives the infinite system:

$$b_n = \sum_{j=0}^{n} \left(\sum_{k=0}^{j} p_k p_{j-k} \right) c_{n-j}, \quad n = 0, 1, 2, \ldots.$$

We get a recursion on the Taylor coefficients of the Blaschke products:

$$c_0 = \frac{b_0}{p_0^2}, \quad c_n p_0^2 + c_{n-1} 2 p_0 p_1 + \cdots + c_0 \left(2 p_0 p_n + 2 p_1 p_{n-1} + \cdots \right) = b_n.$$

However, by the result in Zygmund's book, [107], we have

$$b_n = \sum_{k=0}^{n} q_k r_{n-k}, \quad \sum_{n=0}^{\infty} |q_n|^2 = \sum_{n=0}^{\infty} |r_n|^2 = 1.$$

We deduce the following:

$$\sum_{j=0}^{n} \left(\sum_{k=0}^{j} p_k p_{j-k} \right) c_{n-j} = \sum_{k=0}^{n} q_k r_{n-k}.$$

This shows the following.

Theorem 73.0.181. *Let* $B(z) = \sum_{n=0}^{\infty} c_n z^n$ *be a Blaschke product. Then there are three sequences* $\{p_n\}$, $\{q_n\}$, $\{r_n\}$, *such that*

1. $\sum_{n=0}^{\infty} |p_n|^2 = \sum_{n=0}^{\infty} |q_n|^2 = \sum_{n=0}^{\infty} |r_n|^2 = 1$ *and*

2. $\sum_{j=0}^{n} (\sum_{k=0}^{j} p_k q_{j-k}) c_{n-j} = \sum_{k=0}^{n} q_k r_{n-k}$, *and if the sequence*

3. $\{s_n\}$ *is defined by the recursion*:

$$p_0 s_0 = 1, \quad p_0 s_n + p_1 s_{n-1} + \cdots + p_n s_0 = 0 \quad (n > 0) \text{ then } \limsup |s_n|^{1/n} \leq 1$$

Proof. Only the last assertion (3) on $\{s_n\}$ needs a justification. But this follows by the fact that the analytic function

$$\sqrt{\frac{g(z)}{B(z)}} = \sum_{n=0}^{\infty} p_n z^n$$

has no zeros in U. So,

$$\frac{1}{\sum_{n=0}^{\infty} p_n z^n} = \sum_{n=0}^{\infty} s_n z^n$$

is analytic in U, and hence has a radius of convergence which is at least 1. This radius equals

$$r = \left(\limsup |s_n|^{1/n} \right)^{-1}.$$

$\square$

Chapter 74

The Paper of Newman and Shapiro [79]

We refer to [79, Section 2, p. 253]. The following formula is derived:

$$\exp\left(a\frac{z+1}{z-1}\right) = \sum_{n=0}^{\infty} e^{-a} L_n^{(-1)}(2a) z^n \quad (a > 0),$$

where $L_n^{(-1)}$ denote generalized Laguerre polynomials. Asymptotically, we have

$$e^{-a} L_n^{(-1)}(2a) = \pi^{-1/2}(2a)^{1/4} n^{-3/4} \cos(2\sqrt{2\cdot a \cdot n}) + O(n^{-5/4}).$$

Hence,

$$\exp\left(-\lambda\frac{1+kz}{1-kz}\right) = \sum_{n=0}^{\infty}\{e^{-\lambda} L_n^{(-1)}(2\lambda) k^n\} z^n \quad (\lambda > 0, \, |k| = 1).$$

We recall that the extremals for the Krzyż conjecture A_n normalized by $f(0) > 0$ have the following form:

$$\exp\left(-\sum_{j=1}^{N}\lambda_j\frac{1+k_jz}{1-k_jz}\right) = \prod_{j=1}^{N}\exp\left(-\lambda_j\frac{1+k_jz}{1-k_jz}\right)$$

$$= \prod_{j=1}^{N}\sum_{m=0}^{\infty}\{e^{-\lambda_j} L_m^{(-1)}(2\lambda_j) k_j^m\} z^m$$

$$= \sum_{k=0}^{\infty} \left(\sum_{\substack{m_1+\cdots+m_N=k \\ m_1,\ldots,m_N\geq 0}} \prod_{j=1}^{N} \left\{ e^{-\lambda_j} L_{m_j}^{(-1)}(2\lambda_j) k_j^{m_j} \right\} \right)$$

$$\times z^k. \quad (N \leq n, \lambda_j > 0, |k_j| = 1).$$

In particular, for $k = n$, we get the following coefficient:

$$a_n = \sum_{\substack{m_1+\cdots+m_N=n \\ m_1,\ldots,m_N\geq 0}} \prod_{j=1}^{N} \left\{ e^{-\lambda_j} L_{m_j}^{(-1)}(2\lambda_j) k_j^{m_j} \right\}$$

$$= e^{-\lambda_1-\cdots-\lambda_N} \times \sum_{\substack{m_1+\cdots+m_N=k \\ m_1,\ldots,m_N\geq 0}} \prod_{j=1}^{N} \left\{ L_{m_j}^{(-1)}(2\lambda_j) k_j^{m_j} \right\}.$$

$$(74.0.1)$$

That representation motivates natural questions such as the following: Is it true that for any admissible values of the parameter space, we have

$$\left| \sum_{\substack{m_1+\cdots+m_N=n \\ m_1,\ldots,m_N\geq 0}} \prod_{j=1}^{N} \left\{ L_{m_j}^{(-1)}(2\lambda_j) k_j^{m_j} \right\} \right| \leq 2 \quad (N \leq n, \lambda_j > 0, |k_j| = 1)?$$

$$(74.0.2)$$

If true, then by equation (74.0.1), this would imply that

$$|a_n| \leq e^{-\lambda_1-\cdots-\lambda_N} \cdot 2 = 2 \cdot a_0.$$

The doubling principle would then imply the Krzyż conjecture for the index n. Later on, we will see that the estimate in (74.0.2) cannot be valid as is! If we plug into (74.0.2) the values of the parameters conjectured by Krzyż for the normalized extremal, we obtain

$$\left| \sum_{\substack{m_1+\cdots+m_n=n \\ m_1,\ldots,m_n\geq 0}} \prod_{j=1}^{n} \left\{ L_{m_j}^{(-1)}\left(\frac{2}{n}\right) e^{2\pi i \cdot j \cdot m_j/n} \right\} \right| = 2.$$

We bring some basic facts on the generalized Laguerre polynomials. We refer to [7, pp. 66–67]. Let $\alpha \in (-1, \infty)$. Note that $\alpha \neq -1$.

(a) Rodrigues' formula. Let

$$L_n^{(\alpha)}(x) := \frac{1}{n!e^{-x}x^\alpha}\frac{d^n}{dx^n}\left(e^{-x}x^{\alpha+n}\right).$$

Then $\{L_n^{(\alpha)}\}_{n=0}^\infty$ is a sequence of orthogonal polynomials on $[0,\infty)$ associated with the weight function $w(x) := x^\alpha e^{-x}$. That is, $L_n^{(\alpha)} \in \mathbb{C}_n[x]$ and $\int_0^\infty L_n^{(\alpha)}(x)L_m^{(\alpha)}(x)x^\alpha e^{-x}dx = 0$ for any non-negative integers $n \neq m$.

(b) Normalization. We have

$$\int_0^\infty (L_n^{(\alpha)}(x))^2 x^\alpha e^{-x}dx = \frac{\Gamma(\alpha+n+1)}{n!} \quad \text{and} \quad L_n^{(\alpha)}(0) = \binom{n+\alpha}{n}.$$

(c) Explicit form.

$$L_n^{(\alpha)}(x) = \sum_{m=0}^n \frac{(-1)^m}{m!}\binom{n+\alpha}{n-m}x^m.$$

(d) Differential equation. The function $y = L_n^{(\alpha)}(x)$ satisfies the following ODE

$$x\frac{d^2y}{dx^2} + (\alpha+1-x)\frac{dy}{dx} + ny = 0.$$

(e) Recurrence relation. The sequence $\{L_n^{(\alpha)}(x)\}_{n=0}^\infty$ satisfies

$$(n+1)L_{n+1}^{(\alpha)}(x) = [(2n+\alpha+1)-x]L_n^{(\alpha)}(x) - (n+\alpha)L_{n-1}^{(\alpha)}(x),$$

with $L_0^{(\alpha)}(x) = 1$ and $L_1^{(\alpha)}(x) = -x+\alpha+1$.

(f) Generating function.

$$\sum_{n=0}^\infty L_n^{(\alpha)}(x)z^n = \exp\left(\frac{xz}{z-1}\right)(1-z)^{-\alpha-1}.$$

The case of interest in the Krzyż problem is the endpoint $\alpha = -1$. Not all of the above properties are valid for this value of the parameter α. We have the Rodrigues' formula

$$L_n^{(-1)}(x) = \frac{x}{n!e^{-x}} \frac{d^n}{dx^n} \left\{ x^{n-1} e^{-x} \right\}, \text{ the weight function: } w(x) = x^{-1} e^{-x}.$$

For positive integers $n \neq m$, we have

$$\int_0^\infty L_n^{(-1)}(x) L_m^{(-1)}(x) x^{-1} e^{-x} dx = 0 \quad (n \neq m, \ n, m \in \mathbb{Z}^+),$$

and normalization

$$\int_0^\infty \left(L_n^{-1}(x) \right)^2 x^{-1} e^{-x} dx = \frac{\Gamma(n)}{n!} = \frac{1}{n} \quad (n \in \mathbb{Z}^+).$$

By equation (74.0.1), we have the following:

$$a_n(\lambda_1, \ldots, \lambda_N; k_1, \ldots, k_N) = \sum_{\substack{m_1 + \cdots + m_N = n \\ m_1, \ldots, m_N \geq 0}} \prod_{j=1}^N \left\{ e^{-\lambda_j} L_{m_j}^{(-1)}(2\lambda_j) k_j^{m_j} \right\}.$$

Let us fix an integer $0 \leq m \leq N$. Let us fix one of the $\lambda_k \in \{\lambda_1, \ldots, \lambda_N\}$. Then we note that by the orthogonality and the normalization relations, we have

$$(n \neq m) \int_0^\infty L_n^{(-1)}(2x) L_m^{(-1)}(2x) (2x)^{-1} e^{-(2x)} dx$$

$$= \frac{1}{2} \int_0^\infty L_n^{(-1)} L_m^{(-1)} (2x)^{-1} e^{-(2x)} d(2x) = \frac{1}{2} \cdot 0 = 0,$$

and for $n \neq 0$

$$\int_0^\infty (L_n^{(-1)}(2x))^2 (2x)^{-1} e^{-(2x)} dx$$

$$= \frac{1}{2} \int_0^\infty (L_n^{(-1)}(2x))^2 (2x)^{-1} e^{-(2x)} d(2x) = \frac{1}{2n}.$$

We split the sum that represents a_n by separating the terms that contain λ_k. We have

$$(2\lambda_k)^{-1}e^{-\lambda_k}L_m^{(-1)}(2\lambda_k)a_n(\lambda_1,\ldots,\lambda_N;k_1,\ldots,k_N)$$

$$- \sum_{\substack{m_1+\cdots+m_N=n \\ m_1,\ldots,m_N\geq 0 \\ m_k=m}} (2\lambda_k)^{-1}e^{-\lambda_k}(L_m^{(-1)}(2\lambda_k))^2 k_k^{m_k}$$

$$\times \prod_{\substack{j=1 \\ j\neq k}}^{N}\left\{e^{-\lambda_j}L_{m_j}^{(-1)}(2\lambda_j)k_j^{m_j}\right\}$$

$$+ (2\lambda_k)^{-1}e^{-\lambda_k} \sum_{\substack{m_1+\cdots+m_N=n \\ m_1,\ldots,m_N\geq 0 \\ m_k\neq m}} L_m^{(-1)}(2\lambda_k)$$

$$\times \prod_{j=1}^{N}\{e^{-\lambda_j}L_{m_j}^{(-1)}(2\lambda_j)k_j^{m_j}\} \quad (m>0).$$

Integrating the last identity $\int_0^\infty \ldots d\lambda_k$ gives, in light of the orthogonality and normalization relations, the following:

$$\int_0^\infty (2\lambda_k)^{-1}e^{-\lambda_k}L_m^{(-1)}(2\lambda_k)a_n(\lambda_1,\ldots,\lambda_N;k_1,\ldots,k_N)d\lambda_k$$

$$= \sum_{\substack{m_1+\cdots+m_N=n \\ m_1,\ldots,m_N\geq 0 \\ m_k=m}} \left(\frac{1}{2m}\right)k_k^m \prod_{\substack{j=1 \\ j\neq k}}^{N}\{e^{-\lambda_j}L_{m_j}^{(-1)}(2\lambda_j)k_j^{m_j}\}.$$

$$(74.0.3)$$

We can repeat this integration process for each one of the indices $1,\ldots,N$. Each integration is induced by a different index and the sum of the parameters m is at most n. The result is

$$\int_0^\infty \cdots \int_0^\infty \left(\prod_{j=1}^{N}(2\lambda_j)^{-1}e^{-\lambda_j}L_{m_j}^{(-1)}(2\lambda_j)\right) a_n(\lambda_1,\ldots,\lambda_N;k_1,\ldots,k_N)$$

$$\times d\lambda_1\ldots d\lambda_N = \prod_{j=1}^{N}\left(\frac{1}{2m_j}\right)k_j^{m_j},$$

where $m_1, \ldots, m_N \geq 0$, $m_1 + \cdots + m_N = n$. Hence,

$$\int_0^\infty \cdots \int_0^\infty \left(\prod_{j=1}^N \lambda_j^{-1} e^{-\lambda_j} L_{m_j}^{(-1)}(2\lambda_j) \right) a_n(\lambda_1, \ldots, \lambda_N; k_1, \ldots, k_N) d\lambda_1 \ldots d\lambda_N$$

$$= \prod_{j=1}^N \left(\frac{k_j^{m_j}}{m_j} \right), \quad (m_1, \ldots, m_N \geq 0, \; m_1 + \cdots + m_N = n).$$

This can be written as follows:

$$\int_0^\infty \cdots \int_0^\infty \left(\prod_{j=1}^N \lambda_j^{-1} \right) \left(\prod_{j=1}^N e^{-\lambda_j} L_{m_j}^{(-1)}(2\lambda_j) \overline{k}_j^{m_j} \right)$$

$$\times a_n(\lambda_1, \ldots, \lambda_N; k_1, \ldots, k_N) d\lambda_1 \ldots d\lambda_N$$

$$= \prod_{j=1}^N \left(\frac{1}{m_j} \right), \quad (m_1, \ldots, m_N \geq 0, \; m_1 + \cdots + m_N = n).$$

We sum that identity on all the choices of $m_1, \ldots, m_N \geq 0$, $m_1 + \cdots + m_N = n$ and note that by (74.0.1)

$$\sum_{\substack{m_1 + \cdots + m_N = n \\ m_1, \ldots, m_N \geq 0}} \prod_{j=1}^N (e^{-\lambda_j} L_{m_j}^{(-1)}(2\lambda_j) \overline{k}_j^{m_j}) = \overline{a}_n(\lambda_1, \ldots, \lambda_N; k_1, \ldots, k_N).$$

Hence, we finally obtain

$$\int_0^\infty \cdots \int_0^\infty \left(\prod_{j=1}^N \lambda_j^{-1} \right) |a_n(\lambda_1, \ldots, \lambda_N; k_1, \ldots, k_N)|^2 \, d\lambda_1 \ldots \lambda_N$$

$$= \sum_{\substack{m_1 + \cdots + m_N = n \\ m_1, \ldots, m_N \geq 0}} \prod_{j=1}^N \left(\frac{1}{m_j} \right). \tag{74.0.4}$$

An immediate conclusion is that the integral on the left-hand side is independent of the uni-modular parameters $k_1, \ldots, k_N$. We need to figure out and correct the cases where some $m_j = 0$, for otherwise the expressions we get are meaningless because of the reciprocals $\frac{1}{m_j}$. The origin of that inconvenience lies in the fact that for the generalized Laguerre polynomials $\{L_n^{(-1)}(x)\}_{n \geq 0}$, orthogonality and normalization relations are valid only for indices $n \geq 1$, i.e., natural numbers only. It is easy to compute that $L_0^{(-1)}(x) \equiv 1$. But this implies that the normalization integral $\int_0^\infty (L_0^{(-1)}(x))^2 w(x) dx = \int_0^\infty (L_0^{(-1)}(x))^2 x^{-1} e^{-x} dx = \infty$ for it diverges at the origin $x = 0$. Similarly, $L_0^{(-1)}$ is not orthogonal to some of the other polynomials. For example $\int_0^\infty L_0^{(-1)}(x) L_3^{(-1)}(x) w(x) dx$ is a non-zero real number. That suggests that all of our arguments and calculations above should be performed only for indices $m_k > 0$. This of course will not ruin the condition $m_1 + \cdots + m_N = n$ because not adding the zero indices do not change that sum. Also, multiplying (or not) by $L_m^{(-1)}(2\lambda) k^m$ for the problematic index $m = 0$ is multiplying by the number one, so that does not change the product. Let us gain some practice with the system $\{L_m^{(-1)}\}_m$ by making simple calculations.

Example 74.0.182. The Laguerre polynomials of with parameter -1 satisfy

$$\exp\left(\frac{-xz}{1-z}\right) = \sum_{n=0}^{\infty} L_n^{(-1)}(x) z^n.$$

Let us compute the first few polynomials using that definition. We denote $f(z) = \exp\left(\frac{-xz}{1-z}\right)$. Here, x is fixed. We have $f(0) = 1$. So,

$$f'(z) = f(z) \cdot \frac{-x}{(z-1)^2}. \text{ Hence } f'(0) = -x.$$

$$f''(z) = f(z) \cdot \left(\frac{x^2 + 2x(z-1)}{(z-1)^4}\right). \text{ Hence } \frac{f''(0)}{2} = \frac{1}{2}(x^2 - 2x).$$

$$f'''(z) = f(z) \cdot \left(\frac{-x^3}{(z-1)^6} + \frac{-2x}{(z-1)^5} + \frac{-4x^2}{(z-1)^5} + \frac{-6x}{(x-1)^4}\right).$$

Hence, $\quad \dfrac{f'''(0)}{3!} = \dfrac{1}{6} \cdot (-x^3 + 6x^2 - 6x).$

$$L_0^{(-1)}(x) = 1, \; L_1^{(-1)}(x) = -x, \; L_2^{(-1)}(x) = \frac{1}{2}(x^2 - 2x),$$

$$L_3^{(-1)}(x) = \frac{1}{6} \cdot (-x^3 + 6x^2 - 6x).$$

Let us verify the computation by re-doing it using the Rodrigues' formula.

$$L_n^{(-1)}(x) = \frac{1}{n! e^{-x} x^{-1}} \frac{d^n}{dx^n}\left(e^{-x} x^{n-1}\right).$$

For $n = 0$:

$$\frac{1}{0! e^{-x} x^{-1}} \cdot e^{-x} x^{-1} = 1.$$

For $n = 1$:

$$\frac{1}{1!e^{-x}x^{-1}}(-e^{-x}) = -x.$$

For $n = 2$:

$$\frac{1}{2!e^{-x}x^{-1}}(e^{-x} - e^{-x}x)' = \frac{1}{2!e^{-x}x^{-1}}(-e^{-x} + e^{-x}x - e^{-x}) = \frac{1}{2}(x^2 - 2x).$$

Also, the explicit forms agree with the above calculations. Next, let us compute the first few normalizations:

$$\int_0^\infty (L_n^{(-1)}(x))^2 x^{-1} e^{-x} dx.$$

For $n = 0$, $\int_0^\infty 1^2 \cdot x^{-1} e^{-x} dx$ diverges.

For $n = 1$, $\int_0^\infty (-x)^2 x^{-1} e^{-x} dx = \Gamma(2) = 1$.

For $n = 2$, $\int_0^\infty (\frac{1}{2}(x^2 - 2x))^2 x^{-1} e^{-x} dx = \frac{1}{4}\int_0^\infty (x^3 - 4x^2 + 4x)e^{-x} dx = \frac{1}{4} \times (\Gamma(4) - 4\Gamma(3) + 4\Gamma(1)) = \frac{1}{4}(6 - 8 + 4) = \frac{1}{2}$.

For $n = 3$, $\int_0^\infty (\frac{-x}{6}(6 - 6x + x^2))^2 x^{-1} e^{-x} dx = \frac{1}{36}\int_0^\infty e^{-x}x(36 + 36x^2 + x^4 - 72x + 12x^2 - 12x^3)dx = \frac{1}{36}(36\Gamma(2) + 36\Gamma(4) + \Gamma(6) - 72\Gamma(3) + 12\Gamma(4) - 12\Gamma(5)) = \frac{1}{3}$. These agree with $\int_0^\infty e^{-x}x^{-1}(L_n^{(-1)}(x))^2 dx = \frac{1}{n}$, for $n \geq 1$. Let us check orthogonality with $L_0^{(-1)} \equiv 1$:

With $L_1^{(-1)}(x)$, $\int_0^\infty e^{-x}x^{-1}L_0^{(-1)}(x)L_1^{(-1)}(x)dx = \int_0^\infty e^{-x}x^{-1}(-x)dx = -1$. Not orthogonal.

With $L_2^{(-1)}(x)$, $\int_0^\infty e^{-x}x^{-1}\frac{1}{2}(x^2 - 2x)dx = \frac{1}{2}(\Gamma(2) - 2\Gamma(1)) = 0$. Orthogonal.

With $L_3^{(-1)}(x)$, $\int_0^\infty e^{-x}x^{-1}(\frac{-x}{6}(6 - 6x + x^2))dx = -\frac{1}{6}(6\Gamma(1) - 6\Gamma(2) + \Gamma(3)) = -\frac{1}{6}(6 - 6 + 2) = -\frac{1}{3}$. Not orthogonal.

On the other hand, orthogonality for $n \neq m \geq 1$:

With $L_1^{(-1)}(x)$, $L_3^{(-1)}(x)$, $\int_0^\infty e^{-x}x^{-1}(-x)\frac{-x}{6}(6 - 6x + x^2)dx = \frac{1}{6}(6\Gamma(2) - 6\Gamma(3) + \Gamma(4)) = 0$. Orthogonal.

With $L_2^{(-1)}(x)$, $L_3^{(-1)}(x)$, $\int_0^\infty e^{-x} x^{-1} (\frac{1}{2}(x^2 - x))(\frac{-x}{6}(6 - 6x + x^2)) dx = 0$. Orthogonal.

We saw that the fact that $\{L_n^{(-1)}(x)\}_{n \geq 1}$ is orthogonal (with the weight $e^{-x} x^{-1}$) while $\{L_n^{(-1)}(x)\}_{n \geq 0}$ in not orthogonal and furthermore, the normalization of $L_0^{(-1)}(x)$ diverges, causes complications in deriving integral identities on the Taylor coefficients of inner functions in U.

Chapter 75

A More Accurate Computation Taking into Account the Anomaly of $L_0^{(-1)}(x)$

By equation (74.0.1), we have the following:

$$a_n = \sum_{\substack{m_1+\cdots+m_N=n \\ m_1,\ldots,m_N \geq 0}} \prod_{j=1}^{N} \left\{ e^{-\lambda_j} L_{m_j}^{(-1)}(2\lambda_j) k_j^{m_j} \right\}.$$

Hence, for a fixed index k, $1 \leq k \leq N$ and any positive m_k, $0 < m_k \leq n$, we obtain by orthogonality:

$$\int_0^{\infty} (2\lambda_k)^{-1} e^{-\lambda_k} L_{m_k}^{(-1)}(2\lambda_k) a_n(\lambda_1,\ldots,\lambda_N; k_1,\ldots,k_N) d\lambda_k$$

$$= \sum_{\substack{m_1+\cdots+m_N=n \\ m_1,\ldots,m_N \geq 0}} \left(\frac{1}{2m_k} \right) k_k^{m_k} \prod_{\substack{j=1 \\ j \neq k}}^{N} \left\{ e^{-\lambda_j} L_{m_j}^{(-1)}(2\lambda_j) k_j^{m_j} \right\}.$$

We repeat this integration process for each one of the indices $1,\ldots,N$ for which $m_j > 0$. Each integration is induced by a different such an index j for which $m_j > 0$. The sum of the total number of the positive m_j's is exactly n. The indices i are those for which $m_i = 0$ do not contribute to the total sum $m_1 + \cdots + m_N = n$. The result is for $m_{i_1},\ldots,m_{i_L} > 0$ such that $L \leq N$

and $m_{i_1} + \cdots + m_{i_L} = n$ we have the identity

$$
\int_0^\infty \cdots \int_0^\infty \left(\prod_{j=1}^L \lambda_{i_j}^{-1} e^{-\lambda_{i_j}} L_{m_{i_j}}^{(-1)}(2\lambda_{i_j}) \right)
$$

$$
\times a_n(\lambda_1, \ldots, \lambda_N; k_1, \ldots, k_N) d\lambda_{i_1} \ldots d\lambda_{i_L}
$$

$$
= \prod_{j=1}^L \left(\frac{k_{i_j}^{m_{i_j}}}{m_{i_j}} \right).
$$

This can be written as follows:

$$
\int_0^\infty \cdots \int_0^\infty \left(\prod_{j=1}^L e^{-\lambda_{i_j}} L_{m_{i_j}}^{(-1)}(2\lambda_{i_j}) \overline{k}_{i_j}^{m_{i_j}} \right)
$$

$$
\times a_n(\lambda_1, \ldots, \lambda_N; k_1, \ldots, k_N) \prod_{j=1}^L \lambda_{i_j}^{-1} d\lambda_{i_1} \ldots d\lambda_{i_L}
$$

$$
= \prod_{j=1}^L \left(\frac{1}{m_{i_j}} \right).
$$

We note that the right-hand side of this identity is independent of $k_1, \ldots, k_N$ and also of $\{\lambda_k \,|\, k \in \{1, \ldots, N\} - \{i_1, \ldots, i_L\}\}$. Let us take λ_k for which $k \in \{\{1, \ldots, N\} - \{i_1, \ldots, i_L\}\}$. Let us multiply both sides by $e^{-\lambda_k}$ and integrate $\int_0^\infty \ldots d\lambda_k$. We get the following:

$$
\int_0^\infty \cdots \int_0^\infty e^{-\lambda_k} \left(\prod_{j=1}^L e^{-\lambda_{i_j}} L_{m_{i_j}}^{(-1)}(2\lambda_{i_j}) \overline{k}_{i_j}^{m_{i_j}} \right)
$$

$$
\times a_n(\lambda_1, \ldots, \lambda_N; k_1, \ldots, k_N) \prod_{j=1}^L \lambda_{i_j}^{-1} d\lambda_{i_1} \ldots d\lambda_{i_L}
$$

$$
= \prod_{j=1}^L \left(\frac{1}{m_{i_j}} \right) \int_0^\infty e^{-\lambda_k} d\lambda_k = \prod_{j=1}^L \left(\frac{1}{m_{i_j}} \right).
$$

We note the following fact:

$$
\sum_{\substack{m_1 + \cdots + m_N = n \\ m_1, \ldots, m_N \geq 0}} \prod_{j=1}^N \left\{ e^{-\lambda_j} L_{m_j}^{(-1)}(2\lambda_j) \overline{k}_j^{m_j} \right\} = \overline{a}_n(\lambda_1, \ldots, \lambda_N; k_1, \ldots, k_N).
$$

If we sum all of the identities we have, the result is

$$
\int_0^\infty \cdots \int_0^\infty \sum_{\substack{m_{i_1}+\cdots+m_{i_L}=n \\ m_{i_1},\ldots,m_{i_L}>0 \\ 1\le L\le N}} \left(\prod_{j=1}^L \lambda_{i_j}^{-1}\right)\left(\prod_{j=1}^L e^{-\lambda_{i_j}} L_{m_{i_j}}^{(-1)}(2\lambda_{i_j})\overline{k}_{i_j}^{\,m_{i_j}}\right)
$$

$$
\times a(\lambda_1,\ldots,\lambda_N;k_1,\ldots,k_N)d\lambda_1\ldots\lambda_N
$$

$$
= \sum_{\substack{m_{i_1}+\cdots+m_{i_L}=n \\ m_{i_1},\ldots,m_{i_L}>0 \\ 1\le L\le N}} \prod_{j=1}^L \left(\frac{1}{m_{i_j}}\right),
$$

where in the integral we extended the definition of the m_{i_j} by $m_{i_j}=0$ for $L<j\le N$. So, the sum in the integral is a kind of a weighted sum for $\overline{a}_n(\lambda_1,\ldots,\lambda_N;k_1,\ldots,k_N)$ with weights $\prod_{j=1}^L \lambda_{i_j}^{-1}$ that correspond to the positive elements $m_{i_j}>0$ in the sum that represents n as a sum of N non-negative integers.

Chapter 76

Properties of Generalized Laguerre Polynomials $L_n^{(-1)}(x)$

We quote from the classical book [101]. In Section 5.2 (p. 102), the book offers a generalization of the Laguerre polynomials. Initially, those polynomials $\{L_n^{(\alpha)}(x)\}$ were defined for $\alpha > -1$ by the following conditions of orthogonality and normalization (p. 100):

$$\int_0^\infty e^{-x} x^\alpha L_n^{(\alpha)}(x) L_m^{(\alpha)}(x)\,dx = \Gamma(\alpha+1)\binom{n+\alpha}{n}\delta_{nm}, \quad n,\, m = 0, 1, 2, \dots.$$

$$(76.0.1)$$

In addition, it is required that the coefficient of x^n in the polynomial $L_n^{(\alpha)}(x)$ of degree n has the sign $(-1)^n$. Also, the following short notation is standard, $L_n^{(0)}(x) = L_n(x)$. The analog of Rodrigues' formula is

$$e^{-x} x^\alpha L_n^{(\alpha)}(x) = \frac{1}{n!}\left(\frac{d}{dx}\right)^n (e^{-x} x^{n+\alpha}). \tag{76.0.2}$$

The explicit representation is

$$L_n^{(\alpha)}(x) = \sum_{\nu=0}^{n}\binom{n+\alpha}{n-\nu}\frac{(-x)^\nu}{\nu!}. \tag{76.0.3}$$

Also,

$$L_n^{(\alpha)}(0) = \binom{n+\alpha}{n}, \tag{76.0.4}$$

433

and

$$l_n^{(\alpha)} = \frac{(-1)^n}{n!}$$
(76.0.5)

is the coefficient of x^n in $L_n^{(\alpha)}(x)$. As a generating function,

$$L_0^{(\alpha)}(x) + L_1^{(\alpha)}(x)w + \cdots + L_n^{(\alpha)}(x)w^n + \cdots = (1-w)^{-\alpha-1}\exp\left(-\frac{xw}{1-w}\right).$$
(76.0.6)

Recurrence,

$$L_0^{(\alpha)}(x) = 1, \ L_1^{(\alpha)}(x) = x + \alpha + 1,$$

$$nL_n^{(\alpha)}(x) = (-x + 2n + \alpha - 1)L_{n-1}^{(\alpha)}(x) - (n + \alpha - 1)L_{n-2}^{(\alpha)}(x),$$

$$n = 2, 3, 4, \ldots.$$
(76.0.7)

For the "kernel polynomial" $K_n^{(\alpha)}(x, y)$, we have

$$\Gamma(\alpha + 1)K_n^{(\alpha)}(x, y) = \sum_{\nu=0}^{n}\left\{\binom{\nu + \alpha}{\nu}\right\}^{-1} L_\nu^{(\alpha)}(x)L_\nu^{(\alpha)}(y)$$

$$= (n + 1)\left\{\binom{\nu + \alpha}{\nu}\right\}^{-1}$$

$$\times \frac{L_n^{(\alpha)}(x)L_{n+1}^{(\alpha)}(y) - L_{n+1}^{(\alpha)}(x)L_n^{(\alpha)}(y)}{x - y}.$$
(76.0.8)

The special case $y = 0$ is particularly important:

$$\sum_{\nu=0}^{n} L_\nu^{(\alpha)}(x) = (n + \alpha + 1)L_n^{(\alpha)}(x) - (n + 1)L_{n+1}^{(\alpha)}(x).$$
(76.0.9)

One can deduce by (76.0.3) or (76.0.6),

$$\sum_{\nu=0}^{n} L_\nu^{(\alpha)}(x) = L_n^{(\alpha+1)}(x), \ L_n^{(\alpha)}(x) = L_n^{(\alpha+1)}(x) - L_n^{(\alpha+1)}(x), \quad (76.0.10)$$

$$\frac{d}{dx}L_n^{(\alpha)}(x) = -L_{n-1}^{(\alpha+1)}(x) = x^{-1}\{nL_n^{(\alpha)}(x) - (n + \alpha)L_{n-1}^{(\alpha)}(x)\}.$$
(76.0.11)

Theorem 5.1 ([101]). *Let*

$$J_\alpha(z) = \sum_{\nu=0}^{\infty} \frac{(-1)^\nu \left(\frac{z}{2}\right)^{\alpha+2\nu}}{\nu!\,\Gamma(\nu+\alpha+1)},$$

be the Bessel function of the first kind of order α. Then

$$\sum_{n=0}^{\infty} \left\{ \binom{n+\alpha}{n} \right\}^{-1} L_n^{(\alpha)}(x) L_n^{(\alpha)}(y) w^n$$

$$= \Gamma(\alpha+1)(1-w)^{-1} \exp\left\{ -(x+y)\frac{w}{1-w} \right\} (-xyw)^{-\alpha/2}$$

$$\times J_\alpha\left\{ \frac{2(-xyw)^{1/2}}{1-w} \right\}, \tag{76.0.12}$$

and

$$\sum_{n=0}^{\infty} \frac{L_n^{(\alpha)}(x)}{\Gamma(n+\alpha+1)} = e^w (xw)^{-\alpha/2} J_\alpha\left\{ 2(xw)^{1/2} \right\}. \tag{76.0.13}$$

Now, the generalization:

By means of (76.0.3), the definition of Laguerre polynomials can be extended to arbitrary complex values of α. No reduction in the degree ever occurs. For $n \geq 1$, we have $L_n^{(\alpha)}(0) = 0$ if and only if $\alpha = -k$, k integral, $1 \leq k \leq n$. In this case, $x = 0$ is a zero of precise order k, and from (76.0.3) we obtain

$$L_n^{(-k)}(x) = (-x)^k \frac{(n-k)!}{n!} \sum_{\nu=0}^{n-k} \binom{n}{n-k-\nu} \frac{(-x)^\nu}{\nu!}$$

$$= (-x)^k \frac{(n-k)!}{n!} L_{n-k}^{(k)}(x). \tag{76.0.14}$$

Formula (76.0.3) remains true for arbitrary α.

Chapter 77

Checking the 2-Bound and the Role of the λ-Weighted Moments of the Uni-modular Parameters k_j

The 2-bound was presented in equation (74.0.2), namely, it was asked whether

$$\left| \sum_{\substack{m_1+\cdots+m_N=n \\ m_1,\ldots,m_N \geq 0}} \prod_{j=1}^{N} \left\{ L_{m_j}^{(-1)}(2\lambda_j) k_j^{m_j} \right\} \right| \leq 2 \quad (N \leq n, \lambda_j > 0, |k_j| = 1)\,?$$

That clearly cannot be true. Since the relaxing factors e^{λ_j} were taken out of the equation, it is rather obvious (and we will demonstrate shortly) that there cannot be any (finite) upper bound. But in naively trying to compute the sum above, we will come across the interesting fact that it is always representable as a polynomial in the λ-weighted moments of the k_j. Let us consider the first few cases of n in (74.0.2). We recall that we computed the first few (-1)-Laguerre polynomials:

$$L_0^{(-1)}(x) \equiv 1, \; L_1^{(-1)}(x) = -x, \; L_2^{(-1)}(x) = \frac{1}{2}(x^2 - 2x),$$

$$L_3^{(-1)}(x) = \frac{1}{6}(-x^3 + 6x^2 - 6x),\ldots.$$

The case $n = 1$: Then $N = 1$ and $m_1 = 1$, so we have $|L_1^{(-1)}(2\lambda_1)k_1| = 2\lambda_1$. That already provides a trivial negative answer to the question above ("Is the 2-bound valid?").

However, regardless of that, let us continue the computations of the sum in the left-hand side of (74.0.2).

The case $n = 2$: Then there are two possibilities, $N = 1$ or $N = 2$. If $N = 1$, then $m_1 = 2$, so that we have

$$|L_2^{(-1)}(2\lambda_1)k_1^2| = \frac{1}{2}|((2\lambda_1)^2 - 2\cdot 2\lambda_1)| = 2\lambda_1|\lambda_1 - 1|.$$

If $N = 2$, then there are 3 terms in the sum. These correspond to $m_1 = 0$, $m_2 = 2$ or $m_1 = m_2 = 1$ or $m_1 = 0$, $m_2 = 2$. So, we have

$$|L_2^{(-1)}(2\lambda_1)L_0^{(-1)}(2\lambda_2)k_1^2 + L_1^{(-1)}(2\lambda_1)L_1^{(-1)}(2\lambda_2)k_1k_2$$
$$+L_0^{(-1)}(2\lambda_1)L_2^{(-1)}(2\lambda_2)k_2^2|$$
$$= \left|\frac{1}{2}((2\lambda_1)^2 - 2\cdot(2\lambda_1))k_1^2 + (-2\lambda_1)(-\lambda_2)k_1k_2 + \frac{1}{2}((2\lambda_2)^2 - 2\cdot(2\lambda_2))k_2^2\right|$$
$$= \left|2\lambda_1(\lambda_1 - 1)k_1^2 + 4\lambda_1\lambda_2k_1k_2 + 2\lambda_2(\lambda_2 - 1)k_2^2\right|$$
$$= 2\left|(\lambda_1k_1 + \lambda_2k_2)^2 - (\lambda_1k_1^2 + \lambda_2k_2^2)\right|.$$

The next case $n = 3$ is cumbersome. So, we take just one possibility.

The case $n = N = 3$: The multi-index (m_1, m_2, m_3) runs through the following possibilities:

$$(3,0,0), (2,1,0), (2,0,1), (1,2,0), (1,1,1), (1,0,2), (0,3,0), (0,2,1),$$
$$(0,1,2), (0,0,3).$$

For example, $(3,0,0)$ induces $L_3^{(-1)}(2\lambda_1)L_0^{(-1)}(2\lambda_2)L_0^{(-1)}(2\lambda_3)k_1^3$. The multi-index $(1,0,2)$ induces $L_1^{(-1)}(2\lambda_1)L_0^{(-1)}(2\lambda_2)L_2^{(-1)}(2\lambda_3)k_1k_3^2$, and so on.... Plugging in the expressions of the corresponding Laguerre polynomials and performing the required algebraic manipulations, we obtain, as the final result in this case,

$$2\left\{-(\lambda_1k_1^3 + \lambda_2k_2^3 + \lambda_3k_3^3) + 4(\lambda_1k_1^2 + \lambda_2k_2^2 + \lambda_3k_3^2)(\lambda_1k_1 + \lambda_2k_2 + \lambda_3k_3)\right.$$
$$\left. -\frac{4}{3}(\lambda_1k_1 + \lambda_2k_2 + \lambda_3k_3)^3\right\}.$$

All these demonstrate the following result which probably is well known to experts.

Theorem 77.0.183. *The following sum*

$$\sum_{\substack{m_1+\cdots+m_N=n \\ m_1,\ldots,m_N \geq 0}} \prod_{j=1}^{N} \left\{ L_{m_j}^{(-1)}(2\lambda_j) k_j^{m_j} \right\}$$

is a polynomial in the λ-weighted moments of the k_j, $\lambda_1 k_1^m + \cdots + \lambda_N k_N^m$ of homogeneous degree n in the uni-modular parameters $k_1,\ldots,k_N$, and with rational coefficients.

Coming back towards (maybe) fixing the 2-bound question, we inquire about the following type of questions. Namely, how large can the following expression $|e^{-\lambda} L_m^{(-1)}(2\lambda)|$ be for a fixed $m \in \mathbb{Z}_{\geq 0}$ over the open half line $\lambda > 0$? Since $\lim_{\lambda \to 0} |e^{-\lambda} L_m^{(-1)}(2\lambda)| = 0$, it is clear that the supremum is attained possibly for $\lambda = 0$ (which out of our half a line). We might try to do the obvious and λ-differentiate and equate the derivative to zero. We arrive at

$$2(L_m^{(-1)})'(2\lambda) - L_m^{(-1)}(2\lambda) = 0.$$

By equation (5.1.4) in [101], we have the following:

$$\frac{d}{dx} L_m^{(\alpha)}(x) = -L_{m-1}^{(\alpha+1)}(x) = x^{-1}\{m \cdot L_m^{(\alpha)}(x) - (m+\alpha)L_{m-1}^{(\alpha)}(x)\}.$$

Chapter 78

A Corrected Version
of the Integral Identity

Let us fix an integer $0 < m \le N$ and a positive parameter $\lambda_k \in \{\lambda_1, \ldots, \lambda_N\}$. We have three types of relations which are orthogonality, normalization and a third type that involves $L_0^{(-1)}(x)$. This originates in the fact that the system of Laguerre polynomials $\{L_m^{(-1)}(x)\}_{m \ge 1}$ is orthogonal and with finite normalizations with respect to the weight $w(x) = x^{-1}e^{-x}$. On the other hand, the system $\{L_m^{(-1)}(x)\}_{m \ge 0}$ is not orthogonal and the normalization of $L_0^{(-1)}(x)$ diverges. We summarize the three types of relations in the following:

$$\int_0^\infty L_k^{(-1)}(2x) L_m^{(-1)}(2x)(2x)^{-1}e^{-2x}dx = \begin{cases} 0 & \text{if} \quad 0 < k \ne m \\ \frac{1}{2m} & \text{if} \quad k = m \\ I_m & \text{if} \quad k = 0 \end{cases}.$$

We note that the third type is the result of integrating $L_m^{(-1)}(x)$ using the Laguerre weight $w(2x) = (2x)^{-1}e^{-2x}$, i.e.,

$$I_m = \int_0^\infty L_m^{(-1)}(2x)(2x)^{-1}e^{-2x}dx.$$

This follows by the fact that the zero Laguerre polynomial $L_0^{(-1)}(2x) \equiv 1$ is not necessarily orthogonal to $L_m^{(-1)}(2x)$, $m > 0$. So, we split the representation of the Taylor coefficient $a_n(\lambda_1, \ldots, \lambda_N; k_1, \ldots, k_N)$ into three (and not into two) components:

$$(2\lambda_k)^{-1}e^{-\lambda_k}L_m^{(-1)}(2\lambda_k)a_n(\lambda_1,\ldots,\lambda_N;k_1,\ldots,k_N)$$

$$= \sum_{\substack{m_1+\cdots+m_N=n \\ m_1,\ldots,m_N\geq 0 \\ m_k=m}} (2\lambda_k)^{-1}e^{-2\lambda_k}(L_{m_k}^{(-1)}(2\lambda_k))^2 k_k^{m_k}$$

$$\times \cdot \prod_{\substack{j=1 \\ j\neq k}}^{N}\left\{e^{-\lambda_j}L_{m_j}^{(-1)}(2\lambda_j)k_j^{m_j}\right\}$$

$$+ (2\lambda_k)^{-1}e^{-\lambda_k}\sum_{\substack{m_1+\cdots+m_N=n \\ m_1,\ldots,m_N\geq 0 \\ 0<m_k\neq m}} L_m^{(-1)}(2\lambda_k)\prod_{j=1}^{N}\left\{e^{-\lambda_j}L_{m_j}^{(-1)}(2\lambda_j)k_j^{m_j}\right\}$$

$$+ (2\lambda_k)^{-1}e^{-\lambda_k}\sum_{\substack{m_1+\cdots+m_N=n \\ m_1,\ldots,m_N\geq 0 \\ m_k=0}} L_m^{(-1)}(2\lambda_k)\prod_{j=1}^{N}\left\{e^{-\lambda_j}L_{m_j}^{(-1)}(2\lambda_j)k_j^{m_j}\right\}.$$

Integrating the last identity $\int_0^\infty \ldots d\lambda_k$ gives

$$\int_0^\infty (2\lambda_k)^{-1}e^{-\lambda_k}L_m^{(-1)}(2\lambda_k)a_n(\lambda_1,\ldots,\lambda_N;k_1,\ldots,k_N)d\lambda_k$$

$$= \sum_{\substack{m_1+\cdots+m_N=n \\ m_1,\ldots,m_N\geq 0 \\ m_k=m}} \left(\frac{1}{2m}\right)k_k^m \prod_{\substack{j=1 \\ j\neq k}}^{N}\left\{e^{-\lambda_j}L_{m_j}^{(-1)}(2\lambda_j)k_j^{m_j}\right\}$$

$$+ \sum_{\substack{m_1+\cdots+m_N=n \\ m_1,\ldots,m_N\geq 0 \ \ m_k=0}} I_m \cdot \prod_{\substack{j=1 \\ j\neq k}}^{N}\left\{e^{-\lambda_j}L_{m_j}^{(-1)}(2\lambda_j)k_j^{m_j}\right\}.$$

We can repeat this integration process for each one of the indices $1,\ldots,N$. It is crucial to choose in each step a positive integer $m>0$. Choosing $m=0$ and integrating will lead to divergent integrals. We recall (by (74.0.1)) that

$$a_n = \sum_{\substack{m_1+\cdots+m_N=n \\ m_1,\ldots,m_N\geq 0}} \prod_{j=1}^{N}\left\{e^{-\lambda_j}L_{m_j}^{(-1)}(2\lambda_j)k_j^{m_j}\right\}.$$

Our process consists of multiplying by $(2\lambda_k)^{-1}e^{-\lambda_k}L_m^{(-1)}(2\lambda_k)$ and then integrating $\int_0^\infty \ldots d\lambda_k$. We do that with $m>0$. We repeat doing that till the sum

$\sum m$ equals n. If $m_k = 0$, we get I_m but since $m_1 + \cdots + m_k + \cdots + m_N = n$ ($m_k = 0$) necessarily for some k $m_k \neq m$ and by orthogonality, we get 0. Why don't we get a zero? Exactly when our m's equal respectively only the positive m_j's in $m_1 + \cdots + m_N = n$, this gives

$$\prod_{\substack{j=1 \\ m_j > 0}} \left(\frac{1}{2m_j} \right) k_j^{m_j} \prod_{\substack{j=1 \\ m_j = 0}} e^{-\lambda_j}.$$

All the other terms in

$$\sum_{\substack{m_1 + \cdots + m_N = n \\ m_1, \ldots, m_N \geq 0}} \prod_{j=1}^{N} \{ e^{-\lambda_j} L_{m_j}^{(-1)}(2\lambda_j) k_j^{m_j} \}$$

are annihilated. So, we have

$$\int_0^\infty \cdots \int_0^\infty \left\{ \prod_{\substack{j=1 \\ m_j > 0}}^{N} (2\lambda_j)^{-1} e^{-\lambda_j} L_{m_j}^{(-1)}(2\lambda_j) \right\}$$

$$\times a_n(\lambda_1, \ldots, \lambda_N; k_1, \ldots, k_N) d\lambda_{i_1} \ldots d\lambda_{i_l}$$

$$= \prod_{j=1}^{l} \left(\frac{1}{2m_{i_j}} \right) k_{i_j} m_{i_j} \prod_{s=1}^{N-l} e^{-\lambda_{k_s}} k_{k_s}^0,$$

where $m_1 + \cdots + m_N = n$, $m_1, \ldots, m_N \geq 0$, the positive m's are the following l ones: $m_{i_1}, \ldots, m_{i_l} > 0$ (meaning that $m_{i_1} + \cdots + m_{i_l} = n$ and $m_t = 0$ $\forall t \notin \{i_1, \ldots, i_l\}$). The complementary $N - l$ indices are $k_1, \ldots, k_{N-l}$ ($m_{k_1} = \cdots = m_{k_{N-l}} = 0$). If we multiply both sides by $\prod_{s=1}^{N-l} e^{-\lambda_{k_s}}$ and integrate $\int_0^\infty \cdots \int_0^\infty \ldots d\lambda_{i_1} \ldots d\lambda_{i_{N-l}}$, we obtain

$$\int_0^\infty \cdots \int_0^\infty \left\{ \prod_{j=1}^{l} (2\lambda_{i_j})^{-1} e^{-\lambda_{i_j}} L_{m_{i_j}}^{(-1)}(2\lambda_{i_j}) \right\} \left\{ \prod_{s=1}^{N-l} e^{-\lambda_{k_s}} \right\} a_n d\lambda_1 \ldots d\lambda_N$$

$$= \prod_{j=1}^{l} \left(\frac{1}{2m_{i_j}} \right) k_{i_j}^{m_{i_j}} \cdot \left(\frac{1}{2} \right)^{N-l}. \tag{78.0.1}$$

A second possibility is not to multiply and only integrate $\int_0^\infty \ldots$ $\int_0^\infty \ldots d\lambda_{k_1} \ldots d\lambda_{k_{N-l}}$. This time, we get

$$\int_0^\infty \cdots \int_0^\infty \left\{ \prod_{j=1}^{l} (2\lambda_{i_j})^{-1} e^{-\lambda_{i_j}} L_{m_{i_j}}^{(-1)}(2\lambda_{i_j}) \right\}$$

$$\times a_n(\lambda_1, \ldots, \lambda_N; k_1, \ldots, k_N) d\lambda_1 \ldots d\lambda_N$$

$$= \prod_{j=1}^{l} \left(\frac{1}{2m_{i_j}} \right) k_{i_j}^{m_{i_j}}. \tag{78.0.2}$$

Expanding (78.0.1): Let us divide both sides by $\prod_{j=1}^{l} (k_{i_j}/2)^{m_{i_j}}$, remembering that $1/k_{i_j}^{m_{i_j}} = \overline{k}_{i_j}^{m_{i_j}}$. The result is

$$\int_0^\infty \cdots \int_0^\infty \left(\prod_{j=1}^{l} \lambda_{i_j} \right)^{-1} \left(\prod_{j=1}^{N} e^{-\lambda_j} L_{m_j}^{(-1)}(2\lambda_j) \overline{k}_j^{m_j} \right) a_n d\lambda_1 \ldots d\lambda_N$$

$$= \left(\frac{1}{2} \right)^{N-l} \prod_{j=1}^{l} \left(\frac{1}{m_{i_j}} \right). \tag{78.0.3}$$

Expanding (78.0.2): We apply a similar transformation and obtain

$$\int_0^\infty \cdots \int_0^\infty \left(\prod_{j=1}^{l} \lambda_{i_j} \right)^{-1} \left(\prod_{j=1}^{l} e^{-\lambda_{i_j}} L_{m_{i_j}}^{(-1)}(2\lambda_{i_j}) \overline{k}_{i_j}^{m_{i_j}} \right) a_N d\lambda_1 \ldots d\lambda_N$$

$$= \prod_{j=1}^{l} \left(\frac{1}{m_{i_j}} \right). \tag{78.0.4}$$

Comparing (78.0.3) with (78.0.4), we obtain

$$2^{N-l} \int_0^\infty \cdots \int_0^\infty \left(\prod_{j=1}^{l} \lambda_{i_j} \right)^{-1} \left(\prod_{j=1}^{N} e^{-\lambda_j} L_{m_j}^{(-1)}(2\lambda_j) \overline{k}_j^{m_j} \right) a_n d\lambda_1 \ldots d\lambda_N$$

$$= \int_0^\infty \cdots \int_0^\infty \left(\prod_{j=1}^{l} \lambda_{i_j} \right)^{-1} \left(\prod_{j=1}^{l} e^{-\lambda_{i_j}} L_{m_{i_j}}^{(-1)}(2\lambda_{i_j}) \overline{k}_{i_j}^{m_{i_j}} \right) a_N d\lambda_1 \ldots d\lambda_N$$

$$= \prod_{j=1}^{l} \left(\frac{1}{m_{i_j}} \right).$$

A remarkable feature of these identities is the fact that the left-hand side includes the uni-modular parameters k_j, but the right-hand side is independent of these parameters. The integral averaging on the positive parameters, λ_j, eliminates also the uni-modular parameters k_j.

Chapter 79

$\frac{2}{e}$ Again

We refer to [3]. Here is the introduction of this nice paper: "If $\{z_n\}$ is a sequence (finite or infinite) of complex numbers of modulus less than 1 such that $\sum (1 - |z_n|) < \infty$, then the Blaschke product

$$B(z) = \prod_n \frac{|z_n|}{z_n} \cdot \frac{z_n - z}{1 - \bar{z}_n z},$$

converges uniformly on compact subsets of the unit disc U. We let $\mathcal{B}_\infty$ denote the set of Blaschke products whose zero sequences are infinite. In this paper, we show that

$$\inf_{B \in \mathcal{B}_\infty} \limsup_{r \to 1} (1 - r)^{-1} \int_{-\pi}^{\pi} \left(1 - \left|B(re^{i\theta})\right|\right)^2 \frac{d\theta}{2\pi}$$

$$= \max_{0 \le x < 1} (1 + \sqrt{1 - x})_2 F_1 \left(^{1/2} 2^{1/2}; x\right) = \gamma_0$$

where $_2F_1 \left(^{1/2} 2^{1/2}; x\right)$ is a hypergeometric function. Using one of Gauss' identities for the hypergeometric functions and tables for the complete elliptic integrals, we obtain the estimate $\gamma_0 \ge 0.285$. We have two applications of this result. In [79], it was shown that if $B(z) = \sum_{n=0}^{\infty} a_n z^n \in \mathcal{B}_\infty$ then $\limsup_{n \to \infty} n|a_n| \ge \frac{1}{\pi} = 0.3183\ldots$. We improve this to show that if $B(z) = \sum_{n=0}^{\infty} a_n z^n \in \mathcal{B}_\infty$ then $\limsup_{n \to \infty} n|a_n| \ge \sqrt{\frac{\gamma_0}{2}} \ge 0.37749$. In the other direction, we modify a method of Newman and Shapiro in that same paper to show that if $\epsilon > 0$ is given, there is a $B(z) = \sum_{n=0}^{\infty} a_n z^n \in \mathcal{B}_\infty$ such that $\limsup n|a_n| \le \frac{2}{e} + \epsilon$. We conjecture that if $B(z) = \sum_{n=0}^{\infty} a_n z^n \in \mathcal{B}_\infty$ then $\limsup_{n \to \infty} n|a_n| \ge \frac{2}{e}$". Thus, another $\frac{2}{e}$ conjecture. It is this conjecture we put in the center of attention of this chapter. Let us though proceed with quoting from [3]. "It is well known that if $B(z) = \sum_{n=0}^{\infty} a_n z^n \in \mathcal{B}_\infty$ then $\sum_{n=1}^{\infty} n|a_n|^2 = \infty$. As another application of our main theorem, we improve this to show that if $B(z) = \sum_{n=0}^{\infty} a_n z^n \in \mathcal{B}_\infty$, then $\limsup_{n\to\infty} \sum_{k \in I_n} k|a_k|^2 \geq \frac{\gamma_0}{8} \geq 0.0356$, where $I_n = \{k \mid k \in \mathbb{Z}^+, 2^n \leq k < 2^{n+1}\}$".

It is natural to ask if there is a connection between the following two conjectures on extremal problems.

$$\max\left\{ |a_n| \;\middle|\; f(z) = \sum_{k=0}^{\infty} a_k z^k, \; 0 < |f(z)| \leq 1 \; \forall \, |z| < 1 \right\} = \frac{2}{e},$$

and

$$\inf\left\{ \limsup_{n\to\infty} n|a_n| \;\middle|\; B(z) = \sum_{k=0}^{\infty} a_k z^k \in \mathcal{B}_\infty \right\} = \frac{2}{e}.$$

Let us "play" with some computations related to the above question.

Chapter 80

Non-trivial Blaschke Products

We outline a classical construction. We consider the following inner function:

$$S(z) = \exp\left(-\frac{1+z^n}{1-z^n}\right),$$

where $n \in \mathbb{Z}^+$ is fixed. We compute $\lim_{r \to 1^-} S(re^{i\theta}) = S(e^{i\theta}) = \exp\left(-i\cot\left(\frac{n\theta}{2}\right)\right)$, for every $e^{i\theta} \in \mathbb{T} - \{1, e^{i2\pi/n}, \ldots, e^{i2k\pi/n}, \ldots, e^{i2(n-1)\pi/n}\}$, $0 \leq k \leq n-1$, and $\lim_{r \to 1^-} S(re^{i2k\pi/n}) = 0$. We fix any $\alpha \in U$, $\alpha \neq 0$ and define

$$\Phi(z) = \frac{\alpha - S(z)}{1 - \overline{\alpha}S(z)}, \quad (z \in U).$$

Then, $|\Phi(z)| < 1$, $(z \in U)$, and for every $e^{i\theta} \notin \{1, e^{i2\pi/n}, \ldots, e^{i2(n-1)\pi/n}\}$,

$$\lim_{r \to 1^-} \Phi(re^{i\theta}) = \frac{\alpha - S(e^{i\theta})}{1 - \overline{\alpha}S(e^{i\theta})} \in \mathbb{T}$$

and also $\lim_{r \to 1^-} \Phi(re^{i2k\pi/n}) = \alpha \neq 0$. Hence, $\Phi(z)$ is an inner function and in fact $\Phi(z)$ is a Blaschke product. This follows by the following:

Theorem 1.15 ([71], p. 17). *Let Φ be an inner function for U. Suppose that $\lim_{r \to 1^-} |\Phi(re^{i\theta})| \neq 0$ for all $e^{i\theta} \in \mathbb{T}$. Then Φ is a Blaschke product.*

Remark. The assumption $\lim_{r \to 1^-} |\Phi(re^{i\theta})| \neq 0$ means that either the radial limit $\lim_{r \to 1^-} |\Phi(re^{i\theta})|$ does not exist or it exists but its value is not zero.

The zeros of $\Phi(z)$ are solutions of the equation

$$S(z) = \exp\left(-\frac{1+z^n}{1-z^n}\right) = \alpha.$$

The solutions are given by

$$z_k^n = \frac{\log|\alpha| + i(\arg\alpha + 2k\pi) + 1}{\log|\alpha| + i(\arg\alpha + 2k\pi) - 1}, \quad (k \in \mathbb{Z}).$$

In particular, for $\alpha = \frac{1}{e}$, these zeros are given by,

$$z_k^n = \frac{i \cdot k\pi}{i \cdot k\pi - 1}, \quad (k \in \mathbb{Z}).$$

We note that without using Theorem 1.15, it is not clear that

$$\Phi(z) = \frac{\alpha - \exp\left(-\frac{1+z^n}{1-z^n}\right)}{1 - \overline{\alpha} \cdot \exp\left(-\frac{1+z^n}{1-z^n}\right)}, \quad (\alpha \in U - \{0\}),$$

is a Blaschke product. We note that $\Phi(z) = (T_\alpha \circ S)(z)$ where $T_\alpha(w) = \frac{\alpha - w}{1 - \overline{\alpha}w}$ is the appropriate U-automorphism (sometimes called a Frostman shift). We would like to estimate the Taylor coefficients of $\Phi(z) = \sum_{k=0}^{\infty} a_k z^k$. Our motivation originates in the problem that was raised at the end of Chapter 79. Namely, the functions $S(z)$ are the conjectured extremals (up to rotations) for the first extremal problem (i.e., the Krzyż conjecture). The Frostman shift, applied to these extremals, produces for all the values of the parameter α in the punctured unit disc ($U - \{0\}$) Blaschke products. It is interesting to figure out if those Blaschke products can serve as conjectured extremals for the second extremal problem. If indeed this is the case, then it gives a hope of a connection between the two problems, i.e., via non-trivial Frostman shifts of the unit disc. This estimate of the Taylor coefficients turns out not to be easy. In the sequel we shall outline few possible ways of evaluation. The straightforward way would be probably to represent a_k using integral with respect to the argument θ of $z = re^{i\theta}$. Namely, we arrive at the following standard identity: Let $S(z) = \sum_{k=0}^{\infty} b_k^{(1)} z^k$. Then

$$b_k^{(1)} = \frac{1}{2\pi}\int_0^{2\pi} S(e^{i\theta})e^{-ik\theta}d\theta = \frac{1}{2\pi}\int_0^{2\pi}\exp\left(-i\cot\left(\frac{n\theta}{2}\right)\right)e^{-ik\theta}d\theta$$

$$= \frac{1}{2\pi}\int_0^{2\pi}\exp\left(-i\left(k\theta + \cot\left(\frac{n\theta}{2}\right)\right)\right)d\theta.$$

So

$$b_k^{(1)} = \sum_{j=0}^{\infty}\frac{1}{2\pi j!}\int_0^{2\pi}\left(-i\left(k\theta + \cot\left(\frac{n\theta}{2}\right)\right)\right)^j d\theta.$$

If we use the representations $\Phi(z) = (T_\alpha \circ S)(z)$, we might want the following:

$$T_\alpha(w) = \frac{\alpha - w}{1 - \overline{\alpha}w} = \alpha - (1 - |\alpha|^2)(w + \overline{\alpha}w^2 + \overline{\alpha}^2 w^3 + \overline{\alpha}^3 w^4 + \cdots). \quad (|w| < 1)$$

Hence, $\Phi(z) = \alpha - (1 - |\alpha|^2) \sum_{j=1}^{\infty} \overline{\alpha}^{j-1} S(z)^j$. If we use the notation

$$S(z)^j = \exp\left(-j\frac{1 + z^n}{1 - z^n}\right) = \sum_{l=0}^{\infty} b_l^{(j)} z^{nl},$$

then

$$\Phi(z) = \alpha - (1 - |\alpha|^2) \sum_{j=1}^{\infty} \left(\overline{\alpha}^{j-1} \sum_{l=0}^{\infty} b_l^{(j)} z^{nl}\right)$$

$$= \alpha - (1 - |\alpha|^2) \sum_{l=0}^{\infty} \left(\sum_{j=1}^{\infty} \overline{\alpha}^{j-1} b_l^{(j)}\right) z^{nl}.$$

We might try to express $b_l^{(j)}$ in terms of Laguerre polynomials. Using the generating function, we have

$$\exp\left(\frac{-xw}{1 - w}\right) = L_0^{(-1)}(x) + L_1^{(-1)}(x)w + L_2^{(-1)}(x)w^2 + \cdots.$$

$$\exp\left(-j\frac{1 + z}{1 - z}\right) = \sum_{l=0}^{\infty} e^{-j} L_l^{(-1)}(2j) z^l = \sum_{l=0}^{\infty} b_l^{(j)} z^l,$$

so $b_l^{(j)} = e^{-j} L_l^{(-1)}(2j)$. This gives the following identity: $a_k = 0$ if $n \nmid k$. If $k = n \cdot l$, then

$$a_k = a_{nl} = -(1 - |\alpha|^2) \sum_{j=1}^{\infty} \overline{\alpha}^{j-1} b_l^{(j)}$$

$$= -(1 - |\alpha|^2) \sum_{j=1}^{\infty} \overline{\alpha}^{j-1} e^{-j} L_l^{(-1)}(2j), \quad \text{for } l > 0.$$

Finally, $a_0 = \alpha$. So, for $l \geq 1$, $k = n \cdot l$, we have

$$a_k = -\frac{1 - |\alpha|^2}{\overline{\alpha}} \sum_{j=1}^{\infty} \left(\frac{\overline{\alpha}}{e}\right)^j L_l^{(-1)}(2j).$$

Using equation (5.2.1) in [101, p. 102]:

$$L_l^{(-1)}(2j) = (-2j)\frac{(l-1)!}{l!}\sum_{\nu=0}^{l-1}\binom{l}{l-1-\nu}\frac{(-2j)^\nu}{\nu!}$$

$$= \frac{1}{l}\sum_{\nu=0}^{l-1}\binom{l}{l-1-\nu}\frac{(-2j)^{\nu+1}}{\nu!}.$$

$$a_k = -\frac{1-|\alpha|^2}{l\cdot\overline{\alpha}}\sum_{\nu=0}^{l-1}\binom{l}{l-1-\nu}\frac{1}{\nu!}\left\{\sum_{j=1}^{\infty}\left(\frac{\overline{\alpha}}{e}\right)^j(-2j)^{\nu+1}\right\}.$$

Let us denote $\rho = \frac{\overline{\alpha}}{e}$, then $|\rho| < \frac{1}{e}$. We need $\sum_{j=1}^{\infty}(-2j)^{\nu+1}\rho^j = (-2)^{\nu+1}\sum_{j=1}^{\infty}j^{\nu+1}\rho^j$. Using elementary manipulations on the following geometrical progression $\sum_{j=1}^{\infty}\rho^j x^j = \frac{1}{1-\rho x}$, $(|\rho x| < 1)$, we can compute these sums. The manipulations include the application of three steps: (1) differentiation with respect to x, (2) substituting $x = 1$ to get the formula, (3) multiplying the output of (1) by x. Here are the first few identities:

$$\sum_{j=1}^{\infty}j\rho^j = \frac{(-\rho)(-1)}{(1-\rho)^2}, \quad \sum_{j=1}^{\infty}j^2\rho^j = (-\rho)(-1)\cdot\left\{\frac{-1}{(1-\rho)^2} - \frac{-2}{(1-\rho)^3}\right\},$$

$$\sum_{j=1}^{\infty}j^3\rho^j = (-\rho)(-1)\left\{\frac{1}{(1-\rho)^2} - 3\cdot 2\frac{1}{(1-\rho)^3} + 3\cdot 2\frac{1}{(1-\rho)^4}\right\},$$

$$\sum_{j=1}^{\infty}j^4\rho^j = (-\rho)(-1)\left\{\frac{-1}{(1-\rho)^2} + \frac{2(1+3\cdot 2)}{(1-\rho)^3}\right.$$

$$\left. -\frac{3(3\cdot 2 + 3\cdot 2)}{(1-\rho)^4} + \frac{4\cdot 3\cdot 2}{(1-\rho)^5}\right\},\dots.$$

In general, we extract the recursions as follows:

$$\sum_{j=1}^{\infty}j^k\rho^j x^j = \frac{A_{k,1}}{1-\rho x} + \frac{A_{k,2}}{(1-\rho x)^2} + \frac{A_{k,3}}{(1-\rho x)^3} + \cdots + \frac{A_{k,k+1}}{(1-\rho x)^{k+1}},$$

$$\sum_{j=1}^{\infty}j^{k+1}\rho^j x^{j-1} = \frac{(-\rho)(-1)A_{k,1}}{(1-\rho x)^2} + \frac{(-\rho)(-2)A_{k,2}}{(1-\rho x)^3}$$

$$+ \frac{(-\rho)(-3)A_{k,3}}{(1-\rho x)^4} + \cdots + \frac{(-\rho)(-(k+1))A_{k,k+1}}{(1-\rho x)^{k+2}},$$

$$\sum_{j=1}^{\infty} j^{k+1} \rho^j x^j = (-1) A_{k,1} \frac{(1 - \rho x) - 1}{(1 - \rho x)^2}$$

$$+ (-2) A_{k,2} \frac{(1 - \rho x) - 1}{(1 - \rho x)^3} + \cdots + (-(k+1)) A_{k,k+1}$$

$$\times \frac{(1 - \rho x) - 1}{(1 - \rho x)^{k+2}} \cdot (-l) A_{k,l} \frac{(1 - \rho x) - 1}{(1 - \rho x)^{l+1}} + (-(l+1))$$

$$\times A_{k,l+1} \frac{(1 - \rho x) - 1}{(1 - \rho x)^{l+2}}$$

$$= \cdots \frac{l A_{k,l} - (l+1) A_{k,l+1}}{(1 - \rho x)^{l+1}} \cdots .$$

Finally, we get the following recursions:

$$A_{k+1,l+1} = l A_{k,l} - (l+1) A_{k,l+1}, \quad 1 \le l \le k+1.$$

The initial conditions are clearly

$$A_{1,1} = -1, \quad A_{1,2} = 1.$$

We recall that we are searching for an upper bound of a_n.

Chapter 81

Other Techniques to Estimate the Taylor Coefficients of the Frostman Shifts (the Non-trivial Blaschke Products)

In this chapter, we will apply a useful identity from [79] to estimate the coefficients of

$$\frac{\alpha - S(z)}{1 - \overline{\alpha}S(z)}, \quad (\alpha \in U - \{0\}),$$

where the singular inner functions $S(z)$ are the conjectured extremals for the Krzyż conjecture (up to rotations),

$$S(z) = \exp\left(-\frac{1+z^n}{1-z^n}\right).$$

Here, $n \in \mathbb{Z}^+$ is fixed. We note that the non-zero coefficients must be among a_{nl}. We recall the following from Lemma 2 and its proof in [79, pp. 251–252]. Let $f(z)$ be a Blaschke product with zeros $\{z_n\}$ in U. Let the power series of $f(z)$ be $f(z) = \sum_{n=0}^{\infty} a_n z^n$. Then

$$a_n = \frac{1}{2\pi} \int_0^{2\pi} f(e^{i\theta}) e^{-in\theta} d\theta.$$

By conjugation

$$\overline{a}_n = \frac{1}{2\pi i} \int_{|z|=1} \frac{z^{n-1}}{f(z)} dz,$$

and simple considerations concerning the boundary behavior of $f(z)$ justify the computation of this integral by residues. We get

$$\overline{a}_n = \sum_{k=1}^{\infty} \frac{z_k^{n-1}}{f'(z_k)}.$$

This identity was used also in the proof of Theorem 3 in [3, p. 12]. Namely, if $\{z_n\}$ is a sequence (finite or infinite) of complex numbers of modulus less than 1 such that $\sum(1 - |z_n|) < \infty$, then the Blaschke product

$$B(z) = \prod_{n} \frac{|z_n|}{z_n} \cdot \frac{z_n - z}{1 - \overline{z}_n z},$$

converges uniformly on compact subsets of the unit disc U. Let the power series expansion of this Blaschke product be given by $B(z) = \sum_{n=0}^{\infty} a_n z^n$, $(|z| < 1)$. Then we have

$$\overline{a}_n = \sum_{k=1}^{\infty} \frac{z_k^{n-1}}{B'(z_k)}.$$

Here, we have

$$B'(z_k) = \frac{1}{1 - z_k^2} \prod_{l \neq k} \frac{z_l - z_k}{1 - z_k z_l}.$$

Let us apply this identity to our case,

$$\Phi(z) = \frac{\alpha - \exp\left(-\frac{1+z^n}{1-z^n}\right)}{1 - \overline{\alpha} \exp\left(-\frac{1+z^n}{1-z^n}\right)}, \quad (z \in U),$$

for some fixed $\alpha \in U - \{0\}$. The function $\Phi(z)$ is that Blaschke product on U, with its zero set given by

$$z_k^n = \frac{\log|\alpha| + i(\arg \alpha + 2k\pi) + 1}{\log \alpha + i(\arg \alpha + 2k\pi) - 1}, \quad (k \in \mathbb{Z}).$$

In the particular case where $\alpha = \frac{1}{e}$, these zeros are given by

$$z_k^n = \frac{i \cdot k\pi}{i \cdot k\pi - 1}, \quad (k \in \mathbb{Z}).$$

Since we are interested in $\limsup_{k\to\infty} k|a_k|$, it is clearly sufficient to consider the case $n = 1$. Thus, we will be considering, from now and till the end of the chapter, the following Frostman shift:

$$\Phi(z) = \frac{\alpha - \exp\left(-\frac{1+z}{1-z}\right)}{1 - \overline{\alpha}\exp\left(-\frac{1+z}{1-z}\right)}, \quad (z \in U),$$

where the zeros are given by

$$z_k = \frac{\log|\alpha| + i(\arg\alpha + 2k\pi) + 1}{\log\alpha + i(\arg\alpha + 2k\pi) - 1}, \quad (k \in \mathbb{Z}),$$

and in the particular case $\alpha = \frac{1}{e}$,

$$z_k = \frac{i \cdot k\pi}{i \cdot k\pi - 1}, \quad (k \in \mathbb{Z}).$$

Let us check the Blaschke condition for those later zeros,

$$1 - |z_k| = 1 - \left|\frac{i \cdot k\pi}{i \cdot k\pi - 1}\right| = 1 - \frac{|k|\pi}{\sqrt{1 + k^2\pi^2}} = \frac{1}{1 + k^2\pi^2 + |k|\pi\sqrt{1 + k^2\pi^2}}.$$

Indeed, the Blaschke condition is fulfilled. Let us try and check the $\limsup_{k\to\infty} k|a_k|$ in this particular case. Thus,

$$B(z) = \frac{\frac{1}{e} - \exp\left(-\frac{1+z}{1-z}\right)}{1 - \frac{1}{e}\exp\left(-\frac{1+z}{1-z}\right)} = \frac{1 - \exp\left(-\frac{2z}{1-z}\right)}{e - \exp\left(-\frac{1+z}{1-z}\right)},$$

and

$$B(z) = e^{ic}\prod_{k}\frac{|z_k|}{z_k} \cdot \frac{z_k - z}{1 - \overline{z}_k z}, \quad \text{where } z_k = \frac{i \cdot k\pi}{i \cdot k\pi - 1}.$$

By the identity mentioned above for the Taylor coefficients, we have

$$\overline{a}_n = \sum_{k=-\infty}^{\infty}\frac{z_k^{n-1}}{B'(z_k)}, \quad \text{and } B'(z_k) = \frac{1}{1 - z_k^2}\prod_{l\neq k}\frac{z_l - z_k}{1 - z_k z_l}.$$

We want to avoid infinite products. So we take advantage of the fact that we have a finite explicit representation of $B(z)$ and differentiate it.

$$B'(z) = \frac{d}{dz}\left\{\frac{1 - \exp\left(-\frac{2z}{1-z}\right)}{e - \exp\left(-\frac{1+z}{1-z}\right)}\right\} = \frac{2\left\{\exp\left(\frac{1-3z}{1-z}\right) - \exp\left(-\frac{1+z}{1-z}\right)\right\}}{(1 - z^2)\left\{e - \exp\left(-\frac{1+z}{1-z}\right)\right\}^2}.$$

However, this complicated expression results because we used in a straight-forward manner the basic arithmetical rules of differentiation. A much better (shorter and simpler) option is the use of the chain rule, which we now do. We use $B(z) = (T_\alpha \circ S)(z)$, where $T_\alpha(w) = \frac{\alpha - w}{1 - \overline{\alpha} w}$ and where we choose $\alpha = \frac{1}{e}$. The chain rule gives us $B'(z) = T'_\alpha(S(z)) \cdot S'(z)$. We have for the automorphism

$$T'_\alpha(w) = \left(\frac{\alpha - w}{1 - \overline{\alpha} w} \right)' = \frac{-1 + |\alpha|^2}{(1 - \overline{\alpha} w)^2}.$$

For the singular inner function, we have

$$S'(z) = \exp\left(-\frac{1 + z}{1 - z} \right)' = -\frac{2}{(1 - z)^2} \exp\left(-\frac{1 + z}{1 - z} \right).$$

Combining things, we obtain

$$B'(z) = \frac{-1 + |\alpha|^2}{\left(1 - \overline{\alpha} \exp\left(-\frac{1+z}{1-z} \right) \right)^2} \cdot \frac{-2}{(1 - z)^2} \exp\left(-\frac{1 + z}{1 - z} \right)$$

$$= \frac{2(1 - |\alpha|^2) \exp\left(-\frac{1+z}{1-z} \right)}{(1 - z)^2 \left(1 - \overline{\alpha} \exp\left(-\frac{1+z}{1-z} \right) \right)^2}.$$

In our particular case, $\alpha = \frac{1}{e}$ and the result is

$$B'(z) = \frac{2(1 - \frac{1}{e^2}) \exp\left(-\frac{1+z}{1-z} \right)}{(1 - z)^2 \left(1 - \frac{1}{e} \exp\left(-\frac{1+z}{1-z} \right) \right)^2}.$$

This expression is a bit more convenient to handle than the expression we got without using the chain rule. We use it. Let us plug into $B'(z)$ the value $z = z_k = \frac{i \cdot k\pi}{i \cdot k\pi - 1}$.

$$1 + z_k = \frac{2k\pi i - 1}{k\pi i - 1}, \quad 1 - z_k = \frac{-1}{k\pi i - 1}, \quad -\frac{1 + z_k}{1 - z_k} = 2k\pi i - 1.$$

So,

$$\exp\left(-\frac{1 + z_k}{1 - z_k} \right) = \exp(2k\pi i - 1) = \frac{1}{e},$$

and

$$B'(z_k) = \left(\frac{2e}{e^2 - 1}\right)(1 - k\pi i)^2.$$

We now form the infinite series from [79] to obtain $\bar{a}_n$:

$$\bar{a}_n = \sum_{k=-\infty}^{\infty} \frac{z_k^{n-1}}{B'(z_k)} = \left(\frac{e^2 - 1}{2e}\right) \sum_{k=-\infty}^{\infty} \frac{(k\pi i)^{n-1}}{(k\pi i - 1)^{n+1}}.$$

Chapter 82

Further Computation in the Case $\alpha = \frac{1}{e}$

We would like to compute the following:

$$\left(\frac{e^2 - 1}{2e}\right) \cdot \limsup_{n \to \infty} n \cdot \left|\sum_{k=-\infty}^{\infty} \frac{(k\pi i)^{n-1}}{(k\pi i - 1)^{n+1}}\right|. \qquad (82.0.1)$$

Summing conjugates, we get

$$\frac{(k\pi i)^{n-1}}{(k\pi i - 1)^{n+1}} + \frac{(-k\pi i)^{n-1}}{(-k\pi i - 1)^{n+1}} = 2\Re \frac{(k\pi i)^{n-1}}{(k\pi i - 1)^{n+1}}$$

$$= \left(\frac{-2}{k^2\pi^2}\right) \Re \left(\frac{k\pi i}{k\pi i - 1}\right)^{n+1} \quad (k \in \mathbb{Z}^+).$$

Hence, our computation is of the following:

$$\left(\frac{e^2 - 1}{2e}\right) \cdot \limsup_{n \to \infty} 2n \cdot \left|\sum_{k=1}^{\infty} \left(\frac{1}{k^2\pi^2}\right) \Re \left(\frac{k\pi i}{k\pi i - 1}\right)^{n+1}\right|. \qquad (82.0.2)$$

Let us express $\frac{k\pi i}{k\pi i - 1}$ in polar coordinates. The modulus is (here, $k \in \mathbb{Z}^+$)

$$r_k = \left|\frac{k\pi i}{k\pi i - 1}\right| = \frac{k\pi}{\sqrt{1 + k^2\pi^2}}.$$

For the argument, we use the following:

$$\arg\left(\frac{k\pi i}{k\pi i - 1}\right) = \arg(k\pi i) - \arg(k\pi i - 1) = \frac{\pi}{2} - \left(\frac{\pi}{2} + \theta_k\right) = -\theta_k.$$

A simple geometric sketch gives us

$$\cos\theta_k = \frac{k\pi}{\sqrt{1 + k^2\pi^2}}, \quad \sin\theta_k = \frac{1}{\sqrt{1 + k^2\pi^2}}, \quad \tan\theta_k = \frac{1}{k\pi}.$$

Hence,

$$\frac{k\pi i}{k\pi i - 1} = r_k(\cos\theta_k - i\sin\theta_k) = \frac{k\pi}{\sqrt{1 + k^2\pi^2}}\left(\frac{k\pi}{\sqrt{1 + k^2\pi^2}} - i\frac{1}{\sqrt{1 + k^2\pi^2}}\right).$$

Incidentally,

$$\frac{k\pi i}{k\pi i - 1} = \cos\theta_k(\cos\theta_k - i\sin\theta_k),$$

and so

$$\left(\frac{k\pi i}{k\pi i - 1}\right)^{n+1} = (\cos\theta_k)^{n+1}(\cos(n+1)\theta_k - i\sin(n+1)\theta_k).$$

We can express that in terms of Chebyshev's polynomials. Chebyshev's polynomials of the first kind are

$$\cos m\theta = T_m(\cos\theta).$$

Chebyshev's polynomials of the second kind are

$$\frac{\sin(m+1)\theta}{\sin\theta} = U_m(\cos\theta).$$

$$\left(\frac{k\pi i}{k\pi i - 1}\right)^{n+1} = (\cos\theta_k)^{n+1}(T_{n+1}(\cos\theta_k) - i\sin\theta_k U_{n+1}(\cos\theta_k))$$

$$= (\cos\theta_k)^{n+1}(T_{n+1}(\cos\theta_k) - i\sqrt{1 - \cos 2\theta_k} \cdot U_{n+1}(\cos\theta_k))$$

$$= \left(\frac{k\pi}{\sqrt{1 + k^2\pi^2}}\right)^{n+1}\left(T_{n+1}\left(\frac{k\pi}{\sqrt{1 + k^2\pi^2}}\right)\right.$$

$$\left. - i\frac{1}{\sqrt{1 + k^2\pi^2}}U_{n+1}\left(\frac{k\pi}{\sqrt{1 + k^2\pi^2}}\right)\right).$$

We only need the real part of the expression above. Thus, we aim at computing the following:

$$\left(\frac{e^2 - 1}{2e}\right)\cdot\limsup_{n\to\infty}2n\cdot\left|\sum_{k=1}^{\infty}\left(\frac{1}{k^2\pi^2}\right)\left(\frac{k\pi}{\sqrt{1 + k^2\pi^2}}\right)^{n+1}T_{n+1}\left(\frac{k\pi}{\sqrt{1 + k^2\pi^2}}\right)\right|.$$

$$(82.0.3)$$

We can express it as follows:

$$\left(\frac{e^2-1}{2e}\right) \cdot \limsup_{n\to} n \cdot \left| \sum_{k=1}^{\infty} \left(\frac{1}{k^2\pi^2}\right) \left(\frac{k\pi}{\sqrt{1+k^2\pi^2}}\right)^{n+1} \right.$$

$$\times \left\{ \left(\frac{k\pi}{\sqrt{1+k^2\pi^2}} + i\frac{1}{\sqrt{1+k^2\pi^2}}\right)^{n+1} \right.$$

$$\left. \left. + \left(\frac{k\pi}{\sqrt{1+k^2\pi^2}} - i\frac{1}{\sqrt{1+k^2\pi^2}}\right)^{n+1} \right\} \right|. \qquad (82.0.4)$$

Also,

$$\left(\frac{e^2-1}{2e}\right) \cdot \limsup_{n\to} n \cdot \left| \sum_{k=-\infty}^{\infty} \left(\frac{1}{k^2\pi^2}\right) \left(\frac{k\pi i}{k\pi i - 1}\right)^{n+1} \right|, \qquad (82.0.5)$$

and

$$\left(\frac{e^2-1}{2e}\right) \cdot \limsup_{n\to} n \cdot \left| \sum_{k=-\infty}^{\infty} \left(\frac{1}{k^2\pi^2}\right) \left(1 + \frac{1}{k\pi i - 1}\right)^{n+1} \right|. \qquad (82.0.6)$$

Let us discuss equation (82.0.5) (for example). The infinite series

$$\sum_{k=-\infty}^{\infty} \left(\frac{1}{k^2\pi^2}\right) \left(\frac{k\pi i}{k\pi i - 1}\right)^{n+1},$$

is absolutely convergent for any $n \in \mathbb{Z}^+$, and for all those values of n, it is uniformly bounded, in absolute value, by $\frac{1}{3}$.

$$\sum_{k=-\infty}^{\infty} \left| \left(\frac{1}{k^2\pi^2}\right) \left(\frac{k\pi i}{k\pi i - 1}\right)^{n+1} \right| < \sum_{\substack{k=-\infty \\ k \neq 0}}^{\infty} \frac{1}{k^2\pi^2} = \frac{1}{3}.$$

We recall that

$$\left| \frac{k\pi i}{k\pi i - 1} \right| < 1.$$

Let us consider the following part of our infinite series:

$$\sum_{k=1}^{\infty} \left(\frac{1}{k^2\pi^2}\right) \left(\frac{k\pi i}{k\pi i - 1}\right)^{n+1}.$$

The other part is its complex conjugate and both are bounded from above by $\frac{1}{6}$. Let us fix a natural number $m \in \mathbb{Z}^+$ and an $\epsilon > 0$ and find tails of the series that are bounded from above by $\frac{\epsilon}{m}$. We are looking for a $u(m, \epsilon) > 0$ such that

$$\left| \sum_{k=u(m,\epsilon)+1}^{\infty} \left(\frac{1}{k^2\pi^2} \right) \left(\frac{k\pi i}{k\pi i - 1} \right)^{n+1} \right| < \frac{\epsilon}{m}.$$

A crude estimate of $u(m, \epsilon)$ gives a linear bound in m with a coefficient depending on ϵ.

$$\left| \sum_{k=u(m,\epsilon)+1}^{\infty} \left(\frac{1}{k^2\pi^2} \right) \left(\frac{k\pi i}{k\pi i - 1} \right)^{n+1} \right| < \sum_{k=u(m,\epsilon)+1}^{\infty} \frac{1}{k^2\pi^2}$$

$$\leq \int_{u(m,\epsilon)}^{\infty} \frac{dx}{x^2\pi^2} = \frac{1}{\pi^2} \cdot \frac{1}{u(m,\epsilon)}.$$

That leads to

$$\frac{1}{\pi^2} \cdot \frac{1}{u(m,\epsilon)} = \frac{\epsilon}{m} \Rightarrow u(m,\epsilon) = \left(\frac{1}{\pi^2\epsilon} \right) m.$$

Using that for $m = n$ gives

$$n \cdot \left| \sum_{k=1}^{\infty} \left(\frac{1}{k^2\pi^2} \right) \left(\frac{k\pi i}{k\pi i - 1} \right)^{n+1} \right| \leq n \cdot \left| \sum_{k=1}^{u(n,\epsilon)} \left(\frac{1}{k^2\pi^2} \right) \left(\frac{k\pi i}{k\pi i - 1} \right)^{n+1} \right|$$

$$+ n \cdot \left| \sum_{k=u(n,\epsilon)+1}^{\infty} \left(\frac{1}{k^2\pi^2} \right) \left(\frac{k\pi i}{k\pi i - 1} \right)^{n+1} \right|$$

$$< n \cdot \left| \sum_{k=1}^{u(n,\epsilon)} \left(\frac{1}{k^2\pi^2} \right) \left(\frac{k\pi i}{k\pi i - 1} \right)^{n+1} \right| + \epsilon.$$

We now face a problem of estimating a finite sum:

$$\limsup_{n\to\infty} n \cdot \left| \sum_{k=1}^{u(n,\epsilon)} \left(\frac{1}{k^2\pi^2} \right) \left(\frac{k\pi i}{k\pi i - 1} \right)^{n+1} \right|.$$

As a rough estimate, we will compare that finite sum to the corresponding integral:

$$\int_{1}^{u(n,\epsilon)} \left(\frac{1}{x^2\pi^2} \right) \left(\frac{x\pi i}{x\pi i - 1} \right)^{n+1} dx.$$

We use the following partial fractions:

$$\left(\frac{1}{x^2\pi^2}\right)\left(\frac{x\pi i}{x\pi i - 1}\right)^{n+1} = \frac{A}{x^2} + \frac{B}{x} + \sum_{j=1}^{n+1}\frac{C_j}{(x\pi i - 1)^j}.$$

We easily obtain $A = B = C_1 = 0$, $C_2 = C_{n+1} = -1$ and generally

$$C_j = -\binom{n-1}{j-2}, \quad 2 \leq j \leq n+1.$$

Changing the index j to $j - 2$ gives

$$\left(\frac{1}{x^2\pi^2}\right)\left(\frac{x\pi i}{x\pi i - 1}\right)^{n+1} = -\sum_{j=0}^{n-1}\binom{n-1}{j}\frac{1}{(x\pi i - 1)^{j+2}}.$$

Integrating from $x = 1$ to $x = u(n,\epsilon) = \left(\frac{1}{\pi^2\epsilon}\right)n$ gives

$$\int_1^{u(n,\epsilon)}\left(\frac{1}{x^2\pi^2}\right)\left(\frac{x\pi i}{x\pi i - 1}\right)^{n+1}dx$$

$$= -\sum_{j=0}^{n-1}\binom{n-1}{j}\int_1^{u(n,\epsilon)}\frac{dx}{(x\pi 1 - 1)^{j+2}}$$

$$= \frac{1}{n\pi i}\left\{\left(1 + \frac{1}{(u(n,\epsilon)\pi i - 1)}\right)^n - \left(1 + \frac{1}{(\pi i - 1)}\right)^n\right\}.$$

We conclude that

$$n\cdot\int_1^{u(n,\epsilon)}\left(\frac{1}{x^2\pi^2}\right)\left(\frac{x\pi i}{x\pi i - 1}\right)^{n+1}dx$$

$$= \frac{1}{\pi i}\left\{\left(1 + \frac{1}{(u(n,\epsilon)\pi i - 1)}\right)^n - \left(1 + \frac{1}{(\pi i - 1)}\right)^n\right\}$$

$$\to_{n\to\infty}\frac{1}{\pi i}\left\{\exp\left(\frac{\pi^2\epsilon}{\pi i}\right) - 0\right\} = \frac{e^{-\pi\epsilon i}}{\pi i}.$$

Thus, we finally arrive at

$$\left(\frac{e^2 - 1}{2e}\right)\limsup_{n\to\infty} n\cdot\left|\sum_{k=-\infty}^{\infty}\left(\frac{1}{k^2\pi^2}\right)\left(\frac{k\pi i}{k\pi i - 1}\right)^{n+1}\right|$$

$$= 2\left(\frac{e^2 - 1}{2e}\right)\frac{1}{\pi} = \frac{e^2 - 1}{e\pi}.$$

Remark 82.0.184. We note the following:

$$\frac{e^2 - 1}{e\pi} = 0.7481556\ldots, \quad \text{while} \quad \frac{2}{e} = 0.735759\ldots.$$

The numbers are rather close but still are in agreement with the conjecture of Patrick Ahern and Hong Oh Kim, in [3, p. 3]. The conjecture is that

$$\limsup_{n\to\infty} n|a_n| \geq \frac{2}{e},$$

for any Blaschke product $B(z) = \sum_{n=0}^{\infty} a_n z^n \in \mathcal{B}_\infty$, i.e., with an infinite zero set.

We recall that our computation above was only "a rough estimate". We compared the finite sum we needed to compute with the corresponding integral:

$$\int_1^{u(n,\epsilon)} \left(\frac{1}{x^2\pi^2}\right) \left(\frac{x\pi i}{x\pi i - 1}\right)^{n+1} dx.$$

We now turn and make the computation solid.

Chapter 83

Using the Euler–MacLaurin Formula
for Our Estimate

We are interested in

$$\limsup_{n\to\infty} n \cdot \left| \sum_{k=-\infty}^{\infty} \frac{(k\pi i)^{n-1}}{(k\pi i - 1)^{n+1}} \right|.$$

We reduced the estimation to estimating finite sums by choosing a fixed $\epsilon > 0$ and considering

$$\limsup_{n\to\infty} n \cdot \left| \sum_{k=1}^{(1/\pi^2\epsilon)n} \frac{(k\pi i)^{n-1}}{(k\pi i - 1)^{n+1}} \right|.$$

We will recall the Euler–MacLaurin formula. The general form appears in Proposition 1.3, in [57]. In fact, we can do with the simplest case.

Proposition 1.2 (Euler–MacLaurin formula, [57]). *Let $0 < y < x$ be two real numbers and assume that $f : [y, x] \to \mathbb{R}$ has continuous derivative on $[y, x]$. Then*

$$\sum_{y < n \leq x} f(n) = \int_y^x f(t)dt + \int_y^x (t - [t])f'(t)dt + ([x] - x)f(x) - ([y] - y)f'(y).$$

The general formula involves as many as desired explicit finitely many terms that include the Bernoulli numbers. Also, one has the same formula when the function $f(z)$ is analytic complex valued and the integral is along a line segment. Thus, let us consider our function

$$f(x) = \frac{(x\pi i)^{n-1}}{(x\pi i - 1)^{n+1}},$$

then its derivative is given by

$$f'(x) = \frac{\pi i (x\pi i)^{n-2}(-2x\pi i - (n-1))}{(x\pi i - 1)^{n+2}}.$$

Using the above formula, we have

$$\sum_{k=1}^{u} \frac{(k\pi i)^{n-1}}{(k\pi i - 1)^{n+1}} = \int_{1}^{u} \frac{(x\pi i)^{n-1} dx}{(x\pi i - 1)^{n+1}}$$

$$+ \int_{1}^{u} (x - [x]) \frac{\pi i (x\pi i)^{n-2}(-2x\pi i - (n-1))}{(x\pi i - 1)^{n+2}} dx$$

$$+ ([u] - u) \frac{(u\pi i)^{n-1}}{(u\pi i - 1)^{n+1}}.$$

The first integral on the right-hand side is as follows: We will use partial fractions,

$$(x\pi i - 1)^{-(n+1)} \cdot (1 + (x\pi i - 1))^{n-1} = \sum_{j=0}^{n-1} \binom{n-1}{j} (x\pi i - 1)^{-(j+2)}.$$

$$\int_{1}^{u} \frac{(x\pi i)^{n-1} dx}{(x\pi i - 1)^{n+1}} = \sum_{j=0}^{n-1} \binom{n-1}{j} \int_{1}^{u} (x\pi i - 1)^{-j-2} dx$$

$$= \frac{1}{n\pi i} \left(\sum_{j=0}^{n} \binom{n}{j} \left(\frac{1}{\pi i - 1} \right)^{j} - \sum_{j=0}^{n} \binom{n}{j} \left(\frac{1}{u\pi i - 1} \right)^{j} \right)$$

where we made a use in the identity

$$\binom{n-1}{j} \cdot \frac{1}{j+1} = \frac{1}{n} \cdot \binom{n}{j+1}.$$

Thus, we finally have

$$\int_{1}^{u} \frac{(x\pi i)^{n-1} dx}{(x\pi i - 1)^{n+1}} = \frac{1}{n\pi i} \left(\left(1 + \frac{1}{\pi i - 1} \right)^{n} - \left(1 + \frac{1}{u\pi i - 1} \right)^{n} \right). \quad (83.0.1)$$

The second integral on the left-hand side of our Euler–MacLaurin formula: $(x - [x])\pi i$ multiplies the following rational function:

$$(x\pi i - 1)^{-(n+2)} (1 + (x\pi i - 1))^{n-2} (-2(1 + (x\pi i - 1)) - (n-1))$$

$$= (x\pi i - 1)^{-(n+2)} \left(-2(1 + (x\pi i - 1))^{n-1} - (n-1)(1 + (x\pi i - 1))^{n-2} \right)$$

$$= -2 \sum_{j=0}^{n-1} \binom{n-1}{j} (x\pi i - 1)^{-(j+3)} - (n-1) \sum_{j=0}^{n-2} \binom{n-2}{j} (x\pi i - 1)^{-(j+4)}.$$

Let us integrate $\int_1^u \ldots dx$ separately the two terms above.

$$\sum_{j=0}^{n-1} \binom{n-1}{j} \int_1^u (x\pi i - 1)^{-(j+3)} dx$$

$$= \frac{1}{\pi i} \sum_{j=0}^{n-1} \binom{n-1}{j} \frac{1}{j+2} \left(\frac{1}{\pi i - 1}\right)^{j+2}$$

$$- \frac{1}{\pi i} \sum_{j=0}^{n-1} \binom{n-1}{j} \frac{1}{j+2} \left(\frac{1}{u\pi i - 1}\right)^{j+2}.$$

This time, we will use the following identity:

$$\binom{n-1}{j} \frac{1}{j+2} = \frac{1}{(n+1)n} \binom{n+1}{j+2} (j+1).$$

The first term becomes

$$\frac{1}{\pi i} \sum_{j=0}^{n-1} \frac{1}{(n+1)n} \binom{n+1}{j+2} (j+1) \left(\frac{1}{\pi i - 1}\right)^{j+2}$$

$$= \frac{1}{(\pi i - 1)^2 (n+1)n\pi i} \left(\sum_{j=0}^{n+1} \binom{n+1}{j} (j-1) \left(\frac{1}{\pi i - 1}\right)^{j-2}\right)$$

$$- \frac{1}{(n+1)n\pi i} \left(-\left(\frac{1}{\pi i - 1}\right)^{-2}\right).$$

To compute the last sum, we use the following identity:

$$\frac{d}{dT} \left(\sum_{j=0}^{n+1} \binom{n+1}{j} T^{j-1}\right) = \frac{(1+T)^n}{T^2} (nT - 1).$$

We conclude that

$$\frac{1}{\pi i} \sum_{j=0}^{n-1} \binom{n-1}{j} \frac{1}{j+2} \left(\frac{1}{\pi i - 1}\right)^{j+2} - \frac{1}{\pi i} \sum_{j=0}^{n-1} \binom{n-1}{j} \frac{1}{j+2} \left(\frac{1}{u\pi i - 1}\right)^{j+2}$$

$$= \frac{1}{(n+1)n\pi i} \left(\left(1 + \frac{1}{\pi i - 1}\right)^n \left(\frac{n}{\pi i - 1} - 1\right)\right.$$

$$\left. - \left(1 + \frac{1}{u\pi i - 1}\right)^n \left(\frac{n}{u\pi i - 1} - 1\right)\right).$$

In terms of the integral, we arrived at

$$-2\sum_{j=0}^{n-1}\binom{n-1}{j}\int_1^u (x\pi i - 1)^{-(j+3)}\,dx$$

$$= \frac{-2}{(n+1)n\pi i}\left(\left(1+\frac{1}{\pi i - 1}\right)^n\left(\frac{n}{\pi i - 1}-1\right)\right.$$

$$\left.-\left(1+\frac{1}{u\pi i - 1}\right)^n\left(\frac{n}{u\pi i - 1}-1\right)\right). \qquad (83.0.2)$$

Next, we expand the second integral:

$$\sum_{j=0}^{n-2}\binom{n-2}{j}\int_1^u (x\pi i - 1)^{-(j+4)}\,dx$$

$$= \frac{1}{\pi i}\sum_{j=0}^{n-2}\binom{n-2}{j}\frac{1}{j+3}\left(\frac{1}{\pi i - 1}\right)^{j+3}$$

$$-\frac{1}{\pi i}\sum_{j=0}^{n-2}\binom{n-2}{j}\frac{1}{j+3}\left(\frac{1}{u\pi i - 1}\right)^{j+3}.$$

We will use the following identity:

$$\binom{n-2}{j}\frac{1}{j+3} = \frac{1}{(n+1)n(n-1)}\binom{n+1}{j+3}(j+1)(j+2).$$

So, the first sum is

$$\frac{1}{\pi i}\sum_{j=0}^{n-2}\frac{1}{(n+1)n(n-1)}\binom{n+1}{j+3}(j+1)(j+2)\left(\frac{1}{\pi i - 1}\right)^{j+3}$$

$$= \frac{1}{(n+1)n(n-1)\pi i(\pi i - 1)^3}$$

$$\times\left(\sum_{j=0}^{n+1}\binom{n+1}{j}(j-1)(j-2)\left(\frac{1}{\pi i - 1}\right)^{j-3}\right).$$

To compute this last sum, we will use the following identity:

$$\frac{d^2}{dT^2}\left(\sum_{j=0}^{n+1}\binom{n+1}{j}T^{j-1}\right) = \frac{(1+T)^{n-1}}{T^3}\left(n(n-1)T^2 - 2(n-1)T + 2\right).$$

The quadratic in T on the right-hand side is not factorizable over $\mathbb{R}$. We conclude the following:

$$-(n-1)\sum_{j=0}^{n-2}\binom{n-2}{j}\int_1^u (x\pi i - 1)^{-(j+4)}dx$$

$$= \frac{-1}{(n+1)n\pi i}\left(\left(1+\frac{1}{\pi i - 1}\right)^{n-1}\left((n^2-n)\left(\frac{1}{\pi i - 1}\right)^2\right.\right.$$

$$\left.-2(n-1)\left(\frac{1}{\pi i - 1}\right)+2\right)-\left(1+\frac{1}{u\pi i - 1}\right)^{n-1}$$

$$\left.\times\left((n^2-n)\left(\frac{1}{u\pi i - 1}\right)^2-2(n-1)\left(\frac{1}{u\pi i - 1}\right)+2\right)\right). \qquad (83.0.3)$$

The fourth (and last) element in the Euler–MacLaurin formula that corresponds to our problem is the following:

$$([u]-u)\frac{(u\pi i)^{n-1}}{(u\pi i - 1)^{n+1}} = ([u]-u)\left(\frac{-1}{u^2\pi^2}\right)\left(1+\frac{1}{u\pi i - 1}\right)^{n+1}. \qquad (83.0.4)$$

We recall that we are interested in estimating the following:

$$\left(\frac{e^2-1}{2e}\right)\lim_{\epsilon\to 0^+}\limsup_{n\to\infty} n\cdot\left|\sum_{k=1}^{(1/\pi^2\epsilon)n}\frac{(k\pi i)^{n-1}}{(k\pi i - 1)^{n+1}}\right|.$$

For that purpose, we will form the expressions $I_1(n)$, $I_2(n)$, $I_3(n)$ and $I_4(n)$ and try to separate our estimates into the four parts $\limsup_{n\to\infty} n\cdot I_j(n)$, $j = 1, 2, 3, 4$ hoping that it will solve our problem. We recall that $u = (1/\pi^2\epsilon)n$. We will denote $c := c(\epsilon) = \frac{1}{\pi^2\epsilon}$.

$\boldsymbol{I_1(n)}$):

$$n\cdot I_1(n) = n\cdot\frac{1}{n\pi i}\left(\left(1+\frac{1}{\pi i - 1}\right)^n - \left(1+\frac{1}{u\pi i - 1}\right)^n\right)$$

$$= \frac{1}{\pi i}\left(\left(\frac{\pi i}{\pi i - 1}\right)^n - \left(1+\frac{1}{u\pi i - 1}\right)^n\right).$$

By

$$\left|\frac{\pi i}{\pi i - 1}\right| < 1,$$

we deduce that

$$\left(1 + \frac{1}{\pi i - 1}\right)^n \to_{n \to \infty} 0.$$

Also,

$$\left(1 + \frac{1}{c\pi i n - 1}\right)^n \to_{n \to \infty} e^{1/(c\pi i)}.$$

Thus, we established the following estimate:

$$\lim_{n \to \infty} n \cdot I_1(n) = \frac{i}{\pi} e^{-\pi \epsilon i}.$$

$\boldsymbol{I_2(n))}$:

$$n \cdot I_2(n) = \frac{-2}{(n+1)\pi i} \left(\left(1 + \frac{1}{\pi i - 1}\right)^n \left(\frac{n}{\pi i - 1} - 1\right)\right.$$
$$\left. - \left(1 + \frac{1}{c\pi i n - 1}\right)^n \left(\frac{n}{c\pi i n - 1} - 1\right)\right).$$

Since

$$\left(1 + \frac{1}{\pi i - 1}\right)^n \to_{n \to \infty} 0$$

geometrically, we have

$$\left(1 + \frac{1}{\pi i - 1}\right)^n \left(\frac{n}{\pi i - 1} - 1\right) \to_{n \to \infty} 0.$$

We saw above

$$\left(1 + \frac{1}{c\pi i n - 1}\right)^n \to_{n \to \infty} e^{1/(c\pi i)}.$$

and this clearly implies that

$$-\left(1 + \frac{1}{c\pi i n - 1}\right)^n \left(\frac{n}{c\pi i n - 1} - 1\right) \to_{n \to \infty} -e^{1/(c\pi i)} \cdot \left(\frac{1}{c\pi i} - 1\right).$$

Hence,

$$n \cdot I_2(n) \to_{n \to \infty} 0 \cdot \left(0 - e^{1/(c\pi i)} \cdot \left(\frac{1}{c\pi i} - 1\right)\right) = 0.$$

$I_3(n)$):

$$n \cdot I_3(n) = n \cdot \frac{-1}{(n+1)n\pi i} \left(\left(1 + \frac{1}{\pi i - 1}\right)^{n-1} \left((n^2 - n) \left(\frac{1}{\pi i - 1}\right)^2 \right.\right.$$

$$\left. - 2(n - 1) \left(\frac{1}{\pi i - 1}\right) + 2 \right) - \left(1 + \frac{1}{c\pi i n - 1}\right)^{n-1}$$

$$\times \left((n^2 - n) \left(\frac{1}{c\pi i n - 1}\right)^2 - 2(n - 1) \left(\frac{1}{c\pi i n - 1}\right) + 2 \right)\bigg).$$

We have the following:

$$\left(1 + \frac{1}{\pi i - 1}\right)^{n-1} \left((n^2 - n) \left(\frac{1}{\pi i - 1}\right)^2 - 2(n - 1) \left(\frac{1}{\pi i - 1}\right) + 2 \right) \to_{n \to \infty} 0,$$

because

$$\left(1 + \frac{1}{\pi i - 1}\right)^{n-1} \to_{n \to \infty} 0, \quad \text{geometrically.}$$

Also, using arguments similar to those used for $I_2(n)$, we have

$$-\left(1 + \frac{1}{c\pi i n - 1}\right)^{n-1} \left((n^2 - n) \left(\frac{1}{c\pi i n - 1}\right)^2 - 2(n - 1) \left(\frac{1}{c\pi i n - 1}\right) + 2 \right)$$

$$\to_{n \to \infty} -e^{1/(c\pi i)} \left(\frac{1}{-c^2 \pi^2} - \frac{2}{c\pi i} + 2 \right).$$

Hence,

$$n \cdot I_3(n) \to_{n \to \infty} 0 \cdot \left(-e^{1/(c\pi i)} \left(\frac{1}{-c^2 \pi^2} - \frac{2}{c\pi i} + 2 \right) \right) = 0.$$

$I_4(n)$):

Since the expression $([c\pi i n] - c\pi i n)$ has an absolute value at most 1, we have

$$n \cdot I_4(n) = n \cdot ([c\pi i n] - c\pi i n) \left(\frac{-1}{c^2 \pi^2 n^2} \right) \left(1 + \frac{1}{c\pi i n - 1} \right)^{n+1}$$

$$\to_{n \to \infty} 0 \cdot e^{1/(c\pi i)} = 0.$$

Remark 83.0.185. We ignored in the estimates of the integrals that are related to $I_2(n)$ and $I_3(n)$ the factor $(x - [x])$ within the integrands, since this factor is at most 1 in absolute value and since by now we know that $n \cdot I_2(n)$, $n \cdot I_3(n) \to_{n \to \infty} 0$, in the way they tend to 0 (i.e., even in absolute values).

We concluded the proof of the following

Theorem 83.0.186. *The case $\alpha = 1/e$ in the Frostman shift of*

$$S(z) = \exp\left(-\frac{1+z}{1-z}\right).$$

The following limit exists and equals

$$2\left(\frac{e^2-1}{2e}\right) \cdot \lim_{n\to\infty} n \cdot \left|\sum_{k=1}^{\infty} \frac{(k\pi i)^{n-1}}{(k\pi i - 1)^{n+1}}\right| = \frac{e^2-1}{e\pi}.$$

Chapter 84

Generalizing Theorem 83.0.186

We would like to generalize Theorem 83.0.186, which handles the case $\alpha = \frac{1}{e}$ of the Frostman shift

$$\frac{\alpha - S(z)}{1 - \overline{\alpha}S(z)}, \quad \text{where } S(z) = \exp\left(-\frac{1+z}{1-z}\right).$$

We choose a real and positive parameter $s > 0$. The connection with the Frostman shift parameter is $\alpha = e^{-s}$. Thus, we will deal with the following family of Frostman shifts:

$$\frac{e^{-s} - S(z)}{1 - e^{-s}S(z)}, \quad s > 0.$$

The zeros of a member in this family are exactly the solutions of the equation:

$$\exp\left(-\frac{1+z}{1-z}\right) = e^{-s} \Rightarrow -\frac{1+z}{1-z} = -s + 2\pi i k, \ k \in \mathbb{Z}.$$

So the zeros are exactly given by

$$z_k = \frac{(s-1) - 2\pi i k}{(s+1) - 2\pi i k}, \quad k \in \mathbb{Z}.$$

The Ahern–Kim conjecture in this case, see [3, p. 3], asserts that

$$\limsup_{n \to \infty} n|a_n| \geq \frac{2}{e},$$

for any $s > 0$ for which the Frostman shift,

$$\frac{e^{-s} - S(z)}{1 - e^{-s}S(z)} = \sum_{n=0}^{\infty} a_n z^n,$$

is a Blaschke product, say $B(z)$. Using the identity of Newman and Shapiro in [79], we have

$$\bar{a}_n = \sum_{k=-\infty}^{\infty} \frac{z_k^{n-1}}{B'(z_k)},$$

for the power series coefficients of the Blaschke product $B(z)$ whose zero set is $\{z_k\}_{k=-\infty}^{\infty}$. We differentiate the Blaschke product, which is also a Frostman shift using the chain rule,

$$B'(z) = \frac{d}{dz}\left(\frac{\alpha - S(z)}{1 - \overline{\alpha}S(z)}\right) = \frac{2(1-|\alpha|^2)S(z)}{(1-z)^2(1-\overline{\alpha}S(z))^2}.$$

The zeros satisfy $S(z_k) = \alpha$ and hence,

$$B'(z_k) = \frac{2\alpha}{(1-z_k)^2(1-|\alpha|^2)},$$

and we obtain by the Newman–Shapiro identity mentioned above

$$\bar{a}_n = \left(\frac{1-|\alpha|^2}{2\alpha}\right)\sum_{k=-\infty}^{\infty}(1-z_k)^2 z_k^{n-1}.$$

In our particular family of Frostman shifts, $\alpha = e^{-s}$, $s > 0$ and

$$z_k = \frac{(s-1) - 2\pi ik}{(s+1) - 2\pi ik}, \quad k \in \mathbb{Z},$$

which gives us the following formula:

$$\bar{a}_n = s\left(\frac{e^{2s}-1}{e^s}\right)\sum_{k=-\infty}^{\infty}\frac{(-s+1+2\pi ik)^{n-1}}{(-s-1+2\pi ik)^{n+1}}.$$

The conjecture of Ahern–Kim asserts in this case:

$$2 \cdot \left|\frac{e^{2s}-1}{e^s}\right| \limsup_{n\to\infty} n \left|\sum_{k=-\infty}^{\infty}\frac{(-s+1+2\pi ik)^{n-1}}{(-s-1+2\pi ik)^{n+1}}\right| \geq \frac{2}{e} \quad \text{for } s > 0.$$

Our Theorem 83.0.186 says that for $s = 1$ the $\limsup_{n\to\infty}$ is in fact a limit $\lim_{n\to\infty}$ and the value of this limit is

$$\frac{e^2-1}{e\pi} = 0.748156316383\ldots > 0.73575888234\ldots = \frac{2}{e}.$$

Before proceeding, let us note that for $k = 0$, we have

$$n \cdot 2 \left| \frac{e^{2s} - 1}{e^s} \right| \cdot \left| \frac{1-s}{1+s} \right|^{n+1} \cdot \frac{1}{|1-s|^2} \to_{n \to \infty} 0,$$

because

$$\left| \frac{1-s}{1+s} \right|^{n+1} \to_{n \to \infty} 0 \quad \text{at least geometrically}.$$

Hence, ignoring the element indexed $k = 0$ does not change anything. Next, we note that the following two complementary parts of the infinite series are complex conjugates. This follows by the fact that $\alpha = e^{-s}$, $s > 0$ is a real parameter:

$$\sum_{k=-\infty}^{-1} \frac{(-s+1+2\pi ik)^{n-1}}{(-s-1+2\pi ik)^{n+1}} \quad \text{and} \quad \sum_{k=1}^{\infty} \frac{(-s+1+2\pi ik)^{n-1}}{(-s-1+2\pi ik)^{n+1}}.$$

This has two implications. The first is as follows:

$$\sum_{k=-\infty}^{\infty} \frac{(-s+1+2\pi ik)^{n-1}}{(-s-1+2\pi ik)^{n+1}} = 2 \sum_{k=1}^{\infty} \Re \left\{ \frac{(-s+1+2\pi ik)^{n-1}}{(-s-1+2\pi ik)^{n+1}} \right\}.$$

The second is that it will be sufficient to estimate the following:

$$n \cdot \left| \sum_{k=1}^{\infty} \frac{(-s+1+2\pi ik)^{n-1}}{(-s-1+2\pi ik)^{n+1}} \right|.$$

Taking these last series (for which $k \geq 1$), we have

$$\sum_{k=1}^{\infty} \frac{(-s+1+2\pi ik)^{n-1}}{(-s-1+2\pi ik)^{n+1}} = \sum_{k=1}^{\infty} \frac{1}{(-s+1+2\pi ik)^2} \left(\frac{-s+1+2\pi ik}{-s-1+2\pi ik} \right)^{n+1}.$$

Since $s > 0$, we have

$$\left| \frac{-s+1+2\pi ik}{-s-1+2\pi ik} \right| < 1,$$

and hence,

$$\sum_{k=1}^{\infty} \left| \frac{(-s+1+2\pi ik)^{n-1}}{(-s-1+2\pi ik)^{n+1}} \right| < \sum_{k=1}^{\infty} \frac{1}{|-s+1+2\pi ik|^2}$$

$$= \sum_{k=1}^{\infty} \frac{1}{(1-s)^2 + 4\pi^2 k^2} < \infty.$$

So, these series are absolutely convergent. Let us fix $\epsilon > 0$ and ask for tails of the series that are bounded from above by $\frac{\epsilon}{n}$. We are looking for a $u_\epsilon(n) > 0$ such that

$$\left| \sum_{k=u_\epsilon(n)+1}^{\infty} \frac{(-s+1+2\pi ik)^{n-1}}{(-s-1+2\pi ik)^{n+1}} \right| < \frac{\epsilon}{n}.$$

A crude estimate on $u_\epsilon(n)$ goes as follows:

$$\left| \sum_{k=u_\epsilon(n)+1}^{\infty} \frac{(-s+1+2\pi ik)^{n-1}}{(-s-1+2\pi ik)^{n+1}} \right| < \sum_{k=u_\epsilon(n)+1}^{\infty} \frac{1}{(1-s)^2 + 4\pi^2 k^2}$$

$$\leq \sum_{k=u_\epsilon(n)+1}^{\infty} \frac{1}{4\pi^2 k^2}$$

$$\leq \int_{u_\epsilon(n)}^{\infty} \frac{dx}{4\pi^2 x^2} = \frac{1}{4\pi^2} \cdot \frac{1}{u_\epsilon(n)} \leq \frac{\epsilon}{n}.$$

This leads to

$$\frac{1}{4\pi^2} \cdot \frac{1}{u_\epsilon(n)} = \frac{\epsilon}{n} \Rightarrow u_\epsilon(n) = \left(\frac{1}{4\pi^2 \epsilon} \right) n.$$

This is linear in n, with the coefficient $\frac{1}{4\pi^2 \epsilon}$. Using that, we obtain

$$n \cdot \left| \sum_{k=1}^{\infty} \frac{(-s+1+2\pi ik)^{n-1}}{(-s-1+2\pi ik)^{n+1}} \right| < n \cdot \left| \sum_{k=1}^{u_\epsilon(n)} \frac{(-s+1+2\pi ik)^{n-1}}{(-s-1+2\pi ik)^{n+1}} \right| + \epsilon.$$

Now, we face the problem of estimating a finite sum:

$$\limsup_{n \to \infty} n \cdot \left| \sum_{k=1}^{\left(\frac{1}{4\pi^2 \epsilon} \right) n} \frac{(-s+1+2\pi ik)^{n-1}}{(-s-1+2\pi ik)^{n+1}} \right|.$$

For that purpose, we will use the Euler–MacLaurin identity (a restricted form):

$$\sum_{y < n \leq x} f(n) = \int_y^x f(t)dt + \int_y^x (t - [t])f'(t)dt + ([x] - x)f(x) - ([y] - y)f(y).$$

We take

$$f(x) = \frac{(-s+1+2\pi i x)^{n-1}}{(-s-1+2\pi i x)^{n+1}}.$$

$$f'(x) = \frac{4\pi i(-s+1+2\pi i x)^{n-2}(-2\pi i x + (s-n))}{(-s-1+2\pi i x)^{n+2}}.$$

$$\sum_{k=1}^{u} \frac{(-s+1+2\pi i k)^{n-1}}{(-s-1+2\pi i k)^{n+1}} = \int_{1}^{u} \frac{(-s+1+2\pi i x)^{n-1}}{(-s-1+2\pi i x)^{n+1}}\,dx + \int_{1}^{u} 4\pi i(x-[x])$$

$$\times \frac{(-s+1+2\pi i x)^{n-2}(-2\pi i x + (s-n))}{(-s-1+2\pi i x)^{n+2}}\,dx$$

$$+ ([u]-u)\frac{(-s+1+2\pi i u)^{n-1}}{(-s-1+2\pi i u)^{n+1}}.$$

We now elaborate on each term that appears on the right-hand side.

First integral:
$(-s+1+2\pi i x)^{n-1} = (2+(-1-s+2\pi i x))^{n-1}$ so that the partial fractions identity gives us the following:

$$(-1-s+2\pi i x)^{-(n+1)} \sum_{j=0}^{n-1} \binom{n-1}{j} 2^j (-1-s+2\pi i x)^{(n-1)-j}$$

$$= \sum_{j=0}^{n-1} \binom{n-1}{j} 2^j (-1-s+2\pi i x)^{-(j+2)}.$$

We integrate this $\int_{1}^{u} \ldots dx$. We get

$$\frac{1}{2\pi i} \sum_{j=0}^{n-1} \binom{n-1}{j} 2^j \left(\frac{1}{j+1}\right) \left(\frac{1}{2\pi i - 1 - s}\right)^{j+1}$$

$$- \frac{1}{2\pi i} \sum_{j=0}^{n-1} \binom{n-1}{j} 2^j \left(\frac{1}{j+1}\right) \left(\frac{1}{2\pi i u - 1 - s}\right)^{j+1}.$$

We now use the following identity:

$$\binom{n-1}{j} \cdot \frac{1}{j+1} = \frac{1}{n} \cdot \binom{n}{j+1}.$$

So,

$$\frac{1}{2\pi ni}\sum_{j=0}^{n-1}\binom{n}{j+1}2^j\left(\frac{1}{2\pi i-1-s}\right)^{j+1}$$

$$=\frac{1}{4\pi ni}\left(\sum_{j=0}^{n}\binom{n}{j}\left(\frac{2}{2\pi i-1-s}\right)^{j}-1\right)$$

$$=\frac{1}{4\pi ni}\left(\left(1+\frac{2}{2\pi i-1-s}\right)^{n}-1\right).$$

Similarly, for the second term,

$$\frac{1}{2\pi ni}\sum_{j=0}^{n-1}\binom{n}{j+1}2^j\left(\frac{1}{2\pi iu-1-s}\right)^{j+1}$$

$$=\frac{1}{4\pi ni}\left(\left(1+\frac{2}{2\pi iu-1-s}\right)^{n}-1\right).$$

By $s>0$, we have

$$\left|1+\frac{2}{2\pi i-1-s}\right|^{2}=\frac{(2\pi)^2+(1-s)^2}{(2\pi)^2+(1+s)^2}<1\quad\text{and so}$$

$$n\cdot\frac{1}{4\pi ni}\left(\left(1+\frac{2}{2\pi i-1-s}\right)^{n}-1\right)\to_{n\to\infty}\frac{1}{\pi i}(0-1)=\frac{-1}{4\pi i}.$$

For estimating the second term, we recall that $u=c\cdot n$ $\left(c=\frac{1}{4\pi^2\epsilon}\right)$, so

$$\left(1+\frac{2}{2\pi iu-1-s}\right)^{n}$$

$$=\left\{\left(1+\frac{1}{\pi icn-\left(\frac{1+s}{2}\right)}\right)^{\pi icn-(1+s)/2}\left(1+\frac{1}{\pi icn-\left(\frac{1+s}{2}\right)}\right)^{(1+s)/2}\right\}^{1/(\pi ic)}$$

$$\to_{n\to\infty}\{e\cdot1\}^{1/(\pi ic)}=e^{-i/(\pi c)}.$$

Thus,

$$n\cdot\frac{1}{4\pi ni}\left(\left(1+\frac{2}{2\pi iu-1-s}\right)^{n}-1\right)\to_{n\to\infty}\frac{1}{4\pi i}e^{-i/(\pi c)}-\frac{1}{4\pi i}.$$

We proved the following:

$$n \cdot \int_1^u \frac{(-s+1+2\pi ix)^{n-1}}{-s-1+2\pi ix)^{n+1}} dx \to_{n\to\infty} -\frac{1}{4\pi i} - \left(\frac{1}{4\pi i}e^{-i/(\pi c)} - \frac{1}{4\pi i}\right)$$
$$- \frac{1}{4\pi i}0^{-i/(\pi c)}. \tag{84.0.1}$$

Second integral:
The integrand contains the factor $x - [x]$ which (in principle) turns the single integral into an infinite series the terms of which are integrals between pairs of consecutive integers. However, even though we have those discontinuities at the integers, the nice thing about $x - [x]$ is that $0 \le x - [x] < 1$. Thus, we will start by estimating the integral, where we remove the bounded factor $x - [x]$ from the integrand. So, we are looking at the following integral:

$$4\pi i \int_1^u \frac{(-s+1+2\pi ix)^{n-2}(-2\pi ix + (s-n))}{(-s-1+2\pi ix)^{n+2}} dx$$

$$\times (-s+1+2\pi ix)^{n-2}(-2\pi ix + (s-n))$$

$$= (2 + (-s-1+2\pi ix))^{n-2}(-(n+1) - (-s-1+2\pi ix))$$

$$= \left\{ \sum_{j=0}^{n-2} \binom{n-2}{j} 2^j(-s-1+2\pi ix)^{(-2)-j} \right\} ((n+1)$$

$$+ (-s-1+2\pi ix))(-1)$$

$$= -\sum_{j=0}^{n-2} \binom{n-2}{j} 2^j(-s-1+2\pi ix)^{(n-1)-j}$$

$$- (n+1) \sum_{j=0}^{n-2} \binom{n-2}{j} 2^j(-s-1+2\pi ix)^{(n-2)-j}.$$

We multiply this by $(-s-1+2\pi ix)^{-(n+2)}$:

$$-\sum_{j=0}^{n-2} \binom{n-2}{j} 2^j(-s-1+2\pi ix)^{-(j+3)}$$

$$-(n+1) \sum_{j=0}^{n-2} \binom{n-2}{j} 2^j(-s-1+2\pi ix)^{-(j+4)}.$$

We integrate $\int_1^u \ldots dx$:

$$-\sum_{j=0}^{n-2} \binom{n-2}{j} 2^j \left(\frac{1}{2\pi i}\right) \left(\frac{1}{1-(j+3)}\right) (-s-1+2\pi ix)^{1-(j+3)} \Bigg|_1^u$$

$$= \frac{1}{2\pi i} \sum_{j=0}^{n-2} \binom{n-2}{j} 2^j \left(\frac{1}{j+2}\right) \left\{ \left(\frac{1}{-s-1+2\pi iu}\right)^{j+2} \right.$$

$$\left. - \left(\frac{1}{-s-1+2\pi i}\right)^{j+2} \right\}$$

we make a use in the identity:

$$\binom{n-2}{j} \cdot \frac{1}{j+2} = \frac{1}{(n-1)n} \binom{n}{j+2} (j+1),$$

and continue our chain of equalities:

$$= \frac{1}{2\pi i} \cdot \frac{1}{(n-1)n} \left(\left\{ \sum_{j=2}^{n} \binom{n}{j} (j-1) \left(\frac{2}{-s-1+2\pi iu}\right)^{j-2} \right\} \right.$$

$$\times \left(\frac{1}{-s-1+2\pi iu}\right)^2$$

$$\left. - \left\{ \sum_{j=2}^{n} \binom{n}{j} (j-1) \left(\frac{2}{-s-1+2\pi i}\right)^{j-2} \right\} \left(\frac{1}{-s-1+2\pi i}\right)^2 \right).$$

The last two sums are of the following form:

$$\sum_{j=2}^{n} \binom{n}{j} (j-1) T^{j-2} = \frac{d}{dT} \left\{ \sum_{j=2}^{n} \binom{n}{j} T^{j-1} \right\}$$

$$= \frac{(1+T)^{n-1}}{T^2} ((n-1)T - 1) + \frac{1}{T^2}.$$

For

$$T = \frac{2}{-s-1+2\pi iu},$$

we obtain the following:

$$\frac{1}{2\pi i} \cdot \frac{1}{(n-1)n} \left(\frac{1}{-s-1+2\pi iu}\right)^2$$
$$\times \left\{ \left(1 + \frac{2}{-s-1+2\pi iu}\right)^{n-1} \cdot \left(\frac{-s-1+2\pi iu}{2}\right)^2 \right.$$
$$\left. \times \left(\frac{2(n-1)}{-s-1+2\pi iu} - 1\right) + \left(\frac{-s-1+2\pi iu}{2}\right)^2 \right\}.$$

We recall that $u = cn$ and hence when $n \to \infty$, we have the following estimates:

$$\left(\frac{1}{-s-1+2\pi iu}\right)^2 = \Omega\left(\frac{1}{n^2}\right), \quad \left(1 + \frac{2}{-s-1+2\pi iu}\right)^{n-1} = \Omega\left(1\right),$$
$$\left(\frac{-s-1+2\pi iu}{2}\right)^2 = \Omega\left(n^2\right), \quad \frac{2(n-1)}{-s-1+2\pi iu} - 1 = \Omega\left(1\right).$$

Hence, after multiplying by n, we obtain the following estimate:

$$n \cdot \frac{1}{2\pi i} \cdot \frac{1}{(n-1)n} \left(\frac{1}{-s-1+2\pi iu}\right)^2$$
$$\times \left\{ \left(1 + \frac{2}{-s-1+2\pi iu}\right)^{n-1} \cdot \left(\frac{-s-1+2\pi iu}{2}\right)^2 \right.$$
$$\left. \times \left(\frac{2(n-1)}{-s-1+2\pi iu} - 1\right) + \left(\frac{-s-1+2\pi iu}{2}\right)^2 \right\}$$
$$= \frac{1}{2\pi i} \frac{1}{n-1} \Omega\left(\frac{1}{n^2}\right) \left\{ \Omega\left(1\right) \cdot \Omega\left(n^2\right) \cdot \Omega\left(1\right) + \Omega\left(n^2\right) \right\} \to_{n\to\infty} 0.$$

For

$$T = \frac{2}{-s-1+2\pi i},$$

we obtain the following:

$$\frac{1}{2\pi i} \cdot \frac{1}{(n-1)n} \left(\frac{1}{-s-1+2\pi i}\right)^2$$
$$\times \left\{ \left(1 + \frac{2}{-s-1+2\pi i}\right)^{n-1} \cdot \left(\frac{-s-1+2\pi i}{2}\right)^2 \right.$$
$$\left. \times \left(\frac{2(n-1)}{-s-1+2\pi i} - 1\right) + \left(\frac{-s-1+2\pi i}{2}\right)^2 \right\}.$$

We should note the following:

$$\left| 1 + \frac{2}{-s-1+2\pi i} \right|^2 = \frac{(1-s)^2 + 4\pi^2}{(1+s)^2 + 4\pi^2} < 1, \quad \text{by } s > 0.$$

So,

$$\left(1 + \frac{2}{-s-1+2\pi i} \right)^{n-1} \to_{n\to\infty} 0, \quad \text{geometrically,}$$

and so

$$n \cdot \frac{1}{2\pi i} \cdot \frac{1}{(n-1)n} \left(\frac{1}{-s-1+2\pi i} \right)^2$$

$$\times \left\{ \left(1 + \frac{2}{-s-1+2\pi i} \right)^{n-1} \cdot \left(\frac{-s-1+2\pi i}{2} \right)^2 \right.$$

$$\times \left(\frac{2(n-1)}{-s-1+2\pi i} - 1 \right) + \left(\frac{-s-1+2\pi i}{2} \right)^2 \left. \right\} \to_{n\to\infty} 0.$$

Hence, we proved the following:

$$-n \int_1^u \sum_{j=0}^{n-2} \binom{n-2}{j} 2^j (-s-1+2\pi i x)^{-(j+3)} dx \to_{n\to\infty} 0.$$

Now, we integrate the second sum:

$$-(n+1) \sum_{j=0}^{n-2} \binom{n-2}{j} 2^j \int_1^u (-s-1+2\pi i x)^{-(j+4)} dx$$

$$= \frac{n+1}{2\pi i} \sum_{j=0}^{n-2} \binom{n-2}{j} 2^j \left(\frac{1}{j+3} \right)$$

$$\times \left\{ \left(\frac{1}{-s-1+2\pi i u} \right)^{j+3} - \left(\frac{1}{-s-1+2\pi i} \right)^{j+3} \right\}$$

and we will make a use in the following identity:

$$\binom{n-2}{j} \cdot \frac{1}{j+3} = \frac{1}{(n-1)n(n+1)} \binom{n+1}{j+3} (j+2)(j+1),$$

now we continue the chain of equalities:

$$= \frac{1}{(n-1)n} \cdot \frac{1}{2\pi i} \left(\left\{ \sum_{j=3}^{n+1} \binom{n+1}{j} (j-1)(j-2) \left(\frac{2}{-s-1+2\pi iu} \right)^{j-3} \right\} \right.$$

$$\times \left(\frac{1}{-s-1+2\pi iu} \right)^3 - \left\{ \sum_{j=3}^{n+1} \binom{n+1}{j} (j-1)(j-2) \left(\frac{2}{-s-1+2\pi i} \right)^{j-3} \right\}$$

$$\left. \times \left(\frac{1}{-s-1+2\pi i} \right)^3 \right).$$

The last two sums are of the following form:

$$\sum_{j=3}^{n+1} \binom{n+1}{j} (j-1)(j-2) T^{j-3} = \frac{d^2}{dT^2} \left\{ \sum_{j=3}^{n+1} \binom{n+1}{j} T^{j-1} \right\}$$

$$= \frac{(n+1)n(1+T)^{n-1}}{T} - 2\frac{(n+1)(1+T)^n}{T^2} + \frac{2(1+T)^{n+1}}{T^3} - \frac{2}{T^3}.$$

For

$$T = \frac{2}{-s-1+2\pi iu},$$

we have

$$\frac{1}{(n-1)n} \cdot \frac{1}{2\pi i} \cdot \left(\frac{1}{-s-1+2\pi iu} \right)^3 \cdot \left\{ (n+1)n \left(1 + \frac{2}{-s-1+2\pi iu} \right)^{n-1} \right.$$

$$\times \left(\frac{-s-1+2\pi iu}{2} \right) - 2(n+1) \left(1 + \frac{2}{-s-1+2\pi iu} \right)^n \left(\frac{-s-1+2\pi iu}{2} \right)^2$$

$$\left. + 2 \left(1 + \frac{2}{-s-1+2\pi iu} \right)^{n+1} \left(\frac{-s-1+2\pi iu}{2} \right)^3 - 2 \left(\frac{-s-1+2\pi iu}{2} \right)^3 \right\}.$$

We recall that $u = cn$ and hence, when $n \to \infty$, we have

$$\left(\frac{1}{-s-1+2\pi iu} \right)^3 = \Omega \left(\frac{1}{n^3} \right), \quad \left(1 + \frac{2}{-s-1+2\pi iu} \right)^k = \Omega(1),$$

$$k = n-1, \, n, \, n+1,$$

$$\left(\frac{-s-1+2\pi iu}{2} \right)^k = \Omega(n^k), \, k = 1, \, 2, \, 3.$$

Hence, after multiplying by n, we obtain the following estimate:

$$n \cdot \frac{1}{(n-1)n} \cdot \frac{1}{2\pi i} \cdot \left(\frac{1}{-s-1+2\pi i u}\right)^3 \cdot \left\{ (n+1)n\left(1+\frac{2}{-s-1+2\pi i u}\right)^{n-1}\right.$$

$$\times \left(\frac{-s-1+2\pi i u}{2}\right) - 2(n+1)\left(1+\frac{2}{-s-1+2\pi i u}\right)^n$$

$$\times \left(\frac{-s-1+2\pi i u}{2}\right)^2 + 2\left(1+\frac{2}{-s-1+2\pi i u}\right)^{n+1}\left(\frac{-s-1+2\pi i u}{2}\right)^3$$

$$\left. -2\left(\frac{-s-1+2\pi i u}{2}\right)^3\right\}$$

$$= \frac{1}{n-1}\cdot\frac{1}{2\pi i}\Omega\left(\frac{1}{n^3}\right)\left\{(n+1)n\cdot\Omega(1)\cdot\Omega(n)\right.$$

$$\left. -2(n+1)\cdot\Omega(1)\cdot\Omega(n^2)+\Omega(1)\cdot\Omega(n^3)-\Omega(n^3)\right\}\to_{n\to\infty}0.$$

For

$$T = \frac{2}{-s-1+2\pi i},$$

we have

$$\frac{1}{(n-1)n}\cdot\frac{1}{2\pi i}\cdot\left(\frac{1}{-s-1+2\pi i}\right)^3\cdot\left\{(n+1)n\left(1+\frac{2}{-s-1+2\pi i}\right)^{n-1}\right.$$

$$\times\left(\frac{-s-1+2\pi i}{2}\right)-2(n+1)\left(1+\frac{2}{-s-1+2\pi i}\right)^n\left(\frac{-s-1+2\pi i}{2}\right)^2$$

$$\left. +2\left(1+\frac{2}{-s-1+2\pi i}\right)^{n+1}\left(\frac{-s-1+2\pi i}{2}\right)^3-2\left(\frac{-s-1+2\pi i}{2}\right)^3\right\}.$$

We recall that for $k = n-1,\, n,\, n+1$, we have

$$\left(1+\frac{2}{-s-1+2\pi i}\right)^k \to_{n\to\infty}0, \quad \text{geometrically,}$$

so clearly n times the second sum $\to_{n\to\infty}0$. We proved the following:

$$4\pi i\cdot n\cdot\int_1^u\frac{(-s+1+2\pi i x)^{n-2}(-2\pi i x+(s-n))}{(-s-1+2\pi i x)^{n+2}}dx\to_{n\to\infty}0,$$

and hence, as was explained before,

$$n\cdot\int_1^u 4\pi i(x-[x])\frac{(-s+1+2\pi i x)^{n-2}(-2\pi i x+(s-n))}{(-s-1+2\pi i x)^{n+2}}dx\to_{n\to\infty}0.$$

$$(84.0.2)$$

Finally,

$$([u] - u)\frac{(-s+1+2\pi iu)^{n-1}}{(-s-1+2\pi iu)^{n+1}} = ([u] - u)\left(\frac{1}{-s+1+2\pi iu}\right)^2$$

$$\times \left(1 + \frac{2}{-s-1+2\pi iu}\right)^{n+1}$$

$$= \Omega\left(\frac{1}{n^2}\right) \cdot \Omega(1).$$

So, we have

$$n \cdot ([u] - u) \cdot \frac{(-s+1+2\pi iu)^{n-1}}{(-s-1+2\pi iu)^{n+1}} \to_{n\to\infty} 0. \qquad (84.0.3)$$

By equations (84.0.1), (84.0.2) and (84.0.3), we have

$$4 \cdot \left|\frac{e^{2s}-1}{e^s}\right| \cdot n \left|\sum_{k=1}^{u} \frac{(-s+1+2\pi ik)^{n-1}}{(-s-1+2\pi ik)^{n+1}}\right| \to_{n\to\infty} 4 \cdot \left|\frac{e^{2s}-1}{e^s}\right| \cdot \frac{1}{4\pi} = \left|\frac{e^{2s}-1}{\pi e^s}\right|.$$

Letting $s \to 0+$ and after that $\epsilon \to 0^+$ (in that order) we get the total repeated limit equals 0. In the reverse order, i.e., first $\epsilon \to 0^+$ we obtain

$$4 \cdot \left|\frac{e^{2s}-1}{e^s}\right| \cdot n \left|\sum_{k=1}^{\infty} \frac{(-s+1+2\pi ik)^{n-1}}{(-s-1+2\pi ik)^{n+1}}\right|.$$

That brings us right to the original estimate of the infinite series. We cannot deduce anymore that when $s \to 0^+$ the result is 0. Indeed, it cannot approach 0. Otherwise, we would have $\lim_{n\to\infty} n|a_n| = 0$, which contradicts the results in [79] and [3] which assert that $\lim_{n\to\infty} n|a_n| \geq \frac{1}{\pi}$ and that $\lim_{n\to\infty} n|a_n| \geq \sqrt{\gamma_0/2} \geq 0.37749$, respectively.

Chapter 85

Looking Once More at the Representation of $\lim_{n \to \infty} n|a_n|$ for $\alpha = e^{-s}, \; s > 0$

We have $(\alpha = e^{-s}, \; s > 0)$

$$n\,|a_n| = 2\left(\frac{e^{2s}-1}{e^s}\right) n \left| \sum_{k=-\infty}^{\infty} \frac{(-s+1+2\pi ik)^{n-1}}{(-s-1+2\pi ik)^{n+1}} \right|. \tag{85.0.1}$$

We used the following:

Proposition 1.2 ([57]). *Let* $0 < y < x$ *be two real numbers and assume that* $f : [y,x] \to \mathbb{R}$ *has a continuous derivative on* $[y,x]$. *Then*

$$\sum_{y<n\leq x} f(n) = \int_y^x f(t)dt + \int_y^x (t-[t])f'(t)dt + ([x]-x)f(x) - ([y]-y)f(y).$$

However, we used it on the function,

$$f(x) = \frac{(-s+1+2\pi ix)^{n-1}}{(-s-1+2\pi ix)^{n+1}}, \quad x \in [1,u] = \left[1, \left(\frac{1}{4\pi^2\epsilon}\right)n\right].$$

Thus, $f(x)$ is not a real-valued function. For that particular function, we have

$$\sum_{k=1}^{u} \frac{(-s+1+2\pi ik)^{n-1}}{(-s-1+2\pi ik)^{n+1}} = \int_1^u \frac{(-s+1+2\pi ix)^{n-1}}{(-s-1+2\pi ix)^{n+1}}dx$$

$$+ \int_1^u 4\pi i(x-[x]) \frac{(-s+1+2\pi ix)^{n-2}(-2\pi ix+(s-n))}{(-s-1+2\pi ix)^{n+2}}dx$$

$$+ ([u]-u)\frac{(-s+1+2\pi iu)^{n-1}}{(-s-1+2\pi iu)^{n+1}}. \tag{85.0.2}$$

We computed each one of the three terms on the right-hand side for some fixed $\epsilon > 0$, $s > 0$ when $n \to \infty$. The computations of the first and of the third terms seem to be reasonable. In the computations of the second term, we "waved hands" in that we agreed to ignore from the integrand, the factor $x - [x]$. Maybe that caused problems? Let us compute the second term without ignoring the $x - [x]$ factor. We are dealing with $\int_1^u (x - [x]) f'(x) dx$. We list all the integers between 1 and u (endpoints included): $< 2 < \cdots < k < k+1 < \cdots < N \le u$. Then

$$\int_1^u (x - [x]) f'(x) dx = \int_1^2 + \int_2^3 + \cdots + \int_k^{k+1} + \cdots + \int_{N-1}^N$$
$$+ \int_N^u (x - [x]) f'(x) dx.$$

For $k \le x < k+1$, $[x] = k$ and hence,

$$\int_k^{k+1} (x - [x]) f'(x) dx = \int_k^{k+1} (x - k) f'(x) dx.$$

$$(x - k) \frac{(-s + 1 + 2\pi ix)^{n-2}(-2\pi ix + (s - n))}{(-s - 1 + 2\pi ix)^{n+2}}$$

$$= (x - k)\left\{ -\sum_{j=0}^{n-2} \binom{n-2}{j} 2^j (-s - 1 + 2\pi ix)^{-(j+3)} \right.$$

$$\left. - (n + 1) \sum_{j=0}^{n-2} \binom{n-2}{j} 2^j (-s - 1 + 2\pi ix)^{-(j+4)} \right\}.$$

We can write

$$x - k = \frac{(-s - 1 + 2\pi ix) + (s + 1 - 2\pi ik)}{2\pi i}.$$

Hence,

$$(x - k) f'(x) = \frac{1}{2\pi i}\left\{ -\sum_{j=0}^{n-2} \binom{n-2}{j} 2^j (-s - 1 + 2\pi ix)^{-(j+2)} \right.$$

$$- (n + s + 2 - 2\pi ik) \sum_{j=0}^{n-2} \binom{n-2}{j} 2^j (-s - 1 + 2\pi ix)^{-(j+3)}$$

$$\left. - (n + 1)(s + 1 - 2\pi ik) \sum_{j=0}^{n-2} \binom{n-2}{j} 2^j (-s - 1 + 2\pi ix)^{-(j+4)} \right\}. \quad (85.0.3)$$

The first term is independent of k and so we can integrate $\int_1^u \ldots$ instead of in between consecutive integers followed by summation. Let us compute that

$$\int_1^u \sum_{j=0}^{n-2} \binom{n-2}{j} 2^j (-s - 1 + 2\pi i x)^{-(j+2)} dx$$

$$= \frac{1}{4\pi i (n-1)} \left\{ \left(1 + \frac{2}{-s - 1 + 2\pi i} \right)^{n-1} - \left(1 + \frac{2}{-s - 1 + 2\pi i u} \right)^{n-1} \right\}.$$

We recall that $u = c \cdot n$ $(c = 1/(4\pi^2 \epsilon))$, so

$$\left(1 + \frac{2}{-s - 1 + 2\pi i} \right)^{n-1} = \left(\frac{-s + 1 + 2\pi i}{-s - 1 + 2\pi i} \right)^{n-1} \to_{n\to\infty} 0.$$

$$\left(1 + \frac{2}{2\pi i c n - (s + 1)} \right)^{n-1} \to_{n\to\infty} e^{-i/(\pi c)}.$$

Putting together the sign and the constants, we finally get

$$-\frac{4\pi i}{2\pi i} \int_1^u \sum_{j=0}^{n-2} \binom{n-2}{j} 2^j (-s - 1 + 2\pi i x)^{-(j+2)} dx$$

$$= \frac{i}{2\pi(n-1)} \left\{ \left(1 + \frac{2}{-s - 1 + 2\pi i} \right)^{n-1} - \left(1 + \frac{2}{-s - 1 + 2\pi i u} \right)^{n-1} \right\}.$$

So,

$$n \cdot \left(-\frac{4\pi i}{2\pi i} \right) \int_1^u \sum_{j=0}^{n-2} \binom{n-2}{j} 2^j (-s - 1 + 2\pi i x)^{-(j+2)} dx \to_{n\to\infty} \frac{-i}{2\pi} e^{-i/(\pi c)}.$$

$$(85.0.4)$$

We recall that we already know

$$n \cdot \int_1^u \frac{(-s + 1 + 2\pi i x)^{n-1}}{(-s - 1 + 2\pi i x)^{n+1}} dx \to_{n\to\infty} \frac{i}{4\pi} e^{-i/(\pi c)}. \qquad (85.0.5)$$

$$n \cdot ([u] - u) \frac{(-s + 1 + 2\pi i u)^{n-1}}{(-s - 1 + 2\pi i u)^{n+1}} \to_{n\to\infty} 0. \qquad (85.0.6)$$

So, we are left with the computation of the following:

$$n \cdot 4\pi i \cdot \frac{1}{2\pi i} \left\{ - \sum_{k=1}^{N-1} (n+s+2-2\pi ik) \sum_{j=0}^{n-2} \binom{n-2}{j} 2^j \right.$$

$$\times \int_k^{k+1} (-s-1+2\pi ix)^{-(j+3)} dx - \sum_{k=1}^{N-1} (n+1)(s+1-2\pi ik)$$

$$\left. \times \sum_{j=0}^{n-2} \binom{n-2}{j} 2^j \int_k^{k+1} (-s-1+2\pi ix)^{-(j+4)} dx \right\}$$

$$+ n \cdot 4\pi i \cdot \frac{1}{2\pi i} \left\{ -(n+s+2-2\pi iN) \sum_{j=0}^{n-2} \binom{n-2}{j} 2^j \right.$$

$$\times \int_N^u (-s-1+2\pi ix)^{-(j+3)} dx$$

$$\left. -(n+1)(s+1-2\pi iN) \sum_{j=0}^{n-2} \binom{n-2}{j} 2^j \int_N^u (-s-1+2\pi ix)^{-(j+4)} dx \right\}.$$

If we choose the parameter ϵ to be such that $u \in \mathbb{Z}^+$, the $N = u$ and the second term above is 0. An elaborated computation that uses the elementary facts:

Telescopic series, $\alpha \sum_{k=1}^{N-1} (a_{k+1} - a_k) = \alpha(a_N - a_1)$, and the second-order telescopic series, $\beta \sum_{k=1}^{N-1} k(b_{k+1} - b_k) = \beta((N-1)b_N - \sum_{k=1}^{N-1} b_k)$, give us eventually the total as follows:

$$n \sum_{k=1}^{N-1} \frac{(-s+1+2\pi ik)^{n-1}}{(-s-1+2\pi ik)^{n+1}}, \quad \text{as should be.}$$

Thus, taking into an account (85.0.2), (85.0.5) and (85.0.6), we see that in order to prove the Ahern–Kim conjecture for the family of the Frostman shifts that are parametrized by $s > 0$, we need to prove the following:

$$\lim_{\epsilon \to 0^+} \limsup_{n \to \infty} \left| \frac{i}{4\pi} e^{-i \cdot 4\pi \epsilon} + n \int_1^{(1/(4\pi^2 \epsilon))n} 4\pi i(x - [x]) \right.$$

$$\left. \times \frac{(-s+1+2\pi ix)^{n-2}(-2\pi ix + (s-n))}{(-s-1+2\pi ix)^{n+2}} dx \right| \geq \frac{2}{e}. \quad (85.0.7)$$

Chapter 80

On the Paper [97]

The paper reports an investigation of certain properties of measures on the unit circle $\mathbb{T}$ associated with singular inner functions which omit values in the unit disc U. The results are used to resolve some open questions concerning inner functions. In particular, the author disproves a conjecture of Herrero concerning the structure of the inner functions under the uniform topology of H^∞. Why are we interested in this paper? Because we want to understand the following natural question: **If $f(z)$ is an inner function and if $\alpha \in U$, then the zero set of $f(z) - \alpha$, $z \in U$, which is denoted by $Z(f(z) - \alpha)$, is a Blaschke sequence (unless it is an empty set).** Here, zeros are counted in a manner that their multiplicities are taken into account. Does every Blaschke sequence of U has the form $Z(f(z) - \alpha)$? Of course! Just take $f(z) = B(z)$ to be the Blaschke product associated with our Blaschke sequence and $\alpha = 0$. What if we do not allow $f(z)$ to vanish in U? Thus, we ask: Is it true that every Blaschke sequence of U has the form $Z(S(z) - \alpha)$ for some singular inner function $S(z)$ and some α, $0 < |\alpha| < 1$? The author of this book was confident that the answer is "yes" again. He was surprised to find out in the end, that this is not at all obvious! It is immediate that problems of this sort are closely related to the $U \to U$ surjectivity of certain related functions. In other words, it was useful to understand the set of values omitted by inner functions. This is how we get to read the nice paper [97].

1. Main results.

The author of the paper makes a use of the universal covering surface of a planar domain, and in the uniformization theorem, in order to construct inner functions that are not surjections $U \to U$. This is an effective tool for that purpose. If A is a (relatively) closed subset of U with logarithmic capacity zero, then the universal covering surface of $U - A$ is conformally equivalent to U. If Φ_A is a uniformizer of $U - A$, then Φ_A is an inner function whose image (the author of the paper calls it range) is precisely $U - A$. For the main result, the author assumes $0 \in A$, so the inner function Φ_A does not vanish in U and hence it is a singular inner function.

Theorem I ([97]). *Let A be a closed subset of U of logarithmic capacity zero, $0 \in A$, and let μ be the singular measure on $\mathbb{T}$ associated with the conformal mapping Φ_A of U onto $U - A$.*

(a) *If 0 is an isolated point of A, then μ is discrete, i.e., it consists entirely of point masses.*

(b) *If 0 is a limit point of A, then μ is continuous, i.e., it has no point masses.*

The proofs of parts (a) and (b) require entirely different techniques. A main ingredient is the following interesting mapping theorem:

Theorem II ([97]). *Let F be an analytic function from U into the left half-plane with the property that $\liminf_{r \to 1-} (1 - r)|F(r)| > 0$. Then, for $M > 0$, the disc $\{w \in \mathbb{C} \,|\, |w - F(r)| < M\}$ lies in the image of F for all r sufficiently close to 1, $0 < r < 1$.*

The author explains that his study was originally motivated by certain conjectures in [43]. Later on (Section 5), the author uses his Theorem I to arrive at the following result, which disproves one of Herrero's conjectures in [43].

Theorem III ([97]). *The collection of discrete singular inner functions is not closed in H^∞ (with the uniform topology).*

2. Preliminaries.

2.1. Inner Functions:

A function $f \in H^\infty$ is an inner function if it has unimodular radial limits at almost all (Lebesgue measure) points of $\mathbb{T}$. Every inner function factors uniquely as a product $f(z) = B(z) \cdot S(z)$, $z \in U$, where B is a Blaschke product, formed with the zeros of f, and S is a singular inner function, i.e., one which does not vanish in U. All singular measures discussed are positive, finite, Borel measures on $\mathbb{T}$ which are singular with respect to Lebesgue measure. Each singular measure μ gives rise to a singular inner function S_μ defined by

$$S_\mu(z) = \exp\left(\int_{\mathbb{T}} \frac{z + \xi}{z - \xi} d\mu(\xi)\right), \quad z \in U,$$

and conversely, every singular inner function is a unimodular constant times S_μ for some unique singular μ. In the case that μ is a unit mass concentrated at 1, S_μ is denoted by S_1. Specifically,

$$S_1(z) = \exp\left(\frac{z+1}{z-1}\right).$$

If f is inner and if it factors as $f(z) = e^{i\theta}B(z)S_\mu(z)$, then S_ν is said to be a singular factor of f if $\mu - \nu \geq 0$, so that $S_\mu = S_{\mu-\nu}S_\nu$. Two particular types of singular inner functions will be of interest to the author. Let $f = e^{i\theta}S_\mu$. If μ is discrete (consists entirely of point masses), then f is termed a discrete singular function. If μ has no discrete part, then f is termed a continuous singular inner function. Following Herrero, the author denotes the collections of inner, singular inner, discrete singular inner, and continuous singular inner functions by, $\mathcal{F}$, $\mathcal{F}_s$, $\mathcal{F}_a$, $\mathcal{F}_c$, respectively. For each $\alpha \in U$, Ψ_α denotes the Möbius transformation of U onto itself given by

$$\Psi_\alpha = \frac{z - \alpha}{1 - \overline{\alpha}z}, \quad z \in U.$$

Of course, these are inner functions. Finally, for later reference, the author notes the following properties of inner functions:

(a) $\mathcal{F}$ and $\mathcal{F}_s$ are closed in H^∞.
(b) $\mathcal{F}_a \cup \mathcal{F}_c \neq \mathcal{F}s$ since, in general, a singular measure μ may have both discrete and continuous parts. $\mathcal{F}_a \cap \mathcal{F}_c$ consists of trivial inner functions only (i.e., unimodular constants).
(c) If $f \in \mathcal{F}$ and $g \in H^\infty$, $\|g\|_\infty \leq 1$, then a necessary and sufficient condition for $f \circ g \in \mathcal{F}$ is that $g \in \mathcal{F}$.
(d) For each $f \in \mathcal{F}$, $\|\Psi_\alpha \circ f - \Psi_\beta \circ f\|_\infty \to 0$ as $\alpha \to \beta$ in U.

2.2. **Omitted and exceptional values:**

$\mathcal{A}$ will denote the collection of (relatively) closed subsets of U with logarithmic capacity zero.

(a) Every closed countable subset of U is in $\mathcal{A}$. However, $\mathcal{A}$ also contains uncountable subsets of U. Any closed subset of a set in $\mathcal{A}$ is in $\mathcal{A}$.
(b) If $A \in \mathcal{A}$, then $U - A$ is connected.
(c) $A \in \mathcal{A}$ if and only if $\Psi_\alpha(A) \in \mathcal{A}$ for all $\alpha \in U$.

For each $f \in \mathcal{F}$, we define the exceptional set $E(f)$ and the omitted set $O(f)$ as follows:

$$E(f) = \{\alpha \in U | \Psi_\alpha \circ f \text{ has a non-trivial singular factor}\},$$

$$O(f) = \{\alpha \in U | \Psi_\alpha \circ f \text{ is singular}\}.$$

Clearly, $O(f) \subseteq E(f)$.

2.3. Frostman's construction:

A famous theorem of Frostman states the following: "If f is inner, then $E(f)$ is a set of logarithmic capacity zero." Less well known is his converse: "For each $A \in \mathcal{A}$, there is an inner function f such that $O(f) = E(f) = A$." See [33]. Frostman's construction goes as follows ([14], Chapter 2, Section 8):

For $A \in \mathcal{A}$, let R_A be the universal covering surface of $U - A$ with covering projection $\pi : R_A \to U - A$. By the uniformization theorem, R_A is conformally equivalent to U. Let $\tilde{\Phi}_A : U \to R_A$ be a mapping, which realizes this equivalence. Then $\Phi_A : U \to U$, defined by $\Phi_A = \pi \circ \tilde{\Phi}_A$ is an inner function, and $O(\Phi_A) = E(\Phi_A) = A$. The mapping Φ_A is called an uniformizer of $U - A$. Note that $\tilde{\Phi}_A$, and hence Φ_A, is determined only up to composition with a conformal mapping of U. The results in the paper, though, are independent of the choice of Φ_A, so Φ_A will be some particular choice of a uniformizer of $U - A$.

2.4. Proposition. *Suppose $A \in \mathcal{A}$ and $\Phi \in \mathcal{F}$, then $\Phi = \Phi_A$ if and only if, for each $f \in H^\infty$ with $f(U) \subseteq U - A$, there exists $g \in H^\infty$, $\|g\|_\infty \leq 1$, such that $f = \Phi_A \circ g$.*

2.5. Corollaries.

(a) *If $A \in \mathcal{A}$, $\alpha \in U$, and $B = \Psi_\alpha(A)$, then $\Phi_B = \Psi_\alpha \circ \Phi_A$.*
(b) *If $A \in \mathcal{A}$ and $f \in \mathcal{F}$, then $O(f) \supseteq A$ if and only if there is a $g \in \mathcal{F}$ with $f = \Phi_A \circ g$.*
(c) *If $A = \{0\}$, then $\Phi_A = S_1$.*

To see the third corollary, let $e^{i\theta} S_\mu$ be any singular inner function and we write

$$g(z) = \frac{\log e^{i\theta} S_\mu(z) + 1}{\log e^{i\theta} S_\mu(z) - 1},$$

using any determination of the logarithm. Clearly, $e^{i\theta} S_\mu = S_1 \circ g$.

2.6. Angular derivatives:

Let r denote a real number, $0 < r < 1$. Let $g \in H^\infty$, $\|g\|_\infty \leq 1$. If g has radial limit 1 at the point $1 \in \mathbb{T}$, it can be shown that

$$\lim_{r \to 1^-} \frac{1 - g(r)}{1 - r} = \lim_{r \to 1^-} \frac{|1 - g(r)|}{1 - r} = c,$$

where the common limit c is either real or infinite. Moreover, if $c < \infty$, then also $c = \lim_{r \to 1^-} g'(r)$, and we say that g has the finite angular derivative c at 1. Note that by Schwarz Lemma, $c > 0$. The properties of angular derivatives were developed by Carathéodory, [10, Sections 298 and 299]. Putting all this together, we see that g has radial limit 1 and finite angular derivative c at 1 if and only if

$$\limsup_{r \to 1^-} \frac{|1 - g(r)|}{1 - r} = c < \infty. \tag{86.0.1}$$

This allows the following statement of a result due to Fisher which is central to the proof of Theorem I(b).

2.7. Theorem ([31]). *Let $g \in H^\infty$, $\|g\|_\infty \leq 1$ and let S_μ be the singular inner factor of $S_1 \circ g$. Then*

$$\mu(1) = \liminf_{r \to 1^-} \frac{1 - r}{|1 - g(r)|}.$$

A simple note on surjectivity of Blaschke products.

Before going on with the paper of Kenneth Stephenson, let us recall a well-known result of Frostman which implies that for any inner function $f \in \mathcal{F}$ there exists a sequence $\{B_n\}_n$ of Blaschke products that converges to f uniformly on compact subsets of U. We deduce the following:

Proposition 86.0.187. *Let $f \in \mathcal{F}$ be inner and suppose that there is a sequence of surjective $U \to U$ Blaschke products such that $B_n \to f$ uniformly on compact subsets of U. Then $O(f) \subseteq f(\mathbb{T})$ sequentially, i.e., for each $a \in O(f)$ there is a sequence $\{w_n\}_n \subseteq U$ such that $w_n \to_{n \to \infty} w \in \mathbb{T}$ and $f(w_n) \to_{n \to \infty} a$.*

Proof. For every n, there is some $z_n \in U$ such that $B_n(z_n) - a = 0$ (B_n is a $U \to U$ surjection, by our assumption). Clearly, $\{z_n\}_n$ cannot have a limit point $w \in U$ for then $0 = B_n(z_n) - a \to f(w) - a$ along a subsequence B_{n_k}. This contradicts $a \in O(f)$. Hence, the derived set $\emptyset \neq \{z_n\}'_n \subseteq \mathbb{T}$ which implies the result. $\qquad \square$

Corollary 86.0.188. *Let f be a Blaschke product. Then $O(f) \subseteq f(\mathbb{T})$ sequentially.*

Proof. In Proposition 86.0.187, we take for B_n the sequence of the partial products of f. $\qquad \square$

3. Proof of Theorem I(a).

First, we provide an explanation why the conclusions of Theorem I do not depend on the choice of the uniformizer Φ_A. In [98, Section 5], it was shown that if $f \in \mathcal{F}_c$ and $g \in \mathcal{F}$, then $(f \circ g) \in \mathcal{F}_c$. If $0 \in A$ and if Φ_1 and Φ_2 are any two uniformizers of $U - A$, then there is a conformal mapping Ψ of U such that $\Phi_1 = \Phi_2 \circ \Psi$ and $\Phi_2 = \Phi_1 \circ \Psi^{-1}$. We see that $\Phi_1 \in \mathcal{F}_c$ if and only if $\Phi_2 \in \mathcal{F}_c$. Likewise, Φ_1 has a factor in $\mathcal{F}_c$ if and only if Φ_2 also has a factor in $\mathcal{F}_c$, that is $\Phi_1 \in \mathcal{F}_a$ if and only if $\Phi_2 \in \mathcal{F}_a$.

3.1. **Theorem I(a).** *If $A \in \mathcal{A}$ and 0 is an isolated point of A, then Φ_A is a discrete singular inner function.*

Proof. For convenience, choose Φ_A so that $\Phi_A(0) > 0$. Then $\Phi_A = S_\mu$ for some singular measure μ. Let G consist of those points $\xi \in \mathbb{T}$ for which $\lim_{r \to 1^-} S_\mu(r\xi) = 0$. It is known that G has full μ-measure (Theorem 1.2 in [22]). Therefore, it will suffice to prove that G is at most countable. Choose $\delta > 0$ so that if $B = \{z \in \mathbb{C} | |z| < \delta\}$, then $A \cap \overline{B} = \{0\}$. Let Ω be the open set $\pi^{-1}(B) \subseteq R_A$ with open components Ω_n, $n = 1, 2, \ldots$. The countability of this collection follows from the fact that R_A (the universal covering surface of $U - A$) is separable. Observe that for each n, $R_A - \Omega_n$ is connected. For suppose that a and b are arbitrary points of $R_A - \Omega_n$. Since R_A is connected, we can choose a path $\gamma : [0, 1] \to R_A$ with $\gamma(0) = a$, $\gamma(1) = b$ and consider $\Gamma = \pi \circ \gamma$. Γ is path from $\pi(a)$ to $\pi(b)$ in $U - A$ and, because of the conditions on B, is clearly homotopic in $U - A$ to a path $\tilde{\Gamma}$ from $\pi(a)$ to $\pi(b)$ which misses B. If $\tilde{\gamma}$ is the lifting of $\tilde{\Gamma}$ to R_A (i.e., $\pi\tilde{\gamma} = \tilde{\Gamma}$) with $\tilde{\gamma}(0) = a$, then $\tilde{\gamma}(1) = b$. Thus, the path $\tilde{\gamma}$ lies in $R_A - \Omega_n$ and connects a to b. Fix a point $\xi \in G$. If L is the image on R_A of the radial segment to ξ under the mapping Φ_A, then from some point on, the curve L will lie entirely within one of the components of Ω. Call it $\Omega_{n(\xi)}$. The countability of G follows if we show that the map $\xi \to n(\xi)$ is one-to-one. Defining $V \subseteq U$ by $V = \tilde{\Phi}_A^{-1}(\Omega_{n(\xi)})$, it is enough to prove that $\overline{V} \cap \mathbb{T} = \{\xi\}$. Clearly, $\xi \in \overline{V} \cap \mathbb{T}$. Suppose there exists $\xi_1 \neq \xi$ with $\xi_1 \in \overline{V} \cap \mathbb{T}$. Because Φ_A has radial limits of modulus 1 almost everywhere on $\mathbb{T}$, we can choose $\rho \in \mathbb{T}$ so that ρ and $-\rho$ lie in different arcs of $\mathbb{T} - \{\xi, \xi_1\}$, and so that

$$\lim_{r \to 1^-} |\Phi_A(r\rho)| = 1 = \lim_{r \to 1^-} |\Phi_A(-r\rho)|. \tag{86.0.2}$$

Since V is connected, the choice of ρ ensures the line segment $[\rho, -\rho]$ intersects V, while the condition in equation (86.0.2) ensures that the segment leaves V near both ends. However, this implies that $U - V$ is not connected, contradicting the fact that $\tilde{\Phi}_A(U-V) = R_A - \Omega_{n(\xi)}$ is connected. Therefore, $\overline{V} \cap \mathbb{T} = \{\xi\}$, and the proof is completed. $\square$

3.2. Remarks. When A is not a singleton, the measure μ above is not a "typical" discrete singular measure. Results of Siedel and Lohwater (see [14], Theorems 5.13 and 5.14) imply, for example, that the closed support of μ is a perfect subset of $\mathbb{T}$ with positive logarithmic capacity. Moreover, our proof (Stephenson's proof!) shows that μ has a countable number of points of density, which is strictly stronger than the conclusion that μ is discrete.

4. **Proof of Theorem I(b).**

4.1. Theorem II. *Let F be an analytic function from U into the left half-plane $\{w \in \mathbb{C} | \Re\, w < 0\}$ such that*

$$\liminf_{r \to 1^-}(1 - r)|F(r)| > 0. \tag{86.0.3}$$

Then, for each $M > 0$, the disc $\{w \in \mathbb{C} | |w - F(r)| < M\}$ lies in the range (image) of F for all r sufficiently close to 1.

Proof. First, observe that it is sufficient to prove the conclusion in the case that $M = \frac{1}{3}$. For other values of M, just apply this result to an appropriate dilation F_t of F, $F_t(z) = t \cdot F(z)$, $0 < t < \infty$. Let Ψ denote the Möbius transformation

$$\Psi(z) = \frac{z+1}{z-1}, \quad z \neq 1.$$

Ψ is its own inverse and maps the left half plane conformally onto U. Define $g : U \to U$ by $g = \Psi \circ F$. The condition in equation (86.0.3) on F implies that $\liminf_{r \to 1^-}(1 - r)|1 - F(r)| > 0$, which converts precisely to the condition in equation (86.0.1) on g, so we know that as $r \to 1^-$,

$$|1 - g(r)| \to 0, \ |g'(r)| \to c \quad \text{and}$$
$$\frac{|1 - g(r)|}{1 - r} \to c \text{ for some } c, \ 0 < c < \infty. \tag{86.0.4}$$

Choose r sufficiently near 1 such that $g'(r) \neq 0$, and define the auxiliary functions

$$g_r(z) = (g \circ \Psi_r)(z), \ h_r(z) = \frac{g_r(z) - g(r)}{g'_r(0)}.$$

Then $|g_r'(0)| = |g'(r)|(1 - r^2) \neq 0$, so h_r is well-defined and satisfies:

(1) $h_r(0) = 0$. (2) $h_r'(0) = 1$. (3) $\|h_r\|_\infty \leq \frac{2}{|g_r'(0)|}$.

By a result of Dieudonné (Theorem VI.10 in [102])

$$h_r(U) \supseteq \left\{ w \in \mathbb{C} \;\middle|\; |w| < \frac{|g_r'(0)|}{16} \right\}.$$

Unraveling our auxiliary functions, we see that $g(U)$ contains the disc

$$D_r = \left\{ w \in \mathbb{C} \;\middle|\; |w - g(r)| < |g'(r)|^2(1 - r^2)^2 \frac{1}{16} \right\}$$

and hence that $F(U)$ contains the set $B_r = \Psi(D_r)$. Of course, $F(r) \in B_r$ and B_r is a disc because Ψ is a Möbius transformation. We will be finished if we can show that for r sufficiently near 1, the distance $d_r = \inf\{|F(r) - w| \,|\, w \in \partial B_r\}$ is larger than $\frac{1}{3}$. Fix r temporarily. Let C_1, C_2, and C_3 be circles centered at 1 with radii, respectively,

(i) $|1 - g(r)| + |g'(r)|^2(1 - r^2)^2 \frac{1}{16}$,
(ii) $|1 - g(r)|$, and
(iii) $|1 - g(r)| - |g'(r)|^2(1 - r^2)^2 \frac{1}{16}$.

We see that C_2 passes through $g(r)$, the center of D_r, while C_1 and C_3 are tangent to D_r. The images $\Psi(C_1)$, $\Psi(C_2)$ and $\Psi(C_3)$ are also circles centered at 1, and, of course $\Psi(C_2)$ passes through $F(r)$ and $\Psi(C_1)$ and $\Psi(C_3)$ are tangent to B_r. So, the distance d_r is either the distance from $\Psi(C_2)$ to $\Psi(C_3)$ or from $\Psi(C_2)$ to $\Psi(C_1)$. Furthermore, it is clear that the later distance is smaller. Let

$$\begin{cases} S_1 = 1 - |1 - g(r)| - |g'(r)|^2(1 - r^2)^2 \frac{1}{16}, \\ S_2 = 1 - |1 - g(r)|. \end{cases}$$

Then

$$d_r = |\Psi(S_1) - \Psi(S_2)| = \left| \frac{|g'(r)|^2(1 + r)^2}{8|(1 - g(r))/(1 - r)|^2 + \frac{1}{2}|1 - g(r)||g'(r)|^2(1 + r)^2} \right|.$$

(A long computation.)

Finally, let $r \to 1^-$. The condition in equation (86.0.4) implies $d_r \to \frac{1}{2}$. In particular, for r sufficiently close to 1, $d_r > \frac{1}{3}$ and the proof of Theorem II is complete. $\qquad\square$

4.2. **Theorem.** *Let h be a non-vanishing function in H^∞, $\|h\|_\infty \leq 1$, and assume its singular inner factor has a non-trivial discrete part. Then, for each integer $N \geq 1$, there exists $\delta = \delta(N) > 0$ such that h assumes every w, $0 < |w| < \delta$, at least N times.*

Proof. Let S_μ be the singular inner factor of h, and assume without loss of generality that $\mu(1) > 0$. By Proposition 2.4 and Theorem 2.7, $h = S_1 \circ g$, where

$$\liminf_{r \to 1^-} \frac{1 - r}{|1 - g(r)|} = \mu(1). \tag{86.0.5}$$

$F = \Psi \circ g$ then satisfies the hypotheses of Theorem II. We need one additional property of F. We know from the properties of angular derivatives that equation (86.0.5) implies

$$\lim_{r \to 1^-} \frac{1 - g(r)}{1 - r} = c, \quad 0 < c < \infty.$$

Therefore,

$$\lim_{r \to 1^-} \frac{\Im\, g(r)}{1 - \Re\, g(r)} = 0.$$

But $\frac{\Im\, g(r)}{1 - \Re\, g(r)}$ is the tangent of the angle the line from $g(r)$ to 1 makes with the real axis, so we see that g maps the radial segment to 1 onto an arc which is tangent to the real axis at 1. As for $F = \Psi \circ g$, this means that $|F'(r)| \to \infty$ as $r \to 1^-$, and that, for each θ, $\frac{\pi}{2} < \theta < \pi$, for r sufficiently near 1, $\theta < \arg F(r) < (2\pi - \theta)$. So,

$$\Re\, F(r) \to -\infty \text{ as } r \to 1^-. \tag{86.0.6}$$

If N is any positive integer, then by the conclusion of Theorem II, there exists r_0, $0 < r_0 < 1$, such that for all r, $r_0 \leq r < 1$,

$$F(U) \supseteq \{w \in \mathbb{C} \,|\, |w - F(r)| < 2\pi N\}. \tag{86.0.7}$$

Let $\delta = \exp\{\Re\, F(r_0)\}$. Then equations (86.0.6) and (86.0.7) imply that $\exp(F)$ assumes each w, $0 < |w| < \delta$, at least N times. But $h = \exp(F)$ and the proof is completed. $\qquad\square$

The last theorem is expected to be true because of the following reason: Our "master" example for the factor of h in the theorem, i.e., a non-trivial discrete singular inner function is

$$S_1(z) = \exp\left(\frac{z + 1}{z - 1}\right).$$

Indeed, this function is defined on $\mathbb{C} - \{1\}$ and 1 is an essential singularity. By the theorem of Picard on an isolated essential singularity, any neighborhood of this singularity has an image (by S_1) that covers all of $\mathbb{C} \cup \{\infty\}$ infinitely often except for one exceptional point $z = 0$. However, unlike a "typical" essential singularity which covers $\mathbb{C} \cup \{\infty\}$ usually in a wild fashion, for inner functions such as S_1, the behavior of S_1 at the essential singularity is kind of tame: Namely, the part of a 1-neighborhood within U covers only the portion of the Riemann sphere within U, and moreover mostly a neighborhood of zero for $\lim_{r \to 1^-} \exp\left(\frac{r+1}{r-1}\right) = 0$. The portion of a 1-neighborhood on $\mathbb{T}$ is mapped by S_1 into the unit circle $\mathbb{T}$, because $|S_1(\xi)| = 1$ for $\xi \in \mathbb{T} - \{1\}$, so the portion of the 1-neighborhoods that is "responsible" to cover the vicinity of ∞ lies in the exterior of the closed unit disc, $\mathbb{C} - U$. The geometry of the way the covering of 1-neighborhoods is dominated by S_1, even if S_1 is only a factor of h. In fact, we also expect that h as in the theorem will cover U infinitely often except for, maybe, one point. We were looking at a statement of the following form:

For every Blaschke product $B(z)$, there exists a singular inner function (can be a unimodular constant), $T(z)$, such that $B(z) \cdot T(z)$ is not a $U \to U$ surjection. Suppose that the derived set of the Blaschke sequence $\{\alpha_n\}_{n=1}^{\infty}$ is not $\mathbb{T}$. Say $1 \notin (\{\alpha_n\}_{n=1}^{\infty})'$. Then $B(z) \cdot S_1(z)$ has an isolated singular point at $z = 1$, which is an essential singular point of exactly the same character as $z = 1$ is for $S_1(z)$. Thus, we expect that $B(z) \cdot S_1(z)$ is either a $U \to U$ surjection or that it covers U infinitely often except for a single Picard exceptional value. We insisted that $1 \notin (\{\alpha_n\}_{n=1}^{\infty})'$ in order to avoid the possibility that $z = 1$ is not an isolated singularity of $B(z) \cdot S_1(z)$ in which case we know of no strong results such as Picard's theorem. Thus, at this point, we feel that given a Blaschke product $B(z)$ in order to have a mapping of the form $B(z) \cdot T(z)$, where $T(z)$ is a singular inner function, so that $B(z) \cdot T(z)$ is not a $U \to U$ surjection, we better look at continuous singular inner function, $T(z) \in \mathcal{F}_c$. Certainly, we might want to avoid factors of $T(z)$ which are discrete inner functions where the essential singular points on $\mathbb{T}$ are off $(\{\alpha_n\}_{n=1}^{\infty})'$.

Another possibility might be to consider Blaschke sequences $\{\alpha_n\}_{n=1}^{\infty}$ whose derived set equals $\mathbb{T}$, $(\{\alpha_n\}_{n=1}^{\infty})' = \mathbb{T}$. Taking the corresponding Blaschke product $B(z)$, even $B(z) \cdot S_1(z)$ can be non-surjective $U \to U$ (at least for now).

4.3. Theorem I(b). *If $A \in \mathcal{A}$ and 0 is a limit point of A, then Φ_A is a continuous singular inner function.*

Proof. Assume that Φ_A has a non-trivial discrete factor and apply Theorem 4.2 with $N = 1$. We conclude that Φ_A assumes every value $w \neq 0$ in some neighborhood of 0, contradicting the fact that 0 is a limit point of omitted values of Φ_A. Therefore, Φ_A must be a continuous singular inner function. $\qquad\square$

5. Applications.

5.1. Inner functions in H^∞.

In [43], Herrero conjectured that the collections $\mathcal{F}_a$ and $\mathcal{F}_c$ are closed in H^∞. Using his Theorem I, Stephenson proves that this is not true for $\mathcal{F}_a$.

5.2. Theorem III. $\mathcal{F}_a$ *is not closed in* H^∞.

Proof. Let $A = \{0, \frac{1}{2}, \frac{1}{3}, \frac{1}{4}, \ldots\} \subset U$. Then $A \in \mathcal{A}$ and 0 is a limit point of A. For $n \geq 2$, let $\alpha_n = \frac{1}{n}$ and define $f_n = \Psi_{\alpha_n} \circ \Phi_A$. Each f_n is a singular inner function, and $\|\Phi_A - f_n\|_\infty \to 0$ as $n \to \infty$. By Theorem I(b), $\Phi_A \in \mathcal{F}_c$, so it suffices to prove that $f_n \in \mathcal{F}_a$, $n \geq 2$. But this follows from Theorem I(a) since, by Corollary 2.5(a), f_n is the uniformizer of $U - \Psi_{\alpha_n}(A)$ and 0 is an isolated point of $\Psi_{\alpha_n}(A)$. $\qquad\square$

5.3. Herrero's conjecture is that $\mathcal{F}_c = \overline{\mathcal{F}_c}$ remains open. Stephenson says that his results seem to lend support to this conjecture in the following sense. Consider Theorem 4.2.

5.4. If f is a singular inner function with a non-trivial discrete factor, then there is a $\delta > 0$ such that $\Psi_\alpha \circ f$ fails to be singular if $|\alpha| < \delta$.

That is, f is isolated in H^∞ from any singular inner function of the form $\Psi_\alpha \circ f$. Can it be that this f is isolated from all the singular inner functions (except possibly those with a common factor)? For example, if $f \in \mathcal{F}_a$, must there exist an $\epsilon > 0$ such that $\|f - g\|_\infty \geq \epsilon$ for every $g \in \mathcal{F}_a$ which is not a unimodular constant times f?

5.5. We can also use Theorem I to answer two questions raised by Caughran and Shields. In [16], they asked whether a discrete singular inner function can omit an uncountable number of values in U. The answer is yes, Φ_A is such a function whenever $A \in \mathcal{A}$ is an uncountable set with 0 as isolated point. Seidel proved ([14] Theorem 5.13) that if μ is a singular measure whose closed support in $\mathbb{R}$ has an isolated point, then S_μ omits no value in U other than 0. In an unpublished manuscript, Caughran and Shields suggest that this conclusion may hold under weaker hypothesis that the discrete part of μ has a closed support with an isolated point. The following example shows this to be false: Choose $A \in \mathcal{A}$ to be uncountable with 0 as isolated point. Then Φ_A is a discrete singular inner function, and without loss of generality we may assume $\Phi_A = S_\mu$ with $\mu(1) = 1$. Construct an inner function g so that:

(a) g has radial limit 1 at 1,
(b) g has a finite angular derivative at 1, and
(c) g has a finite angular derivative at no other point of $\mathbb{T}$.

Such a function can be constructed, for example, by a modification of the method of Lemma 5.9 in [98]. Now consider the composition $S_\nu = S_\mu \circ g$. It was shown in [98, Section 5] that ν can have a point mass only at a point where g has a finite angular derivative. Thus, ν has at most one mass point. Since $\mu(1) = 1$, S_1 is a factor of S_μ, so by Theorem 2.7, ν does have a point mass at 1. We conclude that ν has precisely one point mass, yet $S_\nu = S_\mu \circ g$ omits the uncountable set A.

5.6. In [31], Fisher proved that for $h \in H^\infty$, $\|h\|_\infty \leq 1$, the set of α in U for which $\Psi_\alpha \circ h$ has a discrete singular factor is at most countable. Let $\{\rho_j\}_{j=1}^\infty$ be the (countable) collection of masses associated with all the discrete singular factors of $\{\Psi_\alpha \circ h | \alpha \in U\}$. In Theorem 2 in [31], Fisher proves that at most a finite number of masses ρ_j exceed each $\delta > 0$. In a private communication, Fisher asked whether, in fact, $\sum_{j=1}^\infty \rho_j < \infty$. We can answer this in the negative. Note first that if μ is a singular measure, then $\mu(\mathbb{T}) = -\log|S_\mu(0)|$. Using the set A and the notation from the proof of Theorem III (of this paper of Stephenson), let $h = \Phi_A$. Since each $f_n = \Psi_{\alpha_n} \circ h$ is totally discrete, the associated masses sum to $-\log|f_n(0)|$. But as $n \to \infty$, $|f_n(0)| \to |h(0)| \neq 1$, so $\sum_{n=2}^\infty (-\log|f_n(0)|) = \infty$.

6. Comments and questions.

6.1 Let μ be a singular measure with non-trivial discrete part. Define $\delta(\mu)$ to be the largest value of δ for which the conclusion of 5.4 above holds with $f = S_\mu$. What can one say about the map $\mu \to \delta(\mu)$? For instance, what are necessary and sufficient conditions for $\delta(\mu)$ to be less than 1? It is necessary that the closed support of μ be a perfect set of positive capacity in $\mathbb{T}$ (see Remark 3.2), however, this is far from sufficient. Using the methods of Herrero [43] for example, it can be shown that if μ has a point mass at the endpoint of any arc where S_μ is analytic, then $\delta(\mu) = 1$. When $\delta(\mu) < 1$, its size would seem to depend on characteristics of the discrete part of μ. To be specific, perhaps, if we normalize μ in an appropriate way, $\delta\mu$ will necessarily be larger than some absolute constant. This would not be too surprising since the existence of $\delta(\mu)$ depends on Theorem II, which is a result of Bloch type.

6.2. We have been concerned primarily with the omitted value of inner functions. To what extent do our results carry over to exceptional values of inner functions? More generally, the exceptional set $E(h)$ makes sense whenever $\|h\|_\infty \leq 1$. Do our results have analogues in the general case? Consider this result, for example: If $f \in \mathcal{F}$, then the set

$$\{\alpha \in O(f) | \Psi_\alpha \circ f \text{ has a factor in } \mathcal{F}_a\}$$

is at most countable. This follows trivially from our work because this set is discrete, but it also is a special case of Fisher's result quoted in Section 5.6. Does Fisher's result hold because the α's for which $\Psi_\alpha \circ h$ has a discrete singular inner factor form a discrete set?

Chapter 87

More Properties of Blaschke Sequences

Here are few natural questions to ask on Blaschke sequences.

Question (1):

(a) Is it true that for every Blaschke sequence $\{\alpha_n\}_{n=1}^\infty$ there exists a singular inner function $S(z)$ and a number $\alpha \in U - \{0\}$ such that $Z(S(z) - \alpha) = \{\alpha_n\}_{n=1}^\infty$? Here, we denote by $Z(f(z))$ the zero set of a function $f(z)$ analytic in U. If $f \not\equiv 0$, then this is either empty, or a sequence. We order the elements of this sequence by ascending modulus where each α_n is repeated as many times as its multiplicity as a zero of $f(z)$.

(b) In case the answer to 1(a) is positive, can we choose α at will?

Question (2): Is there an inner function $S(z)$, such that for every Blaschke sequence $\{\alpha_n\}_{n=1}^\infty$ there is a $\alpha \in U$ such that $Z(S(z) - \alpha) = \{\alpha_n\}_{n=1}^\infty$? That is, a kind of a universal inner function $S(z)$ that fits all the Blaschke sequences.

Question (3): Is there an inner function $S(z)$ such that for every Blaschke sequence generating the Blaschke product $B(z)$ there is a $\alpha \in U$ and a $c \in \mathbb{R}$ such that

$$\frac{\alpha - S(z)}{1 - \overline{\alpha}S(z)} = e^{ic}B(z)?$$

This is closely related to (2).

The answer to Question (2) is "no".

Proof. For otherwise if $\{\alpha_n\}_{n=1}^\infty$ is a Blaschke sequence, then we can change $\alpha_1 = w$ in U in an arbitrary manner (not changing $\alpha_2, \alpha_3, \ldots$) and

507

still get a Blaschke sequence, $\{w, \alpha_2, \alpha_3, \ldots\}$. Under our assumption, there will correspond a $\alpha(w) \in U$ such that $S(z) - \alpha(w)$ vanishes in U exactly at $\{w, \alpha_2, \alpha_3, \ldots\}$. So, for some $\lambda = \lambda(\alpha_1, \alpha_2)$ depending only on α_1 and α_2, we have

$$\frac{S(z) - \alpha(w_1)}{S(z) - \alpha(w_2)} = \lambda(w_1, w_2)\frac{z - w_1}{z - w_2}.$$

We note that

$$\frac{S - \alpha(w_1)}{S - \alpha(w_2)} = \xi \Rightarrow S - \alpha(w_1) = \xi \cdot (S - \alpha(w_2)) \Rightarrow (1-\xi)S = \alpha(w_1) - \xi\alpha(w_2)$$

and if $\alpha(w_1) \neq \alpha(w_2)$, then $\xi \neq 1$ so

$$S(z) = \frac{\alpha(w_1) - \xi\alpha(w_2)}{1 - \xi}.$$

In our case, this gives

$$S(z) = \frac{(\alpha(w_1) - \lambda\alpha(w_2))z + (-\alpha(w_1)\alpha_2 + \lambda w_1\alpha(w_2))}{(1 - \lambda)z + (-w_2 + \lambda w_1)},$$

which must be a U-automorphism. That, of course, is not possible and this contradiction concludes the proof. $\qquad\square$

A second conclusion: The answer to Question (3) is "no".

The answer to Question (1a) is "yes", if

$$\text{Image}\left\{\prod_n \frac{|\alpha_n|}{\alpha_n} \cdot \frac{\alpha_n - z}{1 - \overline{\alpha}_n z}\right\} \neq U.$$

Proof. For any $\alpha \in U$, we can define the following:

$$S(z) := \frac{\alpha - \prod_n \frac{|\alpha_n|}{\alpha_n}\frac{\alpha_n - z}{1 - \overline{\alpha}_n z}}{1 - \overline{\alpha}\prod_n \frac{|\alpha_n|}{\alpha_n}\frac{\alpha_n - z}{1 - \overline{\alpha}_n z}}.$$

Then $S(z) \in \text{Inn}$ and $Z(S(z) - \alpha) = \{\alpha_n\}_{n=1}^{\infty}$ (multiplicities are taken into account). This follows because $\frac{\alpha - w}{1 - \overline{\alpha}w}$ is an involution and hence,

$$\frac{\alpha - S(z)}{1 - \overline{\alpha}S(z)} = \prod_n \frac{|\alpha_n|}{\alpha_n}\frac{\alpha_n - z}{1 - \overline{\alpha}_n z}.$$

Now, just take

$$\alpha \notin \text{Image}\left\{\prod_n \frac{|\alpha_n|}{\alpha_n}\frac{\alpha_n - z}{1 - \overline{\alpha}_n z}\right\}.$$

$\qquad\square$

So, Question (1) is not answered at this stage for those Blaschke sequences $\{\alpha_n\}_{n=1}^{\infty}$ for which

$$\prod_n \frac{|\alpha_n|}{\alpha_n} \frac{\alpha_n - z}{1 - \overline{\alpha}_n z}$$

maps U onto U. This, naturally, leads to the following question:

(1c) Let $\{\alpha_n\}_{n=1}^{\infty}$ be a Blaschke sequence. Can we find a singular inner function $T(z)$ and a $\alpha \in U$ such that:

$$\alpha \notin \text{Image} \left\{ \prod_n \frac{|\alpha_n|}{\alpha_n} \frac{\alpha_n - z}{1 - \overline{\alpha}_n z} \cdot T(z) \right\}?$$

This is equivalent to

$$\prod_n \frac{|\alpha_n|}{\alpha_n} \frac{\alpha_n - z}{1 - \overline{\alpha}_n z} \cdot T(z),$$

which is not an onto mapping $U \to U$ for some singular inner function $T(z)$. If such a $T(z)$ and α exist then the answer to Question (1)(a) is "YES" with $S(z)$ defined as above. If no such a singular inner function exists, then the set

$$\left(\prod_n \frac{|\alpha_n|}{\alpha_n} \frac{\alpha_n - z}{1 - \overline{\alpha}_n z} \right) \cdot \{\text{singular inner functions including}$$
$$\text{the unimodular constants}\}$$

contains only $U \to U$ surjective mappings. This would say: Any inner function with the zero set $\{\alpha_n\}_{n=1}^{\infty}$ is a $U \to U$ surjection. This is an attribute of the very special Blaschke sequence $\{\alpha_n\}_{n=1}^{\infty}$. In this case, we can take an arbitrary singular inner function $T(z)$ and produce (by composition) the singular inner function

$$T\left(\prod_n \frac{|\alpha_n|}{\alpha_n} \frac{\alpha_n - z}{1 - \overline{\alpha}_n z} \right).$$

Then the inner function

$$\left(\prod_n \frac{|\alpha_n|}{\alpha_n} \frac{\alpha_n - z}{1 - \overline{\alpha}_n z} \right) \cdot T\left(\prod_n \frac{|\alpha_n|}{\alpha_n} \frac{\alpha_n - z}{1 - \overline{\alpha}_n z} \right)$$

is a $U \to U$ surjection. But $w = \prod_n \frac{|\alpha_n|}{\alpha_n} \frac{\alpha_n - z}{1 - \overline{\alpha}_n z}$ is surjective. So, we can eliminate from the condition any dependency on the Blaschke sequence

$\{\alpha_n\}_{n=1}^{\infty}$, for $wT(w)$ must be a $U \to U$ surjection for all the singular inner functions $T(z)$. Let us pause with Question (1) and note first few easier and more elementary properties of Blaschke sequences. The next question is as follows:

Question (4): Let $\{\alpha_n\}_{n=1}^{\infty}$ be a Blaschke sequence and let $z \in U$. Is the sequence

$$\left\{ \frac{|\alpha_n|}{\alpha_n} \frac{\alpha_n - z}{1 - \overline{\alpha}_n z} \right\}_{n=1}^{\infty}$$

a Blaschke sequence? The answer is "YES".

Proof.   First, we note that $\{\alpha_n\}_{n=1}^{\infty}$ is a Blaschke sequence if and only if $\{\alpha_n^2\}_{n=1}^{\infty}$ is a Blaschke sequence. For:

$$\sum_n (1 - |\alpha_n|) < \infty \Rightarrow \sum_n (1 - |\alpha_n|^2) = \sum_n (1 - |\alpha_n|)(1 + |\alpha_n|)$$

$$\leq 2 \sum_n (1 - |\alpha_n|) < \infty$$

$$\Rightarrow \{\alpha_n^2\}_{n=1}^{\infty} \text{ is a Blaschke sequence,}$$

and also,

$$\sum_n (1 - |\alpha_n|^2) < \infty \Rightarrow \sum_n (1 - |\alpha_n|) \leq \sum_n (1 - |\alpha_n|)(1 + |\alpha_n|) < \infty$$

$$\Rightarrow \{\alpha_n\}_{n=1}^{\infty} \text{ is a Blaschke sequence.}$$

Thus, we look at the following:

$$1 - \left| \frac{\alpha_n - z}{1 - \overline{\alpha}_n z} \right|^2 = \frac{1 + |\alpha_n|^2 |z|^2 - |\alpha_n|^2 - |z|^2}{1 + |\alpha_n|^2 |z|^2 - 2\Re\{z\overline{\alpha}_n\}} \leq \frac{(1 - |\alpha_n|^2)(1 - |z|^2)}{(1 - |\alpha_n||z|)^2}.$$

So, (by $1 - |\alpha_n||z| \geq 1 - |z|$):

$$1 - \left| \frac{\alpha_n - z}{1 - \overline{\alpha}_n z} \right|^2 \leq \frac{(1 - |\alpha_n|^2)(1 - |z|^2)}{(1 - |z|)^2} = \left(\frac{1 + |z|}{1 - |z|} \right)(1 - |\alpha_n|^2),$$

$$\Rightarrow \sum_n \left(1 - \left| \frac{\alpha_n - z}{1 - \overline{\alpha}_n z} \right|^2 \right) \leq \left(\frac{1 + |z|}{1 - |z|} \right) \sum_n (1 - |\alpha_n|^2) < \infty. \qquad \square$$

The next question is as follows:

Question (5): The product sequence $\{\prod_{k=1}^{N} \alpha_n^{(k)}\}_{n=1}^{\infty}$ of a finite number N of Blaschke sequences $\{\alpha_n^{(k)}\}_{n=1}^{\infty}$, $k = 1, \ldots, N$, is a Blaschke sequence.

Proof. It is necessary and sufficient to prove the claim for $N = 2$ sequences. Thus, let $\{\alpha_n\}_{n=1}^{\infty}$, $\{\beta_n\}_{n=1}^{\infty}$ be two Blaschke sequences. We use the identity $2(1 - |\alpha_n \beta_n|) = (1 - |\alpha_n|)(1 + |\beta|) + (1 + |\alpha_n|)(1 - |\beta_n|)$ to deduce the following:

$$\sum_n (1 - |\alpha_n \beta_n|) = \frac{1}{2} \sum_n (1 - |\alpha_n|)(1 + |\beta|) + \frac{1}{2} \sum_n (1 + |\alpha_n|)(1 - |\beta_n|)$$

$$\leq \sum_n (1 - |\alpha_n|) + \sum_n (1 - |\beta_n|) < \infty. \qquad \square$$

The similar claim as in (5) is false if the number of sequences $N = \infty$ is infinite.

Example 87.0.189. This example is such that the reasons for not being a Blaschke sequence are not entirely trivial.

$$\{\alpha_n^{(k)}\}_{n=1}^{\infty} = \left\{ 1 - \left(\frac{1}{n}\right)^{1+\frac{1}{k}} \right\}_{n=1}^{\infty} , \quad k = 1, 2, 3, \ldots.$$

All of these are Blaschke sequences $\left(\sum_{n=1}^{\infty} (1 - |\alpha_n^{(k)}|) = \sum_{n=1}^{\infty} \left(\frac{1}{n}\right)^{1+1/k} < \infty\right)$. The product sequence is

$$\left\{ \prod_{k=1}^{\infty} \left(1 - \left(\frac{1}{n}\right)^{1+\frac{1}{k}}\right) \right\}_{n=1}^{\infty} .$$

It satisfies

$$\prod_{k=1}^{\infty} \left(1 - \left(\frac{1}{n}\right)^{1+\frac{1}{k}}\right) < 1 - \left(\frac{1}{n}\right)^{1+\frac{1}{n}} .$$

Hence,

$$\sum_n \left(1 - \prod_{k=1}^{\infty} \left(1 - \left(\frac{1}{n}\right)^{1+\frac{1}{k}}\right)\right) \geq \sum_n \left(\frac{1}{n}\right)^{1+\frac{1}{n}} = \infty. \qquad \square$$

We now return to the more difficult problem posed in (1)(a).

Chapter 88

Blaschke Sequences and the $U \to U$ Surjectivity of Certain Inner Functions

The question we are concerned with is the following:

(1)(a) Is it true that for any Blaschke sequence, $\{\alpha_n\}_{n=1}^{\infty}$, there exist a singular inner function $S(z)$ and a number $\alpha \in U - \{0\}$, such that $Z(S(z) - \alpha) = \{\alpha_n\}_{n=1}^{\infty}$, where multiplicities are taken into account?

We saw that this is closely related to the $U \to U$ surjectivity, of certain inner functions. More precisely, we have the following:

Proposition 88.0.190. *Let $\{\alpha_n\}_{n=1}^{\infty}$ be a Blaschke sequence. The following two statements are equivalent.*

(i) *There is no singular inner function $S(z)$ and a number $\alpha \in U - \{0\}$, such that $Z(S(z) - \alpha) = \{\alpha_n\}_{n=1}^{\infty}$, where multiplicities are taken into account.*

(ii) *The following set of functions*

$$\left(\prod_n \frac{|\alpha_n|}{\alpha_n} \left(\frac{\alpha_n - z}{1 - \overline{\alpha}_n z} \right) \right) \cdot \{singular\ inner\ functions\ including\ the\ unimodular\ constants\}$$

contains only $U \to U$ surjective mappings.

Proof. **(i)$\Rightarrow$(ii):** Let us assume, in order to get a contradiction, that (ii) is false. We will denote the corresponding Blaschke product by

$$B(z) = \prod_n \frac{|\alpha_n|}{\alpha_n} \left(\frac{\alpha_n - z}{1 - \overline{\alpha}_n z} \right).$$

Let $T(z)$ be a singular inner function such that for some $\alpha \in U - \{0\}$, we have $\alpha \notin \text{Image}\{B(z)\cdot T(z)\}$. We construct the corresponding Frostman shift

$$S(z) = \frac{\alpha - B(z) \cdot T(z)}{1 - \overline{\alpha}B(z) \cdot T(z)}.$$

Then $S(z)$ is an inner function that does not vanish anywhere in U. Hence, it is a singular inner function for U. Furthermore, we have the following identity:

$$B(z) \cdot T(z) = \frac{\alpha - S(z)}{1 - \overline{\alpha}S(z)}.$$

Hence, $Z(S(z) - \alpha) = \{\alpha_n\}_{n=1}^{\infty}$ which violates (i).

(ii)$\Rightarrow$(i): Let us assume, in order to get a contradiction, that (i) is false. Let $S(z)$ be a singular inner function, $\alpha \in U - \{0\}$, such that $Z(S(z) - \alpha) = \{\alpha_n\}_{n=1}^{\infty}$. Then by the factorization theorem in $H^{\infty}(U)$, we have

$$\frac{\alpha - S(z)}{1 - \overline{\alpha}S(z)} = B(z) \cdot T(z),$$

for some singular inner function $T(z)$. Thus, the following

$$S(z) = \frac{\alpha - B(z) \cdot T(z)}{1 - \overline{\alpha}B(z) \cdot T(z)},$$

is a singular inner function. But this implies that $\alpha \notin \text{Image}\{B(z) \cdot T(z)\}$. So $B(z) \cdot T(z)$ is not a $U \to U$ surjection. That contradicts (ii). $\qquad\square$

In fact, we can do better than Proposition 88.0.190. We can replace statement (ii) with a statement that contains no reference to Blaschke products. This is a simple and basic statement which makes the following interesting.

Theorem 88.0.191. *The following two statements are equivalent.*

(i) *There is Blaschke sequence which is not the zero set (multiplicities are taken into account) of any function of the form $S(z) - \alpha$, where $S(z)$ is a singular inner function and $\alpha \in U - \{0\}$.*

(ii) *For every singular inner function, $T(z)$, the function $z \cdot T(z)$ is a $U \to U$ surjection.*

Proof. **(i)$\Rightarrow$(ii):** Let us assume, in order to get a contradiction, that (ii) is false. Let $\{\alpha_n\}_{n=1}^{\infty}$ be the Blaschke sequence in (i) and let $B(z)$ be the corresponding Blaschke product:

$$B(z) = \prod_n \frac{|\alpha_n|}{\alpha_n} \left(\frac{\alpha_n - z}{1 - \overline{\alpha}_n z} \right).$$

Then, if $T(z)$ is a singular inner function such that $z \cdot T(z)$ is not a $U \to U$ surjection, then also $B(z) \cdot T(B(z))$ is not a $U \to U$ surjection (Image$\{B(z) \cdot T(B(z))\} \subseteq$ Image$\{z \cdot T(z)\} \subset U$). Say $\alpha \notin$ Image$\{B(z) \cdot T(B(z))\}$ $(\alpha \in U)$. But then

$$S(z) = \frac{\alpha - B(z) \cdot T(B(z))}{1 - \overline{\alpha} B(z) \cdot T(B(z))},$$

is a singular inner function. Moreover,

$$B(z) \cdot T(B(z)) = \frac{\alpha - S(z)}{1 - \overline{\alpha} S(z)}.$$

Hence, $Z(S(z) - \alpha) = \{\alpha_n\}_{n=1}^{\infty}$ which contradicts (i). Now for the inverse implication.

(ii)$\Rightarrow$(i): So, we assume that (ii) holds true but that (i) does not. Then every Blaschke sequence is a $Z(S(z) - \alpha)$. By Proposition 88.0.190 (written in the form (not (i)) $\Leftrightarrow$ (not (ii))), we deduce that for every Blaschke product $B(z)$, the set

$$B(z) \cdot \{\text{singular inner functions including the unimodular constants}\}$$

includes mappings that are not $U \to U$ surjections. In particular, taking $B(z) = z$ will give us a contradiction to the assumption (ii). $\square$

We can state Theorem 88.0.191 in an equivalent way, as follows:

Theorem 88.0.192. *The following two statements are equivalent.*

(i) *Every Blaschke sequence is a zero (multiplicities are taken into account) of a function of the form $S(z) - \alpha$, for some singular inner function $S(z)$ and some number $\alpha \in U - \{0\}$.*

(ii) *There exists a singular inner function $T(z)$, such that $z \cdot T(z)$ is not a $U \to U$ surjection.*

To complete the solution of the problem (1)(a), we need to find if the equivalent statements (i) and (ii) of Theorem 88.0.191 are true or not!

Chapter 89

The Family of Refined Inner Functions

In this chapter, we will answer the question asked at the end of Chapter 88 for a special subclass of the family of singular inner functions. We know of a close connection between

(1) The existence of a Blaschke sequence which is not $Z(S(z) - \alpha)$ for any singular inner function $S(z)$ and any $\alpha \in U - \{0\}$.

and

(2) The $U \to U$ surjectivity of all the inner functions of the form $z \cdot T(z)$, where $T(z)$ is any singular inner function.

(Theorem 88.0.191).
In this chapter, we will establish a connection of (1), (2) with

(3) The boundary behavior of inner functions (not necessarily the singular inner functions).

We recall that a function $f(z) \in H^\infty(U)$ is called an inner function if $\lim_{r \to 1^-} |f(re^{i\theta})| = 1$ for almost all $\theta \in [0, 2\pi)$. This important definition was motivated by Fatou's theorem that for any $g \in H^\infty(U)$, $\lim_{r \to 1^-} g(re^{i\theta})$ exists and is finite for almost all $\theta \in [0, 2\pi)$, and by the well-known theorem of Beurling on invariant $H^2(U)$-subspaces. We single out a special subclass of the family of inner functions.

Definition 89.0.193. An inner function $f(z)$ is called a refined inner function (an R-inner function) if $\lim_{r \to 1^-} f(re^{i\theta}) = 1$ for all $e^{i\theta}$ on $|z| = 1$, except possibly for a set of capacity zero on $|z| = 1$. We will freely use the induced notions: R-singular inner function, R-continuous inner function and an R-Blaschke product and an R-Blaschke sequence. The last notion is merely the zero set of an R-Blaschke product.

Remark 89.0.194. A set which is of capacity zero is of linear measure zero, but not conversely. The middle-thirds set of Cantor is of a positive capacity.

Remark 89.0.195. If g is an R-inner function and f is an R-inner function, then $f \circ g$ is an R-inner function.

Proof. There is a subset C of $\mathbb{T}$ of capacity zero, such that $\lim_{r \to 1^-} |g(re^{i\theta})| = 1$ for all $e^{i\theta} \in \mathbb{T} - C$. Out of those $e^{i\theta}$, we take only those for which $g(re^{i\theta}) = R(r)e^{i\phi(r)}$ so that when $r \to 1^-$, $R(r)e^{i\phi(r)}$ is a non tangential path in U that ends on $e^{i\phi(1)} \in \mathbb{T}$, such that along this path $|f(R(r)e^{i\phi(r)})| \to 1$. This excludes another subset C_1 of $\mathbb{T}$ of capacity zero (because also f is an R-inner function). $\qquad\square$

Remark 89.0.196. If g and f are R-inner functions, then $g \cdot f$ is an R-inner function.

Proposition 89.0.197. *Let $\{\alpha_n\}_{n=1}^{\infty}$ be an R-Blaschke sequence. The following two statements are equivalent.*

(i) *There is no R-singular inner function $S(z)$ and a number $\alpha \in U - \{0\}$, such that $Z(S(z) - \alpha) = \{\alpha_n\}_{n=1}^{\infty}$, where multiplicities are taken into account.*

(ii) *The set*

$$\left(\prod_n \frac{|\alpha_n|}{\alpha_n} \left(\frac{\alpha_n - z}{1 - \overline{\alpha}_n z} \right) \right) \cdot \{R - singular\ inner\ functions\ including\ the\ unimodular\ constants\}$$

contains only $U \to U$ surjections.

Proof. **(i)$\Rightarrow$(ii):** Let us assume, in order to get a contradiction, that (ii) is false. We will denote the corresponding Blaschke product by

$$B(z) = \prod_n \frac{|\alpha_n|}{\alpha_n} \left(\frac{\alpha_n - z}{1 - \overline{\alpha}_n z} \right).$$

Let $T(z)$ be an R-singular inner function such that for some $\alpha \in U - \{0\}$, we have $\alpha \notin \text{Image}\{B(z) \cdot T(z)\}$. We construct the corresponding Frostman shift

$$S(z) = \frac{\alpha - B(z) \cdot T(z)}{1 - \overline{\alpha} B(z) \cdot T(z)}.$$

Then $S(z)$ is an inner function that does not vanish anywhere in U. It is an R-singular inner function by Remark 89.0.196 and alike. Furthermore, we have the following identity:

$$B(z) \cdot T(z) = \frac{\alpha - S(z)}{1 - \overline{\alpha}S(z)}.$$

Hence, $Z(S(z) - \alpha) = \{\alpha\}_{n=1}^{\infty}$ which violates (i).

(ii)$\Rightarrow$(i): Let us assume, in order to get a contradiction, that (i) is false. Let $S(z)$ be an R-singular inner function and let $\alpha \in U - \{0\}$ such that $Z(S(z) - \alpha) = \{\alpha_n\}_{n=1}^{\infty}$. By the $H^{\infty}(U)$ factorization theorem, we have

$$\frac{\alpha - S(z)}{1 - \overline{\alpha}S(z)} = B(z) \cdot T(z),$$

for some singular inner function $T(z)$. Since $S(z)$ and $B(z)$ are R-inner functions, it follows that $T(z)$ is an R-singular inner function. Thus,

$$S(z) = \frac{\alpha - B(z) \cdot T(z)}{1 - \overline{\alpha}B(z) \cdot T(z)}$$

is a singular inner function which implies that $\alpha \notin \text{Image}\{B(z) \cdot T(z)\}$. Thus $B(z) \cdot T(z)$ is not a $U \to U$ surjection. That contradicts (ii). $\qquad\square$

Theorem 89.0.198. *The following two statements are equivalent.*

(i) *There is an R-Blaschke sequence which is not the zero set (multiplicities are taken into account) of any function of the form $S(z) - \alpha$, where $S(z)$ is an R-singular inner function and $\alpha \in U - \{0\}$.*

(ii) *For every R-singular inner function $T(z)$, the function $z \cdot T(z)$ is a $U \to U$ surjection.*

Proof. **(i)$\Rightarrow$(ii):** Let us assume, in order to get a contradiction, that (ii) is false and let $\{\alpha_n\}_{n=1}^{\infty}$ be the R-Blaschke sequence in (i). Let us denote the corresponding R-Blaschke product by

$$B(z) = \prod_n \frac{|\alpha_n|}{\alpha_n} \left(\frac{\alpha_n - z}{1 - \overline{\alpha}_n z} \right).$$

Then, if $T(z)$ is an R-singular inner function such that $z \cdot T(z)$ is not a $U \to U$ surjection, then also $B(z) \cdot T(B(z))$ is not a $U \to U$ surjection. Say, $\alpha \notin \text{Image}\{B(z) \cdot T(z)\}$, $(\alpha \in U)$. But then

$$S(z) = \frac{\alpha - B(z) \cdot T(B(z))}{1 - \overline{\alpha}B(z) \cdot T(B(z))},$$

is an R-singular inner function. Moreover, we have

$$B(z) \cdot T(B(z)) = \frac{\alpha - S(z)}{1 - \overline{\alpha}S(z)}.$$

Hence, $Z(S(z) - \alpha) = \{\alpha_n\}_{n=1}^{\infty}$ which contradicts (i). Now for the inverse implication.

(ii)⇒(i): We assume that (ii) holds true but that (i) does not. Then every R-Blaschke sequence is a $Z(S(z) - \alpha)$, for some R-singular inner function $S(z)$ and some $\alpha \in U - \{0\}$. By Proposition 89.0.197 (written in the contrapositive fashion), we deduce that for every R-Blaschke product, $B(z)$, the set

$$B(z) \cdot \{\text{R} - \text{singular inner functions including the unimodular constants}\},$$

includes mappings that are not $U \to U$ surjections. In particular, we may take $B(z) = z$ and this contradicts (ii). $\square$

We can now conclude the following result, which is a bit surprising.

Theorem 89.0.199. *There is an R-Blaschke sequence which is not the zero set (multiplicities are taken into account) of any function of the form $S(z) - \alpha$, where $S(z)$ is an R-singular inner function and $\alpha \in U - \{0\}$.*

Proof. We will demonstrate the validity of statement (ii) in Theorem 89.0.198. We use the following result from [66]. It also appears in [14]:

Theorem 5.14 ([14], p. 109). *Let $w = f(z)$ be analytic and bounded, $|f(z)| < 1$ in $|z| < 1$, and let the radial limit of the modulus $|f(re^{i\theta})|$ be 1 for all $e^{i\theta}$ on $|z| = 1$, except possibly for a set of capacity zero on $|z| = 1$. Then unless $f(z)$ reduces to a finite Blaschke product or to a constant, every value of $|w| < 1$ is assumed infinitely often by $f(z)$ with at most one exception.*

Thus, let $T(z)$ be an R-singular inner function and consider $z \cdot T(z)$. Then $z \cdot T(z)$ is an R-inner function which is not a finite Blaschke product nor a constant in our case. According to the theorem of Lohwater, it assumes infinitely often every value of $|w| < 1$ with at most one exception. Since $T(z)$ is a singular inner function, $z \cdot T(z)$ assumes $w = 0$ exactly once (in $z = 0$). So, this is the exceptional point mentioned in the theorem of Lohwater. We conclude that $z \cdot T(z)$ is a $U \to U$ surjection. This concludes our proof. $\square$

Problem 89.0.200. Give a constructive characterization of all the R-Blaschke sequences that are not $Z(S(z) - \alpha)$ for some R-singular inner function $S(z)$ and some $\alpha \in U - \{0\}$.

Theorem 89.0.201. *Let $\{z_n\}_{n=1}^{\infty}$ be a Blaschke sequence which equals the zero set $Z(S(z)-\alpha)$, for some singular inner function $S(z)$ and some number $\alpha \in U - \{0\}$. Then*

(i) *There exists a singular inner function $T(z)$ such that*

$$B(z) \cdot T(z) = \frac{\alpha - S(z)}{1 - \overline{\alpha}S(z)}, \quad S(z) = \frac{\alpha - B(z) \cdot T(z)}{1 - \overline{\alpha}B(z) \cdot T(z)},$$

where

$$B(z) = \prod_{n=1}^{\infty} \frac{|z_n|}{z_n} \frac{z_n - z}{1 - \overline{z}_n z}$$

up to multiplication by a uni-modular constant, and with the standard change for $z_n = 0$.

(ii) *$\alpha \notin \operatorname{Image} B(z) \cdot T(z)$.*

(iii)

$$\forall\, z \in U, \ \forall\, n \in \mathbb{Z}^{+}, \ \left| \frac{\alpha - S\left(\frac{z_n-z}{1-\overline{z}_n z}\right)}{1 - \overline{\alpha}S\left(\frac{z_n-z}{1-\overline{z}_n z}\right)} \right| \leq |z| \left| T\left(\frac{z_n - z}{1 - \overline{z}_n z}\right) \right|,$$

equivalently

$$\forall\, z \in U, \ \forall\, n \in \mathbb{Z}^{+}, \ \left| \frac{\alpha - S(z)}{1 - \overline{\alpha}S(z)} \right| \leq \left| \frac{z_n - z}{1 - \overline{z}_n z} \right| |T(z)|.$$

Moreover, equality holds even for one $z \neq z_n$ in U if and only if we have the following identity

$$\frac{\alpha - S(z)}{1 - \overline{\alpha}S(z)} = \lambda \cdot \left(\frac{z_n - z}{1 - \overline{z}_n z} \right) \cdot T(z),$$

for some uni-modular constant $|\lambda| = 1$, in which case the Blaschke product degenerates to

$$B(z) = \lambda \cdot \left(\frac{z_n - z}{1 - \overline{z}_n z} \right).$$

(iv) *$\forall\, n \in \mathbb{Z}^{+}$, $|S'(z_n)| \leq |T(z_n)|$. Moreover, if there is equality even for a single $n \in \mathbb{Z}^{+}$, then we have the same identity as in (iii). In particular, $\forall\, n \in \mathbb{Z}^{+}$, $|S'(z_n)| < 1$.*

Proof. We already know (i) and (ii). For (iii) and (iv), we fix $n \in \mathbb{Z}^+$. Using (i), we define the following function:

$$f(z) = B\left(\frac{z_n - z}{1 - \bar{z}_n z}\right) = \left(1/T\left(\frac{z_n - z}{1 - \bar{z}_n z}\right)\right) \cdot \left(\frac{\alpha - S\left(\frac{z_n - z}{1 - \bar{z}_n z}\right)}{1 - \bar{\alpha}S\left(\frac{z_n - z}{1 - \bar{z}_n z}\right)}\right).$$

We note that since $\mathrm{Aut}(U)$ is a group (with composition of mappings as the binary operation), also,

$$f(z) = B\left(\frac{z_n - z}{1 - \bar{z}_n z}\right),$$

is a Blaschke product. In fact, one can observe the following elementary computation in order to be convinced:

$$\frac{z_1 - \frac{z_2 - z}{1 - \bar{z}_2 z}}{1 - \bar{z}_1 \frac{z_2 - z}{1 - \bar{z}_2 z}} = \left(\frac{1 - \bar{z}_1 z_2}{1 - z_1 \bar{z}_2}\right) \cdot \frac{\left(\frac{z_2 - z_1}{1 - \bar{z}_2 z_1}\right) - z}{1 - \left(\frac{z_2 - z_1}{1 - \bar{z}_2 z_1}\right) z}.$$

By $S(z_n) = \alpha$, we obtain $f(0) = 0$, and since, as was noted above, $f(z)$ is a Blaschke product, we also have $\forall z \in U, |f(z)| \leq 1$. So, we deduce, by the Schwarz Lemma, that $|f(z)| \leq |z|, \forall z \in U$ and $|f'(0)| \leq 1$ and if equality holds for one of these inequalities (even for just one $z \in U - \{0\}$ in the first inequality), then $f(z) = \lambda \cdot z$ for some uni-modular constant $|\lambda| = 1$. Thus,

$$\forall z \in U, \forall n \in \mathbb{Z}^+, \quad \left|\frac{\alpha - S\left(\frac{z_n - z}{1 - \bar{z}_n z}\right)}{1 - \bar{\alpha}S\left(\frac{z_n - z}{1 - \bar{z}_n z}\right)}\right| \leq |z|\left|T\left(\frac{z_n - z}{1 - \bar{z}_n z}\right)\right|, \qquad (89.0.1)$$

with equality if and only if

$$\forall z \in U, \forall n \in \mathbb{Z}^+, \quad \frac{\alpha - S\left(\frac{z_n - z}{1 - \bar{z}_n z}\right)}{1 - \bar{\alpha}S\left(\frac{z_n - z}{1 - \bar{z}_n z}\right)} = \lambda \cdot z \cdot T\left(\frac{z_n - z}{1 - \bar{z}_n z}\right), \qquad (89.0.2)$$

for some uni-modular constant $|\lambda| = 1$. Alternatively, since $w = \frac{z_n - z}{1 - \bar{z}_n z}$ is an involution in the group $(\mathrm{Aut}(U), \circ)$, i.e., $z = \frac{z_n - w}{1 - \bar{z}_n w}$, we have

$$\forall z \in U, \forall n \in \mathbb{Z}^+, \quad \left|\frac{\alpha - S(z)}{1 - \bar{\alpha}S(z)}\right| \leq \left|\frac{z_n - z}{1 - \bar{z}_n z}\right| |T(z)|, \qquad (89.0.3)$$

and equality holds even for a single $z \neq z_n$ in U if and only if we have the following identity:

$$\forall z \in U, \ \forall n \in \mathbb{Z}^+, \ \frac{\alpha - S(z)}{1 - \overline{\alpha}S(z)} = \lambda \cdot \left(\frac{z_n - z}{1 - \overline{z}_n z}\right) \cdot T(z), \qquad (89.0.4)$$

for some uni-modular constant $|\lambda| = 1$. We now differentiate $f(z)$. We use the following elementary computation:

$$\left(\frac{z_n - z}{1 - \overline{z}_n z}\right)'_z = \frac{-1 + |z_n|^2}{(1 - \overline{z}_n z)^2},$$

and the chain rule to get: $f'(z) = \frac{F}{E}$ where

$$F = \frac{1 - |z_n|^2}{(1 - \overline{z}_n z)^2} \left\{ T'\left(\frac{z_n - z}{1 - \overline{z}_n z}\right)\left(1 - \overline{\alpha}S\left(\frac{z_n - z}{1 - \overline{z}_n z}\right)\right)\left(\alpha - S\left(\frac{z_n - z}{1 - \overline{z}_n z}\right)\right) \right.$$
$$\left. + T\left(\frac{z_n - z}{1 - \overline{z}_n z}\right)S'\left(\frac{z_n - z}{1 - \overline{z}_n z}\right)(1 - |\alpha|^2) \right\},$$

and

$$E = \left\{ T\left(\frac{z_n - z}{1 - \overline{z}_n z}\right)\left(1 - \overline{\alpha}S\left(\frac{z_n - z}{1 - \overline{z}_n z}\right)\right) \right\}^2.$$

Hence,

$$f'(0) = \frac{(1 - |z_n|^2)\{T'(z_n)(1 - \overline{\alpha}S(z_n))(\alpha - S(z_n)) + T(z_n)S'(z_n)(1 - |\alpha|^2)\}}{T(z_n)^2(1 - \overline{\alpha}S(z_n))^2}.$$

Since $S(z_n) = \alpha$, we get

$$f'(0) = \frac{S'(z_n)}{T(z_n)}.$$

So, (by $|f'(0)| \leq 1$), we finally get $|S'(z_n)| \leq |T(z_n)|$. $\qquad \square$

Chapter 90

Paper [93] of Wladimir Seidel

The paper was received by the editors on July 17, 1931. It contains results
along the lines of Theorem 5.14 of Lohwater, [14]. It is beautifully written
and is worth studying.

1. **Introduction.** Let $f(z)$ be a regular, analytic function defined in the unit
disc $|z| < 1$. Consider a sequence of points $z_1, z_2, \ldots, z_n, \ldots$ lying in $|z| < 1$
and converging toward the point $z = 1$, a point of discontinuity of $f(z)$. In
general, the sequence of images $w_1 = f(z_1)$, $w_2 = f(z_2), \ldots, w_n = f(z_n), \ldots$
will not converge toward a definite value. We can, however, always select a
subsequence $z_{n_1}, z_{n_2}, \ldots, z_{n_m},$ so that the limit of the second sequence
exists: $\lim_{m \to \infty} f(z_{n_m}) = c$, where c may in some cases be infinite. Such a
value c will be called a cluster value of the function $f(z)$ in the point $z = 1$.
The purpose of the paper is to study the distribution of the cluster values of univalent analytic functions, bounded analytic functions, and certain
intermediate types of analytic functions in a boundary point of their circle
of convergence, which without loss of generality will always be taken to be
the unit circle $|z| = 1$. The point will always be taken to be $z = 1$. There
are essentially three questions with which Seidel will be concerned. By the
way, in this book we, naturally, are more interested in bounded analytic functions. Less in univalent analytic functions. First, the author considers cluster
sets formed by taking all the cluster values of $f(z)$ obtained by approaching
$z = 1$ along a Jordan arc and investigates certain relations between such arcs
and the corresponding cluster sets. This will be done for univalent analytic
functions and for bounded analytic functions. Secondly, sufficient conditions
are investigated that two cluster values corresponding to two sequences of
points $z_1, z_2, \ldots, z_n, \ldots$ and $\xi_1, \xi_2, \ldots, \xi_n, \ldots$ of points in $|z| < 1$ converging
towards $z = 1$ be equal. In the course of this investigation, the author arrives
at a generalization of a theorem of Gross. Finally, the author considers those

types of bounded analytic functions which assume one value infinitely often in $U : |z| < 1$. For these types, we shall see that most of the preceding theorems fail to hold and the author derives alternative theorems for these cases. The investigations are related to the work of Carathéodory on conformal mapping, as well as to the work of Lindelöf, Iversen, and Gross. In this connection, the author mentions an interesting recent (than!) result of Plessner.

Chapter I. Cluster values on curves.

2. Let $f(z)$ be a bounded analytic function in $U : \sup_{z \in U} |f(z)| = M < \infty$. Consider a Jordan arc C all of whose points with the exception of $z = 1$ consist of interior points of the disc U. Joining the end points of C by another Jordan arc D which lies in U and has no points in common with C except its two end points, we obtain a simply connected domain G bounded by the two curves C and D. Let us now set

$$\limsup_{z \to 1} |f(z)| = A, \ z \in C. \tag{90.0.1}$$

Seidel proves the following:

Theorem 1 ([93]). *Let $\{z_i\}$ be an arbitrary sequence of points lying in G and converging toward $z = 1$, for which $\lim_{i \to \infty} |f(z_i)| = \alpha$ exists. Then, $\alpha \leq A$.*

Proof. We can without loss of generality assume that the upper bound M of $|f(z)|$ on U is 1. Let $t = \Phi(z)$ be the function which maps the simply connected domain G on the unit disc $|t| < 1$. Let its inverse be $z = \Psi(t)$. Since G is bounded by a closed Jordan curve, the function $\Psi(t)$ is uniformly continuous in the closed disc $|t| \leq 1$, [81], [11]. Furthermore, the function Ψ can be so chosen that $\Psi(1) = 1$. Our map of the simply connected domain G carries, therefore, the sequence $\{z_i\}$ into a sequence of points $\{t_i\}$ converging toward $t = 1$, for which $\lim_{i \to \infty} |f(\Psi(t_i))| = \alpha$. At the same time, by equation (90.0.1), we have

$$\limsup_{t \to 1} |f(\Psi(t))| = A, \quad |t| = 1. \tag{90.0.2}$$

Let $\epsilon > 0$ be arbitrary small constant. It follows from equation (90.0.2) that there exists an arc γ_1 of the circumference containing the point $t = 1$, such that

$$|f(\Psi(t))| < A + \epsilon, \tag{90.0.3}$$

for any t lying on γ_1. Join the endpoints of γ_1 by a chord σ_1. By a rigid motion of the t-plane, we can always bring it about that the chord σ_1 coincides with (a part of) the real axis. Let us denote the reflection of γ_1 in σ_1 by $\widetilde{\gamma}_1$ and the domain bounded by γ_1 and $\widetilde{\gamma}_1$ by Σ_1. The function $f(\Psi(t))\overline{f}(\Psi(\bar{t}))$ is analytic in Σ_1, and on the two arcs γ_1 and $\widetilde{\gamma}_1$ satisfies the inequality

$$|f(\Psi(t))\overline{f}(\Psi(\bar{t}))| < A + \epsilon.$$

Hence, on the chord σ_1, we have the inequality

$$|f(\psi(t))| < (A + \epsilon)^{1/2}. \tag{90.0.4}$$

However, the same argument may be repeated if the chord σ_1 is replaced by a parallel chord lying in the domain bounded by σ_1 and γ_1. Hence, we see, that the inequality holds in the whole circular domain S_1 bounded by σ_1 and γ_1. We started by assuming that $f(\Psi(t))| < 1$ in S_1 and satisfies the inequality (90.0.3), and we arrived at the inequality (90.0.4) which holds in the circular domain S_1. We may repeat the argument. Consider a chord σ_2 of the unit disc $|t| < 1$ parallel to σ_1 and lying in S_1. Denoting by γ_2 the smaller arc on $|t| = 1$ subtended by σ_2, and by $\widetilde{\gamma}_2$ the reflection of γ_2 in σ_2. We can choose σ_2 in such a manner that the arc $\widetilde{\gamma}_2$ will also lie in S_1. Repeating then the argument for σ_2, we obtain in all points of S_2, the circular domain bounded by σ_2 and $\widetilde{\gamma}_2$, the inequality

$$|f(\Psi(t))| < (A + \epsilon)^{1-2^{-2}}.$$

Repeating the argument n times gives for all the points in the n'th circular domain S_n

$$|f(\Psi(t))| < (A + \epsilon)^{1-2^{-n}}. \tag{90.0.5}$$

Since all but a finite number of points t_i lie in each of the domains S_n, we have

$$\alpha = \lim_{i \to \infty} |f(\Psi(t_i))| \leq \lim_{n \to \infty} (A + \epsilon)^{1-2^{-n}} = A + \epsilon.$$

Since the last inequality holds for all positive ϵ, we obtain $\alpha \leq A$, which completes the proof. $\qquad\square$

Remark 90.0.202. Seidel says that this proof of the theorem, based on a method of Lindelöf, has been kindly suggested to him by Doob.

3. If $f(z)$ is again taken to be the same function as in Theorem 1, it is possible to form the following lower limit:

$$\liminf_{z \to 1} |f(z)| = a, \quad z \in C. \tag{90.0.6}$$

If α is the same number as in Theorem 1, it may be conjectured that $a \leq \alpha$. An example will now be given to show that this inequality is false in general. Consider a sequence of points $\{n_k\}$ lying in the interior of the disc $U : |z| < 1$, converging toward $z = 1$, and having the property that

$$\prod_{k=1}^{\infty} |n_k| > 0. \tag{90.0.7}$$

We form the corresponding Blaschke infinite product:

$$P(z) = \prod_{k=1}^{\infty} \left(\frac{1 - \frac{z}{n_k}}{1 - \overline{n}_k z} \right) |n_k| = \prod_{k=1}^{\infty} \frac{|n_k|}{n_k} \left(\frac{n_k - z}{1 - \overline{n}_k z} \right). \tag{90.0.8}$$

The product (90.0.8) converges uniformly in every bounded domain R of the z-plane which is at a positive distance from the points $z = 1$ and $z = 1/\overline{n}_k$ for $k = 1, 2, \dots$. This follows by the Blaschke condition (90.0.7). (Julia in his recent (than) book, *Principes Géométriques d'Analyse*, Paris, 1930, p. 65, has given a simple proof of the uniform convergence of this Blaschke product in every disc $|z| < \rho < 1$. A simple modification of the proof shows that the convergence holds in the domain R.) In this manner, it is seen that $P(z)$ is analytic in the entire finite plane except in the points $z = 1/\overline{n}_k$, where it has poles, and in the point $z = 1$, and in the interior of the disc, $|z| < 1$ we have $|P(z)| < 1$. In fact, $P(z)$ is also analytic in the point $z = \infty$. This follows by applying the reflection principles of Schwarz according to which the function may be continued analytically beyond the unit circle by means of the functional equation

$$P\left(\frac{1}{\overline{z}}\right) = \frac{1}{\overline{P(z)}}. \tag{90.0.9}$$

Hence, if the origin is not a zero of the function $P(z)$, the point $z = \infty$ is a point of analyticity. We are now in a position to show that if we take $f(z) = P(z)$ and consider the lower limit a in equation (90.0.6), then the relation $a \leq \alpha$ fails to hold always, when α is a cluster value of $P(z)$ in the point $z = 1$. If we choose the unit circle itself, then we have

$$\liminf_{z \to 1} |P(z)| = 1, \quad |z| = 1,$$

whereas $\alpha = 0$ is certainly a cluster value of the function, for $P(n_k) = 0$ for all $k = 1, 2, \ldots$.

4. While the last example shows how complicated the situation may be for bounded analytic functions, we shall see that in the case of equivalent analytic functions the problem can be answered easily.

Theorem 2 ([93]). *Let $f(z)$ be a univalent analytic function defined in the unit disc $|z| < 1$. Let C be a closed Jordan curve which passes through the point $z = 1$ and save for that point lies in the unit disc $|z| < 1$. Let $\{z_i\}$ be a sequence of points lying in the domain G bounded by the curve C and converging toward $z = 1$. Let the limit $\lim_{i \to \infty} f(z_i)$ exist and be equal to α and let S be the cluster set of $f(z)$ in the point $z = 1$ assumed along C. Then, the point α lies in the set S.*

Proof. It is clear that we may restrict our attention to the case of univalent and bounded functions. To see that, if $w = f(z)$ is a general univalent function, it maps the unit disc conformally on a simply connected domain R of the w-plane. By the uniformization theorem, the domain R must have at least two boundary points $w = a$ and $w = b$. It can be easily shown that every simply connected domain R with at least two boundary points can be mapped by a function of the type

$$w' = \left\{ \left(\frac{w - a}{w - b} \right)^{1/2} - c \right\}^{-1},$$

where c is a suitable constant, on a bounded and simply connected domain R'. The following function

$$F(z) = M \cdot \left\{ \left(\frac{f(z) - a}{f(z) - b} \right)^{1/2} - c \right\}^{-1}, \tag{90.0.10}$$

is therefore, univalent and bounded: $|F(z)| < 1$ in the unit disc, mapping it onto a domain R'', when the constant M is suitably chosen. Suppose that Theorem 2 were proved for the case of univalent and bounded functions. We shall prove it for general univalent functions. Solving equation (90.0.10) for $f(z)$ yields

$$f(z) = \frac{b(M + cF(z))^2 - a(F(z))^2}{(M + cF(z))^2 - (F(z))^2}. \tag{90.0.11}$$

Let the cluster set of $F(z)$ along C be S' and $\lim_{i \to \infty} F(z_i) = \alpha'$. According to our assumption, α' lies in the set S', and we wish to show that in this case α will also lie in S. By equation (90.0.10), there corresponds to α one and only one cluster value α' of $F(z)$ which lies in the set S'. By equation (90.0.11), then, α lies in S. We are justified, therefore, in proving Theorem 2 by assuming that the function $f(z)$ is univalent and bounded: $|f(z)| < 1$. Let $z = \Phi(t)$ be a function which maps the domain G, bounded by the closed Jordan curve C, conformally on the disc $|t| < 1$ in such a manner that $z = 1$ corresponds to $t = 1$. By the theorem of Osgood and Carathéodory, $\Phi(t)$ is continuous in the closed disc $|t| \le 1$. This function transforms the points $z_1, z_2, \ldots, z_n, \ldots$ into a set of points $t = t_1$, $t = t_2, \ldots, t = t_n, \ldots$ which lie in the interior of the unit disc $|t| < 1$ and converge toward $t = 1$. On this set of points, the function $f(\Phi(t)) = \Psi(t)$, which, except for $t = 1$, is analytic and bounded everywhere in the disc $|t| \le 1$, converges toward α : $\lim_{n \to \infty} \Psi(t_n) = \alpha$. The cluster set of $\Psi(t)$ in $t = 1$ assumed along the circle $|t| = 1$ is S, the set defined by us in the statement of Theorem 2. We wish to show that α is contained in S or, what is tantamount, that there exists a set of points $T_1, T_2, \ldots$ lying on $|t| = 1$ such that

$$\lim_{n \to \infty} \Psi(T_n) = \alpha. \tag{90.0.12}$$

This last assertion may be proved by the use of Carathéodory's theory of prime ends. By that theory, in particular by Carathéodory's fundamental result, we know that the cluster set of $\Psi(t)$ in every point of $|t| = 1$ is precisely a prime end of the domain R into which the unit disc $|t| < 1$ is mapped by the function $\Psi(t)$. In particular, to the point $t = 1$, there must correspond a prime end E of the domain R. Since $\Psi(t)$ is analytic elsewhere on the circle $|t| = 1$, all remaining prime ends of R are single points. Thus, the assertion about the existence of points $T_1, T_2, \ldots$ with the properties mentioned above and satisfying equation (90.0.12) reduces geometrically to the assertion that there always exists a sequence of boundary points of R, not belonging to the prime end E, which converges toward any pre-assigned point α of E. If this were not the case, there would exist a point $\alpha \in E$ which would not be a limit point of any sequence of boundary points of R not belonging to E. We could then draw a circle C with α as its center and with a radius so small that every boundary point of R, not belonging to E, would lie outside C. Since, however, α as a point of a prime end E, is also a boundary point of R as well, there exists a sequence of interior points of R which converges towards α. There must surely exist at least one interior point p of R which also lies in the interior of C. Consider the line segment $\overline{p\alpha}$. Setting out from p and traveling along this line segment, there will exist a first boundary point q of R which will be reached. This follows by the fact that the boundary of R is a closed point set. Since q lies on the line segment between p and α, it surely lies in C and is, therefore a point of the prime end E. Moreover, q is an accessible point of the prime end E. If there is a second point β of E inside the circle C which does not lie on the straight line through p and α, we can join p and β by a line segment $\overline{p\beta}$ and repeating the argument above (for $\overline{p\alpha}$), arrive at the existence of a second accessible point r of E, different from q. This, however, is a contradiction, for according to a theorem of Carathéodory, a prime end can contain at most one accessible point. The only remaining alternative, then, is that the subset E' of E which lies in the circle C lies wholly on the line segment joining p and α. Since E is a perfect connected set (Carathéodory), E' itself must be a line segment. Let us consider two points α and β of E lying in C. Not every interior point of R, contained in C, will lie on the straight line which contains E'. For if p is such a point, we can construct a circle Γ about p as a center and with a radius so small that Γ will consist only of interior points of R and will lie wholly within C. We, therefore, merely need to choose any point P of Γ which does not lie on the line containing E'. Join P with α and β, respectively, by the line segments L_α and L_β. Since P does not lie on the line containing E', the points α and β will not be the first boundary points of E attained by traversing from P along L_α and L_β, respectively. Hence, α and β are, again, two distinct accessible points of the prime end E, which is a contradiction. This means that to every point α of the prime end E, there exists a sequence $\{p_n\}$ of boundary points of R, not belonging to E, which converges towards α as a limit point. Since $\Psi(t)$ was taken to be analytic everywhere on the boundary circle of the disc $|t| < 1$ except in the point $t = 1$, the result means that there exists a sequence $\{T_n\}$ of points on $|t| = 1$ converging toward $t = 1$ and such that

$$\lim_{n \to \infty} \Psi(T_n) = \alpha.$$

The function $\Psi(t)$ was defined as $f(\Phi(t))$, where $\Phi(t)$ is continuous in the closed disc $|t| \leq 1$. Hence, $\Psi(T_n) = f(z_n)$, where $\{z_n\}$ is a sequence of points on C converging towards $t = 1$ and such that

$$\lim_{n \to \infty} f(z_n) = \alpha.$$

This proves that α lies in S, and there with the theorem. $\qquad\square$

Chapter II. Conditions that cluster values on two sequences be equal.

5. In this section, Seidel investigates the second problem proposed in the Introduction of the paper. Let $f(z)$ be a bounded analytic function:

$$|f(z)| < 1 \quad \text{in the disc } |z| < 1. \tag{90.0.13}$$

Let $z = 1$ be a point of discontinuity of $f(z)$ and let $z_1, z_2, \ldots$ and $z_1', z_2', \ldots$ be two sequences of interior points of the disc converging toward $z = 1$. If the limit $\lim_{n \to \infty} f(z_n) = \alpha$ exists, what conditions are to be imposed upon the second sequence in order that $\lim_{n \to \infty} f(z_n') = \alpha$ should also hold? We recall some facts about the non-Euclidean (hyperbolic) interpretation of the Schwarz Lemma. If we set $w = f(z)$ and consider the w-plane, we see that inequality (90.0.13) means that $f(z)$ assumes only values in the disc $|z| < 1$ that lie in the disc $|w| < 1$. We consider the two unit discs $|z| < 1$ and $|w| < 1$ as carriers of a non-Euclidean (hyperbolic) geometry. In this geometry, angles are measured as in the Euclidean geometry, but the distance between two points z_1 and z_2 is defined as follows: Join the points z_1 and z_2 by a circle orthogonal to $|z| = 1$ and cutting it in the points ξ_1 (next to z_1 on $|z| = 1$) and ξ_2 next to z_2. The non-Euclidean distance $D(z_1, z_2)$ is defined by

$$D(z_1, z_2) = \log \left(\frac{z_1 - \xi_1}{z_1 - \xi_2} \cdot \frac{z_2 - \xi_2}{z_2 - \xi_1} \right),$$

where the expression inside the parentheses is the cross ratio of the four points ξ_1, ξ_2, z_1, z_2. An equivalent expression, involving only z_1 and z_2, is

$$D(z_1, z_2) = \log \left(\frac{1 + \left| \frac{z_1 - z_2}{1 - \bar{z}_1 z_2} \right|}{1 - \left| \frac{z_1 - z_2}{1 - \bar{z}_1 z_2} \right|} \right).$$

The non-Euclidean interpretation of the Lemma of Schwarz [82] can be expressed in the following manner:

The Lemma of Schwarz. *Let $f(z)$ be a bounded analytic function: $|f(z)| < 1$ in the unit disc $|z| < 1$, and z_1 and z_2 two interior points of the unit disc. If $w_1 = f(z_1)$ and $w_2 = f(z_2)$, the relation $D(w_1, w_2) \leq D(z_1, z_2)$ holds for all pairs of interior points z_1 and z_2 of the unit disc $|z| < 1$.*

Remark 90.0.203. Thus, a bounded analytic function is a D-contraction.

6. Seidel proves the following result:

Theorem 3. *Let $f(z)$ be a bounded analytic function: $|f(z)| < 1$, in the unit disc $|z| < 1$ which omits the value α in the unit disc. Let $z_1,\, z_2,\,\dots$ and $z_1',\, z_2',\,\dots$ be two sequences of points of $|z| < 1$ converging toward $z = 1$. If the non-Euclidean distances $D(z_n, z_n')$ are less than a constant M, independent of n: $D(z_n, z_n') < M$ $(n = 1, 2, \dots)$, and if $\lim_{n\to\infty} f(z_n) = \alpha$, then also $\lim_{n\to\infty} f(z_n') = \alpha$.*

Remark 90.0.204. Theorem 3 holds if we allow $f(z)$ to assume the value α a finite number of times in the unit disc. If, however, $f(z)$ assumes the value α in every neighborhood of $z = 1$, then it may be shown by an example that the theorem fails to be true.

7. The example we shall consider shows that if a function has infinitely many zeros in the unit disc, Theorem 3 may fail to hold. Let $0 < t_1 < t_2 < \cdots < t_n < \cdots$ be a sequence of real, positive numbers less than one and converging toward one: $\lim_{n\to\infty} t_n = 1$. Let these numbers satisfy the following Blaschke condition:

$$\prod_{n=1}^{\infty} t_n > 0. \tag{90.0.14}$$

Then, the infinite Blaschke product

$$\Phi(z) = \prod_{n=1}^{\infty} \frac{z - t_n}{1 - t_n z} \tag{90.0.15}$$

represents a bounded analytic function in the unit disc $|z| < 1$ with zeros in the points $z = t_n$ $(n = 1, 2, \dots)$. In particular, if we set

$$t_n = \frac{n! - 1}{n! + 1} \quad (n = 1, 2, \dots), \tag{90.0.16}$$

then equation (90.0.14) is satisfied. We wish to show that Theorem 3 fails to hold for this particular $\Phi(z)$. In order to show this, we consider a second sequence:

$$\tau_n = \frac{(n+1)! - \rho}{(n+1)! + \rho}, \quad 1 < \rho < 2 \quad (n = 1, 2, \dots), \tag{90.0.17}$$

of real positive numbers less than one and converging toward one: $\lim_{n\to\infty} \tau_n = 1$. Furthermore, the following non-Euclidean distances

$$D(\tau_n, t_{n+1}) = \log \rho \tag{90.0.18}$$

are bounded uniformly for all values of n, and, so we shall prove (Seidel not me!),

$$\lim_{n\to\infty} |\Phi(\tau_n)| = \frac{\rho-1}{\rho+1} > 0. \tag{90.0.19}$$

Setting $z = \tau_j$ in equation (90.0.15) and substituting there the values of t_n and τ_j, as given in equations (90.0.16) and (90.0.17), respectively, we obtain

$$\Phi(\tau_j) = \prod_{n=1}^{\infty} \frac{(j+1)! - \rho n!}{(j+1)! + \rho n!}. \tag{90.0.20}$$

Hence,

$$|\Phi(\tau_j)| = \prod_{n=1}^{j-1} \frac{(j+1)! - \rho n!}{(j+1)! + \rho n!} \prod_{n=j}^{j+2} \frac{(j+1)! - \rho n!}{(j+1)! + \rho n!} \prod_{n=j+3}^{\infty} \frac{\rho n! - (j+1)!}{\rho n! + (j+1)!}. \tag{90.0.21}$$

The second factor tends to $\frac{\rho-1}{\rho+1}$ as $j \to \infty$. The limits of the first and third factors are 1 as j becomes infinite. Consider the first factor in the form

$$\exp\left(\sum_{n=1}^{j-1} \log \frac{(j+1)! - \rho n!}{(j+1)! + \rho n!}\right). \tag{90.0.22}$$

Here,

$$\log \frac{(j+1)! - \rho n!}{(j+1)! + \rho n!} = O\left(\log\left(1 + \frac{1}{j^2}\right)\right) = O\left(\frac{1}{j^2}\right) \quad (n = 1, 2, \ldots, j-1).$$

The exponent in equation (90.0.22) is $(j-1)O(j^{-2}) = O(j^{-1}) \to 0$ as $j \to \infty$. As to the last factor in the right side of equation (90.0.21),

$$\exp\left(\sum_{n=j+3}^{\infty} \log \frac{\rho n! - (j+1)!}{\rho n! + (j+1)!}\right), \tag{90.0.23}$$

where the exponent in equation (90.0.23) is

$$O\left(\sum_{n=j+3}^{\infty} n^{-2}\right) = O(j^{-1}) \to 0$$

as $j \to \infty$. This proves equation (90.0.19).

Chapter III. Cluster values of bounded functions.

8. The example given in Section 7 shows that bounded functions assuming one value infinitely often may have entirely different properties as regards their cluster values. These properties are studied in the current section. We return to the function studied in Section 3:

$$p(z) = \prod_{k=1}^{\infty} \frac{1 - \frac{z}{n_k}}{1 - \overline{n}_k z} |n_k|, \quad n_k \to 1, \tag{90.0.24}$$

where the product $\prod_{k=1}^{\infty} |n_k| > 0$. This function, as we have seen, is analytic in the whole z-plane, except the points $z = 1/\overline{n}_k$ $(k = 1, 2, \ldots)$, where $p(z)$ has poles, and the point $z = 1$, which is an isolated essential singularity and limit point of poles. By the Picard theorem, the function $p(z)$ assumes in every neighborhood of the point $z = 1$ every value infinitely often with the exception of at most two values. We possess, however, additional information as to the distribution of the values. The general factor

$$\frac{|n_k|}{n_k} \frac{n_k - z}{1 - \overline{n}_k z}$$

of the product $p(z)$ satisfies the inequality

$$\left| \frac{|n_k|}{n_k} \frac{n_k - z}{1 - \overline{n}_k z} \right| \geq 1 \text{ in } |z| \geq 1.$$

Hence, $|p(z)| \geq 1$ also in $|z| \geq 1$, and only in $|z| \geq 1$. Hence, with the exception of at most two values a and b, every value α, $|\alpha| < 1$, is assumed infinitely often by $p(z)$ in points of the unit disc $|z| < 1$ which converge toward the point $z = 1$. It can be shown, moreover, that at most one value a, $|a| < 1$, can be omitted (or assumed only a finite number of times). For if a is such an exceptional value, then according to equation (90.0.9),

$$p\left(\frac{1}{\overline{z}}\right) = \frac{1}{\overline{p(z)}},$$

which may also be directly verified in the product (90.0.24), the value $\frac{1}{\overline{a}}$ will also be exceptional. If, now, a second value b, $|b| < 1$, were also exceptional, then there would be three values omitted (or only assumed a finite number of times) in the neighborhood of $z = 1$. Hence, we have that with the exception of at most one value a, $|a| < 1$, every value α, $|\alpha| < 1$, is assumed infinitely often by $p(z)$ in points of the unit disc $|z| < 1$, which converge toward the point $z = 1$, and no value β, $|\beta| \geq 1$, is assumed by $p(z)$ in a point of the unit disc $|z| < 1$. If instead of Picard's Theorem, we had used Weierstrass' theorem such that in the neighborhood of an isolated essential singularity, a function approaches every preassigned value, and then applied the same reasoning as before, we would find that the cluster set of the function $w = p(z)$ in the point $z = 1$ is the closed unit circle $|w| \leq 1$.

9. From the function $p(z)$ studied in Section 8, we obtain an interesting example. Let G be an arbitrary simply connected domain lying in the interior of the unit disc $|w| < 1$. A function $w = F(z)$ will be constructed which is analytic and bounded in the unit disc $|z| < 1 : |F(z)| < 1$, and whose cluster set in the point $z = 1$ is precisely the closure $\overline{G}$ of G. Furthermore, every value, save at most one, out of the domain G will be assumed by the function $F(z)$ infinitely often in the unit disc $|z| < 1$. Let $w = \Phi(t)$ be the function which maps the domain G on the unit disc $|t| < 1$ in the t-plane. The function $F(z)$ in question will be given by $F(z) = \Phi(p(z))$.

10. The behavior exemplified by the function $p(z)$ is characteristic of a wider class of functions. We prove the following theorem:

Theorem 4. *Let $w = f(z)$ be a bounded analytic function in the unit disc $|z| < 1 : |f(z)| < 1$. Let $\{n_k\}$ be an infinite sequence of points interior to the unit disc converging toward $z = 1$ in which the function vanishes and let A be an arc of the unit circle, $-\alpha \leq \theta \leq \alpha$, $z = e^{i\theta}$, containing $z = 1$, on which $f(z)$ is continuous except for $z = 1$ and assumes values of modulus 1. Then, $w = f(z)$ assumes every value w, save at most one, of the unit disc $|w| < 1$ infinitely often in the unit disc $|z| < 1$ and assumes no value w, $|w| \geq 1$, in the unit disc. The cluster set of $f(z)$ in $z = 1$ is the closed unit disc $|w| \leq 1$.*

The proof of this theorem is practically the same as in Section 8. We only have to extend $f(z)$ analytically across the arc A by means of the functional equation:

$$f\left(\frac{1}{\overline{z}}\right) = \frac{1}{\overline{f(z)}}.$$

11. By conformal mapping, Theorem 4 is extended as follows:

Theorem 5. *Let $w = f(z)$ be a bounded analytic function in the disc $|z| < 1$, assuming values there which lie in the interior of a domain G bounded by a closed Jordan curve C. Let there be infinitely many zeros $\{n_k\}$ of $f(z)$ in the disc converging toward the point $z = 1$ and let A be an arc of the circle, $-\alpha \leq \theta \leq \alpha$, $z = e^{i\theta}$, containing $z = 1$, on which $f(z)$ is continuous except for $z = 1$ and assumes values which all lie on the curve C. Then, $w = f(z)$*

assumes infinitely often every value w, save at most one, of the domain G, bounded by C, in the unit disc $|z| < 1$. Furthermore, $f(z)$ assumes no value w in the disc $|z| < 1$ which lies on the boundary or in the exterior of G. The cluster set of $f(z)$ in the point $z = 1$ is the closed domain $G \cup C$.

12. The examples studied thus far, of bounded functions with infinitely many zeros, suggest the following alternative to Theorem 3:

Theorem 6. *Let $f(z)$ be a bounded analytic function in the unit disc $|z| < 1$: $|f(z)| < 1$. Let $z_1, z_2, \ldots$ and $z_1', z_2', \ldots$ be two sequences of interior points of the unit disc converging toward $z = 1$, such that the non-Euclidean distances $D(z_n, z_n')$ are less than a positive constant M, independent of n: $D(z_n, z_n') < M$ $(n = 1, 2, \ldots)$. Then, $f(z)$ always has one of the following two properties:*

I. *The cluster sets of $f(z)$ on any two such sequences $\{z_n\}$ and $\{z_n'\}$ are identical.*

II. *The cluster set of $f(z)$ in $z = 1$ contains a disc of the w-plane. Every value from the interior of this disc is assumed infinitely many times by $f(z)$ in $|z| < 1$.*

Proof. If the property I fails to hold for some function $f(z)$, then there must exist a sequence $\{z_n\}$ of interior points of unit disc converging toward the point $z = 1$ on which the function $f(z)$ approaches a value a and a second sequence $\{z_n'\}$ of interior points for which the relation

$$D(z_n, z_n') < M \quad (n = 1, 2, \ldots) \tag{90.0.25}$$

holds and such that on it the function $f(z)$ approaches a value b, different from a. Consider, now, two sets of non-Euclidean circles C_k and D_k of radius $M + \epsilon$, $\epsilon > 0$, and M, respectively, described about the points $z = z_k$ as centers. According to the relation (90.0.25) each circle D_k is contained in the interior of the corresponding circle C_k. Let us now transform the circle C_k by the transformation

$$z = \frac{z_k + m_\epsilon w}{1 + m_\epsilon \bar{z}_k w}, \quad m_\epsilon = \frac{e^{M+\epsilon} - 1}{e^{M+\epsilon} + 1}, \tag{90.0.26}$$

into the circle $|w| < 1$. The function $f(z)$ is thereby transformed into the function

$$\Psi_k(w) = f\left(\frac{z_k + m_\epsilon w}{1 + m_\epsilon \bar{z}_k w}\right). \tag{90.0.27}$$

The functions $\Psi_k(w)$ are defined and analytic in the disc $|w| < 1$ for all k. Furthermore,

$$|\Psi_k(w)| < 1 \quad (k = 1, 2, \ldots), \tag{90.0.28}$$

and

$$\Psi_k(0) = f(z_k) \quad (k = 1, 2, \ldots). \tag{90.0.29}$$

The transformation (90.0.26) carries the circle D_k into the circle

$$|w| \leq \frac{m_0}{m_\epsilon} < 1. \tag{90.0.30}$$

Hence, the images w'_k of the points z'_k satisfy the inequality

$$|w'_k| \leq \frac{m_0}{m_\epsilon} \quad (k = 1, 2, \ldots). \tag{90.0.31}$$

According to (90.0.29) and our assumption, we have

$$\lim_{k \to \infty} \Psi_k(0) = a. \tag{90.0.32}$$

By a theorem of Montel, the family $\{\Psi_k(w)\}$, being uniformly bounded according to (90.0.28), forms a normal family. It is therefore possible to extract a subsequence $\{\Psi_{k_i}(w)\}$ converging uniformly in every closed subdomain of the disc $|w| < 1$, hence, in particular in the circle (90.0.30). The limit function we shall call χ is as follows:

$$\lim_{i \to \infty} \Psi_{k_i}(w) = \chi(w). \tag{90.0.33}$$

The function $\chi(w)$ is analytic and bounded in the disc $|w| < 1 : |\chi(w)| < 1$. From (90.0.32), it follows that $\chi(0) = a$. In order to show that $\chi(w)$ is different from a constant, it must be proved that there exists at least one point of the disc $|w| < 1$ in which $\chi(w)$ is different from a. Denote by w' an arbitrary one of the limit points of the sequence $\{w'_{k_i}\}$. By (90.0.31), it follows that all points as well as all their limit points lie in the disc (90.0.30). Since the convergence (90.0.33) in that disc is uniform, it follows that to each arbitrary positive integer η, there exists a positive integer $k(\eta)$, independent of w, such that

$$|\chi(w'_{k_i}) - \chi(w')| < \frac{\eta}{2} \quad \text{for } k_i > k(\eta). \tag{90.0.34}$$

Furthermore, it follows from the uniform continuity of $\chi(w)$ in the disc (90.0.30) that to the given η there corresponds a positive integer $K(\eta)$, independent of w, such that

$$|\Psi_{k_i}(w'_{k_i}) - \chi(w')| < \frac{\eta}{2} \quad \text{for } k_i > K(\eta). \tag{90.0.35}$$

Adding together the inequalities (90.0.34) and (90.0.35) yields

$$|\chi(w') - \Psi_{k_i}(w'_{k_i})| < \eta$$

for $k_i > k(\eta)$ and $k_i > K(\eta)$, or $\lim_{i\to\infty} \Psi_{k_i}(w'_{k_i}) = \chi(w')$. From the last equation and (90.0.27), it follows that

$$\chi(w') = \lim_{i\to\infty} f(z'_{k_i}) = b,$$

which proves that $\chi(w)$ is not a constant. Hence, there exists a circle $|t-t_0| < \gamma$ such that the function $t = \chi(w)$ assumes in the disc (90.0.30) every value of the circle $|t - t_0| < \gamma$ at least once. It will be shown now that every value t assumed by the function $\chi(w)$ in the disc $|w| < 1$ is a cluster value of the function $f(z)$ in the point $z = 1$. Let t be an arbitrary such value assumed by $\chi(w)$ in some point w_0 of the unit disc. Consider the numbers

$$\chi_{k_i}(w_0) = t_{k_i}. \tag{90.0.36}$$

Let the image point of w_0 by the transformation (90.0.26) be denoted by ξ_k. Then, the point $z = \xi_{k_i}$ is a point of the circle C_{k_i} and we have by (90.0.27) $f(\xi_{k_i}) = t_{k_i}$. Now, $\lim_{i\to\infty} t_{k_i} = t$ by (90.0.33) and (90.0.36). Hence, $\lim_{i\to\infty} f(\xi_{k_i}) = t$, as we set out to prove. Hence, the cluster set of $f(z)$ in the point $z = 1$ contains the disc $|t - t_0| < \gamma$. The last statement of the theorem follows immediately if we observe that, in view of the uniform convergence of the sequence $\{\Psi_{k_i}(w)\}$ to $\chi(w)$ in the disc $|w| < 1$, the equations $\chi(w) = t$, $\Psi_{k_i}(w) = t$ have the same number of roots for sufficiently large values i, t being fixed. $\qquad\square$

Chapter 91

Few Results from the Book by Noshiro on Cluster Sets [78]

We are going to look for covering theorems for analytic functions.

§1. Definitions of Cluster Sets (p. 1).

Let D be an arbitrary domain (of any possible connectivity) with boundary Γ. Let E be a totally disconnected closed set contained in Γ. We suppose that $w = f(z)$ is non-constant, single-valued and meromorphic in D. We associate with every point z_0 of Γ the following sets of values.

(i) **The Cluster Set $C_D(f, z_0)$.** $\alpha \in C_D(f, z_0)$ if there exists a sequence of points $\{z_n\}$ with the following proportion:

$$z_n \in D, \quad \lim_{n\to\infty} z_n = z_0, \quad \lim_{n\to\infty} f(z_n) = \alpha. \tag{91.0.1}$$

If we denote by $\mathfrak{D}_r$ the set of values of $w = f(z)$ in the intersection D_r of D with the circular disc $|z - z_0| < r$, then

$$C_D(f, z_0) = \bigcap_{r>0} \overline{\mathfrak{D}_r}, \tag{91.0.2}$$

where $\overline{\mathfrak{D}_r}$ is the closure of $\mathfrak{D}_r$. Evidently, $C_D(f, z_0)$ is a non-empty closed set. In the particular case where D is a Jordan domain bounded by a simple closed curve, $C_D(f, z_0)$ is either a single point or a continuum. However, this property does not hold in general cases. Take D as the unit circular disc with a radial slit and select a boundary point z_0 ($\neq 0$) on the slit. If $w = f(z)$ is a function mapping D conformally onto $|w| < 1$, then $C_D(f, z_0)$ consists of two points.

541

Remark 91.0.205. Consider the special case where z_0 is an accessible boundary point of D. Then, there exists a path (simple curve) L in D terminating at z_0. Denote by z_r the last point of intersection of L with a circle $C : |z - z_0| = r$ and by L_r the arc $\overparen{z_r, z_0}$ of L. Such an arc is called a last part of L. The intersection D_r of D with $(C) : |z - z_0| < r$ is an open set which consists of at most an enumerable infinite number of connected components. Let Δ_r be the component which contains the last part L_r of L. If we denote by $\mathfrak{D}_r^*$ the value set of $f(z)$ in Δ_r, then $\mathfrak{D}_r^*$ is a domain and, hence, $\overline{\mathfrak{D}}_r^*$ is a continuum. Hence, the set

$$C_D(f, z_0; L) = \bigcap_{r>0} \overline{\mathfrak{D}}_r^* \tag{91.0.3}$$

is either a single point or a continuum. Suppose that L' is another path in D terminating at z_0. If, for every sufficiently small r (> 0), the last parts L_r and L'_r can be joined by a suitable path in D_r, then we say that L and L' are equivalent and define the same accessible boundary point of D at z_0. It is easy to show that if L and L' are equivalent, then

$$C_D(f, z_0; L) = C_D(f, z_0; L'). \tag{91.0.4}$$

Evidently,

$$C_D(f, z_0; L) \subseteq C_D(f, z_0). \tag{91.0.5}$$

In the particular case where D is a Jordan domain,

$$C_D(f, z_0) = C_D(f, z_0; L). \tag{91.0.6}$$

(ii) **The Boundary Cluster Sets $C_\Gamma(f, z_0)$ and $C_{\Gamma-E}(f, z_0)$.** $\alpha \in C_\Gamma(f, z_0)$ (respectively, $C_{\Gamma-E}(f, z_0)$) if there exists a sequence of points $\{\xi_n\}$ of $\Gamma - z_0$ (resp. $\Gamma - z_0 - E$) such that

$$w_n \in C_D(f, \xi_n) \quad \text{for each } n,$$
$$z_0 = \lim_{n\to\infty} \xi_n \quad \text{and } \alpha = \lim_{n\to\infty} w_n;$$

i.e., if M_r denotes the closure of the union $\bigcup_\xi C_D(f, \xi)$ for every ξ of the common part of $\Gamma - z_0$ (resp. $\Gamma - z_0 - E$) and $(C) : |z - z_0| < r$, then $\bigcap_{r>0} M_r$ is $C_\Gamma(f, z_0)$ (resp. $C_{\Gamma-E}(f, z_0)$). Obviously, $C_\Gamma(f, z_0)$ and $C_{\Gamma-E}(f, z_0)$ are closed,

$$C_{\Gamma-E}(f, z_0) \subseteq C_\Gamma(f, z_0) \subseteq C_D(f, z_0), \tag{91.0.7}$$

if $z_0 \in \Gamma - E$ or if $z_0 \in E - E'$, E' denoting the derived set of E, then

$$C_{\Gamma-E}(f, z_0) = C_\Gamma(f, z_0).$$

$C_\Gamma(f, z_0)$ is empty if and only if z_0 is an isolated boundary point; $C_{\Gamma-E}(f, z_0)$ is empty if and only if $z_0 \notin \overline{(\Gamma - E)}$.

(iii) **The Range of Values $R_D(f, z_0)$.** This is defined as the set of values α such that $z_n \in D$, $\lim_{n\to\infty} z_n = z_0$, $f(z_n) = z_0$; i.e.,

$$R_D(f, z_0) = \bigcap_{r>0} \mathfrak{D}_r, \tag{91.0.8}$$

where $\mathfrak{D}_r$ is the value set of $w = f(z)$ in the common part of D and $(C) : |z - z_0| < r$. Accordingly, $R_D(f, z_0)$ is a G_δ set.

(iv) **The Asymptotic Set $A_D(f, z_0)$.** Let z_0 be an accessible boundary point of D. A complex number α is called an asymptotic value of $w = f(z)$ at z_0 if $f(z) \to \alpha$ as $z \to z_0$ along a path in D terminating at z_0. The asymptotic set $A_D(f, z_0)$ is defined as the set of asymptotic values of $f(z)$ at z_0. We define $A_D(f, z_0) = \emptyset$ when z_0 is an inaccessible boundary point, for the sake of convenience.

We shall state a relation between $C_D(f, z_0; L)$ and $C_{\Gamma-E}(f, z_0)$. If z_0 is an accessible boundary point of D defined by a path L in D terminating at z_0 and if z_0 is an accumulation point of $\Gamma - E$, then

$$C_D(f, z_0; L) \cap C_{\Gamma-E}(f, z_0) \neq \emptyset. \tag{91.0.9}$$

To prove this, let $\{\xi_n\}$ be a sequence of points such that $\xi_n \in \Gamma - E$ and $\xi_n \to z_0$. Construct a simple closed curve γ_n passing through ξ_n such that γ_n surrounds z_0 and does not meet E. By a suitable choice of the sequence $\{\gamma_n\}$, we may assume that the diameter of γ_n converges to zero. Let z_n be the last point of intersection of L with γ_n. Then, it is obvious that the component, containing z_n, of the intersection of γ_n with D is a cross-cut of D whose end-points lie in $\Gamma - E$. From this fact follows that for every positive number r, $\overline{\mathfrak{D}}_r^*$ and M_r (defined above) have a point in common and hence (91.0.9) holds.

§1. Compact set of capacity zero and Evans–Selberg's theorem (p. 6).

1. In Noshiro's book [78], "capacity" always means "logarithmic capacity". Let E be a bounded Borel set in the z-plane and μ be a non-negative

completely additive set function defined for the Borel subsets of E. Then μ is called a positive mass-distribution on E. Let μ be a positive mass-distribution on E with total mass unity. Then

$$U^{\mu}(z) = \int_E \log \left| \frac{1}{z - \xi} \right| d\mu(\xi) \tag{91.0.10}$$

is called a (logarithmic) potential of the distribution μ on E. Writing

$$V_{\mu}(E) = \sup_z U^{\mu}(z), \quad V = \inf_{\mu} V_{\mu}(E), \tag{91.0.11}$$

we define the (logarithmic) capacity $C(E)$ of E by

$$C(E) = E^{-V}. \tag{91.0.12}$$

In the case $V = \infty$, we put $C(E) = 0$. Obviously $0 \leq C(E) < \infty$; if $E_1 \subseteq E_2$, then $C(E_1) \leq C(E_2)$, moreover, if there exists a sequence of bounded Borel sets E_n, such that $C(E_n) = 0$ for all n, and if $E = \bigcup_{n=1}^{\infty} E_n$ is bounded, then $C(E) = 0$.

2. Let us consider a domain D, containing $z = \infty$ in its interior, with boundary Γ. We suppose that E is a compact set complementary to D. Let $\{D_n\}$ be an exhaustion of D such that each D_n is bounded by a finite number of simple closed analytic curves Γ_n and such that $\overline{D}_n \subset D_{n+1}$ $(n = 1, 2, \ldots)$. Denote by $g_n(z, \infty)$ Green's function of D_n with pole at $z = \infty$. Since $\{g_n(z, \infty)\}$ is a monotone increasing sequence, the limit is either a finite function $g(z, \infty)$ in D except for $z = \infty$ or a constant ∞. In the former case, $g(z, \infty)$ is called Green's function of D with pole at $z = \infty$ and in the latter case, we say that there is no Green's function of D. It is well known that there is no Green's function of D if and only if $C(E) = 0$.

3. Let D_0 be an arbitrary Jordan domain bounded by a simple closed analytic curve Γ_0 such that $\overline{D}_0 \subset D_1$. For simplicity, we put $G = D - \overline{D}_0$, $G_n = D_n - \overline{D}_0$. We denote by $\omega_n(z) = \omega(z, \Gamma_n, G_n)$ the harmonic measure with boundary values 0 on Γ_0 and 1 on Γ_n respectively. Since $\{\omega_n(z)\}$ is monotonically decreasing, this sequence converges uniformly on any compact set in G (Harnack's theorem); we denote the limiting function by $\omega(z) = \omega(z, \Gamma, G)$. Evidently, $\omega(z)$ is harmonic on $G \cup \Gamma_0$. Since $\omega_n(z) = 0$ on Γ_0, it follows by Schwarz's Principle of Reflection, that $\omega(z)$ is also harmonic on Γ_0. $\omega(z) = 0$ on Γ_0 and $0 \leq \omega(z) < 1$ in G. By the minimum principle, if $\omega(z)$ vanishes at some point of G, then $\omega(z) \equiv 0$. If $\omega(z, \Gamma, G) \equiv 0$, then we say that E is of absolute harmonic measure zero. If Γ contains

a non-degenerate continuum, then $\omega(z, \Gamma, G) > 0$. Accordingly, if Γ is of absolute harmonic measure zero, then Γ (and therefore E) is totally disconnected. Furthermore, Γ is of absolute harmonic measure zero if and only if $C(E) = 0$.

Remark 91.0.206. Letting z_i $(i = 1, 2, \ldots, n)$ vary on a compact set E we denote by V_n the maximum value of the quantity $V(z_1, z_2, \ldots, z_n) = \prod_{k<\lambda}^{1,\ldots,n} |z_k - z_\lambda|$. Then $V_n^{1/\binom{n}{2}}$ is monotonically decreasing and converges to a limit $\tau(E)$ which is named by Fekete, the transfinite diameter of E. It is known that $C(E) = \tau(E)$.

4. **Remark 91.0.207.** Let E be a compact set. If for any positive number ϵ, we can cover E by a sequence of circular discs K_n of radius r_n such that $\sum r_n < \epsilon$, then we say that E is of linear measure zero. Similarly, we define E to be of logarithmic measure zero, by replacing $\sum r_n < \epsilon$ by $\sum \left(\log^+ \frac{1}{r_n}\right)^{-1} < \epsilon$. It is known that if E is of logarithmic measure zero, then $C(E) = 0$. If $C(E) = 0$, then E is of linear measure zero. Their converses are not true.

5. **Evans–Selberg theorem.** Evans and Selberg have proved independently the following theorem.

Theorem 1 ([28, 94]). *Let E be a compact set of capacity zero. Then there exists a positive mass distribution μ on E with total mass unity, such that its potential*

$$u(z) = \int_E \log \left| \frac{1}{z - \xi} \right| d\mu(\xi) \tag{91.0.13}$$

is positively infinite at every point of E and at no other points.

Proof. Given n points $a_1, a_2, \ldots, a_n$ on E, we form polynomial $p(z) = (z - a_1)(z - a_2) \ldots (z - a_n)$. Denote by $\overline{M}_n$ the maximum modulus of $p(z)$, letting z vary on E, i.e., $\overline{M}_n = \max_{z \in E} |p(z)|$, and by M_n the greatest lower bound of $\overline{M}_n$, letting the points $a_1, a_2, \ldots, a_n$ vary on E, i.e., $M_n = \inf \overline{M}_n$. Then M_n is the minimum of $\overline{M}_n$, in other words, by a suitable choice of $a_1^0, a_2^0, \ldots, a_n^0$ on E, there exists a polynomial $T_n(z) = (z - a_1^0)(z - a_2^0) \ldots (z - a_n^0)$ with maximum modulus M_n. Remembering the definition of the transfinite diameter $\tau(E)$ of E and the relation $\tau(E) = C(E)$, we denote by V_n the maximum of $V(z_1, z_2, \ldots, z_n) = \prod_{k<\lambda}^{1,\ldots,n} |z_k - z_\lambda|$, letting z_i $(i = 1, 2, \ldots, n)$ vary on E. Let V_{n+1} be attained by $n+1$ points $b_1, b_2, \ldots, b_{n+1}$ on

E. From the identity $V_{n+1} = V(b_1, b_2, \ldots, b_{n+1}) = |(b_1 - b_2)(b_1 - b_3) \ldots (b_1 - b_{n+1})| \cdot V(b_2, \ldots, b_{n+1})$ follows

$$|(b_1 - b_2)(b_1 - b_3) \ldots (b_1 - b_{n+1})| \geq M_n, \qquad (91.0.14)$$

for otherwise there would exist a point $b_1' \in E$ such that $V(b_1, b_2, \ldots, b_{n+1}) < V(b_1', \ldots, b_{n+1})$. By a cyclic change of suffices of b in (91.0.14), we have

$$V_{n+1} \geq M_n^{\frac{n+1}{2}} \quad \text{and} \quad V_{n+1}^{1/\binom{n+1}{2}} \geq M_n^{1/n}, \quad \text{hence follows}$$

$$\lim_{n \to \infty} V_{n+1}^{1/\binom{n+1}{2}} = \lim_{n \to \infty} M_n^{1/n} = 0,$$

as $\tau(E) = C(E) = 0$. Consider the function

$$u_n(z) = -\log |T_n(z)|^{1/n}$$

$$= \frac{1}{n} \left(\log \left| \frac{1}{z - a_1^0} \right| + \log \left| \frac{1}{z - a_2^0} \right| + \cdots + \log \left| \frac{1}{z - a_n^0} \right| \right).$$

$u_n(z)$ is a potential defined by a certain distribution of equal point masses on E with total mass unity and for every point z on E, $u_n(z) \geq m_n$ where $m_n = -\log M_n^{1/n}$. Since $m_n \to \infty$, we can find a sequence of integers $\{n_j\}$ such that $m_{n_j} \geq 2^j$ $(j = 1, 2, \ldots)$. Put $U_j(z) = 2^{-j} u_{n_j}(z)$ $(j = 1, 2, \ldots)$. Then $U_j(z)$ is a potential of distribution of equal point masses on E with total mass 2^{-j} and evidently $U_j(z) > 1$ on E. Consider finally the function

$$u(z) = \sum_{j=1}^{\infty} U_j(z) = \lim_{\nu \to \infty} \sum_{j=1}^{\nu} U_j(z).$$

Then $u(z)$ is a required potential. In fact, it is a potential of positive mass-distribution on E with total mass unity and hence of the form (91.0.13). At every point z of E, $u(z) = +\infty$ as $u(z) \geq \sum_{j=1}^{\nu} U_j(z) \geq \nu$ for all ν. If $z \in E^c$ and if z has a distance ρ from E, then clearly $u(z) \leq \log \frac{1}{\rho}$. $\qquad \square$

Remark 91.0.208. The potential $u(z)$, in Theorem 1 will be called an Evans–Selberg's potential. For a given compact set of capacity zero, Evans–Selberg's potential is not unique [39].

6. We state some properties of Evans–Selberg's potential $u(z)$. Clearly, $u(z)$ is harmonic outside E except for $z = \infty$ and its boundary value at every point of E is $+\infty$. In the neighborhood of $z = \infty$, $u(z)$ is of the form

$$u(z) = -\log|z| - \omega(z), \tag{91.0.15}$$

where $\omega(z) = \int_E \log\left|1 - \frac{\xi}{z}\right| d\mu(\xi)$ is harmonic at $z = \infty$. Let $v(z)$ be its conjugate harmonic function and put

$$w(z) = u(z) + iv(z). \tag{91.0.16}$$

Then the function $w(z)$ is many-valued and regular outside E except for $z = \infty$, the infinity being a logarithmic singularity. However, the derivative $w'(z) = u_x(z) - iu_y(z)$ is obviously single-valued and regular throughout the domain E^c, $z = \infty$ being a simple zero-point of $w'(z)$, and has a singularity at every point of E. Consequently, the many-valuedness of $w(z)$ arises only in its imaginary part by some additive constants. It can be shown that the level curve $\Gamma_\lambda : u(z) = \lambda$ $(-\infty < \lambda < \infty)$ consists of a finite number of simple closed curves surrounding E, by the minimum principle of harmonic function, and that $\lambda - u(z)$ is none other than Green's function $g(z, \infty)$ in the exterior of Γ_λ. Thus, we see that if there are p closed curves of Γ_λ, then $w'(z)$ has $p - 1$ finite zero-points in the exterior of Γ_λ and moreover, that

$$\int_{\Gamma_\lambda} dv(z) = \int_{\Gamma_\lambda} \frac{\partial u}{\partial n} ds = 2\pi, \tag{91.0.17}$$

where ds denotes the arc length and n the inner normal, [40].

Remark 91.0.209. Extensions of the Evans–Selberg theorem and related theorems have been obtained by Rudin [89], Ugaeri [103], Hong [47], and Inoue [51].

On a result of Lehto

In [78, p. 80], Noshiro quotes a result of Lehto from [65]. In Noshiro's book, the following is the result:

Theorem 15 ([65]). *Let $\Psi_1(\theta)$ and $\Psi_2(\theta)$ be arbitrary measurable functions in the interval $0 \leq \theta \leq 2\pi$. Then, there exists a function $f(z)$, regular in $|z| < 1$, such that $f(e^{i\theta}) = \Psi_1(\theta) + i\Psi_2(\theta)$ almost everywhere, where $f(e^{i\theta})$ denotes the radial limit of $f(z)$ at $e^{i\theta}$. i.e., $f(e^{i\theta}) = \lim_{r\to 1-} f(re^{i\theta})$.*

The remarkable thing about this result of Lehto is the fact that the two measurable functions are arbitrary. A regular function is orientation preserving. Within its domain of definition, the real part and the corresponding imaginary part satisfy the Cauchy–Riemann equations. Thus, it seems that the statement of Theorem 15 is false.

A counter-example.

Let $\Psi_1(\theta)$ and $\Psi_2(\theta)$ be two measurable functions in the interval $0 \leq \theta \leq 2\pi$ and let $f(z)$ be the regular function in $|z| < 1$ that satisfies a.e. $f(e^{i\theta}) = \Psi_1(\theta) + i\Psi_2(\theta)$. Let $g(z)$ be the regular function in $|z| < 1$ that satisfies a.e. $g(e^{i\theta}) = \Psi_1(\theta) - i\Psi_2(\theta)$. If $\Psi_1(\theta)^2 + \Psi_2(\theta)^2 = 1$, $0 \leq \theta \leq 2\pi$ then $f(e^{i\theta})g(e^{i\theta}) \equiv 1$ a.e. By a well-known theorem of Riesz $f(z)g(z) \equiv 1$, $|z| < 1$, so that $g(z) = 1/f(z)$. If $f(z)$ has a zero within $|z| < 1$, then $g(z)$ cannot be regular in $|z| < 1$. For instance, the choices $\Psi_1(\theta) = \cos\theta$, $\Psi_2(\theta) = \sin\theta$ generate $f(z) = z$ and $g(z) = 1/z$. Thus, maybe, the correct form of Theorem 15 should include the restriction that $(\Psi_1(\theta), \Psi_2(\theta))$, $0 \leq \theta \leq 2\pi$ are measurable and the map they generate is orientation preserving. Or, maybe, one should include in the term "regular function" also meromorphic functions.

Some facts on the boundary behavior of functions regular in $|z|<1$.

(1) We briefly recall that an inner function $f(z)$ in $|z| < 1$ is a $U \to U$ surjective mapping if

$$\min_{a \in U} \lim_{r \to 1^-} \left\{ \frac{1}{2\pi i} \oint_{|z|=r} \frac{f'(z)}{f(z) - a} dz \right\} > 0.$$

For some values of $a \in U$, the limit can be $+\infty$. Even the minimum itself over all $a \in U$ can be $+\infty$.

(2) **Spirals** ([78] pp. 75–76).

Theorem 7. *Let $f(z)$ be the regular function in $U : |z| < 1$ defined by*

$$f(z) = \prod_{j=1}^{\infty} \left\{ 1 - \left(\frac{z}{1 - n_j^{-1}} \right)^{n_j^2} \right\}, \quad n_j = 3^j \ (j = 1, 2, \ldots).$$

Then there exists a set of spirals σ_t $(0 \leq t < 1)$ in $|z| < 1$, each of which approaches $\partial U : |z| = 1$ asymptotically, such that every point of U belongs to one, and only one, of these spirals, and for every t,

$$\lim_{\substack{|z| \to 1^- \\ z \in \sigma_t}} f(z) = \infty.$$

Setting $g(z) = \frac{1}{f(z)}$ we have the following.

Corollary 3. *There exists a non-constant function $g(z)$, meromorphic in U with the following property: There is a set of spirals σ_t $(0 \leq t < 1)$ in U, each of which approaches ∂U asymptotically, such that every point of U belongs to one, and only one, of these spirals and, for every t,*

$$\lim_{\substack{|z| \to 1^- \\ z \in \sigma_t}} g(z) = 0.$$

(3) **Chords** ([78] pp. 73–74).

Theorem 6. *There exists a function $f(z)$ regular in U such that for every θ in the interval $0 \leq \theta \leq 2\pi$, $f(z) \to \infty$ as $z \to e^{i\theta}$ along all chords $\rho(\phi)$ of U ending at $e^{i\theta}$ except, perhaps a set of values ϕ of measure zero in the open interval $\left(-\frac{\pi}{2}, \frac{\pi}{2}\right)$. In fact, we can take the function $f(z)$ of Theorem 7 (in part 2) on spirals.*

Let $\mathcal{M}$ be a measurable set of positive measure on ∂U.

Theorem 4. *Let $w = f(z)$ be meromorphic in U. Suppose that for every point $e^{i\theta} \in \mathcal{M}$, there exist two chords Λ_1, Λ_2, terminating at $e^{i\theta}$, on which $f(z)$ is bounded. Then, for almost every point $e^{i\theta} \in \mathcal{M}$, $f(z)$ has a finite angular limit or $f(z)$ assumes ∞ infinitely often in any Stolz angle $\Delta(\theta)$ with vertex at $e^{i\theta}$.*

Corollary 1. *Let $f(z)$ be regular in the unit disc U. Suppose that for every point $e^{i\theta} \in \mathcal{M}$, there are two chords Λ_1, Λ_2 terminating at $e^{i\theta}$ on which $f(z)$ is bounded. Then, $f(z)$ has a finite angular limit at almost every point $e^{i\theta}$ of $\mathcal{M}$.*

Theorem 5. *Let $f(z)$ be meromorphic in the unit disc U. Suppose that for $e^{i\theta} \in \mathcal{M}$, there are two chords Λ_1, Λ_2 terminating at $e^{i\theta}$ along which $f(z)$ converges to zero as $z \to e^{i\theta}$. Then either $f(z) \equiv 0$ or $f(z)$ takes every value γ different from zero infinitely often in any Stolz angle $\Delta(\theta)$ for almost every $e^{i\theta} \in \mathcal{M}$.*

(4) $R^*(f, e^{i\theta})$ ([78] p. 72).

Here is a result from [72]: Let $w = f(z)$ be meromorphic in the unit U. If there are two chords Λ_1, Λ_2 terminating at $e^{i\theta}$ such that $\gamma \notin C_{\Lambda_1}(f, e^{i\theta}) \cup C_{\Lambda_2}(f, e^{i\theta})$, then we say that the value γ belongs to $R^*(f, e^{i\theta})$. $R^*(f, e^{i\theta})$ (which is always an open set) may be empty.

Example 91.0.210. Let $\Psi(w)$ be a meromorphic function for which the cluster set at $w = \infty$ on every simple curve which tends to infinity is total. Consider the function $f(z) = \Psi\left(\frac{1+z}{1-z}\right)$. Then $R^*(f, 1) = \emptyset$ [4]. The example is in page 1072 of that paper.

Theorem 3. *Let $w = f(z)$ be meromorphic in the unit disc U. Let $\mathcal{M}$ be a measurable set of positive measure on ∂U such that for every $e^{i\theta} \in \mathcal{M}$, $R^*(f, e^{i\theta})$ contains the number γ. Then, for almost every point $e^{i\theta}$ of $\mathcal{M}$, either $f(z)$ has an angular limit or $f(z)$ assumes the value γ infinitely often in any Stolz angle $\Delta(\theta)$ with vertex $e^{i\theta}$.*

(5) A result of Lusin and Privalov, and of Plessner ([78] p. 72).

Theorem 2 ([68]). *Let $w = f(z)$ be meromorphic in the unit disc U. Let $\mathcal{M}$ be a set of positive measure on ∂U. Suppose that $f(z)$ has angular limit γ at every point $e^{i\theta} \in \mathcal{M}$, where γ is some fixed complex number. Then, $f(z)$ reduces to a constant.*

The basic result of that type is due to Plessner [83].

Theorem 1. *Let $w = f(z)$ be meromorphic in the unit disc U. Then, almost every point $e^{i\theta}$ of ∂U is either a Fatou point or a Plessner point.*

(6) Inner functions (Functions of class (U) in Seidel's sense) ([78] pp. 32–47).

Any inner function $f(z)$ in U has a unique representation $f(z) = B(z) \cdot g(z)$, where

$$B(z) = \prod_k \frac{\overline{a}_k}{a_k} \cdot \frac{a_k - z}{1 - \overline{a}_k z}$$

is the Blaschke product generated by the zeros of $f(z)$ ($\sum_k (1 - |a_k|) < \infty$) and $g(z)$ is a singular inner function. We consider the function $h(z) = -\log g(z)$, selecting a definite branch of the logarithm. Then $h(z)$ is a single-valued regular function in U. Furthermore, $\Re h(z) \geq 0$ in U. By Herglotz theorem, we have

$$h(z) = \frac{1}{2\pi} \int_0^{2\pi} \frac{e^{i\theta} + z}{e^{i\theta} - z} d\mu(\theta) + i\gamma,$$

where $\mu(\theta)$ is monotonically non-decreasing function of θ in $0 \leq \theta \leq 2\pi$ for which $\mu(\theta)$ has a derivative. By the fact that $g(z)$ is inner, the radial limit of $\Re h(z)$ at $e^{i\theta}$, and therefore $\mu'(\theta)$, is equal to zero a.e. in $0 \leq \theta \leq 2\pi$.

Theorem 1 ([78] p. 32). *Let $w = f(z)$ be an inner function. Then*

$$f(z) = e^{-i\gamma} B(z) \exp\left\{ -\frac{1}{2\pi} \int_0^{2\pi} \frac{e^{i\theta} + z}{e^{i\theta} - z} d\mu(\theta) \right\}, \qquad (91.0.18)$$

where $B(z)$ is the Blaschke product extended over the zero-points of $f(z)$, $\mu(\theta)$ is a monotone non-decreasing function of θ in $0 \le \theta \le 2\pi$ whose derivative $\mu'(\theta)$ is equal to zero a.e. in $0 \le \theta \le 2\pi$ and γ is a real constant.

Theorem 2 ([78] p. 33). *If $f(z)$ is a non-constant inner function and if $f(z)$ is not a Blaschke product, then $f(z)$ admits 0 as its radial limit.*

Remark 91.0.211. There exists a Blaschke product which admits 0 as its radial limit. Frostman constructed the following:

$$B(z) = \prod_{k=1}^{\infty} \frac{\left(1 - \frac{1}{k^2}\right) - z}{1 - \left(1 - \frac{1}{k^2}\right) z},$$

which has the radial limit 0 at $z = 1$.

Theorem 3 ([78] p. 34). *Let $f(z)$ be a non-constant inner function. If $f(z) \ne \alpha$ $(|\alpha| < 1)$ in U, then there exists at least one radius $\theta = \phi$ such that*

$$\lim_{r \to 1^-} f(re^{i\phi}) = \alpha.$$

Lemma 1 ([78] p. 34). *Let $z = z(\xi)$ be a function regular and bounded: $|z(\xi)| < 1$ in the unit disc $|\xi| < 1$ such that $z(0) = 0$. Let E_ξ be a set on $|\xi| = 1$ such that for every $e^{i\theta} \in E_\xi$ the radial limit $z(e^{i\theta})$ is of modulus unity. Denote by E_z the set of values $z(e^{i\theta})$ for all $e^{i\theta} \in E_\xi$. Then $m_* E_\xi \le m^* E_z$, where $m_* E_\xi$ and $m^* E_z$ denote the interior measure of E_ξ and the exterior measure of E_z, respectively.*

Corollary ([78] p. 34). *If we omit the assumption $z(0) = 0$ in Lemma 1 (above), we can assert that $m^* E_z > 0$ provided $m_* E_\xi > 0$.*

Lemma 2 ([78] p. 34). *Let E be a closed set of linear measure zero on ∂U. Then, there exists a function $u(z)$ (an analogue of Evans–Selberg's potential) such that $u(z)$ is positive and harmonic on $U \cup (\partial U - E)$ and the boundary value of $u(z)$ at every point of E is $+\infty$.*

Obviously, the function $w = \chi(z) = e^{-(u(z)+iv(z))}$ which is an analogue of Evans–Selberg's function, is regular and bounded $|\chi(z)| < 1$ on $U \cup (\partial U - E)$ and has the boundary value 0 at every point of E. We note that we are not assuming that the capacity of E is zero.

The following is remarkable: Denote by Γ_λ the level curve $u(z) = \lambda$ (inf $u < \lambda < \infty$) suppose that $\int_{\Gamma_\lambda} dv(z)$ is bounded. Then E must be of capacity zero. See [64].

Remark 91.0.212. Let $f(z)$ be an inner function. Then if $f(z)$ has a singularity at $z_0 = e^{i\theta_0}$, the cluster set $C_U(f, z_0)$ of $f(z)$ at z_0 is the closed disc $|w| \leq 1$ (a result of Seidel).

Theorem 4 ([78] p. 35). *Let $w = f(z)$ be a non-constant inner function and (c) be any disc $|w - \alpha| < \rho$ lying inside $|w| < 1$, whose boundary may be tangent to $|w| = 1$. Denote by Δ any connected component of the inverse image of (c) under $w = f(z)$ and by $z = z(\xi)$ a function which maps $|\xi| < 1$ onto the simply connected domain Δ in a one-to-one conformal manner. Then, the function*

$$W = F(\xi) = \frac{1}{\rho}\{f(z(\xi)) - \alpha\},$$

is also an inner function.

Theorem 5 ([78] p. 36). *(An extension of Iversen's theorem). Let $w = f(z)$ be an inner function and $z = \phi(w)$ be its inverse function defined in the unit disc $|w| < 1$. Then, for any disc $(c): |w - \alpha| < \rho$ lying inside $|w| < 1$ and for any element $e(w, w_0)$ of $z = \phi(w)$ with center w_0 lying in (c), it is possible to find a suitable path joining $w = w_0$ and $w = \alpha$ inside (c) along which the element $e(w, w_0)$ can be continued analytically except perhaps at $w = \alpha$.*

Theorem 6 ([78] p. 37). *Let $w = f(z)$ be a non-constant inner function. Then,*

(i) *the set of all the radial limits $f(e^{i\theta})$ contains $|w| = 1$.*

(ii) *if $w = f(z)$ omits a value α ($|\alpha| < 1$) in $|w| < 1$, then there exists at least one point $e^{i\theta}$ such that $\lim_{r \to 1-} f(re^{i\theta}) = \alpha$.*

(iii) *if $w = f(z)$ omits at least one singularity on $|z| = 1$ and if $f(z)$ assumes α only finitely often in $|z| < 1$, the same assertion as in (ii) holds. This is a result of Seidel.*

Remark 91.0.213. Ohtsuka has constructed an inner function which admits every value of modulus < 1 as a radial limit.

Theorem 7 ([78] p. 37). *(Seidel–Frostman's theorem). Let $w = f(z)$ be an inner function. Suppose that $f(z)$ has a singularity at $z_0 = e^{i\theta_0}$. Then, every value of $U_w : |w| < 1$ is assumed by $f(z)$ infinitely often in any neighborhood of z_0 except perhaps for a set of values of capacity zero, i.e., $U_w - R_U(f, z_0)$ is at most of capacity zero.*

Theorem 15 ([78] p. 46). *(Lohwater). Let $f(z)$ be meromorphic function of bounded type in U, and let $\lim_{r \to 1^-} |f(re^{i\theta})| = 1$ a.e. on an open arc A of $|z| = 1$. If $f(z)$ has a singularity at a point z_0 of A, then every value of modulus 1 which does not belong to $R_U(f, z_0)$ is an asymptotic value of $f(z)$ at some point of each arc of A containing the point z_0.*

Remark 91.0.214. Lohwater asked whether Theorem 15 holds in the case of a meromorphic function of unbounded type. It is difficult to give an answer.

Theorem 16 ([78] p. 46). *Let $f(z)$ be meromorphic in $|z| < 1$. Suppose that $C_{\Lambda_\theta}(f, e^{i\theta})$ lies on $\partial U_w : |w| = 1$ for almost every $e^{i\theta}$ belonging to an open arc A of ∂U containing $z_0 = e^{i\theta_0}$, where Λ_θ is an arbitrary curve in $|z| < 1$ ending at $e^{i\theta}$. Let $f(z)$ have a singularity at z_0. If $f(z)$ has an exceptional value α ($|\alpha| = 1$) in a neighborhood of z_0, then α is an asymptotic value along a path terminating at some point arbitrarily near z_0, under an additional condition that*

(i) $\mathcal{E} = \{e^{i\theta} | \alpha \notin C_{\Lambda_\theta}(f, e^{i\theta})\}$ *is everywhere dense on A.*

Remark 91.0.215. Suppose that $f(z)$ has three exceptional values in a neighborhood of z_0. Then it is known that the radial limit $f(e^{i\theta})$ exists at a dense subset of A (Collingwood–Cartwright). Hence, in Theorem 16, the additional condition (i) can be replaced by the following:

(ii) $f(z)$ has three exceptional values in a neighborhood of z_0.

Condition (i) can also be replaced by (Theorem 10),

(iii) there exists a point β such that $\beta \notin R_U(f, z_0)$, $|\beta| = 1$.

Theorem 17 ([78] p. 47). *Let $f(z)$ be regular in U. Suppose that $C_{\Lambda_\theta}(f, e^{i\theta})$ lies on $\partial U_w : |w| = 1$ for almost every $e^{i\theta}$ belonging to A. Let $f(z)$ have a singularity at $z_0 = e^{i\theta_0}$. If $f(z)$ has an exceptional value α*

($|\alpha| = 1$) *in a neighborhood of z_0, then α is an asymptotic value of $f(z)$ at some point of each arc of A containing z_0.*

Theorem 18 ([78] pp. 47–48). *Let $f(z)$ be meromorphic in U. Suppose that $\lim_{r \to 1^-} |f(re^{i\theta})| = 1$ almost everywhere on an arc A ($\{e^{i\theta}|\theta_1 < \theta < \theta_2\}$) of $|z| = 1$. Let $f(z)$ have a singularity at $z_0 = e^{i\theta_0}$, ($\theta_1 < \theta_0 < \theta_2$). If $f(z)$ has an exceptional value α ($|\alpha| = 1$) at z_0, then α is an asymptotic value arbitrarily near z_0, provided that $\sum_k (1 - |b_k|) < \infty$ where b_k ($k = 1, 2, \ldots$) are β-points of $f(z)$ and $|\beta| \neq 1$.*

Chapter 92

Indestructible Blaschke Products

We recall the following problem: Let $\{z_n\}_{n=1}^{\infty}$ be an infinite Blaschke sequence. Can we find a singular inner function $S(z)$ (in U) and a complex number $\alpha \in U - \{0\}$ such that $Z(S(z) - \alpha) = \{z_n\}_{n=1}^{\infty}$? If we denote the Blaschke product generated by the Blaschke sequence above, by

$$B(z) = \prod_{n=1}^{\infty} \frac{|z_n|}{z_n} \cdot \frac{z_n - z}{1 - \overline{z}_n z},$$

then the answer to the question above is affirmative if and only if we can find a singular inner function $T(z)$ (in U), so that the inner function $B(z) \cdot T(z)$ is not onto U. For if $B(z) \cdot T(z) \neq \alpha$ for some fixed $\alpha \in U$, and for all $z \in U$, then the α-Frostman shift of $B(z) \cdot T(z)$, i.e.,

$$S(z) = \frac{\alpha - B(z) \cdot T(z)}{1 - \overline{\alpha} B(z) \cdot T(z)}$$

is a singular inner function. But we have

$$B(z) \cdot T(z) = \frac{\alpha - S(z)}{1 - \overline{\alpha} S(z)}$$

which implies that $Z(\alpha - S(z)) = Z(B(z) \cdot T(z)) = \{z_n\}_{n=1}^{\infty}$. Also, the opposite implication is true. If $Z(S(z) - \alpha) = \{z_n\}_{n=1}^{\infty}$ for some singular inner function $S(z)$ and some $\alpha \in U - \{0\}$, then also

$$Z\left(\frac{\alpha - S(z)}{1 - \overline{\alpha} S(z)}\right) = \{z_n\}_{n=1}^{\infty},$$

and so the inner function $\frac{\alpha - S(z)}{1 - \overline{\alpha} S(z)}$ has the factorization

$$\frac{\alpha - S(z)}{1 - \overline{\alpha} S(z)} = B(z) \cdot T(z)$$

555

for some singular inner function $T(z)$. But then

$$S(z) = \frac{\alpha - B(z) \cdot T(z)}{1 - \overline{\alpha}B(z) \cdot T(z)}$$

is a singular inner function. So, $\forall z \in U$, $\alpha - B(z) \cdot T(z) \neq 0$. This means that α does not belong to the image of U under $B(z) \cdot T(z)$. Thus, $B(z) \cdot T(z)$ is not onto U. This means that the Blaschke sequence $\{z_n\}_{n=1}^{\infty}$ is not a zero set $Z(S(z) - \alpha)$, where $S(z)$ is some singular inner function if and only if $B(z) \cdot T(z)$ is onto U for every singular inner function $T(z)$. Let us first consider examples in which $T(z) \equiv 1$, i.e., surjective Blaschke products. We recall the following:

Definition 92.0.216. An infinite Blaschke product $B(z)$ is called an indestructible Blaschke product if

$$B_a(z) = \frac{B(z) - a}{1 - \overline{a}B(z)}$$

is also a Blaschke product for every $a \in U$.

We clearly have the following proposition.

Proposition 92.0.217. *If $B(z)$ is an indestructible Blaschke product, then $B(z)$ maps U onto U.*

Proof. Let $a \in U$. Then $B_a(z)$ is a Blaschke product and in particular it has zeros. Let $B_a(z_0) = 0$. Then $B(z_0) = a$. $\qquad\square$

Proposition 92.0.218. *If $B(z)$ is an indestructible Blaschke product, then for every $a \in U$ $B_a(z)$ is also an indestructible Blaschke product.*

Proof. Let us denote $T_a(z) = \frac{z-a}{1-\overline{a}z}$. Then $T_a(B(z)) = B_a(z)$ is a Blaschke product. Let $b \in U$. Then

$$T_b \circ T_a(z) = \frac{1 + \overline{a}b}{(1 + \overline{a}b)} T_{\frac{b+a}{1+b\overline{a}}}(z).$$

Hence, for any $b \in U$, we have

$$T_b(B_a(z)) = T_b(T_a(B(z))) = (T_b \circ T_a)\left(B(z) = \frac{1 + \overline{a}b}{(1 + \overline{a}b)} T_{\frac{b+a}{1+b\overline{a}}}(B(z))\right)$$

which is a Blaschke product (because $B(z)$ is an indestructible Blaschke product and $\left|\frac{1+\overline{a}b}{(1+\overline{a}b)}\right| = 1$). $\qquad\square$

Proposition 92.0.219. *If $B(z)$ is an indestructible Blaschke product and $a \in U$, then there are infinitely many solutions of the equation $B(z) = a$ in U.*

Proof. These are exactly the zeros of the (infinite) Blaschke product $B_a(z)$. $\square$

We end this chapter by pointing out at a simple fact regarding the construction of new surjective inner functions from a given such function. The bottomline is that we may replace the Blaschke product factor of the given surjective inner function by any positive integral power of this Blaschke product without losing the property of $U \to U$ surjectivity.

Proposition 92.0.220. *Let $\{z_k\}_k$ be a Blaschke sequence such that for every singular inner function in U, $S(z)$, the function*

$$S(z) \cdot \prod_k \left(\frac{|z_k|}{z_k}\right) \left(\frac{z_k - z}{1 - \overline{z}_k z}\right) = S(z) \cdot B(z),$$

is onto U. Then for every natural number $N \in \mathbb{Z}^+$, also $S(z) \cdot B(z)^N$ is onto U.

Proof. Since for any singular inner function $S(z)$ also $S(z)^{1/N}$ is a singular inner function, it follows that the function $S(z)^{1/N} \cdot B(z)$ is onto U. The monomial function $U \to U$, $w \to w^N$ is an onto function. Finally, the composition of two $U \to U$ surjections is a $U \to U$ surjection. Hence, indeed, the function $\left(S(z)^{1/N} \cdot B(z)\right)^N$ is onto U. $\square$

Corollary 92.0.221. *For the Blaschke product $B(z)$ of Proposition 92.0.220, the set*

$$\left\{B(z)^N \cdot S(z) \,|\, S(z) \text{ is a singular inner function in } U; N \in \mathbb{Z}^+\right\},$$

which contains only surjections $U \to U$, if closed for multiplication of functions.

Proof. If $B(z)^N \cdot S(z)$ and $B(z)^M \cdot T(z)$ belong to the set above, i.e., $M, N \in \mathbb{Z}^+$, and $S(z)$ and $T(z)$ are singular inner functions in U, so that both $B(z)^N \cdot S(z)$ and $B(z)^M \cdot T(z)$ are $U \to U$ surjections, then $(B(z)^N \cdot S(z))(B(z)^M \cdot T(z)) = B(z)^{M+N} \cdot (S(z)T(z))$ and since $S(z)T(z)$ is a singular inner function the result follows by Proposition 92.0.220. $\square$

Finally, we add one more fact on Blaschke sequences, which might be useful.

Remark 92.0.222. Let $\{z_k\}_k$ be a Blaschke sequence and let $N \in \mathbb{Z}^+$. Then $\{z_k^{1/N}\}_k$ is a Blaschke sequence. Here, $z_k^{1/N}$ stands for the totality of the N roots of z_k, i.e., if $z_k = |z_k|e^{i\phi_k}$ then $z_k^{1/N} = \{|z_k|^{1/N}e^{i(\phi_k+2\pi j)/N}\}_{j=0}^{N-1}$. For by the assumption that $\{z_k\}_k$ is a Blaschke sequence it follows (equivalently) that $\sum_k(1-|z_k|) < \infty$. Now, we have

$$1 - |z_k| = (1 - |z_k|^{1/N})(1 + |z_k|^{1/N} + |z_k|^{2/N} + \cdots + |z_k|^{(N-1)/N}),$$

so that

$$1 - |z_k|^{1/N} = \frac{1 - |z_k|}{(1 - |z_k|^{1/N})(1 + |z_k|^{1/N} + |z_k|^{2/N} + \cdots + |z_k|^{(N-1)/N})}.$$

Also, by our definition

$$\sum_k(1 - |z_k|^{1/N}) = \sum_k \sum_{j=0}^{N-1}(1 - ||z_k|^{1/N}e^{i(\phi_k+2\pi j)/N}|) = N\sum_k(1 - |z_k|^{1/N}).$$

By $\lim_{k\to\infty}|z_k| = 1$ (if $\{z_k\}_k$ is infinite, the only case of interest), we have $|z_k| \geq \frac{1}{2}$ except for finitely many z_k's. So,

$$1 + |z_k|^{1/N} + |z_k|^{2/N} + \cdots + |z_k|^{(N-1)/N}$$

$$\geq 1 + \left(\frac{1}{2}\right)^{1/N} + \cdots + \left(\frac{1}{2}\right)^{(N-1)/N} \geq 1 + (N-1)\frac{1}{2}.$$

Hence,

$$1 - |z_k|^1 N \leq \frac{1 - |z_k|}{1 + (N-1)\frac{1}{2}}.$$

So,

$$N\sum_k(1 - |z_k|^{1/N}) \leq \frac{N}{1 + \frac{1}{2}(N-1)} \cdot \sum_k(1 - |z_k|) < \infty.$$

We included in the last two sums only those z_k's for which $|z_k| \geq \frac{1}{2}$. This leaves out only finitely many z'_ks.

Chapter 93

Contribution of Kari Astala, Planar Quasiconformal Mappings in [23]

Reading the contribution of Kari Astala, Planar Quasiconformal Mappings; Deformations and Interactions, pp. 33–54, he discusses in Section 3 the topic of holomorphic motions and its equivalence to planar quasiconformal mappings:

Definition 3.1 (p. 40). A function $\Phi : \Delta \times A \to \overline{\mathbb{C}}$ (Here $\Delta = \{z \in \mathbb{C} | |z| < 1\}$) is called a holomorphic motion of a set $A \subseteq \overline{\mathbb{C}}$ if

  (i) for any fixed $a \in A$, the map $\lambda \to \Phi(\lambda, a)$ is holomorphic in Δ,
  (ii) for any fixed $\lambda \in \Delta$, the map $a \to \Phi_\lambda(a) = \Phi(\lambda, a)$ is an injection, and
  (iii) the mapping Φ_0 is the identity on A.

Note in particular that no assumptions are made on the set A or on the *á priori* continuity of Φ_λ; this will be obtained later as a consequence. This flexibility gives us a wide variety of examples. The typical ones come of course from complex dynamical systems: Holomorphic deformations of parameters of a dynamical system:

(a) Let Γ be Kleinian group, a discrete group of Möbius transformations with limit set $L(\Gamma) \neq \overline{\mathbb{C}}$. We assume Γ has no elliptic elements. Then changing the coefficients of the group elements in a holomorphic manner gives a holomorphic motion of the limit set, as long as the new groups Γ_λ obtained are discrete, the natural identification $(gh)_\lambda = g_\lambda h_\lambda$ remains an isomorphism and accidental parabolics are formed.

(b) Let R be a rational function which is hyperbolic, i.e., $|(R^n)'(z)| \geq c \cdot k^n$, with $k > 1$, uniformly on the Julia set $J(R)$, $n \in \mathbb{Z}^+$. Then a (small) holomorphic change of the coefficients of R produces a family of rational functions with Julia sets moving holomorphically.

Here is another type of motion: Holomorphic deformations of quasiconformal mappings: If we use the motion f^μ for the quasiconformal mapping which has (compactly supported) complex dilatation μ and is normalized by $f(z) = z + O(|z|^{-1})$ at ∞. Consider the mapping $\lambda \to f^{\lambda\mu}(z)$, $\lambda \in \Delta$. Then $\Phi(\lambda, z) = f^{\lambda\mu}(z)$ is a holomorphic motion.

The crucial point here is that these seemingly unrelated examples are actually just different aspects of one and the same phenomenon.This was realized first in [69] in their study of stability and geometric rigidity in the complex dynamical systems of rational functions. We formulate their result in the following form:

Theorem 3.4 (p. 41). *Let* $\Phi : \Delta \times A \to \overline{\mathbb{C}}$ *be a holomorphic motion of a set* $A \subseteq \overline{\mathbb{C}}$. *Then*

(i) Φ *is jointly continuous in* $\Delta \times A$,
(ii) Φ *extends to a holomorphic motion of the closure* $\overline{A}$, *and*
(iii) *if* $A = \mathbb{C}$, *then the mappings* Φ_λ *are quasiconformal with*

$$K(\Phi_\lambda) \leq \frac{1 + |\lambda|}{1 - |\lambda|}.$$

Proof. By normalizing with a Möbius transformation, we may assume that all Φ_λ's fix ∞. Consider then the different points $x, y, z \in \mathbb{C}$. The function

$$h(\lambda) = \frac{\Phi_\lambda(x) - \Phi_\lambda(y)}{\Phi_\lambda(x) - \Phi_\lambda(z)}$$

is holomorphic in the unit disk and omits the three values $0, 1$ and ∞. Since $\overline{\mathbb{C}} - \{0, 1, \infty\}$ admits a hyperbolic metric ρ and since holomorphic mappings do not increase hyperbolic distances,

$$\rho(h(\lambda), h(0)) \leq \rho(\lambda, 0) = \log \frac{1 + |\lambda|}{1 - |\lambda|}.$$

On the other hand, in the hyperbolic metric, $\overline{\mathbb{C}} - \{0, 1, \infty\}$ is complete or $\rho(z, w) \to \infty$ when z is fixed and $w \to 0$. This means that we can interpret the inequality in the form

$$\left| \frac{\Phi_\lambda(x) - \Phi_\lambda(y)}{\Phi_\lambda(x) - \Phi_\lambda(z)} \right| \leq \eta \left(\left| \frac{x - y}{x - z} \right| \right) \tag{93.0.1}$$

for some continuous strictly increasing function $\eta = \eta_\lambda$ with $\eta_\lambda(0) = 0$. The first two conclusions (i) and (ii) follow from this uniform estimate and the compactness properties of analytic functions. As for the third, (93.0.1) says (by definition) that each Φ_λ is quasi-symmetric, a property equivalent to quasi-conformality for mappings of $\overline{\mathbb{C}}$. Therefore, the Φ_λ's have complex dilatations μ_{Φ_λ} and these depend holomorphically on λ. Here, $\|\mu_{\Phi_\lambda}\|_\infty \leq 1$ and $\mu_{\Phi_0} = \mu_{\mathrm{id}} = 0$. Hence, the last claim follows from Schwarz lemma (for mappings with values in L^∞). $\qquad\square$

It is natural to ask if the motion extends not only to the closure of A but to a motion in the whole complex plane. After a number of partial answers, the picture was completed by Slodkowski who, using methods from several complex variables, proved the following generalized λ-lemma.

Theorem 3.5 (p. 41). *Any holomorphic motion Φ of any set $A \subseteq \overline{\mathbb{C}}$ extends to a holomorphic motion of $\overline{\mathbb{C}}$.*

Combining the third example $(f^{\mu\lambda})$ with Theorems 3.4 and 3.5, we see that quasi-conformal mappings, with a choice of holomorphic deformation to identity, are precisely the same as holomorphic motions, "Planar quasi-conformal mappings"$\equiv$"Holomorphic motions". In particular, this explains and clarifies the role of quasi-conformality in complex dynamics or even complex analysis in general. These results show that whenever we have (any) structure defined in terms of the complex plane and we perturb the parameters of the structure in a holomorphic manner, then quasi-conformal mappings and phenomena will necessarily appear.

Let us pause here with Astala's paper and involve our interest in the Krzyż problem. Among the functions in $\{f \in H(U) | 0 < |f(z)| < 1, |z| < 1\}$ that maximize $|a_n|$ in $f(z) = a_0 + \cdots + a_n + \cdots$ for some fixed $n \geq 1$, there is a function of the following form:

$$f(z) = \exp\left(-\sum_{j=1}^{N} \lambda_j \frac{1 + k_j z}{1 - k_j z}\right),$$

where $N \leq n$, and for $j = 1, \ldots, N$, $\lambda_j > 0$ and $|k_j| = 1$ and for $1 \leq j_1 \neq j_2 \leq N$, $k_{j_1} \neq k_{j_2}$. For certain results, we viewed that form of the extremal functions as being one in a parametrized family of functions. We thought of the parameter space as $\mathbb{R}_{>0}^N \times \mathbb{T}^N$ which is composed of $(\lambda_1, \ldots, \lambda_N; k_1, \ldots, k_N\}$. We used fixed-points results and especially the Borsuk–Ulam Theorem in order to derive those results. Can we view the parametrization as a complex parametrization? In particular, we think of the space as being $\mathbb{C}^N$ or better $(\mathbb{C}^*)^N$ which is composed of

$$(w_1, \ldots, w_N) = (\lambda_1 k_1, \lambda_2 k_2, \ldots, \lambda_N k_N\}.$$

Thus, for $j = 1, \ldots, N$, $w_j \in \mathbb{C}^*$, where $|w_j| = \lambda_j$ and $\frac{w_j}{|w_j|} = k_j$.

Questions:

Does this scheme give us an holomorphic parametrization with values in the unit ball of $H^\infty(U)$?

Is there a way to view the scheme as a kind of motion and obtain results parallel to the one-dimensional theory of holomorphic motions of Mañé–Sad–Sullivan?

Chapter 94

Variational Methods on B, Convergent Series and Boundary Behavior of Certain Holomorphic Functions

In searching for variations for $B_n(p)$, let $f \in H^p(U)$. We look for $g \in H(U)$ such that $f \cdot g \in H^p(U)$. Thus, we have $\int_0^{2\pi} |f(e^{i\theta})|^p d\theta < \infty$ and we want $\int_0^{2\pi} |f(e^{i\theta})g(e^{i\theta})|^p d\theta$. If $g \in H^\infty(U)$ that works, for

$$\int_0^{2\pi} |f(e^{i\theta})g(e^{i\theta})|^p d\theta \le \int_0^{2\pi} |f(e^{i\theta})|^p d\theta \cdot \|g\|_\infty^p < \infty.$$

What comes to mind is the following theorem.

Theorem (Hölder). *Let $f(x)$ and $g(x)$ be positive continuous functions on the interval $[a,b]$. If $p > 1$ and $\frac{1}{p} + \frac{1}{q} = 1$, then*

$$\int_a^b f(x)g(x)dx \le \left(\int_a^b f(x)^p dx \right)^{1/p} \left(\int_a^b g(x)^q dx \right)^{1/q}$$

and

Theorem (Reverse Hölder inequality). *Let $p > 1$ and $\frac{1}{p} + \frac{1}{q} = 1$. If $f(x)$ and $g(x)$ are non-negative, continuous functions and $f(x)^{1/p}g(x)^{1/q}$ is integrable on $[a,b]$, then*

$$\left(\int_a^b f(x)^p dx \right)^{1/p} \left(\int_a^b g(x)^q dx \right)^{1/q} \le \int_a^b S \left(\frac{Y \cdot f(x)^p}{X \cdot g(x)^q} \right) f(x)g(x)dx,$$

where $X = \int_a^b f(x)^p dx$, $Y = \int_a^b g(x)^q dx$ and

$$S(h) = \frac{h^{1/(h-1)}}{e \cdot \log h^{1/(h-1)}}, \quad h \ne 1.$$

Next, we mention a simple reduction to L^2 in the following sense: The condition $\int_0^{2\pi} |h(e^{i\theta})|^p d\theta < \infty$ (here, we think of $h = f \cdot g$) can be re-formulated as a L^2 condition: $\int_0^{2\pi} |h(e^{i\theta})^{p/2}|^2 d\theta < \infty$. Thus, if $h(z)^{p/2} = \sum_{k=0}^\infty a_k z^k$, then $\sum_{k=0}^\infty |a_k|^2 < \infty$. We want a very effective g so that $h^{p/2} \in H^2(U)$ while $\forall \epsilon > 0$, $h^{(p/2)-\epsilon} \notin H^2(U)$. This leads to a calculus problem on series with non-negative terms:

$$(*) \quad \begin{cases} \text{Find a } l^1 \text{ series } \sum_{k=0}^\infty \alpha_k < \infty, \ \alpha_k \geq 0 \, \forall \, k, \text{ such that } \forall \, \epsilon > 0 \text{ we} \\ \text{have } \sum_{k=0}^\infty \alpha_k^{1-\epsilon} = \infty. \end{cases}$$

Remark 94.0.223. We note that the problem with $+\epsilon$ instead of $-\epsilon$ is easy. That is, if $\sum_{k=0}^\infty \alpha_k < \infty$ ($\alpha_k \geq 0 \, \forall \, k$), then $\forall \, \epsilon > 0$, we have $\sum_{k=0}^\infty \alpha_k^{1+\epsilon} < \infty$. For example, assuming sharp inequalities $\alpha_k > 0 \, \forall \, k$, we can use the limit comparison test:

$$\lim_{k \to \infty} \frac{\alpha_k^{1+\epsilon}}{\alpha_k} = \lim_{k \to \infty} \alpha_k^\epsilon = 0 < 1.$$

The problem $(*)$ we posed is a bit more sophisticated but not hard. A solution of problem $(*)$:

Constructing an l^1 series $\sum_{k=0}^\infty \alpha_k < \infty$, $\alpha_k \geq 0 \, \forall \, k$, such that $\forall \, \epsilon > 0$ we have $\sum_{k=0}^\infty \alpha_k^{1-\epsilon} = \infty$.

The idea is simple. We take two sequences $\{M_k\}$ and $\{m_k\}$ such that $\lim_{k \to \infty} M_k = \lim_{k \to \infty} m_k = \infty$ so that $M_k \gg m_k$, $\sum_{k=0}^\infty \frac{1}{M_k} = \infty$ but with a little help from $\{m_k\}$, $\sum_{k=0}^\infty \frac{1}{M_k m_k} < \infty$. If we make a clever choice, then in

$$\sum_{k=0}^\infty \left(\frac{1}{M_k m_k} \right)^{1-\epsilon} = \sum_{k=0}^\infty \left(\frac{1}{M_k} \right)^{1-\epsilon} \left(\frac{1}{m_k} \right)^{11-\epsilon},$$

the sequence $\frac{1}{M_k^{1-\epsilon}}$ is so large, that even after multiplying it by $\frac{1}{m_k^{1-\epsilon}}$ we still have a divergent series $\sum_{k=0}^\infty \left(\frac{1}{M_k m_k} \right)^{1-\epsilon}$ that blows to infinity. Here is a concrete example:

$$\sum \frac{1}{n(\log n)^2} < \infty$$

by the integral test $\int^\infty \frac{dx}{x(\log x)^2} = -\frac{1}{\log x} \Big|^\infty < \infty$. Here, $M_k = k$, $m_k = (\log k)^2$ and indeed $M_k \gg m_k \to \infty$. But for any $\epsilon > 0$, we have

$$\sum \left\{ \frac{1}{n(\log n)^2} \right\}^{1-\epsilon} = \infty$$

because $\sum \frac{1}{n^{1-\epsilon}} = \infty$ and $\frac{1}{n^{1-\epsilon}}$ is so large, that even after multiplying it by $\frac{1}{(\log n)^{2(1-\epsilon)}}$ the resulting series blows to infinity.

This demonstrates the existence of an holomorphic function in $\{\Re z > 0\}$ that has a bizarre boundary behavior: We suppose that $\sum a_n < \infty$ ($\forall n,\ a_n > 0$) such that $\forall \epsilon >$ we have $\sum a_n^{1-\epsilon} = \infty$. We define a kind of Dirichlet's series by

$$f(z) = \sum a_n^{1+z}, \quad \Re z > 0.$$

Of course, $f(z) \in H(\Re z > 0)$. We note that

$$a_n^{1+z} = a_n^{1+x+iy} = a_n^{1+x} a_n^{iy} = a_n^{1+x} \cos(y \log a_n) + a_n^{1+x} \sin(y \log a_n).$$

Clearly, $|a_n^{1+z}| = a_n^{1+\Re z}$. Also, $\sum a_n^{1+iy}$ is absolutely convergent ($\sum a_n < \infty$). But if $\epsilon > 0$, then $\sum a_n^{1-\epsilon} = \infty$ which means that $f(z)$ is defined on $\Re z \geq 0$ (including the boundary $\Re z = 0$, but not outside that domain, i.e., in $\Re z < 0$). For example, on the negative x-axis, $\{x + i \cdot 0 \mid x < 0\}$ the defining series blows up to ∞. So $f(z) \in H(\Re z > 0) \cap C(\Re z \geq 0)$ but the y-axis, $\{0 + i \cdot y \mid y \in \mathbb{R}\}$ is a "natural boundary".

Remark 94.0.224. $\forall y \in \mathbb{R}$ both $\sum a_n^{1+x} \cos(y \log a_n)$ and $\sum a_n^{1+x} \times \sin(y \log a_n)$ are absolutely convergent for $0 \leq x$, and $y \in \mathbb{R}$. However,

$$f(iy) = \sum a_n (\cos(y \log a_n) + \sin(y \log a_n))$$

are not points of holomorphy. f is only continuous there but cannot be analytically extended to (a non-empty part of) $x < 0$. Certainly, the original Dirichlet's series we used is not capable of that.

Chapter 95

Surjective Inner Functions of the Form $z \cdot T(z)$, Where $T(z)$ is a Singular Inner Function

We recall that our motivating question: "Is every Blaschke sequence the zero set of some $S(z) - \alpha$, for some singular inner function $S(z)$ and some number $\alpha \in U$?" is closely related to the question: "Is every function of the form $z \cdot T(z)$ a $U \to U$ surjection, where $T(z)$ is a singular inner function?".

Here is a natural exercise to think about: Is the inner function

$$z \exp\left(-\frac{1+z}{1-z}\right)$$

a $U \to U$ surjection? Here is one possible solution.

Solution. The answer is affirmative, i.e., $z \exp\left(-\frac{1+z}{1-z}\right)$ is a $U \to U$ surjection. We recall Theorem 5.14 in [14, p. 109]:

Let $w = f(z)$ be analytic and bounded, $|f(z)| < 1$ in U, and let the radial limit of the modulus $|f(re^{i\theta})|$ be 1 for all $e^{i\theta}$ on $|z| = 1$, except possibly for a set of capacity zero on $|z| = 1$. Then, unless $f(z)$ reduces to a finite Blaschke product or to a constant, every value of $|w| < 1$ is assumed infinitely often by $f(z)$, with at most one exception.

The function $f(z) = \exp\left(-\frac{1+z}{1-z}\right)$ satisfies the assumption in Theorem 5.14. It is not a finite Blaschke product or a constant and it never assumes the value $w = 0$. So, $z \cdot f(z)$ also satisfies the assumption in Theorem 5.14. It is not a finite Blaschke product or a constant and it assumes the value $w = 0$ exactly once (for $z = 0$). Hence, it is a $U \to U$ surjection, by Theorem 5.14. $\qquad\square$

We note that the solution of this exercise follows immediately from the following proposition:

Proposition 95.0.225. *Let $T(z)$ be an R-singular inner function. Then $z \cdot T(z)$ is a $U \to U$ surjection.*

Proof. This was proved within the proof of our Theorem 89.0.199. $\qquad\square$

In Proposition 95.0.225, $T(z)$ can be a unimodular constant, in which case $z \cdot T(z)$ is clearly a $U \to U$ surjection. The proof of Proposition 95.0.225 (which is embedded within the proof of Theorem 89.0.199) is almost identical to the solution of the exercise on the $U \to U$ surjectivity of $z \exp\left(-\frac{1+z}{1-z}\right)$. We just replace in that solution the function $f(z) = \exp\left(-\frac{1+z}{1-z}\right)$ by the function $T(z)$ (which is an R-singular inner function. This is the only property of $f(z)$ we need in the solution we gave).

Remark 95.0.226. Let us denote by SInn, the subfamily of the inner functions Inn which consists of all the singular inner functions. Let R $-$ SInn be the subfamily of SInn which consists of all the refined singular inner functions.

Proposition 95.0.227. *Suppose* R $-$ SInn *is dense in* SInn, *in the sense that for each $S(z) \in$ SInn there exists a sequence $f_n(z) \in$ R $-$ SInn such that $(f_n)_{n\to\infty}S$ uniformly on compact subsets of U. Then for each $T(z)$ in* SInn, *the function $z \cdot T(z)$ is a $U \to U$ surjection. Thus, there is a Blaschke sequence which is not a zero set of a function of the form $S(z) - \alpha$ for some singular inner function $S(z)$ and some number $\alpha \in U - \{0\}$.*

Proof. To prove that $z \cdot T(z)$ is a $U \to U$ surjection, we take any $a \in U$, and any r such that $|a| < r < 1$ and assume that $z \cdot T(z) \neq a$ for $|z| = r$. We need to show that

$$\frac{1}{2\pi i} \oint_{|z|=r} \frac{(z \cdot T(z))'}{z \cdot T(z) - a} dz \in \mathbb{Z}^+.$$

By the density assumption, there is a sequence $f_n(z) \in$ R $-$ SInn such that $(f_n)_{n\to\infty}S$ uniformly on compact subsets of U. Hence, $(f_n')_{n\to\infty}S'$ uniformly on compact subsets of U and hence

$$\frac{(z \cdot f_n(z))'}{z \cdot f_n(z) - a}_{n\to\infty} \to \frac{(z \cdot T(z))'}{z \cdot T(z) - a}$$

uniformly on $|z| = r$ (we may assume that $\forall\, n \in \mathbb{Z}^+$, $z \cdot f_n(z) - a \neq 0$ for $|z| = r$). Hence,

$$\lim_{n \to \infty} \frac{1}{2\pi i} \oint_{|z|=r} \frac{(z \cdot f_n(z))'}{z \cdot f_n(z) - a} dz = \frac{1}{2\pi i} \oint_{|z|=r} \frac{(z \cdot T(z))'}{z \cdot T(z) - a} dz.$$

But for n large enough, it follows by Proposition 95.0.225 that

$$\frac{1}{2\pi i} \oint_{|z|=r} \frac{(z \cdot f_n(z))'}{z \cdot f_n(z) - a} dz \in \mathbb{Z}^+.$$

Hence

$$\frac{1}{2\pi i} \oint_{|z|=r} \frac{(z \cdot T(z))'}{z \cdot T(z) - a} dz \in \mathbb{Z}^+.$$

$\square$

Remark 95.0.228. Proposition 95.0.227 implies that if any Blaschke sequence is the zero set of a function of the form $S(z) - \alpha$ for some singular inner function $S(z)$ and some number $\alpha \in U - \{0\}$, then

$$\overline{\mathrm{R} - \mathrm{SInn}} \subsetneq \mathrm{SInn}$$

where the closure on the left-hand side is with respect to the topology of uniform convergence on compact subsets of U.

Remark 95.0.229. By a theorem of Fatou, we know that if $S(z)$ is an inner function then the radial limit of the modulus $S(re^{i\theta})|$ is 1 for all $e^{i\theta}$ on $|z| = 1$ except possibly for a set of linear measure zero on $|z| = 1$. Thus, the above density assumption might be related to the question of representation of a set of linear measure zero on $|z| = 1$, which has the supremum of a non-decreasing sequence of subsets of capacity zero on $|z| = 1$.

Chapter 96

More Facts on Blaschke Sequences Which are $Z(S(z) - \alpha)$, For Some Singular Inner Function $S(z)$ and Some Number $\alpha \in U - \{0\}$

Theorem 96.0.230. *Let $\{z_n\}_{n=1}^{\infty}$ be a Blaschke sequence which equals the zero set $Z(S(z) - \alpha)$, for some singular inner function $S(z)$ and some number $\alpha \in U - \{0\}$. Then*

(i) *There exists a singular inner function $T(z)$ such that*

$$B(z) \cdot T(z) = \frac{\alpha - S(z)}{1 - \overline{\alpha}S(z)}, \quad S(z) = \frac{\alpha - B(z) \cdot T(z)}{1 - \overline{\alpha}B(z) \cdot T(z)}$$

where up to a multiplication by a unimodular constant

$$B(z) = \prod_{n=1}^{\infty} \frac{|z_n|}{z_n} \cdot \frac{z_n - z}{1 - \overline{z}_n z}.$$

(ii) $\alpha \notin \operatorname{Im} B(z) \cdot T(z)$.

(iii) $\forall\, z \in U,\ \forall\, n \in \mathbb{Z}^+$,

$$\left| \frac{\alpha - S\left(\frac{z_n - z}{1 - \overline{z}_n z}\right)}{1 - \overline{\alpha}S\left(\frac{z_n - z}{1 - \overline{z}_n z}\right)} \right| \le |z| \left| T\left(\frac{z_n - z}{1 - \overline{z}_n z}\right) \right|.$$

Equivalently, $\forall\, z \in U,\ \forall\, n \in \mathbb{Z}^+$,

$$\left| \frac{\alpha - S(z)}{1 - \overline{\alpha}S(z)} \right| \le \left| \frac{z_n - z}{1 - \overline{z}_n z} \right| \cdot |T(z)|.$$

Moreover, equality holds for one $z \in U$, $z \neq z_n$ if and only if we have an identity

$$\frac{\alpha - S(z)}{1 - \overline{\alpha}S(z)} = \lambda \cdot \left(\frac{z_n - z}{1 - \overline{z}_n z}\right) \cdot T(z),$$

in which case

$$B(z) = \lambda \left(\frac{z_n - z}{1 - \overline{z}_n z}\right),$$

for some unimodular constant $|\lambda| = 1$.

(iv) $\forall n \in \mathbb{Z}^+$, $|S'(z_n)| \leq |T(z_n)|$. *Moreover, if there is equality even for a single $n \in \mathbb{Z}^+$, $|S'(z_n)| < 1$.*

Proof. We already know (i) and (ii). For (iii) and (iv), we fix $n \in \mathbb{Z}^+$. Using (i), we define the following function:

$$f(z) = B\left(\frac{z_n - z}{1 - \overline{z}_n z}\right) = \left(\frac{1}{T\left(\frac{z_n - z}{1 - \overline{z}_n z}\right)}\right) \cdot \left(\frac{\alpha - S\left(\frac{z_n - z}{1 - \overline{z}_n z}\right)}{1 - \overline{\alpha}S\left(\frac{z_n - z}{1 - \overline{z}_n z}\right)}\right).$$

We note that since $\mathrm{Aut}(U)$ is a group (with composition of mappings as the binary operation), also

$$B\left(\frac{z_n - z}{1 - \overline{z}_n z}\right)$$

is a Blaschke product. In fact, one can perform the following elementary computation to be convinced:

$$\frac{z_1 - \frac{z_2 - z}{1 - \overline{z}_2 z}}{1 - \overline{z}_1 \frac{z_2 - z}{1 - \overline{z}_2 z}} = \left(\frac{1 - \overline{z}_1 z_2}{1 - z_1 \overline{z}_2}\right) \cdot \frac{\left(\frac{z_2 - z_1}{1 - \overline{z}_2 z_1}\right) - z}{1 - \left(\frac{z_2 - z_1}{1 - \overline{z}_2 z_1}\right) z}.$$

By $S(z_n) = \alpha$, we obtain $f(0) = 0$, and since as noted above, $f(z)$ is a Blaschke product, we also have $\forall z \in U$, $|f(z)| \leq 1$. So, we deduce, by the Schwarz Lemma, that $|f(z)| \leq |z|$, $\forall z \in U$ and $|f'(0)| \leq 1$ and if equality holds for one of these inequalities (for just one $z \in U - \{0\}$ in the first inequality) then $f(z) = \lambda z$ for some unimodular constant $|\lambda| = 1$. Thus,

$$\left|\frac{\alpha - S\left(\frac{z_n - z}{1 - \overline{z}_n z}\right)}{1 - \overline{\alpha}S\left(\frac{z_n - z}{1 - \overline{z}_n z}\right)}\right| \leq |z|\left|T\left(\frac{z_n - z}{1 - \overline{z}_n z}\right)\right|, \quad \forall z \in U, \tag{96.0.1}$$

with equality (as explained) if and only if

$$\frac{\alpha - S\left(\frac{z_n - z}{1 - \bar{z}_n z}\right)}{1 - \bar{\alpha} S\left(\frac{z_n - z}{1 - \bar{z}_n z}\right)} = \lambda \cdot z \cdot T\left(\frac{z_n - z}{1 - \bar{z}_n z}\right), \quad (|\lambda| = 1). \tag{96.0.2}$$

Alternatively, since $w = \frac{z_n - z}{1 - \bar{z}_n z}$ is an involution in the group $(\mathrm{Aut}(U), \circ)$, i.e., $z = \frac{z_n - w}{1 - \bar{z}_n w}$ we have

$$\left|\frac{\alpha - S(z)}{1 - \bar{\alpha} S(z)}\right| \leq \left|\frac{z_n - z}{1 - \bar{z}_n z}\right| \cdot |T(z)|, \tag{96.0.3}$$

and equality holds for even a single $z \neq z_n$ in U if and only if

$$\frac{\alpha - S(z)}{1 - \bar{\alpha} S(z)} = \lambda \cdot \left(\frac{z_n - z}{1 - \bar{z}_n z}\right) \cdot T(z), \quad \forall z \in U, \text{ (some } |\lambda| = 1). \tag{96.0.4}$$

We now differentiate $f(z)$. We use the following elementary

$$\left(\frac{z_n - z}{1 - \bar{z}_n z}\right)'_z = \frac{-1 + |z_n|^2}{(1 - \bar{z}_n z)^2},$$

and the chain rule to get

$$f'(z) = \frac{\frac{1 - |z_n|^2}{(1 - \bar{z}_n z)^2} \{D\}}{A}$$

where

$$B = T'\left(\frac{z_n - z}{1 - \bar{z}_n z}\right)\left(1 - \bar{\alpha} S\left(\frac{z_n - z}{1 - \bar{z}_n z}\right)\right)\left(\alpha - S\left(\frac{z_n - z}{1 - \bar{z}_n z}\right)\right)$$
$$+ T\left(\frac{z_n - z}{1 - \bar{z}_n z}\right) S'\left(\frac{z_n - z}{1 - \bar{z}_n z}\right)(1 - |\alpha|^2),$$

and

$$A = \left\{T\left(\frac{z_n - z}{1 - \bar{z}_n z}\right)\left(1 - \bar{\alpha} S\left(\frac{z_n - z}{1 - \bar{z}_n z}\right)\right)\right\}^2.$$

Hence,

$$f'(0) = \frac{(1 - |z_n|^2)\{T'(z_n)(1 - \bar{\alpha} S(z_n))(\alpha - s(z_n)) + T(z_n) S'(z_n)(1 - |\alpha|^2)\}}{T(z_n)^2(1 - \bar{\alpha} S(z_n))^2}.$$

Since $S(z_n) = \alpha$, this reduces to the following simple identity:

$$f'(0) = \frac{S'(z_n)}{T(z_n)}.$$

So, we deduce that

$$\left| \frac{S'(z_n)}{T(z_n)} \right| \leq 1,$$

i.e., $|S'(z_n)| \leq |T(z_n)|$. $\qquad\square$

$$\text{Chapter 97}$$

On Theorem 4 in the Paper of Seidel [93]

The paper was received by the editors on July 17, 1931.

Theorem 4 [93]. *Let $w = f(z)$ be a bounded analytic function in the unit disc $|z| < 1 : |f(z)| < 1$. Let $\{n_k\}$ be an infinite sequence of points interior to the unit disc converging toward $z = 1$ in which the function vanishes and let A be an arc of the unit circle, $-\alpha \leq \theta \leq \alpha$, $z = e^{i\theta}$, containing $z = 1$, on which $f(z)$ is continuous except for $z = 1$ and assumes values of modulus 1. Then, $w = f(z)$ assumes every value w, save at most one, of the unit disc $|w| < 1$ infinitely often in the unit disc $|z| < 1$ and assumes no value w, $|w| \geq 1$, in the unit disc. The cluster set of $f(z)$ in $z = 1$ is the closed unit disc $|w| \leq 1$.*

Motivated by this result of Seidel we make the following:

Definition 97.0.231. Let $w = f(z)$ be a bounded analytic function in the unit disc in the following sense: $|f(z)| < 1 \; \forall \, z \in U = \{z \in \mathbb{C} | |z| < 1\}$. We say that $f(z)$ belongs to the Seidel class in $z = 1$ if there is some $0 < \alpha < \pi$ such that the corresponding arc of the unit circle $\{e^{i\theta} | -\alpha \leq \theta \leq \alpha\}$ containing $z = 1$ is such that $f(z)$ is continuous at every point that belongs to it except, maybe, $z = 1$ and assumes values of modulus 1. This Seidel class will be denoted by $\mathrm{Seidel}_1(U)$. Index 1 indicates that it is the Seidel class in $z = 1$. A similar definition applies for the Seidel class in $z_0 = e^{i\theta_0}$. This Seidel class will be denoted by $\mathrm{Seidel}_{z_0}(U)$.

Proposition 97.0.232. *Let $z_0 \in \mathbb{T} = \{e^{i\theta} | 0 \leq \theta < 2\pi\}$.*

1. *If $f, g \in \mathrm{Seidel}_{z_0}(U)$, then $f \cdot g \in \mathrm{Seidel}_{z_0}(U)$.*

2. *An infinite Blaschke product*

$$B(z) = z^m \left(\prod_{k=1}^{\infty} \frac{|a_k|}{a_k} \frac{a_k - z}{1 - \overline{a}_k z} \right)$$

belongs to $\mathrm{Seidel}_{z_0}(U)$ if and only if the derived set of the zeros of $B(z)$ in U does not contain z_0, i.e., $z_0 \notin \{a_k\}'_k$, or z_0 is an isolated point of $\{a_k\}'_k$.

3. $f \in \mathrm{Seidel}_{z_0}(U) \Leftrightarrow \forall\, b \in U, \frac{b-f(z)}{1-\overline{b}f(z)} \in \mathrm{Seidel}_{z_0}(U)$.

Proof.

1. By $f \in \mathrm{Seidel}_{z_0}(U)$, there exists a number $0 < \alpha_f < \pi$ such that $f(z)$ is continuous on $\{z = e^{i\theta} | \theta_0 - \alpha_f \leq \theta \leq \theta_0 + \alpha_f\} - \{z_0 = e^{i\theta_0}\}$ and assumes values of modulus 1. By $g \in \mathrm{Seidel}_{z_0}(U)$, we have a number $0 < \alpha_g < \pi$ such that $g(z)$ is continuous on $\{z = e^{i\theta} | \theta_0 - \alpha_g \leq \theta \leq \theta_0 + \alpha_g\} - \{z_0 = e^{i\theta_0}\}$. Hence, if $\alpha = \min\{\alpha_f, \alpha_g\}$, then $0 < \alpha < \pi$ and $f(z)g(z)$ is continuous on $\{z = e^{i\theta} | \theta_0 - \alpha \leq \theta \leq \theta_0 + \alpha\} - \{z_0 = e^{i\theta_0}\}$ and assumes there only unimodular values. Thus, indeed, $f \cdot g \in \mathrm{Seidel}_{z_0}(U)$.

2. We will use the following well-known facts:

   (a) $B(z)$ has a singular point at every point of the derived set of the zeros of $B(z)$, $\{a_k\}'_k$. It is analytic at every point of $\mathbb{T} - \{a_k\}'_k$.

   (b) $\{a_k\}'_k$ is a closed subset of $\mathbb{T}$.

   By (a), $B(z)$ is continuous at every point of $\mathbb{T} - \{a_k\}'_k$ and only there. By (b), the set $\mathbb{T} - \{a_k\}'_k$ is an open subset of $\mathbb{T}$. Hence, indeed, $B(z) \in \mathrm{Seidel}_{z_0}(U) \Leftrightarrow z_0 \notin \{\{a_k\}'_k\}'$. In fact we can take for α, the distance along $\mathbb{T}$ from $\{z_0\}$ to the closed set $\{a_k\}'_k$.

3. If $f \in \mathrm{Seidel}_{z_0}(U)$ and $b \in U$, then $g(z) = \frac{b-f(z)}{1-\overline{b}f(z)} \in \mathrm{Seidel}_{z_0}(U)$ and we have by the involution property $f(z) = \frac{b-g(z)}{1-\overline{b}g(z)}$ which shows the other direction. $\qquad\square$

Remark 97.0.233. It follows from the proof of part 2 of Proposition 97.0.232, that

$$B(z) = z^m \prod_{k=1}^{\infty} \frac{|a_k|}{a_k} \frac{a_k - z}{1 - \overline{a}_k z} \in \mathrm{Seidel}_{z_0}(U)$$

if and only if there is no subsequence $\{a_{k_i}\}_i$ of $\{a_k\}_k$ such that $z_0 = \lim_{i \to \infty} a_{k_i}$ or if there is such a subsequence, then there is a number $0 < \alpha < \pi$, such that $\forall\, \theta \in \{\phi | \theta_0 - \alpha \leq \phi \leq \theta_0 + \alpha\} - \{\theta_0\}$, $z_0 = e^{i\theta_0}$, there is no subsequence $\{a_{k_j}\}_j$ of $\{a_k\}_k$ such that $e^{i\theta} = \lim_{j \to \infty} a_{k_j}$.

Another useful fact which is clear is worth mentioning here is as follows:

Proposition 97.0.234. *Let $f(z)$, $g(z)$ be bounded analytic functions in the unit disc:* $|f(z)|, |g(z)| < 1 \ \forall z \in U$. *Let $z_0 \in \mathbb{T}$. Then we have*

$$f, \ f \cdot g \in \mathrm{Seidel}_{z_0}(U) \Rightarrow g \in \mathrm{Seidel}_{z_0}(U).$$

In fact, we can take $\alpha_g = \min \alpha_f, \alpha_{f \cdot g}$.

Definition 97.0.235. An infinite Blaschke sequence $\{a_k\}_k$ will be called a Seidel–Blaschke sequence in the point $z_0 \in \mathbb{T}$, if the corresponding Blaschke product $B(z) \in \mathrm{Seidel}_{z_0}(U)$. Equivalently, $z_0 \notin \{\{a_k\}_k'\}'$, where $\{a_k\}_k$ is the infinite Blaschke sequence.

We are ready to state and prove a result.

Theorem 97.0.236. *Let $B(z)$ be an infinite Blaschke product such that $B(z) \in \mathrm{Seidel}_{z_0}(U)$. Then for any singular inner function $S(z)$ such that $S(z) \in \mathrm{Seidel}_{z_0}(U)$, the function $B(z) \cdot S(z)$ assumes every value w, save at most one, of the unit disc $|w| < 1$ infinitely often in the unit disc $|z| < 1$ and assumes no value w, $|w| \geq 1$, in the unit disc. The cluster set of $B(z) \cdot S(z)$ in $z = z_0$ is the closed unit disc $|w| \leq 1$.*

Proof. By $B(z), S(z) \in \mathrm{Seidel}_{z_0}(U)$ and by Proposition 97.0.232 (1), it follows that $B(z) \cdot S(z) \in \mathrm{Seidel}_{z_0}(U)$. Now, the conclusions follow by Theorem 4 of Seidel [93]. $\qquad\qquad \square$

So, this simple conclusion of Theorem 4 of Seidel tells us that the "at most one exceptional value property" is mostly determined by the infinite Blaschke product factor of $f(z) \in \mathrm{Seidel}_{z_0}(U)$. Multiplying this $B(z)$ by any z_0-reasonable singular inner function, $S(z)$, i.e., $S(z) \in \mathrm{Seidel}_{z_0}(U)$ does not affect the "at most one exceptional value property".

Having noticed that, let us inquire (naturally) if the following similar principle is true.

The Principle S:
Let $B(z)$ be an infinite Blaschke product and let $S(z)$ be a singular inner function. Then $B(z)$ is a $U \to U$ surjection if and only if $B(z) \cdot S(z)$ is a $U \to U$ surjection.

Or, if this is not true in that generality, to what extent is it true?

If this is true, then it answers our question, namely: Which Blaschke sequences are $Z(S(z) - \alpha)$ for some singular inner function $S(z)$ and some number $\alpha \in U$? The answer is, those Blaschke sequences for which the corresponding Blaschke product is not a $U \to U$ surjective mapping.

Sketchy ideas:

We try to understand the conjecture above which was labeled "The Principle S" and in case it is not valid in that generality we will try to figure out how to add assumptions so that the modified conjecture will still be interesting. Our approach will be as follows: We will construct a homotopy chain from $B(z)$ to $B(z) \cdot S(z)$, within the family Inn, such that all the chain elements will be inner functions sharing the same zero set, i.e., $Z(B(z))$. We will make an effort to show that the homotopy we constructed preserves the property of "being a $U \to U$ surjection". For t real, $0 \le t \le 1$, we define $F_t(z) = B(z) \cdot S(z)^t$. Then we have $F_0(z) = B(z)$ and $F_1(z) = B(z) \cdot S(z)$. Since $S(z)$ is a singular inner function, it can be represented as follows: $S(z) = \exp(-p(z))$, where $p(z) \in H(U)$, $\Re p(z) \ge 0$ for all $z \in U$ and $\lim_{r \to 1^-} \Re p(re^{i\theta}) = 0$ a.e. in $0 \le \theta < 2\pi$. Hence for any $0 \le t \le 1$ we have $S(z)^t = \exp(-tp(z))$ which is also a singular inner function. We now inquire what are the implications of the requirement that $B(z) \cdot S(z)^t$ is a $U \to U$ surjective mapping given that $F_1(z)$ or $F_0(z)$ is such a mapping. Let us fix a number $a \in U^*$. Then $a \in \mathrm{Im}\{B(z) \cdot S(z)^t\}$, $z \in U$, if and only if

$$f(r,t) := \frac{1}{2\pi} \oint_{|z|=r} \frac{(B(z) \cdot S(z)^t)'}{B(z) \cdot S(z)^t - a} dz \in \mathbb{Z}^+$$

for a fixed $0 \le t \le 1$ and any $r < 1$ which is close enough to 1. By the way, the values of $f(r,t)$, for any $0 \le t \le 1$, $0 < r < 1$ belong to $\mathbb{Z}_{\ge 0}$. Moreover, we assume at the moment that we can find a function $r = r(t)$, such that $0 < r(t) < 1$, $f(r(t),t) = N(t) \in \mathbb{Z}^+$, and $f(r(t),t)$ is a continuous function of t, where t varies in the closed unit interval, $0 \le t \le 1$. If, indeed, finding such a function $r(t)$ is possible, then $N(t)$ is a continuous function of t which assumes only values which are positive integers. Hence $N(t) \equiv N \in \mathbb{Z}^+$ must be a constant function on $0 \le t \le 1$. This implies that $f(r(0),0) = f(r(1),1) = N$, and hence proves that $a \in \mathrm{Im}\{B(z)\}$ if and only if $a \in \mathrm{Im}\{B(z) \cdot S(z)\}$. Since $a \in U$ is fixed, but arbitrary, this, in turn proves that $B(z)$ is a $U \to U$ surjective mapping, if and only if $B(z) \cdot S(z)$ is a $U \to U$ surjective mapping. Of course some how in proving that scheme we will have to make a use in the fact that $B(z)$ is a Blaschke product and $S(z)$ is a singular inner function. Thus everything depends on our ability to construct a function $r = r(t)$ on $0 \le t \le 1$ such that $0 < r(t) < 1$ and such that the function of t, defined by

$$f(r(t),t) = \frac{1}{2\pi i} \oint_{|z|=r(t)} \frac{(B(z) \cdot S(z)^t)'}{B(z) \cdot S(z)^t - a} dz$$

is a positive continuous function of t. Probably, it is sufficient to construct a continuous function $r = r(t)$, $0 \le t \le 1$ with the desired properties, i.e., $0 < r(t) < 1$ and $f(r(t), t) > 0$. For we hope that the continuity of $r(t)$ implies the continuity of $f(r(t), t)$. That seems reasonable. We will try to prove $\{r(t) \text{ continuous}\} \Rightarrow \{f(r(t), t) \text{ is continuous}\}$ when we get there. So, once again, let us a t, $0 \le t \le 1$ and try to understand the implication of the requirement:

$$\operatorname{Im}\left\{B(z) \cdot S(z)^t\right\} = \operatorname{Im}\left\{B(z) \cdot S(z)\right\}$$

or (and this will be our choice)

$$\operatorname{Im}\left\{B(z) \cdot S(z)^t\right\} = \operatorname{Im}\left\{B(z)\right\}.$$

Let $B(z_0) \cdot S(z_0) = \alpha$ for some $|z_0| < 1$ and some $|\alpha| < 1$. We can avoid the case $\alpha = 0$, for then z_0 is a zero of the Blaschke product $B(z)$ in which case $B(z_0) \cdot S(z_0)^t = B(z_0) \cdot S(z_0) = B(z_0) = 0$. We ask if we can find a number z_1, $|z_1| < 1$ such that

$$B(z_1) \cdot S(z_1)^t = B(z_0) \cdot S(z_0) = \alpha \ne 0.$$

At this point, we do not see why this (finding z_1) is possible. However, if z_0 is fixed then for t close enough to 1 we can indeed find such a z_1. So when we fix z_0, there is a $t = t(z_0)$ (in fact many of them) close enough to 1, such that the equation

$$B(z_1) \cdot S(z_1)^t = \alpha(= B(z_0) \cdot S(z_0)) \tag{97.0.1}$$

can be solved for $z_1 = z_1(t)$. We note that we can always take $z_1(1) = Z_0$. There are couple of problems we face. Firstly, we constructed $t(z_0)$ for a fixed z_0 (our $t(z_0)$ should be close to 1). So, t is not arbitrary. Secondly, the existence of a solution $z_1(t)$ is implicit (provided that $t = t(z_0)$ is close enough to 1). If we will be able to show that for a fixed z_0, we can, in fact push $t = t(z_0)$ down, from being in the vicinity of 1 to being, as close as we please to 0, then we will overcome our two problems. So given z_0 (or better α) we know that equation (97.0.1) is solvable for z_1, provided that $t(z_0)$ is close enough to 1. We seek for a continuation process for $t = t(z_0)$ downwards to 0. Since the inner function $B(z) \cdot S(z)^t = B(z) \cdot \exp(-tp(z))$ is an open mapping in the complex variable z and since for $t = t(z_0)$, we have a solution z_1 to the equation (97.0.1): $B(z_1) \cdot \exp(-tp(z_1)) = \alpha$, we ask, if it is possible that equation (97.0.1) will not be solvable in z_1 if we drop slightly the value of t. That is, is it possible that $\forall \epsilon > 0$ (small) we have no solution z to

$$B(z) \cdot \exp(-(t - \epsilon)p(z)) = \alpha = B(z_1) \cdot \exp(-tp(z_1))?$$

We note that this is a question about the α-level curve (in t):

$$B(z) \cdot \exp(-tp(z)) \equiv \alpha$$

where $z = z(t)$. The question is, can this level curve have a positive lower bound $t^* > 0$? That is, we can solve for $z = z(t)$ the equation

$$B(z(t)) \cdot \exp(-tp(z(t))) \equiv \alpha \quad \text{for} \ \ t^* < t \leq 1, \qquad (97.0.2)$$

but not for $0 \leq t \leq t^*$? (at least not near t^*, $t^* - \delta < t \leq t^*$). Since $B(z) \cdot \exp(-tp(z))$ is an open mapping in the complex variable z and since the mapping is also continuous and open in the real variable t this can happen only if our solution $z(t)$ of equation (97.0.2) bumps into the boundary of U, $\partial U = \mathbb{T}$ when t approaches t^* from above:

$$\lim_{t \to (t^*)+} |z(t)| = 1.$$

In other words, the level curve of α, $z(t)$ that started at z_0 went over the boundary ∂U at time t^*. This does not mean at all that we can not find another part of the α-level curve that will make it all the way down to $t = 0$. It might be that there is another point z_1' that is out of the first α-level curve such that when we solve equation (97.0.1) starting from $z_1'(1) = z_0'$ then the continuation process for t downwards will never halt at a positive distance from 0 in which case we will have

$$B(z_1'(0)) = \alpha$$

which means that $\alpha \in \text{Im}\{B(z)\}$ as well as $\alpha \in \text{Im}\{B(z) \cdot S(z)\}$. If this could be done for any $\alpha \in U$ then $B(z)$ and also $B(z) \cdot S(z)$ are both $U \to U$ surjective mappings. At this point, it seems that the geometrical and the topological arguments do not give us definite answers. There is though a refinement based on the geometrical and the topological arguments:

Let us go back to the scenario where we can solve equation (97.0.2) for $0 < t^* < t \leq 1 : B(z(t)) \cdot \exp(-tp(z(t))) = \alpha$. We know that in this situation we have

$$\lim_{t \to (t^*)+} |z(t)| = 1.$$

Let us assume that, in fact, the limit $\lim_{t\to(t^*)+} z(t) = z(t^*)$ exists. So $|z(t^*)| = 1$. Substituting the limiting value in equation (97.0.2), we obtain

$$B(z(t^*)) \cdot \exp(-t^* p(z(t^*))) = \alpha.$$

We recall that $\Re p(e^{i\theta}) = 0$ a.e. $0 < \theta < 2\pi$. If $z(t^*)$ is a point for which $\Re p(z(t^*)) = 0$ then $\exp(-t^* p(z(t^*))) = e^{i\theta_0}$ for some $0 \le \theta_0 < 2\pi$. Hence, $B(z(t^*)) = e^{-i\theta_0}\alpha$. In other words, this does not prove that $\alpha \in \mathrm{Im}\{B(z)\}$ but only that some rotation of α, $e^{-i\theta_0}\alpha \in \mathrm{Im}\{B(z)\}$. Now, we based our last result on the extra assumption that $\lim_{t\to(t^*)+} z(t) = z(t^*)$ exists. We only have $\lim_{t\to(t^*)+} |z(t)| = 1$. However, we can overcome this obstacle. We have $\lim_{t\to(t^*)+} |z(t)| = 1$. By a compactness argument (a cluster point argument), we can extract a sequence $\{t_n\}$, $t_n \to (t^*)^+$ such that the sequential limit $\lim_{n\to\infty} z(t_n) + z_s(t^*)$ exists and, of course $|z_s(t^*)| = 1$. This would give us (on substituting into equation (97.0.2)):

$$B(z_s(t^*)) \cdot \exp(-t^* p(z_s(t^*))) = \alpha \quad (|z_s(t^*)| = 1).$$

So, if $\Re p(z_s(t^*)) = 0$, we obtain the same conclusion:

$$e^{-i\theta_0}\alpha \in \mathrm{Im}\,\{B(z)\} \quad \text{for} \quad \text{some} \quad 0 \le \theta_0 < 2\pi.$$

In terms of the original singular inner function, the assumption $\Re p(z_s(t^*)) = 0$ is equivalent to

$$\lim_{n\to\infty} S(z(t_n)) = S(z_s(t^*)) \quad \text{and} \quad S(z_s(t^*)) = e^{i\theta_0}$$

a uni-modular value of the boundary limit of $S(z)$. In other words the cluster set of $S(z)$ at the point $z_s(t^*) \in \mathbb{T}$ includes a uni-modular cluster value:

$$e^{i\theta_0} \in C_U(S(z); z_s(t^*)).$$

This, of course, is self evident. It only means that for any $\alpha \in U$, $\mathrm{Im}\{B(z)\}$ must contain points on the circle $|w| = |\alpha|$. The alternative to that would have been that $\{w \,|\, |w| = \alpha\} \cap \mathrm{Im}\{B(z)\} = \emptyset$. This implies that $\mathrm{Im}\{B(z)\}$ is disconnected, an absurd!

After presenting a geometrical and topological approach to our problem, we now take an analytic course.

The analytic approach.

We would like to find a piece of the α-level curve

$$B(z(t)) \cdot \exp(-tp(z(t))) \equiv \alpha \quad \text{where} \quad 0 \le t \le 1. \tag{97.0.3}$$

That is, the parameter t varies on the full unit interval. We recall that we have the following initial condition $z(1) = z_0$. The parameter t is continued till t^*, $t^* < t \leq 1$. We would like to show that in fact t^* goes as down as possible, $t^* = 0$. The idea is simply to rewrite the α-level curve problem as an initial value problem. We assume (as we may) that $z'(t)$ exists whenever $z(t)$ is defined. We differentiate (97.0.3) with respect to t, ∂t:

$$z'(t)B'(z(t))\exp(-tp(z(t))) + B(z(t))\left\{-p(z(t)) - tz'(t)p'(z(t))\right\}$$
$$\times \exp(-tp(z(t))) = 0, \quad z(1) = z_0.$$

$$z'\left\{B'(z) - tB(z)p'(z)\right\} - B(z)p(z) = 0, \quad z(1) = z_0.$$

$$\frac{dz}{dt} = \frac{B(z)p(z)}{B'(z) - tB(z)p'(z)}, \quad z(1) = z_0.$$

Before going on with the ODE approach, let us make the following:

Remark 97.0.237. Our starting point was the equation of the α-level curve for the function below of the real variable $t \leq 1$:

$$\begin{cases} B(z(t)) \cdot \exp(-tp(z(t)) \equiv \alpha \in U^*, \\ z(1) = z_0. \end{cases}$$

Let us solve that equation for $t = t(z)$. Solving it for $z = z(t)$ seems to be hard. Taking logarithm of both sides of the equation gives us

$$\log B(z) - t \cdot p(z) = \log \alpha,$$

$$t \cdot p(z) = \log \frac{B(z)}{\alpha}.$$

So,

$$\begin{cases} t(z) = \frac{1}{p(z)} \log \frac{B(z)}{\alpha} \text{ lies inside } [0,1] \\ t(z_0) = 1 \end{cases}.$$

Here,

$$B(z) = z^m \prod_k \frac{|\alpha_k|}{\alpha_k} \cdot \frac{\alpha_k - z}{1 - \overline{\alpha}_k z},$$

for some $m \in \mathbb{Z}_{\geq 0}$ and a Blaschke sequence $\{z_k\}$. Also, $\Re p(z) \geq 0 \ \forall z \in U$ where $\lim_{r \to 1^-} \Re p(re^{i\theta}) = 0$ a.e. $0 \leq \theta < 2\pi$.

One way to express the condition that $t(z)$ is real and positive is by writing $t(z) = |t(z)|$ which, in terms of our analytic functions, means

$$\frac{1}{p(z)} \log \frac{B(z)}{\alpha} = \frac{1}{|p(z)|} \left| \log \frac{B(z)}{\alpha} \right|.$$

Returning to our ODE and its initial value, we recall the following:

$$\frac{dz}{dt} = \frac{B(z)p(z)}{B'(z) - tB(z)p'(z)}, \quad z(1) = z_0.$$

To understand when will we have a global solution $z = z(t)$ on the full unit interval $[0,1]$, let us denote the function on the right-hand side by

$$g(t, z) = \frac{B(z)p(z)}{B'(z) - tB(z)p'(z)}$$

and based on standard IVP theory try to inquire when is $g(t, z)$ Lipschitz in the variable z. This might be an overshoot for as is well known the Lipschitz condition assures both existence and uniqueness. At least now we are not confident about uniqueness. For example, we can imagine that the α-level curve can have bifurcations. At any rate, trying to manipulate $g(t, z_2) - g(t, z_1)$, we get that it equals the following expression:

$$\frac{\left(\frac{B'(z_1)}{B(z_1)} \frac{1}{p(z_1)} - \frac{B'(z_2)}{B(z_2)} \frac{1}{p(z_2)} \right) - t \left(\frac{p'(z_1)}{p(z_1)} - \frac{p'(z_2)}{p(z_2)} \right)}{\left(\frac{B'(z_2)}{B(z_2)} - t\frac{p'(z_2)}{p(z_2)} \right) \left(\frac{B'(z_1)}{B(z_1)} - t\frac{p'(z_1)}{p(z_1)} \right)}.$$

Trying to manipulate $\frac{\partial g}{\partial z}$ we get that it equals the following expression:

$$\frac{-p(z)W(B(z), B'(z)) + B(z)B'(z)p'(z) + B(z)^2 W(p(z), p''(z))t}{(B'(z) - tB(z)p'(z))^2},$$

where we used the standard notation:

$$W(B(z), B'(z)) = \begin{vmatrix} B(z) & B'(z) \\ B'(z) & B''(z) \end{vmatrix} \quad \text{and} \quad W(p(z), p'(z)) = \begin{vmatrix} p(z) & p'(z) \\ p'(z) & p''(z) \end{vmatrix}.$$

We see that trying to find a global solution $z = z(t)$, $0 \le t \le 1$, via inquiring the Lipschitz property in $g(t, z)$ with respect to the variable z lead to complicated estimates. On the other hand, if we think of $t = t(z)$ as a solution which is a function of z we are led to the following ODE:

$$\frac{dt}{dz} = \frac{B'(z) - tB(z)p'(z)}{B(z)p(z)}$$

This time, we naturally define

$$h(z,t) = \frac{1}{g(t,z)} = \frac{B'(z) - tB(z)p'(z)}{B(z)p(z)},$$

and ask about the Lipschitz property of $h(z,t)$ with respect to the variable t. The computations are much easier. The algebraic calculation gives

$$h(z,t_2) - h(z,t_1) = \frac{p'(z)}{p(z)}(t_1 - t_2).$$

The differential calculation gives

$$\frac{\partial h}{\partial t} = -\frac{p'(z)}{p(z)}.$$

Both expressions are simpler. It makes sense to review some of the essential results on global solutions of ODE's of the first order where the independent variable is real and varies within the unit interval $0 \leq t \leq 1$.

Chapter 98

Global Existence Theorems for ODE's and IVP's

The source of this first part of the chapter is as follows:

www.math.washington.edu > crs > 555notes > exist

Let $\mathbb{F}$ be $\mathbb{R}$ or $\mathbb{C}$.

Proposition. *On an interval $I \subseteq \mathbb{R}$ containing t_0, $x = x(t)$ is a solution of $x' = f(t, x)$, $x(t_0) = x_0$ (where $f(t, x)$ is continuous) with $x \in C^1(I)$ if and only if x is a solution of the integral equation $x(t) = x_0 + \int_{t_0}^{t} f(s, x(s))ds$ on I with $x \in C(I)$.*

Definition. Let $I \subseteq \mathbb{R}$ be an interval and $\Omega \subseteq \mathbb{F}^n$. We say that $f(t, x)$ mapping $I \times \Omega$ into $\mathbb{F}^n$ is uniformly Lipschitz continuous with respect to x if there is a constant L (called a Lipschitz constant) for which $\forall t \in I$ $\forall x, y \in \Omega$, $|f(t, x) - f(t, y)| \leq L \cdot |x - y|$ ($|\cdot|$ is the l^2-norm). We say that f is in (C, Lip) on $I \times \Omega$ if f is continuous on $I \times \Omega$ and f is uniformly Lipschitz continuous with respect to x on $I \times \Omega$.

Theorem (Local Existence and Uniqueness for the integral equation for Lipschitz f). *Let $I = [t_0, t_0 + \beta]$ and $\Omega = \overline{B_r(x_0)} = \{x \in \mathbb{F}^n \,|\, |x - x_0| \leq r\}$, and suppose $f(t, x)$ is in (C, Lip) on $I \times \Omega$. Then there exists $\alpha \in (0, \beta]$ for which there is a unique solution of the integral equation*

$$x(t) = x_0 + \int_{t_0}^{t} f(s, x(s))ds \tag{98.0.1}$$

in $C(I_\alpha)$ where $I_\alpha = [t_0, t_0 + \alpha]$. Moreover, we can choose α to be any positive number satisfying $\alpha \leq \beta$, $\alpha \leq \frac{r}{M}$, and $\alpha < \frac{1}{L}$, where $M = \max_{(t,x) \in I \times \Omega} |f(t, x)|$ and L is a Lipschitz constant for f in $I \times \Omega$.

If we set aside the uniqueness part, then the condition $\alpha < \frac{1}{L}$ is unnecessary.

Theorem (Picard Local Existence for (98.0.1) for Lipschitz). *Let $I = [t_0, t_0 + \beta]$, and suppose $f(t, x)$ is in (C, Lip) on $I \times \Omega$, $\Omega = \overline{B_r(x_0)} = \{x \in \mathbb{F}^n \,|\, |x - x_0| \le r\}$. Then there exists a solution $x_*(t)$ of the integral equation (98.0.1) in $C(I_\alpha)$, where $I_\alpha = [t_0, t_0 + \alpha]$, $\alpha = \min(\beta, r/M)$ and $M = \max_{(t,x)\in I\times\Omega} |f(t, x)|$.*

Theorem (Picard Global Existence for (98.0.1) for Lipschitz). *Let $I = [t_0, t_0 + \beta]$, and suppose $f(t, x)$ is in (C, Lip) on $I \times \mathbb{F}^n$. Then there exists a solution $x_*(t)$ of the integral equation (98.0.1) in $C(I)$.*

Corollary. *The solution $x_*(t)$ of (98.0.1) satisfies*

$$|x_*(t) - x_0| \le \frac{M_0}{L}\left(e^{L(t-t_0)} - 1\right)$$

for $t \in I$ (global) or $t \in I_\alpha$ (local), where $M_0 = \max_{t\in I} |f(t, x_0)|$ (global), or $M_0 = \max_{t\in I_\alpha} |f(t, x_0)|$ (local).

Theorem (Cauchy–Peano Existence Theorem). *Let $I = [t_0, t_0 + \beta]$ and $\Omega = \overline{B_r(x_0)}$, and suppose $f(t, x)$ is continuous on $I \times \Omega$. Then there exists a solution $x_*(t)$ of the integral equation (98.0.1), in $C(I_\alpha)$ where $I_\alpha = [t_0, t_0 + \alpha]$, $\alpha = \min(\beta, r/M)$ and $M = \max_{(t,x)\in I\times\Omega} |f(t, x)|$. Thus, $x_*(t)$ is a C^1 solution on I_α of: $x' = f(t, x), x(t_0) = x_0$.*

Remark. In the last theorem, there is no Lipschitz assumption.

Theorem (Gronwall's Inequality-differential form). *Let $I = [t_0, t_1]$. Suppose $a : I \to \mathbb{R}$ and $b : I \to \mathbb{R}$ are continuous, and suppose $u : I \to \mathbb{R}$ is in $C^1(I)$ and satisfies $u'(t) \le a(t)u(t) + b(t)$ for $t \in I$, and $u(t_0) = u_0$. Then*

$$u(t) \le u_0 e^{\int_{t_0}^t a} + \int_{t_0}^t e^{\int_s^t a} \cdot b(s)ds.$$

Theorem (Gronwall's Inequality-integral form). *If ϕ, ψ, $\alpha \in C(I)$ are real-valued with $\alpha \ge 0$ on I, and*

$$\phi(t) \le \psi(t) + \int_{t_0}^t \alpha(s)\phi(s)ds \quad \text{for } t \in I,$$

then

$$\phi(t) \le \psi(t) + \int_{t_0}^t e^{\int_s^t \alpha} \cdot \alpha(s) \cdot \psi(s)ds.$$

In particular, if $\psi(t) \equiv c$ (a constant), then

$$\phi(t) \le c e^{\int_{t_0}^{t} \alpha}.$$

Construction of solutions.

We consider two kinds of results:

(1) Local continuation (continuation at a point — no Lipschitz condition is assumed).
(2) Global continuation (for locally Lipschitz f).

(1) Continuation at a point:

Suppose that $x(t)$ is a solution of the differential equation $x' = f(t,x)$ on an interval I and that f is continuous on some subset $S \subseteq \mathbb{R} \times \mathbb{F}^n$ containing $\{(t, x(t)) | t \in I\}$. Note: no Lipschitz condition is assumed.

Case 1: I is closed at the right-hand side, i.e., $I = (-\infty, b]$, $[a, b]$ or $(a, b]$. Assume further that $(b, x(b))$ is in the interior of S. Then the solution can be extended (by the Cauchy–Peano Existence Theorem) to an interval with right-hand side $b + \beta$ for some $\beta > 0$.

Case 2: I is open at the right-hand side, i.e., $I = (-\infty, b)$, $[a, b)$ or (a, b) with $b < \infty$. Assume further that $f(t, x(t))$ is bounded on $[t_0, b)$ for some $t_0 < b$ with $[t_0, b) \subseteq I$, say $|f(t, x(t))| \le M$ on $[t_0, b)$. Remarks about this assumption are as follows:

(1) If this is true for some $\widetilde{t_0} \in I$, it is true for all $t_0 \in I$, where M depends on t_0. The reason is that in the non-trivial case $t_0 < \widetilde{t_0}$, $f(t, x(t))$ is continuous on $[t_0, \widetilde{t_0}]$. So, the assumption is a condition on the behavior of $f(t, x(t))$ near $t = b$.
(2) The assumption can be restated with a slightly different emphasis: for some $t_0 \in I$, $\{(t, x(t)) | t_0 \le t < b\}$ stays within a subset of S on which f is bounded. For example, $\{(t, x(t)) | t_0 \le t < b\}$ stays within a compact subset of S, this condition is satisfied.

The integral equation (98.0.1)

$$x(t) = x_0 + \int_{t_0}^{t} f(s, x(s))ds$$

holds for $t \in I$. In particular, for $t_0 \le \tau \le t < b$,

$$|x(t) - x(\tau)| = \left| \int_\tau^t f(s, x(s))ds \right| \le \int_\tau^t |f(s, x(s))|ds \le M \cdot |t - \tau|.$$

Thus, for any sequence $t_n \uparrow b$, $\{x(t_n)\}$ is a Cauchy sequence. This implies that $\lim_{t \to b-} x(t)$ exists; call it $x(b^-)$. So $x(t)$ has a continuous extension from I to $I \cup \{b\}$. If in addition $(b, x(b^-))$ is in S, then (98.0.1) holds on $I \cup \{b\}$ as well, so $x(t)$ is a C^1 solution of $x' = f(t, x)$ on $I \cup \{b\}$. Of course, if now in addition $(b, x(b^-))$ is in the interior of S, we are back in Case 1 and can extend the solution $x(t)$ a little beyond $t = b$.

Case 3: I is closed at the left hand side — similar to Case 1.

Case 4: I is open at the left hand side — similar to Case 2.

(2) Global continuation:
Now, suppose that $f(t, x)$ is continuous on an open set $\mathcal{D} \subseteq \mathbb{R} \times \mathbb{F}^n$ and suppose f is locally Lipschitz continuous with respect to x on $\mathcal{D}$. For example, if f is C^1 with respect to x in $\mathcal{D}$, i.e., $\frac{\partial f_i}{\partial x_j}$ exists and is continuous on $\mathcal{D}$ for $1 \le i, j \le n$, then f is locally Lipschitz continuous with respect to x on $\mathcal{D}$. For brevity, we will write $f \in (C, Lip_{loc})$ on $\mathcal{D}$. Let $(t_0, x_0) \in \mathcal{D}$. We want to continue in the time variable t solutions of the initial value problem $x' = f(t, x)$, $x(t_0) = x_0$. Part of being a solution is that $(t, x(t)) \in \mathcal{D}$ (we are assuming that f is only defined in $\mathcal{D}$). We know of local existence of solutions and uniqueness of solutions on any interval. Define

$$T_+ = \sup\{t > t_0 | \text{there exists a solution of the initial}$$

$$\text{value problem on } [t_0, t)\}.$$

By uniqueness, two solutions must agree on their common interval of definition so there exists a solution on $[t_0, T_+)$. define T_- similarly. So (T_-, T_+) is the maximal interval of existence of the solution of the initial value problem. It is possible that $T_+ = \infty$ or (and) $T_- = -\infty$. Note that the maximal interval (T_-, T_+) is open: if the solution could be extended to T_+ (or to T_-), then since $\mathcal{D}$ is open, the results above on continuation at a point imply that the solution could be extended beyond T_+ (or T_-), contradicting the definition of T_+ (or of T_-). The ideal situation would be $T_+ = \infty$ and $T_- = -\infty$, in which case the solution exists for all time t. Another "good" situation is if $f(t, x)$ is not defined for $t \ge T_+$. For example if $a(t) = \frac{1}{1-t}$ (which blows up at $t = 1$) and $x'(t) = a(t)$, we don't expect the solution to exist beyond $t = 1$. Here, if $t_0 = 0$ and $\mathcal{D} = (-\infty, 1) \times \mathbb{R}$, then $T_+ = 1$.

Other, less desirable behavior occurs for $x' = x^2$, $x(0) = x_0 > 0$, $t_0 = 0$, and $\mathcal{D} = \mathbb{R} \times \mathbb{R}$. The solution $x(t) = x_0(1 - x_0 t)^{-1}$ blows up at time $T_+ = x_0^{-1}$ (note that $T_- = -\infty$). Observe that $x(t) \to \infty$ as $t \to (T_+)^-$. So the solution does not just "stop" in the interior of $\mathcal{D}$. This is the general behavior in this situation.

Theorem. *Suppose $f \in (C, LIP_{loc})$ on an open set $\mathcal{D} \subseteq \mathbb{R} \times \mathbb{F}^n$. Let $(t_0, x_0) \in \mathcal{D}$, and let (T_-, T_+) be the maximal interval of existence of the solution of the initial value problem $x' = f(t, x)$, $x(t_0) = x_0$. Given a compact set $K \subset \mathcal{D}$, there exists a $T < T_+$ for which $(t, x(t)) \notin K$ for $t > T$.*

The source of the second part of the chapter is as follows:

math.iitb.ac.in/siva/ma41707/ode91.pdf

Theorem 3.34. *Let $f : \mathbb{I} \times \mathbb{R}^n \to \mathbb{R}^n$ be continuous. Assume that there exist two continuous functions $h, k : \mathbb{I} \to \mathbb{R}_+$ (non-negative real-valued) such that $\|f(x, \overline{y})\| \leq k(x)\|\overline{y}\| + h(x)$, $\forall (x, \overline{y}) \in \mathbb{I} \times \mathbb{R}^n$. Then for every initial data $(x_0, \overline{y}_0) \in \mathbb{I} \times \mathbb{R}^n$, IVP has at least one global solution.*

Theorem 3.36 (continuous dependence). *Let $\Omega \subseteq \mathbb{R}^n$ be a domain and $\mathbb{I} \subseteq \mathbb{R}$ be an interval containing the point x_0. Let $\mathbb{J}$ be a closed and bounded subinterval of $\mathbb{I}$, such that $x_0 \in \mathbb{J}$. Let $\overline{f} : \mathbb{I} \times \omega \to \mathbb{R}^n$ be a continuous function. Let $\overline{y}(x; x_0, \overline{y}_0)$ be a solution on $\mathbb{J}$ of the initial value problem*

$$\overline{y}' = \overline{f}(x, \overline{y}), \quad \overline{y}(x_0) = \overline{y}_0.$$

Let S_α denote the α-neighborhood of the graph of $\overline{y}$, i.e.,

$$S_\alpha = \{(x, \overline{y}) \mid \|\overline{y} - \overline{y}(x; x_0, \overline{y}_0)\| \leq \alpha, \ x \in \mathbb{J}\}.$$

Suppose that there exists an $\alpha > 0$ such that $\overline{f}$ satisfies Lipschitz condition with respect to the variable $\overline{y}$ on S_α. Then the solution $\overline{y}(x; x_0, \overline{y}_0)$ depends continuously on the initial values and on the vector field $\overline{f}$. That is, given $\epsilon > 0$, there exists a $\delta > 0$ such that if $\overline{g}$ is continuous on S_α and the inequalities

$$\|\overline{g}(x, \overline{y}) - \overline{f}x, \overline{y})\| \leq \epsilon \ \text{on} \ S_\alpha, \quad \|\xi - \overline{y}_0\| \leq \delta$$

are satisfied, then every solution $\overline{z}(x; x_0, \xi)$ of the IVP

$$\overline{z}' = \overline{g}(x, \overline{z}), \quad \overline{z}(x_0) = \xi$$

exists on all of $\mathbb{J}$, and satisfies the inequality

$$\|\overline{z}(x; x_0, \xi) - \overline{y}(x; x_0, \overline{y}_0)\| \leq \epsilon, \quad x \in \mathbb{J}.$$

Bibliography

[1] Aleksandrov, A. B., Anderson, J. M., and Nicolau, A., Inner functions, Bloch spaces and symmetric measures, *Proc. London Math. Soc.* **79**(2) (1999) 318–352.

[2] Ahern, P. R. and Clark, D. N., On inner functions with H^p-derivative, *Michigan Math. J.* **21**(2) (1974) 115–127.

[3] Ahern, P. and Kim, H. O., On the mean boundary behavior and the Taylor coefficients of an infinite Blaschke product, *Michigan Math. J.* **31** (1984) 3–14.

[4] Bagemhil, F. and Seidel, W., A general principle involving Baire category, with applications to function theory and other fields, *Proc. Nat. Acad. Sci. (Wash.)* **39** (1953) 1068–1075.

[5] Beurling, A., On two problems concerning linear transformations in Hilbert space, *Acta Math.* **81** (1949) 239–255.

[6] Brown, E. J., On a coefficient problem for nonvanishing H^r functions, *Complex Variables Theory and Applications* **4** (1985) 253–265.

[7] Borwein, P. and Erdélyi, T., *Polynomials and Polynomial Inequalities*, Graduate Texts in Mathematics, Vol. 161 (Springer, New York, 1995).

[8] Brown, E. J. and Walker, B. J., A coefficient estimate for nonvanishing $\mathbb{H}^p$ functions, *Rocky Mountain J. of Math.* **18**(3) (1988) 707–718.

[9] Constantin, A., *Fourier Analysis Part I — Theory, London Mathematical Society*, Student Texts, Vol. 85 (Cambridge University Press, Cambridge, 2016).

[10] Carathéodory, C., *Theory of Functions*, Vol. 2 (Chelsea reprint, New York, 1966), reprint.

[11] Carathéodory, C., ber die gegenseitige Beziehung der Ränder bei der konformen Abbildung des Inneren einer Jordanschen Kurve auf einen Kreis, *Math. Ann.* **73** (1913) 305–320.

[12] Carleson, L., Interpolation by Bounded Analytic Functions and the Corona Problem, *Annals of Math.* **76**(3) (1962) 547–559.

[13] Carmona, J. J. and Donaire, J. J., The converse of Fatou's theorem for Zygmund measures, *Pacific J. Math.* **191**(2) (1999) 207–222.

[14] Collingwood, E. F., F.R.S. and Lohwater, A. J., *The Theory of Cluster Sets*, (Cambridge University Press, Cambridge, 1966).

[15] Craizer, M., Entropy of inner functions, *Israel J. Math.* **74**(2) (1991) 129–168.

[16] Caughran, J. G. and Shields, A. L., Singular inner factors of analytic functions, *Michigan Math. J.* **16** (1969) 409–410.

[17] Cullen, M., Derivatives of singular inner functions, *Michigan Math. J.* **18**(3) (1971) 283–287.

[18] Danielyan, A. A., Rubel's problem on bounded analytic functions *Annales Academiæ Scientiarum Fennicæ Mathematica*, **41** (2016) 813–816.

[19] Détraz, Étude de spectre d'algebras de fonctions analytiques sur le disque unité, *C. R. Acad. Sci. Paris* **269** (1969) 833–835.

[20] Dugundji, J., *Topology* (Allyn and Bacon, Boston, 1966).

[21] Duren, P. L., *Univalent Functions*, A series of Comprehensive Studies in Mathematics, Vol. 259 (Springer, New York, 1983).

[22] Duren, P. L., *Theory of H^p Spaces* (Dover Publications, Inc., Mineola, 2000).

[23] Duren, P., Heinonen, J., Osgood, B. and Palka, B. (eds.), *Quasiconformal Mappings and Analysis, A Collection of Papers Honoring F. W. Gehring* (Springer Science plus Business Media, LLC, 1997).

[24] Duren, P. L., Romberg, B. W. and Shields, A. L., Linear functionals on H^p spaces with $0 < p < 1$, *J. Reine Angew. Math.* **238** (1969) 32–60.

[25] Duren, P. L. and Schuster, A., *Bergmen Spaces, Mathematical Surveys and Monographs*, Vol. 100 (American Mathematical Society, 2004).

[26] Dyakonov, K. M., Smooth functions and co-invariant subspaces of the shift operator, *St. Petersburg Math. J.* **4**(5) (1993) 933–959.

[27] Egerváry, E. und Szász, O., Einige Extremalprobleme im Berliche der trigonometrischen Polynome, *Math. Z.* **27** (1928) 641–692.

[28] Evans, G. C., Potentials and positively infinite singularities of harmonic functions, *Monatsh. Math. Phys.* **43** (1936) 419–424.

[29] Fejes, L., Two inequalities concerning trigonometric polynomials, *J. London Math. Soc.* **14** (1939) 44–46.

[30] Farmer, W. D. and Gorkin, P., Approximation by polynomials and Blaschke products having all zeros on a circle, *Illinois J. Math.*, Vol. **55**(3) (2011) 1105–1118.

[31] Fisher, S., The singular set of a bounded analytic function, *Michigan Math. J.* **20** (1973) 257–261.

[32] Frostman, O., Über den Kapazitätsbegriff und einen Satz von R. Nevanlinna, *Meddel. Lunds Univ. Mat. Sem.* **1** (1934).

[33] Frostman, O., Potentiel d'equilibre et capacité des ensembles avec quelques applications ã la theorie des fonctions, *Meddel. Lunds Univ. Mat. Sem.* **3** (1935) 1–118.

[34] Garnett, J. B., *Bounded Analytic Functions* (Academic Press, New York, 1981).

[35] Gelfand, I. M. and Fomin, S. V., *Calculus of Variations*, Rev. Engl. Edn. Transl. Ed. Richard A. Silverman (Prentice-Hall, Inc. Englewood Cliffs, 1963).

[36] Gamelin, T. W. and Garnett, J., Uniform approximation to bounded analytic functions, *Rev. Unión Mat. Argent.* **25** (1970) 87–94.

[37] Goluzin, G. M., *Geometric Theory of Functions of a Complex Variable*, Translations of Mathematical Monographs, Vol. 26, 2nd edn. (American Mathematical Society, 1969).

[38] Guillory, C. and Sarason, D., Division in $H^\infty + C$, *Michigan Math. J.* **28** (1981) 173–181.

[39] Hällström, G. AF, Zur Berrch-nung der Boderordnung oder Bodenyperordnung eindeutiger Funktionen, *Ann. Acad. Sci. Fenn. A.I.* **193** (1955) 1–16.

[40] Hällström, G. AF, Über meromorphe Funktionen mit mehrfach zusa mmen hängenden Existenzgebieten, *Acta Acad. Abo. Math. Phys.* **12** (1939) 5–100.

[41] Hatcher, A., *Algebraic Topology* (Cambridge University Press, 2001).

[42] Heinz, M., On a class of conformal metrics, *Nagoya Math. J.* **21** (1962) 1–60.

[43] Herrero, D., Inner functions under uniform topology, *Pacific J. Math.* **51** (1974) 167–175.

[44] Hardy, G. H., Littlewood, J. E., Pólya, G., *Inequalities*, 2nd edn. (Cambridge University Press, 1973).

[45] Hoffman, K., *Banach Spaces of Analytic Functions* (Prentice Hall, Englewood Cliffs, 1962).

[46] Hoffman, K., Bounded analytic functions and Gleason parts, *Ann. Math.* **86**(1) (1967) 74–111.

[47] Hong, I., On positively infinite singularities of a solution of the equation $\Delta u + k^2 u = 0$, *Kodai Math. Sem. Rep.* **8** (1956) 8–12.

[48] Horowitz, Ch., Coefficients of nonvanishing functions in H^∞, *Israel J. Math.* **30** (1978) 285–291.

[49] Hummel, J. A., Scheinberg, S. and Zalcman, L., A coefficient problem for bounded nonvanishing functions, *J. Anal. Math.* **31** (1977) 169–190.

[50] Hui, S., Blaschke products and nonvanishing bounded analytic functions separating sequences, *J. Math. Anal. Appl.* **136** (1988) 521–528.

[51] Inoue, M., Positively infinite singularities of solutions of linear elliptic partial differential equations, *J. Inst. Polytech., Osaka City Univ.* **8** (1957) 43–50.

[52] Izuchi, K., Factorization of Blaschke products, *Michigan Math. J.* **40** (1993) 53–75.

[53] Izuchi, K. and Izuchi, Y., Inner functions and division in Douglas algebras, *Michigan Math. J.* **33** (1986) 435–443.

[54] Ivrii, O., Prescribing inner parts of derivatives of inner functions, preprint (2017), arXiv:1702.00090.

[55] Khavinson, S. Ya., Foundations of the theory of extremal problems for bounded analytic functions and various generalizations of them, *Amer. Math. Soc. Transl.* **129** (1986) 1–56.

[56] Khavinson, S. Ya., Foundations of the theory of extremal problems for bounded analytic functions with additional conditions, *Amer. Math. Soc. Transl.* **129** (1986) 64–114.

[57] De Koninck, J.-M. and Luca, F., *Analytic Number Theory, Exploring the Anatomy of Integers*, Graduate Studies in Mathematics, Vol. 134 (American Mathematical Society, Providence, Rhode Island, 2012).

[58] Kolesnikov, S. V., On sets of non-existence of radial limits of bounded analytic functions, *Russ. Acad. Sci. Sb. Math.* **81** (1995) 477–485; *Mat. Sb.* **185** (1994) 91–100.

[59] Koosis, P., *Introduction to H_p Spaces*, 2nd edn., corrected and augmented with two appendices by V. P. Havin, St. Petersburg (Leningrad) State and McGill universities (Cambridge University Press, 1998).

[60] Korenblum, B., Cyclic elements in some spaces of analytic functions, *Bull. Amer. Math. Soc.* **5** (1981) 317–318.

[61] Krzyż, J., Coefficient problem for bounded nonvanishing functions, *Ann. Polon. Math.* **20** (1968) 314.

[62] Kraus D. and Roth O., Critical points, the Gauss curvature equation and Blaschke product, preprint (2011), arXiv: 11094711v2[math.CV].

[63] Kraus, D. and Roth, O., Strong submultiplicativity of the Poincaré metric, *J. Anal.* **24**(1) (2016) 39–50.

[64] Kuramochi, Z. and Kuroda, T., A note on the set of logarithmic capacity zero, *Proc. Jpn. Acad.* **30** (1954) 566–569.

[65] Lehto, O., On the first boundary value problem for functions harmonic in the unit circle, *Ann. Acad. Sci. Fenn. I. A.* **210** (1955) 1–26.

[66] Lohwater, A. J., The boundary behaviour of mermorphic functions, *Ann. Acad. Sci. Fenn. Ser. A. I.* **250/22** (1958) 6.

[67] Loomis, L. H., The converse of the Fatou theorem for positive harmonic functions, *Trans. Amer. Math. Soc.* **53** (1943) 239–250.

[68] Lusin, N. and Privaloff, J., Sur l'unicité et la multiplicité des fonctions analytiques, *Ann. Sci. École Norm. Sup.* **3**(42) (1935) 143–191.

[69] Mañé, R., Sad, P. and Sullivan, D., On the dynamics of rational maps, *Ann. Sci. École Norm. Sup.* **16** (1983) 193–217.

[70] Munkers, J., *Topology*, 2nd edn. (Prentice Hall, Upper Saddle River, 2000).

[71] Mashreghi, J., *Derivatives of Inner Functions, Fields Institute Monographs*, Vol. 31 (The Fields Institute for Research in Mathematical Sciences, Springer, 2013).

[72] Meier, K. E., Über Mengen von Randwerten, eromorpher Funktionen, *Comment. Math. Helvet.* **30** (1955) 224–233.

[73] Milovanović, G. V., Mitrinović, D. S. and Rassias, Th. M., *Topics in Polynomials: Extremal Problems, Inequalities, Zeros* (World Scientific, Singapore, 1994).

[74] Mortini, R., Factorization for bounded analytic functions in the unit disk and an application to prime ideals, *Math. Scand.* **99**(2) (2006) 168–174.

[75] Nevanlinna, R., *Analytic Functions* (Springer, Berlin, 1970). A revised translation from German, of *Eindeutige analytische Funktionen*, 2nd edn. (Grund lehren der mathematischen Wissenschaften, Vol. 46). Die Grundleheren der mathematischen Wissenschften in Eingeldarstellungen, Band 162 (1953).

[76] Nevanlinna, R., Über die Randwerte von analytischen Funktionen, *Comment. Math. Helv.* **2** (1930) 236–252.

[77] Nevanlinna, R., Sur un principe général de l'analyse, *C. R. Acad. Sci. Paris* **199** (1934) 548–550.

[78] Noshiro, K., Cluster Sets, Ergebnisse der Mathematik und Ihrer Grenzgebiete, Neue Folga, Heft **28** (Springer, Berlin, 1960).

[79] Newman, D. J. and Shapiro, H. S., The Taylor coefficients of inner functions, *Michigan Math. J.* **9** (1962) 249–255.

[80] Oyma, K., Approximation by interpolating Blaschke products, *Math. Scand.* **79** (1996) 255–262.

[81] Osgood, W. F. and Taylor, E. H., Conformal transformations on the boundaries of their regions of definition, *Trans Amer. Math. Soc.* **14** (1913) 277–298.

[82] Pick, G., ber eine Eigenschaft der konformen Abbildung kreisförmiger Bereiche, *Math. Ann.* **77** (1915) 1–6.

[83] Plessner, A., Über das Verhalten analytischer Funktionen am Rande ihres Definitionsbereiches, *J. Reine Angew. Math.* **158** (1927) 219–227.

[84] Peretz, R. and Rassias, T. M., Some remarks on theorems of M. Marden concerning composite polynomials, *J. Compl. Var.* **18** (1992) 85–89.

[85] Pólya, G. and Szegö, G., *Problems and Theorems in Analysis II* (Springer, 1976).

[86] Pommerenke, Ch., *Boundary Behaviour of Conformal Maps* (Springer-Verlag, Berlin, 1992).

[87] Ponce, A. C., *Elliptic PDEs, Measures and Capacities: From the Poisson Equation to Nonlinear Thomas-Fermi Problems*, EMS Tracts in Mathematics, Vol. 23 (European Mathematical Society, 2016).

[88] Roberts, J. W., Cyclic inner functions in the Bergman spaces and weak outer functions in H^p, $0 < p < 1$, *Illinois J. Math.* **29** (1985) 25–38.

[89] Rudin, W., Positive infinities of potentials, *Proc. Amer. Math. Soc.* **2** (1951) 967–969.

[90] Rudin, W., *Real and Complex Analysis* (McGraw-Hill, New York, 1974).

[91]   Rudin, W., Tauberian theorems for positive harmonic functions, *Nederl. Akad. Weteusch. Proc. Ser. A* **81** (1978) 376–384.

[92]   Sarason, D., Algebras of functions on the unit circle, *Bull. Amer. Math. Soc.* **79** (1973) 286–299.

[93]   Seidel, Wladimir, On the cluster values of analytic functions, *Trans. Amer. Math. Soc.* **34** (1932) 1–21.

[94]   Selberg H. L., Über die ebenen Punktmengen von der Kapazität Null, *Anh. Norske Videnskaps-Akad. Oslo I Math.-Natar.* **10** (1937).

[95]   Solynin, A. Yu., Ordering of sets, hyperbolic metrics, and harmonic measures, *J. Math. Sci.* **95**(3) (1999) 2256–2266.

[96]   Solynin, A. Yu., Elliptic operators and Choquet capacities, *J. Math. Sci.* **166**(2) (2010) 210–213.

[97]   Stephenson, K., Omitted values of singular inner functions, *Michigan Math. J.* **25** (1978) 91–100.

[98]   Stephenson, K., Isometries of the Nevanlinna class, *Indiana Univ. Math. J.* **26** (1977) 307–324.

[99]   Szegö, G., Koeffizientenabschätzungen bei ebenen und räumlichen harmonischen Entwicklungen, *Math. Ann.* **96** (1926–1927) 601–632.

[100]   Szegö, G., *Orthonormal Polynomials*, 4th edn. (American Mathematical Society, Providence RI, 1975).

[101]   Szegö, G., *Orthogonal Polynomials*, American Mathematical Society, Colloquium Publications, Vol. **XXIII** (American Mathematical Society, Providence, Rhode Island, 1967), (1939, 1st edn., 1958, rev edn.).

[102]   Tsuji, M., *Potential Theory in Modern Function Theory* (Maruzen Company Ltd., Tokyo, 1959).

[103]   Ugaeri, On the general potential and capacity, *Jpn. J. Math.* **20** (1950) 37–42.

[104]   Vitushkin, A. G., Analytic capacity of sets and problems in approximation theory, *Usp. Mat. Nauk.* **22** (1967) 141–199.

[105]   Vitushkin, A. G., Analytic capacity of sets and problems in approximation theory, *Russ. Math. Surveys* **22** (1967) 139–200.

[106]   Wiener, N., Discontinuous boundary conditions and the Dirichlet problem, *Trans. Amer. Math. Soc.* **25**(3) (1923) 307–314.

[107]   Zygmund, A., *Trigonometric Series*, Vol. I, 2nd edn. (Cambridge University Press, Cambridge, 1959).